AF248956

HOST DEFENSES TO INTRACELLULAR PATHOGENS

ADVANCES IN EXPERIMENTAL MEDICINE AND BIOLOGY

Recent Volumes in this Series

HOST DEFENSES TO INTRACELLULAR PATHOGENS

Edited by

Toby K. Eisenstein
Temple University School of Medicine
Philadelphia, Pennsylvania

Paul Actor
Smith, Kline, and French Laboratories
Philadelphia, Pennsylvania

and

Herman Friedman
University of South Florida
College of Medicine
Tampa, Florida

PLENUM PRESS • NEW YORK AND LONDON

Library of Congress Cataloging in Publication Data

Main entry under title:

Host defenses to intracellular pathogens.

(Advances in experimental medicine and biology; v. 162)
"Proceedings of a conference on host defenses to intracellular pathogens, held June
10–12, 1981, at the Franklin Plaza Hotel, in Philadelphia, Pennsylvania" — Verso of t.p.
Includes bibliographical references and index.
1. Immune response — Congresses. 2. Micro-organisms, Pathogenic — Congresses. I.
Eisenstein, Toby K. II. Actor, Paul. III. Friedman, Herman, 1931–   . IV. Series.
[DNLM: 1. Immunity, Cellular — Congresses. 2. Infection — Immunology — Congresses.
W1 AD559 v. 162 / QW 700 C748h 1981]
QR186.H67 1983   599′.0295   82-25957
ISBN 0-306-41259-4

Proceedings of a Conference on Host Defenses to Intracellular Pathogens, held June
10–12, 1981, at the Franklin Plaza Hotel, in Philadelphia, Pennsylvania

©1983 Plenum Press, New York
A Division of Plenum Publishing Corporation
233 Spring Street, New York, N.Y. 10013

Printed in the United States of America

PREFACE

    The subject matter of this volume was the basis for a confer-
ence held in Philadelphia in June, 1981, and is an important one
in the contemporary area of how the host interacts with microorganisms. In conception, it grew out of a graduate course
entitled, "The Infectious Process," which has been taught in the
Department of Microbiology and Immunology at Temple University
School of Medicine during the past twelve years. This course has
explored the broad areas of mechanisms of microbial pathogenesis
and host resistance by in-depth consideration of selected models of
experimental infection and immunity, as well as the clinical
literature. It is noteworthy that there is no adequate text for
this material, as the subject matter naturally crosses a wide
spectrum of traditional disciplinary lines, encompassing topics
as diverse as the mechanisms of action of bacterial toxins, the
role of complement and antibody in phagocytosis, and the importance
of cross-reacting bacterial polysaccharide antigens in vaccine
development. A major portion of the course has always considered
"cellular immunity" as it applies to host defenses to intracellular
pathogens. It is in this area that the necessity for amalgamation
of information from different disciplines is most evident, for one
must be intimately concerned with the interactions between the
microbe and the phagocyte, both before and after specific immune
recognition. To understand the delicate balances at issue in this
type of infectious process, one must bring to bear knowledge of
the biochemistry of killing strategies of phagocytes, of how
microbes can escape the usually lethal armamentarium of the phagocyte, as well as the ways in which the immune system modifies macrophage effector function. This conference, and these Proceedings,
are meant to bridge the interdisciplinary gap, in an effort to
bring relevant information from immunology, pathology, microbiology,
and infectious diseases together, to examine intracellular
infections by pathogenic and opportunistic organisms. Clearly, with
the explosion of information in basic and applied immunology, such
an endeavor will suffer from a lack of completeness, for the
advances in this rapidly expanding field threaten to outpace our
ability to apply the information to understanding resistance to
infection. Ironically, this situation is the reverse of that seen
in the preantibiotic era, when the study of experimental infectious

diseases laid the foundation for many of our basic concepts concerning the immune system, including the protective value of immune sera, the microbicidal nature of complement plus antibody, and delayed hypersensitivity as a hallmark of exposure to microbes.

The majority of the organisms we selected for scrutiny are clearly classified as intracellular pathogens, either obligate or facultative. However, a few of those chosen may fall into a somewhat questionable category with respect to an intracellular criterion. Model systems are presented for some of the fungi which address the question of the relative role of humoral versus cellular immunity in contributing to the overall status of host defenses. We feel that the lessons to be learned from these model systems are highly relevant to the study of traditional intracellular pathogens, and for that reason these organisms have also been included in this volume. _Legionella_ species are emerging as newly recognized, facultative intracellular organisms, and the inclusion of a paper on this topic is a harbinger of this fast growing area of investigation.

The conference on which this volume is based was divided into certain specific areas for discussion. The first session dealt with microbicidal mechanisms of leukocytes in host defenses. In that session the oxygen-independent as well as -dependent mechanisms of microbicidal activities of leukocytes were considered. Questions concerning the role and nature of the antimicrobial activity of these cells, especially macrophage function, and phagolysosome fusion, were discussed in detail. The second session dealt with the immune system and phagocytic cell function. The effects of lymphokines in host defenses, as well as the role of interferon, were particularly emphasized in this session. Various immunologic and non-immune factors activating macrophages for killing of microbes were discussed. Immunosuppression as a virulence factor and as a manifestation of chronic infection of the host with intracellular pathogens was a major theme of this session, and of session three, which focused on host defenses to intracellular bacteria. In particular, much new and exciting information about the immunoregulatory disturbances resulting during tuberculosis and leprosy infection were highlighted. There was extensive discussion concerning the importance of humoral versus cellular responses to diverse species of bacteria such as _Listeria_ and _Salmonella_, as well as consideration of the genetics of natural resistance to these organisms. The fourth session dealt with immune mechanisms of resistance to parasites and fungi, and stressed the distinct but interrelated role of cells versus antibody in infections as diverse as those caused by organisms in these two groups. The final session dealt with mechanisms of anti-viral immunity, involving the role of several immune parameters in a variety of infections, such as persistent viruses and leukemia viruses. Major topics included the role of T cells, histocompatibility genes, and immune surveillance

mechanisms in antiviral immunity, as well as immunosuppressive
effects of various viral infections. Hybridoma antibodies as
probes in various model systems was a recurrent theme.

It is quite apparent from the wide variety of topics discussed
in this conference, that there is an ever-increasing interest in
the important relationship between the immune system and host
defenses to various microorganisms, especially those which preferentially replicate within host cells. It is anticipated that by
focusing attention on this very rapidly evolving field, more
investigators from different disciplines will continue to provide
new information concerning these host-parasite relationships.
Certainly, it is clear that there is much to be gained by applying
lessons learned in model systems with one pathogen to systems
employing other organisms. We hope that the reader will take the
time to study the entire volume, rather than just individual
chapters, and will come away with a sense of the dynamic and
exciting interactions presently taking place across traditional
disciplinary lines in the study of host defense mechanisms.

The importance of the host-immune, as well as non-immune,
systems in protection against intracellular infections, and the
equally important concept that the microbial pathogens can themselves affect host immunity, will undoubtedly provide a rich
source of material for future conferences. It is evident that
each new step in unraveling the intricate inner workings of the
macrophage and the complexities of the immune system brings us
that much closer to understanding the molecular and cellular
basis for host defense against infectious disease.

Toby K. Eisenstein

Paul Actor

Herman Friedman

ACKNOWLEDGEMENTS

     The editors are indebted to the Eastern Pennsylvania Branch
of the American Society for Microbiology under whose sponsorship
this conference was held, and to the members of the organizing
committee whose hard work made the symposium possible.

     We would also like to thank Dr. Joseph Pagano, the branch
President for his encouragement and enthusiastic support of this
endeavor. The Pennsylvania Department of Health-Bureau of
Laboratories generously handled mailings and registrations.

     We are also grateful to Temple University School of Medicine,
Hahnemann Medical College, Thomas Jefferson Medical College, and
the School of Medicine of the University of Pennsylvania for their
sponsorship. The Department of Microbiology and Immunology of
Temple University was particularly supportive of this conference
and this volume by providing office services.

     This conference would not have been possible without the
generous financial support of Smith, Kline and French Laboratories
whose contribution we gratefully acknowledge.

     The excellent proof-reading skills of Linda Hampton, Connie
Gabor, and Peggy Cavota are appreciated. Finally, many thanks to
Mrs. Judy Trachtman for her beautiful camera-ready copy of the
manuscripts.

ORGANIZING COMMITTEE

Chairpersons

Toby K. Eisenstein
Temple University School of.
Medicine
Philadelphia, PA

Herman Friedman
University of South Florida
College of Medicine
Tampa, FL

Committee

Paul Actor
Smith, Kline and French
Laboratories
Philadelphia, PA

Richard L. Crowell
Hahnemann Medical College and
Hospital
Philadelphia, PA

Josephine Bartola
Bureau of Laboratories
Pennsylvania Department of
Health
Lionville, PA

Sarah F. Grappel
Smith, Kline and French
Laboratories
Philadelphia, PA

Paul H. Saluk
Hahnemann Medical College and
Hospital
Philadelphia, PA

Walter Ceglowski
Temple University School of
Medicine
Philadelphia, PA

Robert R. Strauss
Albert Einstein Medical Center
Northern Division
Philadelphia, PA

Frank M. Collins
Truedau Institute, Inc.
Saranac Lake, NY

CONTENTS xiii

INTRODUCTION AND HISTORICAL PERSPECTIVE

Toby K. Eisenstein
Department of Microbiology and Immunology
Temple University School of Medicine
Philadelphia, PA   19140

Various microorganisms have evolved specialized, and in many cases, unique mechanisms for evading host defenses. For organisms such as pneumococci and <u>Haemophilus</u>, the presence of a capsular coat allows the organism to evade phagocytosis and multiply unchecked in the extracellular fluid. The ingestion of the pneumococcus by the pursuing polymorphonuclear leucocyte (PMN) results in the subsequent death of this pathogen in the hydrogen peroxide-filled atmosphere of the phago-lysosome. Minute amounts of antibody directed towards the antigenic determinants of the capsular polysaccharide render the pneumococcus an easy target for the PMN. Thus, excellent protection against <u>Streptococcus pneumoniae</u> infection is provided for non-immune animals receiving injections of immune serum.

In the case of the "intracellular pathogens," the topic of the symposium on which this volume is based, there is a much different situation. These organisms are in many cases inadequately handled by the PMN, the body's first line of defense, and infection results in a mononuclear cell infiltrate characterized by lymphocytes and macrophages. Even these cells, a second line of defense, may prove inadequate to contain the bacterial invasion. Max Lurie, in his classic experiments on immunity to tuberculosis, used the anterior chamber of the rabbit eye as his culture vessel (1). When he injected <u>Mycobacteria</u> into the eye chamber of a normal rabbit along with macrophages obtained from a second normal rabbit, the bacilli were phagocytized. Instead of being killed, the organisms found the intracellular environment conducive to their growth. How these facultative intracellular pathogens manage to escape the onslaught of the macrophage armamentarium of enzymes, peroxides and superoxides is still a topic of much active investigation.

Studies on the mechanism of immunity to intracellular pathogens remained largely unclarified during the first half of the twentieth century, as numerous investigators were unable to transfer immunity to the tubercle bacillus with serum. Yet, it was recognized as far back as 1891 by Koch (2) that infection with M. tuberculosis sensitized the host to give a delayed-type hypersensitivity reaction to culture filtrates of the organism. Von Pirquet (3) recognized that other infectious agents could stimulate similar kinds of skin reactions and referred to them as the allergy of infection. It was noted that this type of "allergy" was different from other hypersensitivities which occurred in closer temporal sequence to application of the elicitin. Yet, the immunologic basis for this characteristic reaction was unknown until 1945, when Chase (4) was able to transfer tuberculin hypersensitivity in guinea pigs with peritoneal exudate cells. This observation laid the foundation for a concept of "cellular immunity" in which immune reactivity resided in cells, but not in serum. Considerable confusion reigned concerning the mechanisms involved, and popular concepts attributed the immunological specificity of the reactions to cell-bound antibodies. Furthermore, the relationship between delayed hyper- sensitivity and immunity was hotly debated (5).

A major conceptual advance came in 1964 when Mackaness carried out a remarkable series of in vivo studies using several bacteria, all of which had the capacity to grow in macrophages (6). These included Brucella, Listeria, and BCG, the attenuated Myco- bacterium. He showed that at various stages of infection with one of these organisms there was cross-protection against one of the other organisms which was antigenically unrelated. Furthermore, the onset of immunity correlated with the ability to elicit delayed hypersensitivity. In a most cleverly designed experiment he tested the ability of Listeria monocytogenes to grow in vivo in 1) groups of mice which had received BCG 14 weeks earlier; 2) those which had received the primary BCG infection but were given a second injection of BCG 3 days before the Listeria infection; or 3) control mice. He found that growth of Listeria was inhibited only in the animals which had received both doses of BCG. The hypothesis was formulated that there was a require- ment for an antigenically specific elicitation of immunity to one intracellular pathogen, which could manifest itself as nonspecific resistance. The immunological mechanism behind these observations was clarified in 1969 by Mackaness (7), who proposed that specifically sensitized lymphocytes could activate macrophages to express nonspecific, enhanced bactericidal properties. David (8) and Bloom and Bennett (9) had already demonstrated in vitro that lymphocytes sensitized to antigen, could elaborate soluble factors, when re-exposed to the antigen that caused inhibition of macrophage migration. Furthermore, it was recognized that the ability of lymphocytes to generate the factor was immunologically specific, and correlated with the ability of the donor animal to give a

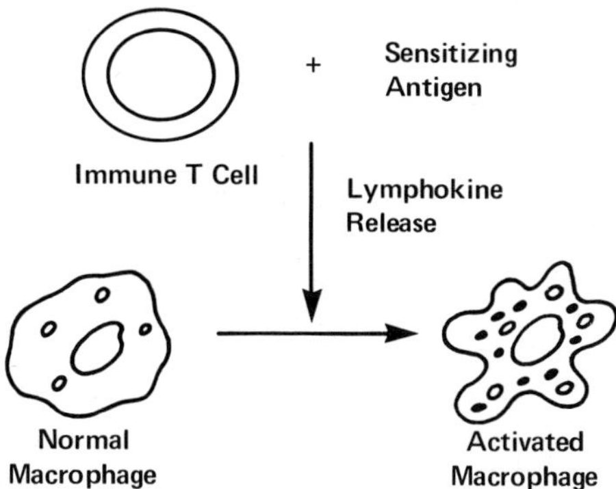

Figure 1. Cellular Immunity.

delayed hypersensitivity reaction _in vivo_. Mackaness unified
these concepts by showing that splenic lymphocytes from _Listeria_-
immune mice could transfer both anti-_Listeria_ immunity and delayed
hypersensitivity to non-immune recipients (7). The antibacterial
immunity was shown to result from activated macrophages, and
Mackaness proposed that the soluble factors described above could
be the mediators from lymphocytes to activate macrophages.

These are some of the experiments which form the basis for
the working model (Fig. 1) which is still being tested with regard
to mechanisms of immunity to intracellular bacterial infection.
Central to this theory is the premise that the macrophage is the
essential effector cell in coping with microorganisms of this
type, and that in order for it to cope effectively, it must be
raised to a state of enhanced bactericidal capacity. This state
of "cellular immunity" is accomplished _via_ the T cell, which when
it meets the specific antigen to which it has been sensitized, is
triggered to relase lymphokines, such as macrophage activating
factor. The normal macrophage is then converted into an "immune"
or "angry" macrophage, a cell which is a match for this special
breed of microbe that finds the intracellular environment so
compatible with its lifestyle. Within this model framework,
we can perhaps hope to understand better the host-parasite
relationship in infections of this type.

REFERENCES

1. Lurie, M.B. (1964). Resistance to tuberculosis: Experimental
        studies in native and acquired defensive mechanisms.
        Harvard University Press, Cambridge, Mass.

2. Koch, R. (1891). Fortsetzung der mittheilungen uber ein
      heilmittel gegen tuberkulose. Deutsche Med. Wochenschr.
      17:101.
3. Von Pirquet, C. (1907). Klinische studien uber vakzination und
      vaksinale allergie. Deuticke, Leipsig.
4. Chase, M.W. (1945). The cellular transfer of cutaneous hyper-
      sensitivity to tuberculin. Proc. Soc. Exp. Biol. Med.
      59:134.
5. Rich, A.R. (1951). The pathogenesis of tuberculosis. Charles
      Thomas, Springfield, Ill.
6. Mackaness, G.B. (1964). The immunological basis of acquired
      cellular resistance. J. Exp. Med. 120:105.
7. Mackaness, G.B. (1969). The influence of immunologically
      committed lymphoid cells on macrophage activity _in vivo_.
      J. Exp. Med. 129:973.
8. David, J.R. (1966). Delayed hypersensitivity _in vitro_: its
      mediation by cell-free substances formed by lymphoid
      cell-antigen interaction. Proc. Natl. Acad. Sci. USA
      56:72.
9. Bloom, B.R. and B. Bennett. (1966). Mechanism of a reaction
      in vitro associated with delayed-type hypersensitivity.
      Science 153:80.

OXYGEN INDEPENDENT MICROBICIDAL MECHANISMS OF HUMAN POLYMORPHONUCLEAR LEUKOCYTES

John K. Spitznagel and Noburu Okamura

Emory University School of Medicine, Atlanta, GA
and Tokyo Medical and Dental University School
of Medicine, Tokyo, Japan

INTRODUCTION

Human polymorphonuclear leukocytes (PMN) have several anti-microbial systems that can be viewed as belonging to two groups. One group depends upon oxidative processes. These include the superoxide anion (O_2^-), the myeloperoxidase-chloride-hydrogen peroxide system (MPHC1), and free hydroxyl radicals ($\cdot$OH) (1). The other group (Table 1) functions independently of oxidative processes and includes increased hydrogen ion concentrations, various cationic proteins, cathepsin G, lysozyme, and apolactoferrin (2). The oxidative bactericidal processes are complex and depend upon soluble enzymes or cofactors present in the cytosol and upon enzymes located on or in membranes. The nonoxidative processes appear to depend solely upon proteins found within the PMN granules (3). Although there is indirect evidence that favors the dominance under various conditions of one or the other of these antimicrobial systems, it is unclear, at present, which

Table 1. Oxygen Independent Antibacterial Components

Cationic Proteins

    Elastase
    Cathepsin G
    Other Azuorphil Granule Proteins

Lysozyme

Lactoferrin

---

mechanisms are mainly responsible for, or crucial in, the bactericidal activity of PMN in the various environments found in the
body. For example, accumulating evidence suggests that oxygen-
independent antimicrobial systems operate _in vivo_ in human PMN
as well as in those of chickens and rabbits (2). But these experiments generally do not permit assessment of the relative contributions of the oxygen-independent and the oxygen-dependent systems.
The general assumption is that the oxygen-dependent systems are
dominant and the oxygen-independent systems provide a backup for
them. However, it is clear that while most environments in the
host are aerobic, relatively anaerobic environments exist in normal,
and particularly in abnormal hosts. It would, therefore, be an
advantage for PMN to have oxygen-independent antimicrobial systems.

We were led to compare the two systems as a result of experiments performed by Rest (3), and Modrzakowski (4) with oxygen-
independent systems isolated from PMN. They reported that _Salmon-
ella typhimurium_ LT2 and its outer membrane mutants showed an
ordered resistance to the bactericidal activity of PMN granule
extracts that decreased as the carbohydrate content of the outer
membrane lipopolysaccharide (LPS) decreased. The LT2 and its
mutants were, however, equally susceptible to the human MPHC1
bactericidal system (5). It seemed to us that, if the bactericidal
granule proteins can operate independently of O_2 in intact meutro-
phil phagolysosomes, the _S. typhimurium_ LT2 and its LPS mutants
might show ordered differences in resistance to killing during
phagocytosis by PMN in anaerobic conditions, just as they have
shown in the _in vitro_ studies. The ordering should resemble that
seen with _in vitro_ oxygen-independent killing due to granule
proteins. We, therefore, adapted PMN to anaerobic conditions,
tested their antimicrobial activities against these _Salmonella_
strains, and compared them with activities of matched neutrophils
working in aerobic conditions.

MATERIALS AND METHODS

The methods are summarized in Table 2. We prepared PMN from
fresh heparinized normal human blood by dextran and Ficoll-Hypaque
sedimentation (6). For anaerobic cultures of PMN we modified
slightly and used the method of Baehner et al.(7). For opsonization we used serum lacking late complement components to avoid
killing the serum sensitive mutants. Human serum deficient in C8
due to a genetically determined fault was the gift of Dr. Steven
Shore of the Center for Disease Control, Atlanta, Georgia. The
serum we used most often was normal human serum that was depleted
of C6, with an affinity column having goat antihuman C6 as ligand.
The anti-C6 was the gracious gift of Dr. Hans Mueller-Eberhard,
Scripp's Clinic, La Jolla, California. PMN were suspended in
Gey's solution and added to tissue culture multi-well dishes at a

Table 2. Methods

PMN monolayers incubated aerobically or anaerobically

Pre-opsonized radiolabeled bacteria

Phagocytosis and killing of bacteria by PMN

Immunofluorescent staining

Determination of superoxide anion production

---

concentration of 1×10^6 to 2×10^6 cells/ml in a total volume of
1 ml for each well. The cells were allowed to adhere for 1, 2,
or 5 hours at 37°C either in a 5% CO_2 incubator for aerobic phagocytosis and killing, or in a Virginia Polytechnic Institute (VPI)
anaerobic chamber for anaerobic phagocytosis and killing. The
chamber contained a gas mixture of nitrogen, hydrogen and
carbon dioxide at concentrations of 85, 10, and 5%, respectively.
It also contained a catalyst to remove residual oxygen. Stringent
anaerobiasis was obtained in the chamber and could be checked with
methylene blue indicator and agar plates that contained resazurin.

The _Salmonella typhimurium_ strains used included LT2, his 642,
HN202, SL1004, SL1181, and TA2168, all kindly supplied by Dr.
Hiroshi Nikaido of the University of California at Berkeley.
Figure 1 shows the LPS structure of the strains we used in this
experiment. The numbers to the right of the figure indicate the
relative antimicrobial potency of granule proteins for the different strains. Bacteria were periodically checked with crystal
violet and LPS phages for their degree of roughness. They were
inoculated into Luria broth containing ^{14}C-labeled glucose and
incubated at 37°C with aeration until log-phase growth was achieved.
They they were washed three times with saline and suspended in 10%
serum in Gey's solution and incubated 40 min at 37°C for opsonization. The bacteria were then washed once in saline and resuspended
in anaerobic Gey's solution for inoculation into PMN culture.

After the preparation of PMN monolayers we added various
amounts of live ^{14}C-labeled _Salmonella typhimurium_ strains to the
cell monolayers, which were then incubated for different periods
of time under either aerobic or anaerobic conditions at 37°C.
All non-living materials for anaerobic experiments were conditioned
by storage in the VPI anaerobic chamber for 24 hours and washed out
with bubbling nitrogen. PMN were conditioned for 1, 2, or 5 hours
in the anaerobic chamber and then inoculated with bacteria.
Paired PMN cultures were prepared simultaneously and conditioned
aerobically for similar time periods.

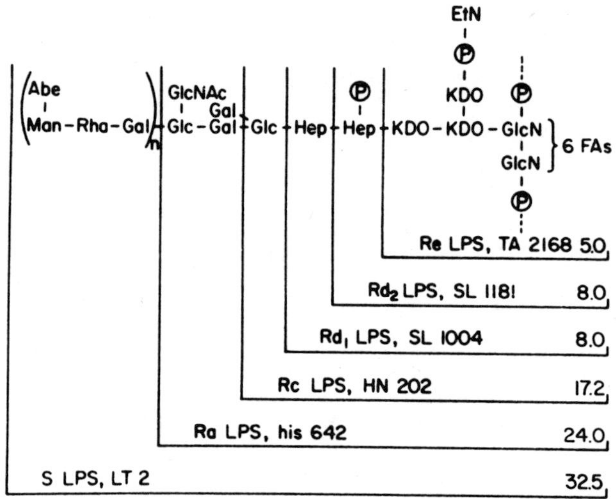

Figure 1. Lipopolysaccharide chemotypes of <u>Salmonella</u> <u>typhimurium</u>
LT2 and its outer membrane mutants used in these experi-
ments. The numbers to the right of the figure indicate
the E.D. 50 of azurophil granule protein for each of the
strains represented in the figure. Abe signifies
abequose; man – mannose; Rha – rhamnose; Gal – galactose;
glcNac – N acetyl glucosamine; Glc – glucose; Hep –
heptose; EtN – ethanolamine; KDO – ketodeoxyoctonate;
GlcN – glucosamine; P – phosphate.

Phagocytosis was terminated by removing the incubation medium,
followed by two washes with phosphate buffered saline. The cell
pellet was scraped from the well and suspended in 1 ml of cold
0.5% Triton X-100 in saline. Triton X-100 at this concentration
disrupted PMN but did not affect bacterial viability. We then
counted ^{14}C activity in a liquid surface plate method. In anaerobic experiments, all the procedures were done in the anaerobic
chamber.

We also performed immunofluorescent staining, in order to
distinguish between actual ingestion and simple attachment of
bacteria to the cell surface. We used goat antibody against human
C'3 and FITC-labeled rabbit IgG antibody against goat IgG (Hyland
Labs). We found that staining of intact, phagocytizing, unfixed
cell monolayers revealed no fluorescence. However, stains of
methanol fixed cells that can no longer exclude antibody revealed
many bacteria. Under these conditions of studying phagocytosis,
most of the cell-associated bacteria were evidently intracellular
and not merely adherent to the cell surface (8).

Although strict anaerobiasis was obtained in the anaerobic chamber, there was a possibility that PMN had available trace amounts of oxygen; hence precursors of oxygen–dependent antimicrobial mechanisms might be already dissolved in their extracellular medium. If this were so they would have available oxygen–dependent metabolism in the anaerobic chamber as they had in an aerobic atmosphere. We decided, therefore, to measure superoxide generation (superoxide dismutase inhibitable reduction of ferricytochrome c (90μM)) by PMN incubated in both aerobic and anaerobic conditions, and to use this measurement made by a modification of the method of Babior as an index of anaerobicity for the PMN (9). As shown in Table 3, PMN incubated in aerobic conditions generated much more superoxide than cells incubated under anaerobic conditions. We used as a positive control, opsonized zymosan, a potent

Table 3. Effect of Anaerobiasis on Superoxide Anion Generation by Human Neutrophils

Stimulus	Ferricytochrome c Reduction[a]			
	Aerobic for		Anaerobic for	
	1 h	5 h	1 h	5 h
None (resting)	1.5	0	0	0
Opsonized LT2 (1×10^8)	8.4	7.2	0	0
Opsonized zymosan (1 mg)	33.0	13.3	13.1	0

[a] Superoxide dismutase-inhibitable cytochrome c reduction by PMN incubated aerobically or anaerobically (nanomoles reduced/ 15 min/ 2.5×10^6 PMN)

stimulus for superoxide generation. Even the opsonized zymosan did not stimulate superoxide generation by anaerobic PMN, although it stimulated vigorous superoxide production among aerobic PMN. Moreover, opsonized zymosan stimulated much more vigorous O_2^- production than did Salmonella LT2. In order to confirm that all O_2^- produced would be detected (10), matched neutrophil suspensions were treated in several experiments with cytochalasin B and their O_2^- production in response to opsonized zymosan was measured with cytochrome c under aerobic and anaerobic conditions. As an additional control, we documented that neutrophils incubated anaerobically even for 5 hours and then returned to an aerobic environment again generated as much superoxide anion as did neutrophils simply incubated aerobically during the same period.

Table 4. Phagocytosis of Bacteria by Aerobic and Anaerobic
          Neutrophils

Bacteria[a]	Aerobic PMN[b]	Anaerobic PMN[b]
LT2	4.4 ± 2.7 (3)	5.9 ± 5.5 (5)
his-642	6.4 ± 1.0 (3)	8.1 ± 5.2 (5)
HN202	11.6 ± 5.3 (3)	12.5 ± 3.7 (5)
SL1004	7.3 ± 0.6 (3)	5.5 ± 1.2 (5)
SL1181	12.7 ± 2.5 (3)	21.6 ± 10.1 (5)
TA2168	8.6 ± 1.8 (3)	9.4 ± 5.5 (4)

[a] Ratio of bacteria to cell; 10 : 1

[b] Per cent of bacteria ingested

RESULTS

Having established conditions for anaerobic culturing of PMN
we then assessed the effect of such anaerobiasis on their phagocytic and bactericidal capacities. The PMN were adapted 1 h, 2 h,
or 5 hr in aerobic and anaerobic conditions and subsequent bactericidal experiments were done for an additional 1 h. Table 4 shows
the rate of ingestion by neutrophils held anaerobically for 1 hour
and then challenged with a ratio of bacteria to cells of 10:1. We
also examined phagocytosis and bactericidal capacity of PMN with
ratios of bacteria to cells of 1:1 and 100:1. We found that the
ratio of bacteria to cells did not have any effect on the degree of
intraleukocytic killing (experiments not shown). Although the
phagocytic rate of PMN for each bacterial strain seemed to differ,
there were similarities in phagocytosis of each strain by aerobic
and anaerobic PMN. Table 5 shows the bactericidal capacity of
these aerobic and anaerobic neutrophils with ratios of bacteria
to cells of 10:1. As the carbohydrate content of LPS decreased

Table 5. Bactericidal Capacity of Aerobic and Anaerobic Neutrophils

Bacteria[a]	Aerobic PMN[b]	Anaerobic PMN[b]
LT2	16.4 ± 9.2 (3)	11.8 ± 6.2 (5)
his-642	7.0 ± 0.4 (3)	6.0 ± 5.3 (5)
HN202	5.2 ± 3.4 (3)	0.5 ± 0.3 (5)
SL1004	3.1 ± 1.3 (3)	0.4 ± 0.7 (5)
SL1181	5.7 ± 7.5 (3)	0.6 ± 0.7 (5)
TA2168	3.5 ± 2.8 (3)	0.1 ± 0.1 (4)

[a] Ratio of bacteria to cell; 10 : 1

[b] Per cent of viable bacteria divided by the total number
    of bacteria that were ingested

from parent LT2 to mutant and from mutant to mutant, the bacteria
became increasingly less resistant to the bactericidal activity of
both aerobic and anaerobic PMN.

In order to test antimicrobial phagocytosis under maximally
stringent anaerobiasis the bactericidal capacity of PMN incubated
and adapted anaerobically for 5 hours was examined with bacteria
adapted to anaerobiasis for up to 2 h before being applied to the
monolayers. Again, the percent of killing due to anaerobic PMN
showed similar results compared with that due to aerobic PMN.
However, the phagocytosis rate by anaerobic PMN was frequently
somewhat less than that by aerobic PMN.

Phagocytosis and killing by both aerobic and anaerobic PMN
were examined with time. The results shown in Figure 2 are from
a representative experiment showing the time course of aerobic
phagocytosis and killing. At increasing times of incubation the
percent of phagocytosis and killing increased as expected. However, most of the bacteria seemed to be killed within 30 min.
Figure 3 shows anaerobic phagocytosis and killing. These PMN
were held anaerobically for 2 hr prior to endocytosis. We see
that under these conditions of anaerobiasis the rate of ingestion
of bacteria was actually somewhat greater than in the aerobic
cultures. Antimicrobial action was very rapid and thorough. It
has been observed by other workers (7) that anaerobic PMN may be
hyperphagocytic. In any event, there was little impairment of
PMN function.

In subsequent experiments done with PMN in suspension and
conditioned in anaerobic conditions for 5 hours (Table 6) the
parent _Salmonella_ _typhimurium_ was well phagocytized and promptly
killed anaerobically as well as aerobically. The only difference
was a slight reduction in uptake and killing under anaerobic
conditions compared with aerobic conditions.

DISCUSSION

We have explored mechanisms of oxygen-independent bactericidal
activity using PMN adapted to anaerobic atmospheres. Under these
conditions oxidative bactericidal mechanisms would be markedly
inhibited. We have succeeded in achieving culture conditions
sufficiently anaerobic that superoxide anion production could no
longer be detected as superoxide dismutase inhibitable reduction
of ferricytochrome c. Under these conditions, phagocytosis and
killing were essentially unchanged from aerobic conditions. We
conclude that oxygen-independent systems were functioning.

Similar evidence of anaerobicity in PMN cultures has been
achieved by Baehner et al. (7). We as well as Baehner found it

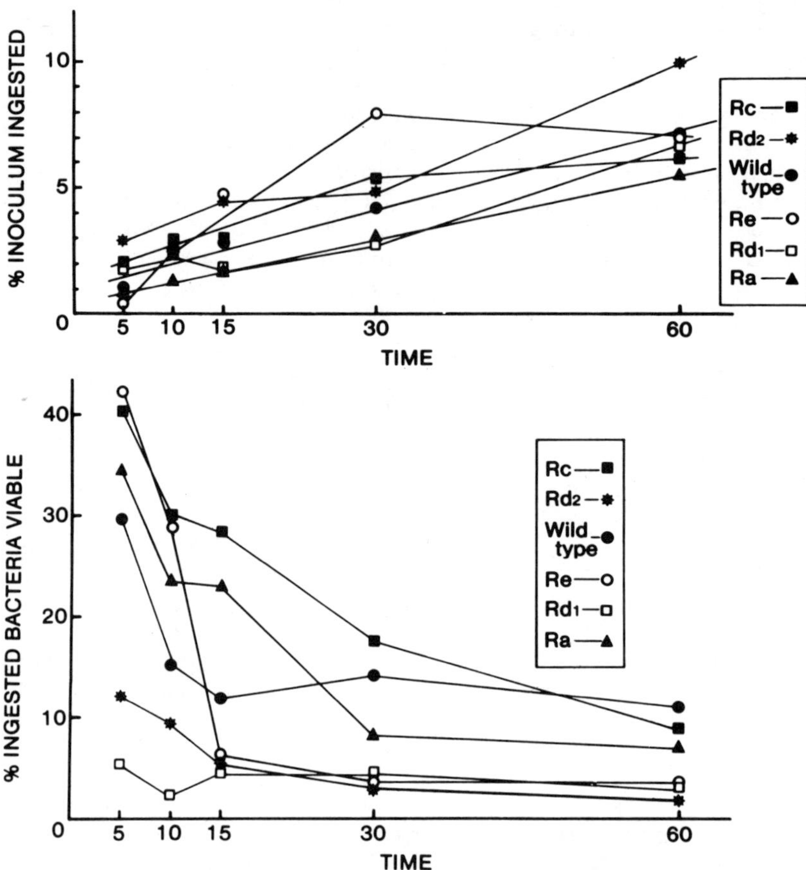

Figure 2. Time course of ingestion (top) and killing (bottom) of *Salmonella typhimurium* LT2 and its mutants. PMN/bacterium ratio = 1/10. The chemotype designations are in accordance with the usage in Fig. 1. The PMN were incubated 2 hours in 5% CO_2 in air before being challenged with the bacteria. The bacteria were preopsonized with C6 depleted normal human serum. Other conditions were as described in Methods and Materials.

necessary to maintain the PMN under stringently anaerobic conditions for at least 5 hours to reproducibly achieve this result. Baehner did not investigate the microbicidal capacity of anaerobic PMN but studied the role of O_2^- in leukocytic reduction of nitroblue tetrazolium NBT. He found NBT reduction apparently failed under strict anaerobiasis. He also found cytochrome c failed to be reduced in that system. We first evaluated the reduction of

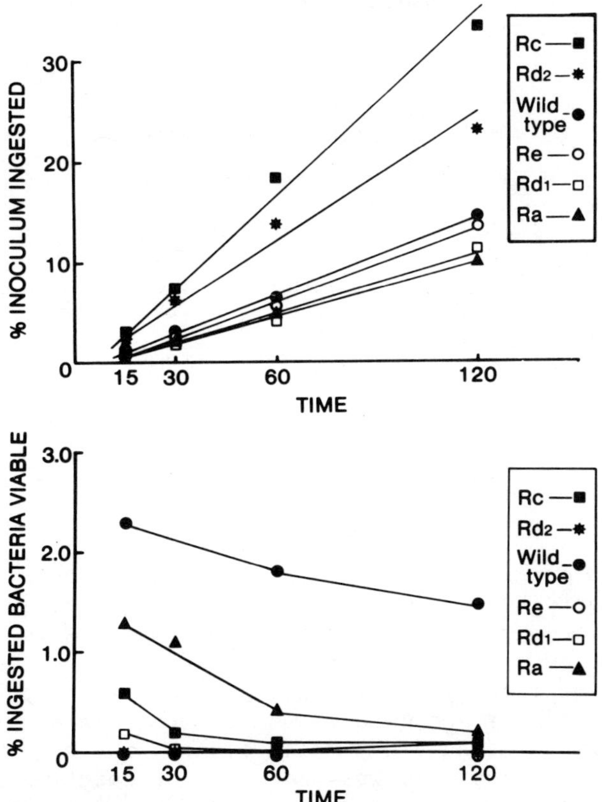

Figure 3.   Time course of ingestion (top) and killing (bottom) of
Salmonella typhimurium LT2 and its outer membrane
mutants. PMN/bacterium ratio = 1/10. The PMN were
incubated 2 hours in a VPI anaerobic chamber before
they were challenged with preopsonized bacteria.
Conditions otherwise as for Fig. 2. This experiment
was performed independently of the one in Fig. 2.

cytochrome c with PMN challenged to ingest opsonized _Salmonella_
LT2 and its mutants and found that it failed under anaerobic
conditions. Because the _Salmonella_ appeared a weak stimulus for
O_2^- production we further evaluated the same function with PMN
challenged with opsonized zymosan. The PMN challenged with zymosan
under anaerobic conditions failed to reduce cytochrome c. This
failure was especially striking because under aerobic conditions
opsonized zymosan elicits vigorous production of O_2^- by PMN.

In further experiments we assessed production of O_2^- among
suspended PMN treated with cytochalasin B and challenged with
opsonized zymosan. Root et al. (5) had shown earlier that

Table 6. Killing of Salmonella LT2 in Suspended Polymorphs Under Aerobic and Anaerobic Conditions

| | SUPERNATANT | | LEUKOCYTE PELLET | |
Time	Aerobic[a]	Anaerobic[b]	Aerobic	Anaerobic
0	2.4×10^6 [c]	3.1×10^6	3.3×10^5	1.2×10^6
60	2.0×10^4	1.1×10^5	2.2×10^4	3.3×10^4
120	5.0×10^3	1.5×10^4	4.4×10^3	1.6×10^4

[a] Cells conditioned 5 hours in 5% CO_2 in air.

[b] Cells conditioned 5 hours in anaerobic chamber.

[c] Viable bacteria per ml.

cytochalasin B maximizes O_2^- release. Anaerobic conditions completely suppressed O_2^- production in response to zymosan among cytochalasin B treated PMN in suspension. We concluded that under these conditions we had achieved stringent anaerobiasis. Nevertheless, we readily confirmed that under aerobic conditions cytochalasin B plus zymosan increased the release of O_2^- by PMN to a greater degree than did zymosan alone. The significance of this degree of anaerobiasis was placed in perspective by our observation that in all cases O_2^- production due to opsonized Salmonella LT2 was both less vigorous and much more easily suppressed than that due to opsonized zymosan. In other words it seemed likely that by achieving conditions sufficient to inhibit O_2^- production in response to opsonized zymosan in cytochalasin B treated PMN, one had substantially transcended the conditions sufficient to deprive PMN phagocytizing Salmonella LT2 of the capacity to respond and elaborate O_2^-.

Kossack et al. (11) recently reported that oxygen metabolism stimulated in human PMN by virulent Salmonella typhi was substantially less than that stimulated by avirulent S. typhi. Interestingly the two S. typhi strains were killed equally well by human PMN and these authors speculate that oxygen-independent mechanisms may account for intraleukocytic killing of Salmonella in PMN.

Our results confirmed that the PMN killed Salmonella efficiently under all of the anaerobic conditions we employed in this study. In this respect killing was of the same order of magnitude whether the PMN had been conditioned for 1, 2 or 5 hours in the anaerobic chamber. Moreover, the Salmonella mutants showed

ordered sensitivity to killing that decreased with the transition
from deep rough outer mutants to smooth parent Salmonella LT2.
This fulfilled our prediction that if oxygen-independent
mechanisms were responsible for intraleukocytic killing, the
mutants and the parent would show ordered resistance. This
would resemble the characteristic order observed in their capacity
to resist killing, in vitro, by PMN granule proteins as observed
by Rest et al. (3). The earlier observations of Rest et al. that
Salmonella mutants and parent were equally sensitive to the antimicrobial effects of the MPHCl system would provide independent
support for our hypothesis.

Curiously the percentages of Salmonella LT2 and its mutants
killed were rather similar under anaerobic and aerobic conditions.
The number of bacteria actually killed tended to be less, under the
most severely anaerobic conditions, e.g. 5-6 hours preincubation
under these conditions the numbers of bacteria ingested were
somewhat less. Thus killing was somewhat impaired under the
most stringent anaerobic conditions. However, killing was substantial in spite of this difference.

Our present results differed to some degree from those of
Mandell (12) who pioneered investigation of antimicrobial action
with PMN. With PMN in suspension and contained in individually
deoxygenated flasks he found that Staphylococcus aureus and several
enterobacteraceae, including S. typimurium, were not killed as
readily by anaerobic PMN as they were by aerobic PMN. However,
Staphylococcus epidermidis, Streptococcus faecalis, Viridans
streptococci, Pseudomonas aeruginosa and anaerobic bacteria were
killed equally by anaerobic and aerobic PMN. We cannot at present
account for these discrepancies but can only point to the substantial differences in the systems we have used. Mandell's PMN
were in suspension in individual deoxygenated flasks and the
bacteria used were from static rather than log phase cultures.
Rest has shown the importance of log phase cultures in antimicrobial experiments with granule proteins (3). Log phase enterobacteraceae are more uniformly sensitive and less resistant,
generally, to the granule proteins. It is noteworthy in fact that
Mandell found anaerobic PMN killed 75% of the inoculum of S.
typhimurium while the aerobic PMN killed 95%. Thus Mandell's
anaerobic cells did kill a substantial proportion of the Salmonella.

The present work has significance beyond the immediate question
concerning the role of potentially oxygen-independent antimicrobial
systems for the killing capacities of PMN. PMN are probably called
upon to function under anaerobic conditions along the gingival
margin where fastidiously anaerobic bacteria are known to flourish
and in a variety of abscesses -- for example abscesses of the
liver where Sabbaj et al. (13) have shown that fastidiously
anaerobic Bacteroides fragilis are significant pathogens. Even

in these situations it appears that there may be a balance relatively
favorable to the host who is often able to contain the organisms
for long periods. It is nevertheless, still relevant to query how
much PMN contribute to the balance between host and parasite under
conditions of liver abscesses. To what extent their oxygen-
independent systems contribute to antimicrobial action they bring
to bear in tissue, remains to be explored in future studies.

SUMMARY

The resistance of <u>Salmonella</u> <u>typhimurium</u> to the anaerobic
antimicrobial action of human neutrophil granulocytes decreased
as the lipopolysaccharide chain length in the outer membrane of
the bacterial decreased.

At this point our results support the hypothesis that oxygen-
independent antimicrobial mechanisms are operative in the killing
of <u>Salmonella</u> <u>typhimurium</u> LT2 and its mutants by PMNs. Indeed
they may comprise a substantial component of such killing under
the conditions we employed.

ACKNOWLEDGMENTS

This work was supported by USPH grant AI 16624.

REFERENCES

1. Klebanoff, S.J. (1975). Antimicrobial leukocytes. Semin.
        Hematol. 12:117-142.
2. Spitznagel, J.K. (1980). Oxygen-independent antimicrobial
        systems in polymorphonuclear leukocytes, in The Reticulo-
        endothelial System: A Comprehensive Treatise. Bio-
        chemistry and Metabolism. Sbarra, A.J. and Strauss, R.,
        eds., New York: Plenum Press. p. 355.
3. Rest, R.F., Cooney, M.H. and Spitznagel, J.K. (1978). Bacter-
        icidal activity of specific and azurophil granules from
        human neutrophil studies with outer membrane mutants of
        <u>Salmonella</u> <u>typhimurium</u> LT2. Infect. Immun. 19:131.
4. Modrzakowski, M.C. and Spitznagel, J.K. (1979). Bacteri-
        cidal activity of fractionated granule contents from
        human polymorphonuclear leukocyte antagonism of granule
        cationic proteins by lipopolysaccharide. Infect. Immun.
        25:597.
5. Rest, R.F. and Spitznagel, J.K. (1977). Myeloperoxidase
        $Cl^--H_2O_2$ bactericidal system;effect of bactericidal
        membrane structure and growth conditions. Infect. Immun.
        19:1110.

6. Boyum, A. (1968). Isolation of mononuclear cells and granulo-
    cytes from human blood isolation of mononuclear cells by
    one centrifugation and of granulocytes by combining cen-
    trifugation and sedimentation at 1 g. Scand. J. Clin.
    Lab. Invest. 21 (Suppl. 97):77.

7. Baehner, R. L., Boxer, L. A. and Davis, J. (1976). The bio-
    chemical basis of nitroblue tetrazolium reduction in
    normal human and chronic granulomatous disease poly-
    morphonuclear leukocytes. Blood, 48:309.

8. Pryzwansky, K.B., MacRae, E.K., Spitznagel, J.K. and Cooney,
    M.H. (1979). Early degranulation of human neutrophils;
    immunocytochemical studies of surface and intracellular
    phagocytic events. Cell 18:1025.

9. Babior, B.M., Kipnes, R.S. and Curnutte, J.T. (1973). Bio-
    logical defense mechanisms in the production by leuko-
    cytes of superoxide a potential bactericidal agent.
    J. Clin. Invest. 52:741.

10. Root, R.K. and Metcalf, J.A. (1977). H_2O_2 release from human
    granulocytes during phagocytosis: Relationship to super-
    oxide anion formation and cellular catabolism of H_2O_2
    studies with normal and cytochalasin B treated cells.
    J. Clin. Invest. 60:1266.

11. Kossack, R.E., Guerrant, R.L., Densen, P., Schaddlin, J. and
    Mandell, G.L. (1981). Diminished neutrophil oxidative
    metabolism after phagocytosis of virulent Salmonella
    typhimurium. Infect. Immun. 31:674-678.

12. Mandell, G.L. (1974). Bactericidal activity of aerobic and
    anaerobic polymorphonuclear neutrophils. Infect. Immun.
    9:337.

13. Sabbaj, J., Sutter, V.L. and Finegold, S.M. (1972). Anaerobic
    pyogenic liver abscess. Ann. Int. Med. 77:629.

OXIDATIVE METABOLISM OF LEUKOCYTES AND ITS RELATIONSHIP TO

BACTERICIDAL ACTIVITY

Lawrence R. DeChatelet, Pamela S. Shirley and Linda C. McPhail

The Department of Biochemistry, The Bowman Gray School of Medicine, Winston-Salem, North Carolina 27103

INTRODUCTION

During the course of phagocytosis, normal polymorphonuclear leukocytes undergo remarkable alterations in oxidative metabolism which are insensitive to cyanide or azide. These events, collectively referred to as the respiratory burst, are listed in Table 1. Although neutrophils are equipped with a variety of bactericidal weapons, including non-oxidative processes such as hydrolytic enzymes and cationic proteins, several lines of evidence suggest that the respiratory burst plays a major role in the killing of

Table 1. The Respiratory Burst

I.   Increased oxygen consumption
II.   Increased glucose oxidation via the hexose monophosphate shunt
III.   Generation of hydrogen peroxide
IV.   Generation of superoxide anion
V.   Generation of chemiluminescence
VI.   Reduction of tetrazolium dyes

many bacteria. Bacteria are ingested well under anaerobic conditions but most are not killed efficiently in this situation (1). Further, cells obtained from patients with chronic granulomatous disease fail to elicit a normal respiratory burst and are unable to adequately kill many types of microorganisms (2). Patients with this disease are highly susceptible to severe pyogenic bacterial infections which are frequently life-threatening.

Although there is a general consensus that the respiratory burst is crucial for normal bactericidal activity, there is, as yet, no clear definition of the molecular events involved in the killing process. There is, however, no scarcity of candidates as indicated by the list of postulated bactericidal mechanisms in Table 2.

The first major group of suggested mechanisms invokes some reaction of hydrogen peroxide while the second suggests that free radi-

Table 2. Summary of Postulated Oxidative Bactericidal Mechanisms
           of Human Neutrophils

   I. H_2O_2-Dependent

       A. Myeloperoxidase-mediated

          1. Iodination

          2. Aldehyde formation

          3. Peptide cleavage

       B. Myeloperoxidase-independent

          1. Lipid peroxidation

          2. Interaction with ascorbate

  II. Reactive Forms of Oxygen

       A. Superoxide anion (O_2^-)

       B. Singlet oxygen ($^1O_2$)

       C. Hydroxyl radical (OH·)

---

cal intermediates of oxygen metabolism are involved. The hydrogen peroxide-dependent mechanisms may, in turn, be subdivided into two classes depending upon whether myeloperoxidase (MPO) is involved. Klebanoff (3) has clearly shown that the combination of MPO with H_2O_2 and a halide (either Cl^- or I^-) is bactericidal in vitro and that ^{125}I is covalently attached to bacteria. This iodination reaction might be expected to result in bacterial death. On the other hand, if chloride is employed as the cofactor, aldehydes are generated from free amino acids and these might be bactericidal (4). There is some evidence that an analogous reaction can occur within protein chains on the surface of the bacteria, resulting in the cleavage of peptide bonds (5). Recent evidence has indicated that stable choramine derivatives of bacterial components might be formed by this system (6). Although the MPO-mediated reactions are attractive and well described, their importance is undermined somewhat by the discovery that MPO-deficient cells show a relatively modest bactericidal defect in vitro and such patients are not plagued with the severe bacterial infections which afflict patients with chronic granulomatous disease (7). To circumvent this problem, several mechanisms which require H_2O_2 but are independent of MPO have been postulated. Shohet has suggested that the peroxidation of unsaturated fatty acids such as arachidonate might be crucial (8); however, normal bacteria contain almost no polyunsaturated fatty acids, so this explanation is not likely. Miller has demonstrated that ascorbate in concert with H_2O_2 forms a bactericidal system which acts synergistically with lysozyme (9). The significance of this reaction in the intact cell is questionable, however, since Stankova et al. found normal bactericidal activity in scorbutic guinea pigs (10).

With the observation by Babior (11) that phagocytizing neutrophils generate significant quantities of the highly reactive superoxide anion, attention was focused in the direction of free radicals as bactericidal agents. The observation that neutrophils generate chemiluminescence during phagocytosis (12) suggested the possible production of singlet oxygen while other evidence has accumulated to indicate that the hydroxyl radical is likewise generated by the cell (13). Although some inhibitor studies have suggested these radicals might be involved in bactericidal activity (14), their true role is still very hazy. We are left with the knowledge that oxidative processes are crucial to the cell, but have no certainty regarding the species involved or the molecular basis of bacterial cell damage.

The mechanism of initiation of the respiratory burst has been the subject of much recent work and will be the topic of the present communication. Most work currently revolves about the activation of a reduced pyridine nucleotide oxidase. Babior has presented evidence for the involvement of the following reactions (15):

$$1. \quad NAD(P)H + 2O_2 \xrightarrow{\text{Oxidase}} NAD(P) + 2O_2^- + 2H^+$$

$$2. \quad 2O_2^- + 2H^+ \longrightarrow H_2O_2 + O_2$$

This mechanism would immediately explain the increased oxygen consumption and production of superoxide anion which accompany phagocytosis. The oxidized form of the pyridine nucleotide would stimulate the hexose monophosphate shunt, while the spontaneous dismutation of superoxide anion would generate hydrogen peroxide. The superoxide anion could likewise engage in secondary reactions which would give rise to chemiluminescence and is capable of reducing nitroblue tetrazolium dye to the insoluble blue formazan. Thus, these reactions could readily explain all the phenomena associated with the respiratory burst as listed in Table 1. The two equations listed represent the simplest possible sequence. It is also possible that the reaction is initiated by activation of a specific oxidase and then amplified by a non-enzymatic chain reaction involving the superoxide anion. Curnutte et al. have demonstrated that such a chain reaction occurs in the presence of Mn^{2+}, but the physiologic significance of such a Mn^{2+}-catalyzed chain is questionable (16).

MATERIALS AND METHODS

Oxidase activity may be assayed in a variety of ways. Our laboratory has employed assays for the oxidized form of the nucleotide (NAD or NADP) with either a sensitive isotopic or fluorometric procedure. Details of the methods have been previously published (17,18). Briefly, reduced pyridine nucleotide is incubated in 0.1 M MES (2-[N-morpholino]ethane sulfonic acid) buffer, pH 6.0 in the presence of 1 mM KCN. Reaction is initiated by the addition of 0.10 mg of a crude granule fraction derived from a homogenate of human neutrophils. The cells are incubated at 37° for three minutes in buffer (resting cells) or in the presence of opsonized zymosan (phagocytizing cells). The reaction is stopped by the addition of an equal volume of cold 0.68 M sucrose and the cell suspensions vigorously homogenized. Debris is removed by centrifugation at 500 g and the supernate is then centrifuged at 27,000 g. The 27,000 g pellet is referred to as the "granule fraction" for convenience, recognizing that it contains a variety of material from the cell. Careful analysis has revealed no significant differences in the content or specific activity of eight different enzymes in the granule fractions derived from resting as opposed to phagocytizing cells. The crude granule fractions are suspended in 0.34 M sucrose, and the protein content of each is adjusted to a concentration of 1.0 mg/ml for use in the various

Table 3. NADPH Oxidase Activity in Granules Isolated from Resting and Phagocytizing Cells[a]

Conditions	NADPH Oxidase nmol NADP/30 min/0.10 mg protein
Spontaneous Oxidation of NADPH	4.68 ± 0.22 (58)
Granule Fraction from Resting Cells	3.70 ± 0.32 (40)
Granule Fraction from Phagocytizing Cells	14.65 ± 0.94 (64)

[a] The isotopic assay was employed in these experiments at an NADPH concentration of 0.17 mM. Values represent the mean ± S.E. The number of individual experiments is given in parentheses. Each value was determined in triplicate in each separate experiment.

assays. Controls are routinely run in the absence of cellular material to account for the spontaneous oxidation of the reduced nucleotide.

RESULTS AND DISCUSSION

Table 3 illustrates results compiled over a period of years in which as many as 64 separate granule fractions were assayed for NADPH oxidase activity by the isotopic assay procedure. Whereas, the granule preparation from resting cells showed no activity above that of the spontaneous at this level of substrate, activity was consistently observed in the granule fraction from cells which had previously been exposed to opsonized zymosan. These data indicate that the enzyme is activated by phagocytosis.

Table 4 compares data employing cells from normal donors and patients with CGD. These experiments were paired, i.e. control and CGD cells were run in parallel on the same day. Nine separate patients with documented CGD were studied on as many as fourteen separate occasions. The data employing normal cells are consistent with those previously summarized in Table 3. The resting granule fraction in normal cells shows no activity above the spontaneous value, while appreciable activity is observed using the granule fraction from cells exposed to opsonized zymosan. In contrast, no

Table 4. NADPH Oxidase Activity in Granules Isolated from CGD and
          Normal Cells[a]

| Conditions | NADPH Oxidase | |
| | nmol NADP/30 min/0.10 mg protein | |
	Normal	CGD
Spontaneous Oxidation of NADPH	5.45 ± 0.82 (12)	5.45 ± 0.82 (12)
Granule Fraction from Resting Cells	4.76 ± 0.79 (14)	3.48 ± 0.27 (12)
Granule Fraction from Phagocytizing Cells	16.98 ± 3.48 (14)	4.42 ± 0.96 (14)

[a]The isotopic assay was employed with an NADPH concentration
of 0.17 mM. Values represent the mean ± S.E. The number of
separate paired experiments is given in parentheses. Each
value was determined in triplicate in each experiment.

activity above the spontaneous value is observed with either the
resting or phagocytizing granule fraction from CGD cells. It should
be emphasized that these experiments employed an NADPH concentration
of 0.17 mM, which is near the estimated physiological concentration
of 0.05-0.10 mM (19). These data indicate that the NADPH oxidase
is somehow defective in CGD. The defect appears to lie in an inability to activate the enzyme by phagocytosis.

There has been considerable controversy in the literature
regarding the nature of the reduced pyridine nucleotide as substrate, with some favoring NADH (20,21) and others NADPH (22,23).
Accordingly, we examined the activity of granule fractions from
normal cells towards both substrates by means of the fluorometric
assay. A compilation of numerous experiments is illustrated in
Fig. 1. Oxidase activity was determined as a function of substrate
concentration. With either NADH or NADPH, the granule fraction
from phagocytizing cells showed increased activity over that
observed in a granule fraction from resting cells. The kinetics
of the reaction with the resting granule fraction were non-linear
with either substrate and suggest the possibility that activation
might involve an allosteric transition in the enzyme, as has been
previously hypothesized (24). The marked similarity in the shape
of the curves and the activation by phagocytosis of activity towards
both substrates suggest that a single enzyme is involved which can
oxidize either NADH or NADPH. The fact that significantly greater
oxidase activity is observed with NADPH suggests that this would
be the preferred physiologic substrate.

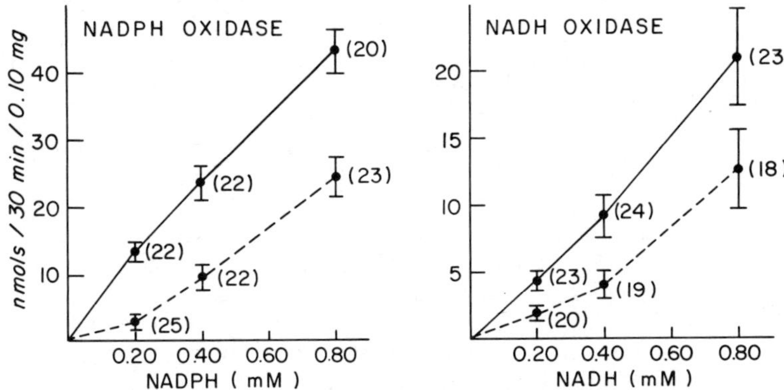

Figure 1. Reduced pyridine nucleotide oxidase(s) as a function of substrate concentration. Values are the mean ± S.E. for the number of experiments given in parentheses. Each value was determined in triplicate in each experiment. Dotted line - granule fraction obtained from resting cells; Solid line - granule fraction from cells exposed to opsonized zymosan.

The respiratory burst may be stimulated in intact cells not only by phagocytosis but also by various soluble stimuli such as phorbol myristate acetate (PMA) (25). If NADPH oxidase were the initiator of the respiratory burst, then it should be activated by such stimuli. Fig. 2 compares the NADPH oxidase activity in granule

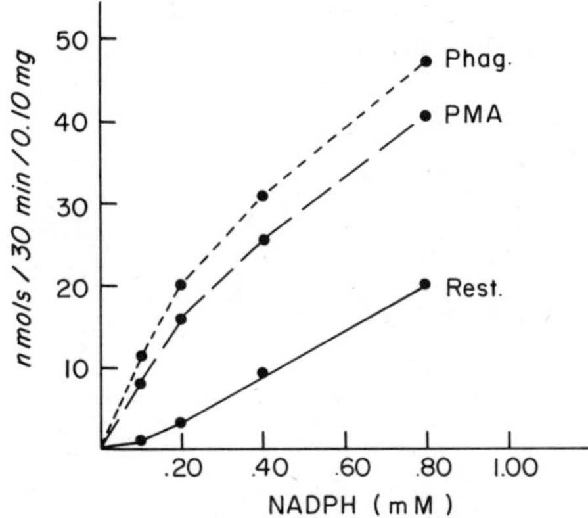

Figure 2. NADPH oxidase activity in granule preparations from resting cells and cells activated by exposure to phorbol myristate acetate or opsonized zymosan. Each value represents the mean of closely agreeing triplicate determinations.

fractions derived from resting cells to those obtained from cells
stimulated with either PMA or phagocytosis of opsonized zymosan.
It is apparent from these data that both soluble and particulate
stimuli, which trigger the respiratory burst in intact cells,
activate NADPH oxidase to a similar degree.

Experiments described to this point have involved activation
of the enzyme in intact cells. We next isolated a granule fraction
from normal resting cells and attempted to activate the enzyme in
vitro by various manipulations. We observed that dialysis of a
granule fraction from resting cells resulted in a marked activation
of enzyme activity similar to that observed in a granule fraction
isolated from phagocytizing cells (Fig. 3). Oxidase activity toward
both NADH and NADPH was increased following dialysis. As before,
the activity with NADPH was significantly greater than that with
NADH. These data also are consistent with a single enzyme which
preferentially oxidizes NADPH but also shows activity toward NADH.
In spite of numerous efforts, we were able to demonstrate the
presence of an inhibitor in the dialysate, so the mechanism of
activation in this case is presently unknown.

All of the assays described have measured the production of the
oxidized form of pyridine nucleotides as a measure of oxidase
activity. It is possible that the cell possesses a number of enzymes
capable of oxidizing NAD(P)H which might not necessarily be involved
in the repiratory burst. Because of this, an independent measure
of oxidase activity was desirable. Iyer et al. (26) first demonstrated the generation of H_2O_2 by phagocytizing neutrophils by means
of formate oxidation. In this reaction, catalyzed by the enzyme
catalase, ^{14}C-formate is oxidized to $^{14}CO_2$ in the presence of H_2O_2.
This assay had previously been employed only with intact cells under
resting and phagocytizing conditions. We adapted the assay to
measure H_2O_2 production by a granule fraction in the presence of

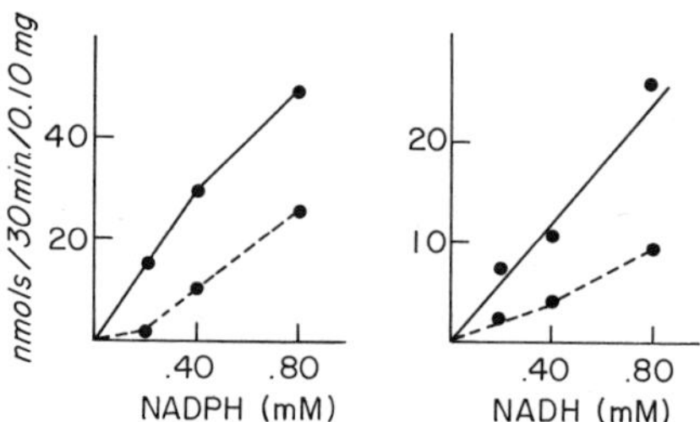

Figure 3. Activation of pyridine nucleotide oxidase by dialysis.
           Each value represents the mean of triplicate determina-
           tions. Data are from a single experiment which is repre-
           sentative of at least five separate experiments. Dotted
           line - resting granule fraction assayed immediate; Solid
           line - same fraction assayed after overnight dialysis.

reduced pyridine nucleotides. Each incubation flask contained
0.10 M MES buffer, pH 6.0; 1400 units catalase; 0.40 mM NAD(P)H;
and 1 µCi (0.3 mM) ^{14}C-formate. Reaction was generally initiated
by the addition of 0.20 mg of granule fraction, and $^{14}CO_2$ released
during the course of the reaction was trapped in a center well
containing potassium hydroxide. Preliminary experiments employing
glucose-glucose oxidase as a source of H_2O_2 demonstrated linearity
between enzyme concentration and $^{14}CO_2$ released from $H^{14}COOH$,
indicating that the assay was a sensitive and accurate indicator
of H_2O_2 production.

Results obtained with this assay were consistent with those
obtained previously measuring oxidized nucleotide as illustrated
in Table 5. Granule fractions from phagocytizing cells always
produced more H_2O_2 than corresponding fractions from resting cells,
with either NADH or NADPH as substrate. Higher activity was
consistently observed in the presence of NADPH than NADH, in agreement with the results shown in Fig. 1. These data reinforce the
suggestion that a single enzyme may be responsible for oxidation
of either nucleotide. Of considerable interest is the observation
that superoxide dismutase inhibits H_2O_2 production from either
nucleotide. This observation is further documented in Table 6.
The generation of H_2O_2 by a granule fraction from phagocytizing
cells is shown to be dependent upon the addition of NADPH. It is
inhibited by substances which react with superoxide anion (SOD and
cytochrome c) and markedly stimulated by the presence of Mn^{2+} which
has previously been demonstrated to catalyze a non-enzymatic chain
reaction (16). The data are consistent with a mechanism of initiation of the respiratory burst which is initiated by a specific
enzyme and propagated by a free radical chain reaction. The simplest

Table 5. Formate Oxidation as a Measure of Hydrogen Peroxide
Production [a]

| | C.P.M. in $^{14}CO_2$ | | | |
| | NADH | | NADPH | |
Granule Fraction Source	−SOD	+SOD	−SOD	+SOD
Resting Cells	2,631	930	6,850	1,606
Phagocytizing Cells	9,851	2,239	25,073	6,576

[a]The experiment is one which is representative of six
separate experiments. Each value was the mean of closely
agreeing triplicate determinations. SOD was added, where
indicated, to a final concentration of 0.7 µg/ml.

Table 6. Effect of Various Compounds on Formate Oxidation by a
          Granule Fraction for Phagocytizing Cells[a]

Description	C.P.M. in $^{14}CO_2$
− NADPH	1,575
+ NADPH	44,980
+ SOD (0.7 µg/ml)	12,593
+ cytochrome c (0.03 mM)	11,700
+ Mn^{2+} (0.5 mM)	219,547

[a]The data are from one experiment which is representative
of two separate experiments. Each value was determined in
triplicate.

such sequence is illustrated below:

$$1. \quad 2O_2 + NADPH \xrightarrow{Oxidase} 2O_2^- + NADP^+ + H^+$$

$$2. \quad O_2^- + NADPH + H^+ \longrightarrow H_2O_2 + NADP \cdot$$

$$3. \quad NADP \cdot + O_2 \longrightarrow NADP^+ + O_2^-$$

$$4. \quad O_2^- + O_2^- + 2H^+ \longrightarrow O_2 + H_2O_2$$

The first reaction is the initiation step catalyzed by NAD(P)H
oxidase which results in the production of O_2^-. Reactions 2 and 3
are non-enzymatic propagation steps which generate both H_2O_2 and
NADP and serve to regenerate O_2^- to continue the sequence.
Reaction 4 involves the dismutation of two molecules of superoxide
anion and serves as a chain termination step. Such a chain reaction
has been postulated by others (27,28) and would permit the generation of a substantial quantity of reactive oxygen products by
activation of only a small number of enzyme molecules. The significance of this mechanism in the intact cells is, however, not yet
known.

SUMMARY

Although human neutrophils possess a wide variety of bactericidal mechanisms, they appear to rely heavily on the generation of
active metabolites of oxygen, including hydrogen peroxide and

superoxide anion for the lethal event. This is most clearly observed
in the case of patients with chronic granulomatous disease (CGD)
whose cells fail to produce these substances during phagocytosis
and show a markedly impaired ability to kill many types of bacteria.
The mechanism of the oxidative changes was investigated by assaying
for reduced pyridine nucleotide oxidase activity. The activity
appeared to be increased by phagocytosis since a granule fraction
from phagocytizing cells showed greater activity than one from
resting cells. This increased activity was observed with either
NADH or NADPH as substrate, but was more pronounced with the latter.
No activation toward either substrate was observed when a fraction
from CGD cells was employed as the source of enzyme. Soluble
stimuli such as phorbol myristate acetate (which cause respiratory
burst activity in intact cells) likewise activate NADPH oxidase.
The enzyme may be activated _in_ _vitro_ by dialysis of a resting granule fraction against either sucrose or distilled water. These
results were confirmed by assaying directly for H_2O_2 production,
utilizing the oxidation of radiolabelled formate to $^{14}CO_2$. This
reaction was dependent upon the presence of reduced pyridine
nucleotide and was greater in fractions from phagocytizing than
resting cells. NADPH was a better substrate than NADH and the
reaction was inhibited by substances which trap superoxide anion.
These results suggest that the respiratory burst in human neutrophils is initiated by an oxidase which preferentially utilizes
NADPH. The oxidase appears to be defective in CGD and the mechanism
for H_2O_2 production likely involves an initial enzymatic reaction
followed by a non-enzymatic chain reaction.

ACKNOWLEDGMENTS

     The research reported from my laboratory was supported by
NIH grant AI-10732. The excellent editorial assistance of
Ms. Gwen Charles is gratefully acknowledged.

REFERENCES

1. McRipley, R. J., and Sbarra, A.J. (1967). J. of Bacteriol.
        94:1417.
2. Holmes, B., Page, A.R., and Good, R.A. (1967). J. Clin.
        Invest. 46:1422.
3. Klebanoff, S.J. (1967). J. Exp. Med. 126:1063.
4. Strauss, R.R., Paul, B.B., Jacobs, A.A., and Sbarra, A.J.
        (1971). Infect. Immun. 3:595.
5. Selvaraj, R.J., Paul, B.B., Strauss, R.R., Jacobs, A.A., and
        Sbarra, A.J. (1974). Infect. Immun. 9:255.
6. Thomas, E.L. (1979). Infect. Immun. 23:522.
7. Klebanoff, S.J., and Hamon, C.B. (1972). J. Reticuloendothel.
        Soc. 12:170.

8. Shohet, S.B., Pitt, J., Baehner, R.L., and Poplack, D.G.
      (1974). Infect. Immun. 10:1321.
9. Miller, T.E. (1969). J. Bacteriol. 98:949.
10. Stankova, L., Gerhardt, N.B., Nagel, L.,and Bigley, R.H. (1975).
      Infect. Immun. 12:252.
11. Babior, B.M., Kipnes, R.S., and Curnutte, J.T. (1973). J. Clin.
      Invest. 52:421.
12. Allen, R.C., Stjernholm, R.L., and Steele, R.H. (1972).
      Biochem. and Biophys. Res. Commun. 47:679.
13. Rosen, H., and Klebanoff, S.J. (1979). J. Clin. Invest. 64:
      1725.
14. Johnston, R.B., Jr., Keele, B.B., Jr., Misra, H.P., Lehmeyer,
      J.E., Webb, L.W., Baehner, R.L. and Rajagopalan, K.V. (1975).
      J. Clin. Invest. 55:1357.
15. Babior, B.M., Curnutte, J.T., and McMurrich, B.J. (1976).
      J. Clin. Invest. 58:989.
16. Curnutte, J.T., Karnovsky, M.L., and Babior, B.M. (1976).
      J. Clin. Invest. 57:1059.
17. DeChatelet, L.R., McPhail, L.C., Mullikin, D., and McCall, C.E.
      (1975). J. Clin. Invest. 55:714.
18. Iverson, D., DeChatelet, L.R., Spitznagel, J.K., and Wang, P.
      (1977). J. Clin. Invest. 59:282.
19. DeChatelet, L.R., McPhail, L.C., Mullikin, D., and McCall,
      C.E. (1974). Infect. Immun. 10:528.
20. Baehner, R.L., and Karnovsky, M.L. (1968). Science 162:1277.
21. Segal, A.W., and Peters, T.J. (1977). Clin. Sci. and Mol.
      Med. 52:429.
22. Patriarca, P., Cramer, R., Moncalvo, S., Rossi, F., and
      Romeo, D. (1971). Arch. of Biochem. and Biophys. 145:255.
23. Hohn, D.C., and Lehrer, R.I. (1975). J. Clin. Invest. 55:707.
24. DeChatelet, L.R., Shirley, P.S., McPhail, L.C., Iverson, D.B.,
      and Doellgast, G.J. (1978). Infect. Immun. 20:398.
25. DeChatelet, L.R., Shirley, P.S., and Johnston, R.B., Jr.
      (1976). Blood 47:545.
26. Iyer, G.Y.N., Islam, D.M.F., and Quastel, J.H. (1961).
      Nature 192:535.
27. Patriarca, P., Dri, P., Kakinuma, K., Tedesco, F., and Rossi,
      F. (1975). Biochim. et Biophys. Acta 285:380.
28. Bellavite, P., Berton, G., and Dri, P. (1980). Biochim.et
      Biophys. Acta 591:434.

SOME PARADOXES OF MACROPHAGE FUNCTION

Mayer B. Goren

Department of Molecular and Cellular Biology, National
Jewish Hospital and Research Center; and Department of
Microbiology and Immunology, University of Colorado
Health Sciences Center, Denver, Colorado   80206

INTRODUCTION

This paper, dealing with several curious paradoxes in macrophage function, is principally in the nature of an editorial
review, although some recent, as yet unpublished, work from our
laboratory is described. In its preparation, I was at first
dismayed to realize that it raises so many questions, but answers
very few. Still the purpose of such an effort is to arouse
interest in what I believe are provocative issues, with the hope
that their examination and analysis might stimulate some productive
research, whether in my own group or by others. Both consequences
are manifestations of the creative process that drives us all, and
therefore they bring both participatory and vicarious satisfaction.

The paradoxes relate principally to the enzyme myeloperoxidase,
and to various expressions of its functionality; to manipulation of
phagosome-lysosome fusion; and to the contributions of the secondary
lysosomal system to intracellular digestion. Specific questions
that are addressed concern 1) the limited functionality so far
demonstrated for granulocyte myeloperoxidase that is acquired by
resident peritoneal macrophages through endocytosis; 2) the
nature of peroxidatic activity that is reputedly acquired by
macrophages when they are stimulated to a tumoricidal level of
activity; 3) the failure of polyanionics sequestered in secondary
lysosomes to inhibit digestion of endocytosed substrates despite
evidence that the polyanionics antagonize fusion of the secondary
lysosomes with endocytic vacuoles on the one hand; or on the
other, directly antagonize the activity of lysosomal enzymes; and

4) the apparent Janus-faced behavior of lysosomotropic weak bases
upon sequestration in lysosomes.

THE PEROXIDASE PARADOX

     Polymorphonuclear leukocytes, eosinophils, promononocytes
and monocytes (except for rabbit monocytes (1)) are more or less
abundantly endowed with the granule enzyme myeloperoxidase (MPO)--
a substance that has been almost unassailably implicated in the
bactericidal activity of these cells (2-4). But macrophages,
which are believed to be the functionally mature cells of the
mononuclear phagocyte system (MPS) are essentially devoid of this
enzyme (5,6). The dilemma has stimulated much of the activity to
delineate alternative oxygen-dependent (7) and oxygen-independent
(8,9) killing mechanisms that might be implicated in the bactericidal capacities of these cells.

     Klebanoff has convincingly shown in many years of study that
in neutrophils and monocytes, H_2O_2 generated in the oxidative burst
that accompanies phagocytosis, very likely participates with MPO
and halides (iodide, chloride) to generate potent microbicidal
agents. With iodide as the cofactor, an important expression of
the Klebanoff system which is conveniently measured and that is
believed to contribute to microbial killing, is the iodination of
bacterial protein, readily demonstrated in the presence of radioiodide. Therefore, the denial to the macrophage of MPO, reputedly
a major antimicrobial armament, is what I have earlier referred to
as the "peroxidase paradox" (10). But, more recently I have been
constrained to wonder if this indeed is a paradox or, instead, is
no more than an alluring but misleading alliterative expression.
Does accumulating evidence suggest that so-called "resident" macrophages are less active, poorer in functions, and perhaps of lesser
importance than accord with that prominence with which we have
endowed them? Some of the evidence that we will consider suggests
that this may be so.

     Elicited macrophages are the interesting exception to the MPO-
famished mononuclear phagocyte. With the peritoneal exudate cell
as an example, this heterogeneous, mostly peroxidase-positive
population may be considered to consist of somewhat altered
resident macrophages on one hand; and on the other, of monocytes
and monocyte-derived MPS cells that were recruited to the site from
the peripheral blood. The stimulus for the elicitation first
brings to the site what I have referred to as "suicidally-programmed
granulocytes" (10). It is reasonable that the various MPS cells,
including the resident macrophages and those which were recruited
to the site, participate together in scavenging operations to clean
up the debris of the short-lived granulocytes; and abundant evidence
has been documented that this is indeed the case (by Atwal, 11;

Daems, 12,13; Nichols and Bainton, 1; vanFurth et al., 5; and
others). It seems teleologically appealing then that resident
macrophages might, by this primitive scavenging, _reacquire_--and
differentiating monocytes might _replenish_ a lysosomal form of
myeloperoxidase, which they obtain from ingesting the dead and
dying granulocyte population. The probable validity of this
interpretation is abundantly supported in numerous illustrations
provided by the various investigators cited immediately above
(see 1,13). And the acquired granulocyte lysosomal enzymes are
logically considered then as circulating within the mononuclear
cells' secondary lysosomal system. It of course is also reasonably
anticipated that other enzymes, relatively restricted to granulocytes, might also be acquired by macrophages-monocytes via
scavenging: lactoferrin, and the special bactericidal cationic
proteins, for example. If we have faith that secondary lysosomes
contain the digestive hydrolases and other enzymes, and that the
lysosomes are capable of fusion with fresh endocytic vacuoles,
then the granulocyte-derived enzymes that were acquired by
scavenging should be functional, at least to a degree.

    Our initial efforts to test this notion in a model system
were recently published (14). We fed effete senescent MPO-rich
human neutrophils or neutrophil granules to resident unelicited
mouse peritoneal macrophages in culture; and we selected Klebanoff
iodination as the test system for functionality. In confirmation
of earlier results by Simmons and Karnovsky (6) we first established that the resident mouse macrophages are entirely incapable
of fixing radioiodide during phagocytosis of zymosan. On the
other hand, identical monolayers phagocytosing zymosan in the
presence of either effete neutrophils, neutrophil granules or
granule lysates fixed significant ^{125}I into a TCA-precipitable
form. Under optimal conditions (Fig. 1), the macrophages did so
as efficiently as an equivalent number of functional human neutrophils. Under these conditions, we were providing the macrophages
the MPO equivalent of about two neutrophils per macrophage (14).
Iodination was most efficient when all of the necessary components
were present simultaneously. But it was also effective, albeit
diminished, if the macrophages were first pulse-fed a "meal" of
neutrophil granules or neutrophil debris and iodination assessed
immediately afterward.

    Our observations in testing the system were consistent with
the interpretation that azide-inhibitable peroxidase (probably
MPO) provided by the neutrophils participated with H_2O_2 generated
by the phagocytosing macrophages to oxidize and to fix radioiodide.

    Our additional findings raised several paradoxical questions,
however, about the nature of this process and of the lysosomal
system. In experiments on pulse feeding of intact senescent
neutrophils to macrophages, the measured peroxidase activity

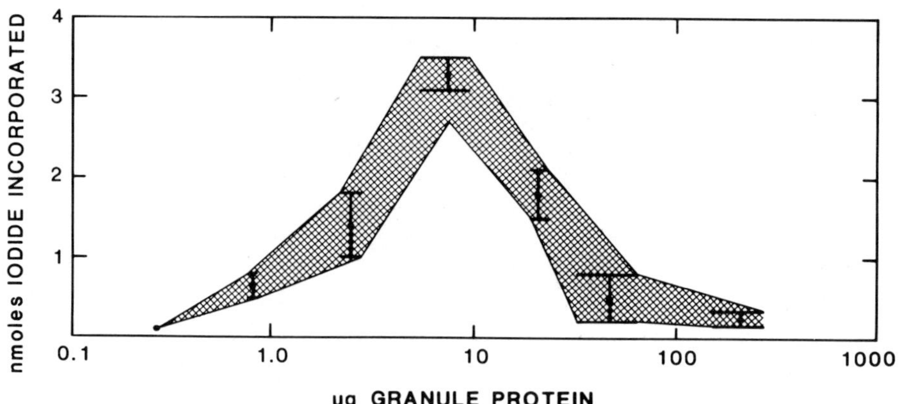

Figure 1.   Fixation of iodide by phagocytosing monolayers as a
            function of neutrophil granule dosage. Graph gives
            standard errors of the mean for both µg granule protein
            and iodine fixed. Reproduced from Heifets, et al. (14)
            with permission of the Journal of the Reticuloendo-
            thelial Society.

incorporated by the cells exhibited a curious cryptic or latent
quality, suggesting that it increased for a time after ingestion,
despite the fact that peroxidase was probably being degraded by
lysosomal hydrolases during the same time. Further, following
pulse feeding of either granules or neutrophils, the temporal
decline in iodinating activity was much greater than would be
expected from the decay of peroxidase activity as measured
enzymatically. Figure 2 shows that the former decayed abruptly,
with a $t_{1/2}$ estimated to be about 1.5 hr. By comparison, as
described in (14), the (hydrolytic) destruction of enyzmatic
activity as measured by the rate of oxidation of ortho-dianisidine
proceeded with a $t_{1/2}$ of about 6 to 9 hours.

The fixation of iodine during phagocytosis of opsonized
zymosan has for some time been viewed as a valid measurement for
assessing one expression of neutrophil function (see especially
Klebanoff and Clark (15)). I find it surprising however that
bactericidal activity may, but often does not, correlate with
iodide fixation (see 16); and my earlier interpretation (10) that
viewed the phenomena as correlated was clearly wrong. Indeed this
was forcefully conveyed to us by the results of our own recent
experiments that compared and found no difference in the bactericidal activity of control populations of macrophages (incapable of
iodine fixation) with that of similar macrophages provided with
an optimum quantity of neutrophil-derived granules and under
conditions commensurate with achieving maximum fixation of iodide.

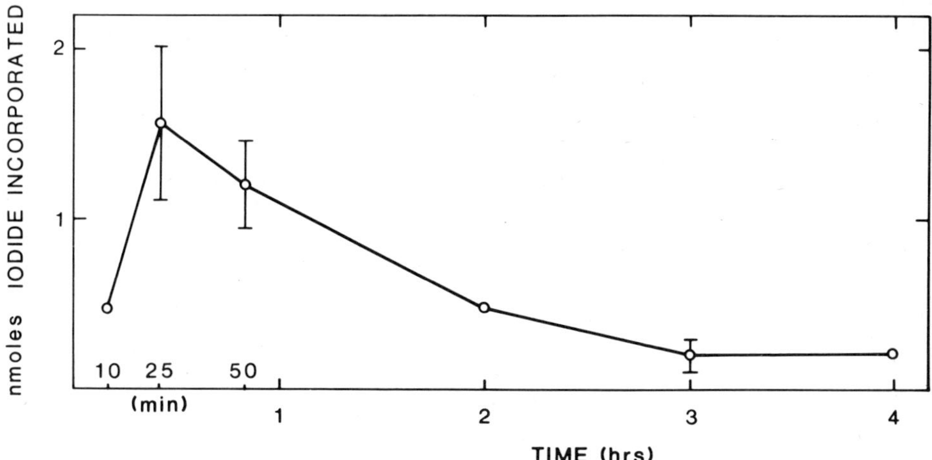

Figure 2.   Decay of "postprandial" iodinating capacity when macrophages were fed with neutrophil granules (15 µg protein)
for various periods of time. Quantitation of iodine
fixation during a subsequent 1.5 hr period followed
immediately after the granule pulse.

Insofar as the logistics of such experiments allowed (and the
complexities should not be dismissed) we found no enhancement of
microbicidal capacity for targets such as S. cerevisiae, C. albicans, or a temperature-sensitive E. coli mutant kindly provided
to us by Dr. Anne Morris Hooke (17). These results are particularly
disappointing when it is recognized that the comparison is made
not between populations that exhibit merely some incremental
difference in iodinating capacity, even a two-to-five-fold
difference. Instead we compared macrophage populations that are
entirely incapable of iodination with populations that fix iodide
as efficiently as fully functional neutrophils--and yet this
profound difference had no measurable effect on microbicidal
activity. We are as yet unable to explain this truly paradoxical
situation.

Lehrer's studies of some years ago seemed to implicate monocyte MPO in killing of Candida species (8). More recent studies
by Sasada and Johnston indicated that elicited peritoneal cells
are more candidacidal than their resident counterparts (18); and
this activity was interpreted as due to a more prolific release
of H_2O_2 by the elicited cells. Whether their content of peroxidase also contributed was not tested, but it seems likely that
for recently recruited monocytes, their content of peroxidase
should participate (vide Lehrer). If this premise is valid, does
the location of the peroxidase in monocyte primary lysosomes

confer or provide a functional quality or accessibility that is
denied to secondary lysosomes in which ingested peroxidase is
sequestered? Can this MPO and other neutrophil granule components
such as lactoferrin and cationic proteins that an elicited macrophage has acquired by a primitive scavenging activity be viewed as
potential true armaments? If so, what more is required to render
them usable? If they are not usable then our concept of the
functions of secondary lysosomes seems to require drastic revisions.

PEROXIDASE AND ITS RELATION TO TUMOR CYTOTOXICITY

     The development of tumor cytotoxicity by macrophages is often
viewed as an expression of a high level of "activation" (19-22).
Nontumoricidal, lesser levels of "stimulation" are developed by
elicitation of (e.g. peritoneal exudate) cells following injection
of casein, thioglycollate broth, or other irritants.

     By comparison with the relatively quiescent status of the
freshly lavaged "resident" peritoneal macrophage, the elicited cell
possesses a significant degree of stimulation and is "primed" for
much greater promotion along the path that Hibbs (19) has designated
the "activation continuum". Excepting elicitation with thioglycollate, which is relatively ineffective in priming for subsequent
development of tumoricidal activity [Ruco and Meltzer (21); compare
however with Hibbs et al. (19)] peritoneal exudate cells obtained
following any one of a number of stimuli may be promoted in vitro
to a tumoricidal level by exposure to one or more of several
stimulants: lipopolysaccharide (19); lymphokine-containing media
generated from exposure of normal lymphocytes to T-cell mitogens,
or of specifically sensitized lymphocytes to the sensitizing
antigen (19-21); or certain serum components (19). There is,
however, considerable evidence that normal resident, non-elicited
peritoneal macrophages can also be stimulated in vitro to a
tumoricidal level: by exposure to larger amounts of lipopolysaccharide (19,23) or to double stranded RNA (23); or, as recited
before, to lymphokines with macrophage-activating properties (20).

     In studies by Ruco and Meltzer (21), later carefully analyzed
by Karnovsky and Lazdins (22) an intriguing feature of the activation process was discerned: invariably the tumoricidal cells were
peroxidase positive. This was taken as evidence of the relative
immaturity of these cells, and was observed principally in
activated macrophages that had been obtained in a two-stage
stimulation process: elicitation (with, for example, fetal calf
serum, latex beads in saline, or starch); followed by in vitro
exposure of the cell harvest to appropriate lymphokines (21).
The elicitation step is of course responsible for the recruitment
of blood derived monocytes--hence the designation of "immaturity"
of the cells. Cytotoxicity was also inducible in harvested

<u>resident</u> mouse peritoneal macrophages by exposing them to super-
natants of PPD-stimulated BCG-immune spleen cell cultures.
"Lymphokine activated tumoricidal macrophages were evident within
4 hr of culture in active supernatants, reached optimal levels by
8 to 12 hr and progressively decreased thereafter" (20). The
concurrent recognition of peroxidase-positivity in these activated
macrophages is not directly addressed in this latter publication,
and whether this was indeed observed was therefore of compelling
interest to Karnovsky and Lazdins in their analysis of the
phenomenon (and to this writer);

"If normal resident or elicited mouse macrophages were
exposed <u>in vitro</u> to supernatants of BCG-sensitized lymphocytes
treated with PPD <u>in vitro</u>, cytotoxicity toward the tumor cells
developed (28). Three matters are notable: first, elicited macrophages were more effective in the system than were normal macrophages; second, the effect persisted for only a short time (20 hr
of culture), and loss thereof was not reversible; third, tumor
cytotoxicity was again correlated with peroxidase positivity.
(The authors are grateful to Dr. Monte Meltzer for information
on this point.)" (22).

It is implicit, however, from a study of the succeeding Ruco
and Meltzer publication (21) that the authors consider the peroxidase positivity not to derive as a consequence of the activation
process <u>per se</u>, but rather that in harvests of either <u>resident</u> or
<u>elicited</u> cells peroxidase positivity is merely associated with
recently arrived or even differentiating monocytes (and offer
evidence in support) and that it is this subpopulation alone (the
relatively immature component) that may ultimately be stimulated
to the tumoricidal level. (We are grateful to Dr. Meltzer for
confirming this interpretation). However, in their studies with
elicited cells no distinction was drawn between MPO-positive
monocytes and macrophages that are MPO-positive as a result of
endocytic activities in the inflammatory focus.

The coexistence of tumor cytotoxicity with peroxidase positivity may be no more than coincidence; and the peroxidase content
may contribute but insignificantly to the expression of tumoricidal
activity--in which event the peroxidase may be simply the marker
that identifies the cell population which can respond appropriately
to the stimulating signals. Alternatively, the peroxidase
(probably MPO) may play a functional role in tumor cell cyto-
toxicity--and once again we are confronted with the question of
whether endocytically acquired MPO can contribute significantly
to this behavior. Certainly there is considerable <u>in vitro</u>
derived evidence that the MPO-H_2O_2-halide system has tumoricidal
activity (24,25). Indeed additional evidence is being gotten in
current collaborative studies kindly described to me by Dr.
Klebanoff and by Dr. Carl Nathan (studies in progress) and with
whose permission the studies are summarized.

Purified MPO derived from horse eosinophils and adsorbed onto chromium-labeled mouse lymphoma cells renders them highly suscept- ible to extracellular killing by BCG-activated macrophages. Un- elicited macrophages alone do not kill these MPO-coated tumor cells, but do so when stimulated (as with PMA) to release H_2O_2 into the medium. The adsorption of the cationic MPO to bacterial surfaces also notably increases its bactericidal potential (see 26-28; see also 29). Although it may be argued that this system appears somewhat artificial, the influx of polymorphs or eosinophils into a focus of infection will quite certainly result in the release of MPO into the milieu, and in its subsequent (even if partial) adsorption.

We are nevertheless constrained to ask whether the activation of macrophages to a tumoricidal level is _always_ accompanied by peroxidase positivity. Several systems have been described--by Alexander and Evans (23) and by Hibbs (19)--that differ in important respects from those reported by Ruco and Meltzer, and which appear to afford legitimate targets for testing the potential universality of the peroxidase phenomenon. The most important difference is in the age of the cells and the time course of activation: in the Ruco-Meltzer system "the capacity of normal resident macrophages to be _activated_ by lymphokines _in vitro_ progressively decreased and was absent by 20 hr in culture" (20). In the studies by Evans and Alexander, non-elicited macrophages first cultured for 24 hr were then activated with either double stranded RNA or endotoxin; and they maintained their antitumor activity for several days. In studies alluded to by Hibbs et al. (19) resident perito- neal macrophages were maintained in culture 48 to 72 hr to develop the "stimulated macrophage" which may then be promoted along the activation continuum with a combination of lymphokines and LPS. In the time periods described it is unlikely that a significant population of the normal resident cells that were MPO-positive (monocytes) at harvest would persist as such. If in these systems peroxidase positivity is indeed found to develop concomitantly with development of tumoricidal capacity by such cell populations, then it would indeed be a quite remarkable and intriguing finding. The recognition and identification of its _origin_ would be of compelling interest.

POLYANIONS, WEAK BASES AND SECONDARY LYSOSOMAL FUNCTION

The evidence to be considered in this section raises disturb- ing questions about the functions and the functionality of macro- phage secondary lysosomes--a theme which in fact characterizes this article as a whole. Among the amazingly diverse (and often antipodal) influences that polyanionic substances have upon cellular functions (30), their sequestration in macrophage

secondary lysosomes allows for their use as tools for manipulating
and probing the behavior of these organelles and their influence
on important intracellular events. Various studies by Hart (31-
33), by Goren (10,34) and by Kielian (35) have implicated polyanionic agents in inducing fusion dysfunction in macrophage
secondary lysosomes so that the delivery of their contents to
phagocytic vacuoles is antagonized. Some agents for which this
behavior has been documented include suramin (31), the mycobacterial
sulfatides, dextran sulfate (34) and poly-D-glutamic acid (32,33,
35; see also 10). However, some contradictory evidence obtained
in ultrastructural examination of cells labeled with electron-
opaque markers for secondary lysosomes denies the evidence
obtained from vital fluorescence microscopy with acridine orange
as lysosomal label (see below). This has raised concerns about
the validity of this fusion antagonism concept (36-38). Although
our efforts for resolving the disparate electron microscopy data
will continue, the consistent evidence from Hart (32,33) the
stereological analyses by Kieliean and Cohn (39) and our own recent
observations (summarized below) are persuasive that the original
"polyanion hypothesis" is probably correct. In that event,
manipulation of the fusion function of secondary lysosomes can
allow dissection of their contribution to specific intracellular
events, to macrophage bactericidal activity, and to intracellular
digestion.

Fluorescent lysosomal labels

     As alluded to in the preceding discussion, phagosome-lysosome
fusion behavior in macrophages has been conveniently studied by
fluorescence microscopy utilizing acridine orange (AO), as the
lysosomal label and viable (32-34) or killed yeasts (compare 32
and 37); or zymosan (35,39) as phagocytic targets. Weak bases
such as AO freely penetrate biological membranes and become
trapped within lysosomes by the lower pH within these organelles.
Sequestration may be enhanced by a proton pump that de Duve has
invoked to account for the high concentration of the weak bases
that can actually accumulate in lysosomes (40). If these AO-
labeled lysosomes fuse with phagosomes the fusion process is
revealed by the colorful influx of dye into these compartments.
Absence of dye transfer, clearly evident in cells exposed to
sufficient polyanions, has been taken as evidence that the fusion
process is inhibited (31,34). We have stressed, however, that AO
is so avidly bound by polyanions that failure to transfer AO to a
phagosome is by no means a reliable indicator of fusion inhibition
(37,38). This is because of the unrestricted traffic through
membranes that the weak bases enjoy. These dyes, even if delivered
to a phagosome, will in the end accumulate where they are most
efficiently trapped--in our experience in polyanion-laden lysosomes.

As a very interesting example that demonstrates this fact, a
control monolayer was labeled with acridine orange in the usual
manner (5 µg/ml, 15 min) and allowed to phagocytize viable yeasts.
After 1.5 hr incubation most of the yeasts were intensely colored
with dye. This monolayer was then left in Hanks medium <u>side-by-</u>
<u>side</u> with a monolayer that had not been exposed to AO but which
had endocytosed mycobacterial sulfatide (SL) and sequestered this
(micellar) polyanion in secondary lysosomes. After 2 hrs of this
side-by-side incubation the AO was essentially quantitatively
transferred to the SL monolayer. The cells of the "donor" monolayer were killed in the process! This may be explained from the
following consideration: the original high accumulation of AO
in the lysosomes required the provision of neutralizing acid
(proton pump?) to trap the dye. When the AO was "siphoned off"
and redistributed to the SL monolayer it could only leave its
original sequestration sites (lysosomes, phagolysosomes) as the
free base and thus would leave behind in these orangelles the
neutralizing acid. It seems likely that the plummeting pH
probably caused lysosomal disruption and killed the cells.

Lissamine-rhodamine as Lysosomal Label

To circumvent the artifacts that may be introduced because of
the variable intracellular traffic that the AO exhibits, we have
sought alternative sensitive lysosomal lables that do not suffer
this deficiency. We have recently found the highly charged
Lissamine Rhodamine (LR) (Fig. 3) to be such a substance (<u>vide</u> 41).
It is highly unlikely that the fluor can freely traverse membranes;
more likely it is probably taken up by "piggy back" endocytosis in
the form of complexes with proteins in the culture medium (40),
and once incorporated in the secondary lysosomal system, remains
there. As will be described in a separate publication, we have
exposed macrophage monolayers to either high concentrations (2 mg/
ml) of LR for several hours, or to lower concentrations (100-400
µg/ml) for up to a week without evidence of overt toxicity.

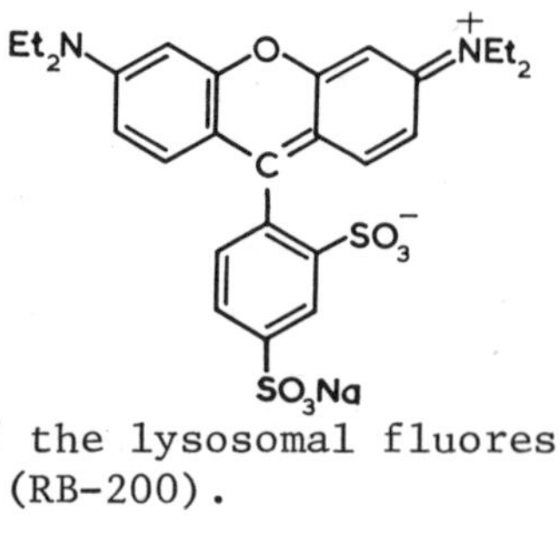

Figure 3.   Structure of the lysosomal fluorescent label Lissamine
             Rhodamine B (RB-200).

As targets for phagocytosis we use killed (boiled) S. cervisiae to
which fluorescein isothiocyanate was conjugated. With combinations
of exciter and barrier filters appropriate for LR, dye delivered to
yeast phagosomes is easily recognizable; while the total number of
yeasts phagocytosed is revealed by exciter and barrier filters
appropriate for fluorescein (41). Figure 4 compares the levels of
fusion seen with control macrophage monolayers and with monolayers
exposed for 5 days to poly-D-glutamic acid. The panorama that is
revealed is not as esthetically exhilarating as that seen with
acridine orange, but we suggest that it may be more trustworthy,
not only because the marker is not free to redistribute between
various membrane bounded organelles, but also because it suggests
what should in fact be expected in biological systems: that fusion
or its abolition are not an all-or-none phenomenon, but that there
are shades of gray. Indeed the stereologic analyses reported by
Kielian and Cohn support this notion (35,39). In our hands
fluorescence microscopy with LR cells shows that in many cells,
very low levels of dye have been delivered to phagosomes even in
polyanion-treated cells (rather poorly reproducible in black and
white renderings--see Fig. 4C). In contrast, the AO technique
simply shows nonfusion in the form of phagosomes that are substantially black spaces against a background of intensely fluorescing
lysosomes (compare 31,33,34,37).

   With this LR labeling technique we have confirmed the induction
of fusion dysfunction by mycobacterial sulfatide, by poly-D-glutamic
acid; and by relatively high concentrations of ammonium ion--a
curious inhibition of the function reported recently by Hart and
Young (42). This latter observations confronts us with still
another paradox. It has been known for a number of years that
accumulation of relatively high levels of NH_4^+ and some other weak
bases (chloroquine) in lysosomes antagonizes lysosomal digestive
processes (see 43,44). This has been interpreted as due to a
raising of the lysosomal pH sufficiently to inhibit the acid hydrolases (45). Hart and Young have reported that certain lysosomotropic weak bases (quinacrine, chloroquine) can relieve the
fusion dysfunction that is imposed by polyanionics (probably by
neutralizing the polyanions), and that these weak bases in fact
promote PL fusion in normal macrophages (33). Why then does NH_4^+
ion act in a totally different manner so that its accumulation
results in fusion inhibition? There is as yet no explanation for
this behavior.

   Although our own evidence with either LR or AO labeled macrophages confirms the block to PL fusion imposed by accumulation of
NH_4^+ ion as reported by Hart and Young (42), we have not been able
to confirm, even with AO-labeled cells, that chloroquine will
alleviate or relieve the block induced in cells previously exposed
to either mycobacterial SL or to poly-D-glutamic acid (compare(37)).

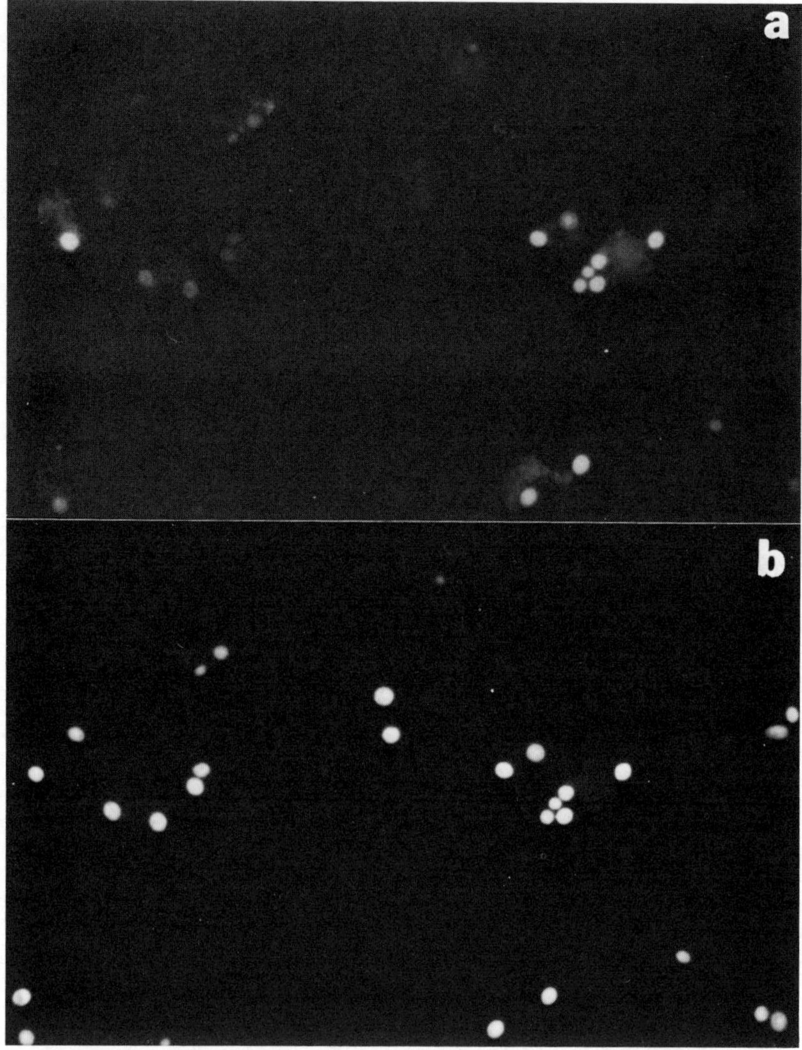

Figure 4. Macrophage monolayers, labeled [fluorescent label
           Lissamine Rhodamine B (RB-200)] and washed, were pulsed
           with opsonized heat-killed baker's yeast that was
           conjugated with fluorescein isothiocyanate. a and b
           show the same field. a) Transfer of RB 200 to yeast
           phagosomes in control monolayer (fluorescence excitation
           at about 530 nm. b) Total of internalized yeasts
           revealed by fluorescein fluorescence (excitation at
           about 400 nm).

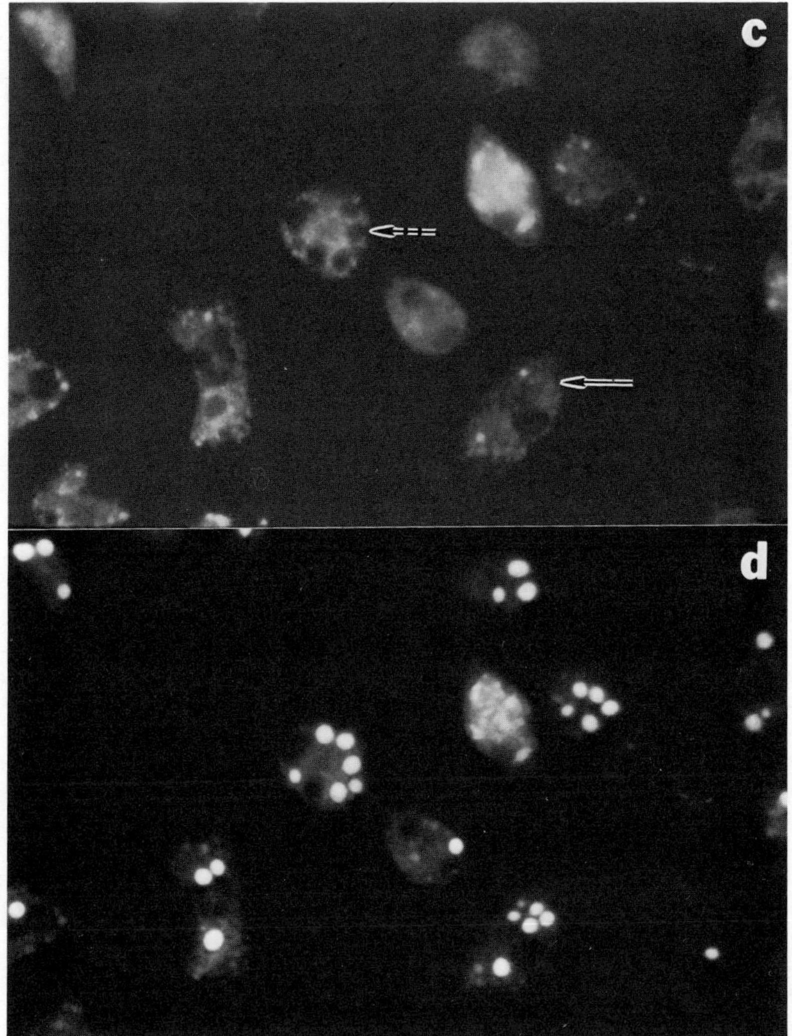

Figure 4. In c and d, the monolayer was preexposed to poly-D-
glutamic acid for 5 days. c) Low levels of RB 200 have
been transferred to some yeast phagosomes (solid arrow)
and some "black holes" are evident where the dye transfer (hence fusion) is especially limited (broken arrow).
d) Same field as c with excitation for fluorescein to
reveal the total of phagocytosed yeasts.

Polyanionics and Intracellular Digestion

A consideration of two properties of appropriately structured polyanionics leads to the deduction that their lysosomal accumulation should result in serious interference with intracellular digestion. First, polyanionics have been demonstrated to inhibit directly the enzymatic activities of a number of lysosomal (and other) hydrolases (46-48); and secondly, as discussed immediately preceding, they induce antagonism to PL fusion--at least for secondary lysosomes. It must be interjected at this point that the lysosomotropic weak bases very likely accumulate in _both_ primary and secondary lysosomes by a process of permeation followed by trapping, whereas the accumulation of polyanionics very likely is via pinocytosis--either "piggyback", or following adsorption to cytoplasmic membrane, which is internalized during endocytic processes. Fusion of these endosomes with either primary or secondary lysosomes sequesters the polyanionics in what are operationally defined as secondary lysosomes. The mycobacterial sulfatides may be a possible exception: from electron microscopic evidence (34,37) it seems probable that lipid droplets of SL may directly penetrate phagosomal and lysosomal membranes so that _fusion_ of SL-containing pinosomes with lysosomes may not be required for their accumulation in lysosomes. However, how can the lysosomal sequestration of other, non-lipidy, polyanionics be explained?--yet another paradox. If membrane bounded organelles that enclose the polyanionics are inhibited from fusion with other vesicles, how do polyanion-containing pinosomes deliver their contents to secondary lysosomes? Do pinosomes, because of their submicroscopic size escape the stricture to fusion that applies to larger vesicles? Again, the answer is not yet known.

Although the lysosomal accumulation of weak bases demonstrably interferes with intracellular digestion (43,44) our recent studies (38, and to be published) surprisingly show that in cells that have accumulated various polyanionics, digestion of endocytosed radiolabeled substrates proceeds at the _same rate_ as in control macrophages. The nature of the substrate has no evident influence: we have used ^{32}P-labeled viable or heat-killed _S_. _cerevisiae_, ^{125}I-keyhole limpet hemocyanin (38), whose processing by macrophages had earlier been studied by Calderon and Unanue (49); and we also tested ^{125}I KLH-anti KLH immune precipitate, and ^{125}I-labeled zymosan. For any of these methods, the degradation and release of radiolabel to the incubation medium were no different in controls as compared with macrophage monolayers exposed to various polyanionics sufficient to inhibit PL fusion as determined, for example, with lissamine-rhodamine labeled cells. Figure 5 shows the time course of release of radiolabel from macrophages that had ingested ^{125}I-KLH. Clearly the rates are identical for control monolayers and for those exposed to sulfatides. Moreover, the data from the earlier study by Calderon and Unanue are

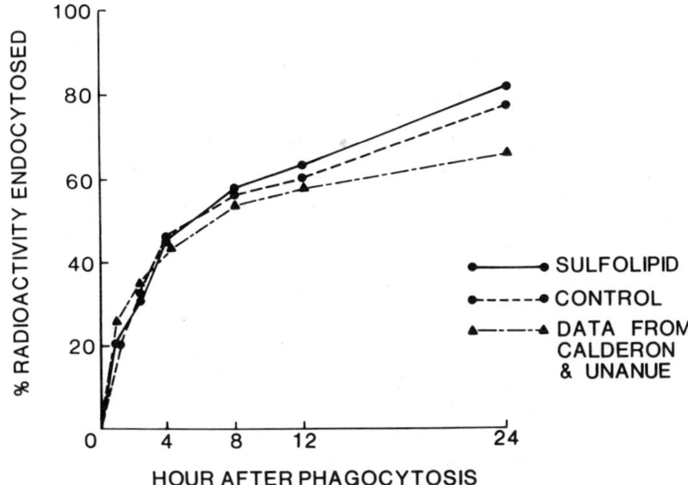

Figure 5.   Digestion of ^{125}I-KLH by control and sulfolipid-exposed
            macrophages as estimated from release of radiolabel
            during 24 hr.   Data from Calderon and Unanue (49) are
            included for comparison.

essentially superimposable on our own.   The results of our most
recent studies are illustrated in Figure 6, which compares the
release of radiolabel (^{125}I) from iodinated zymosan after ingestion

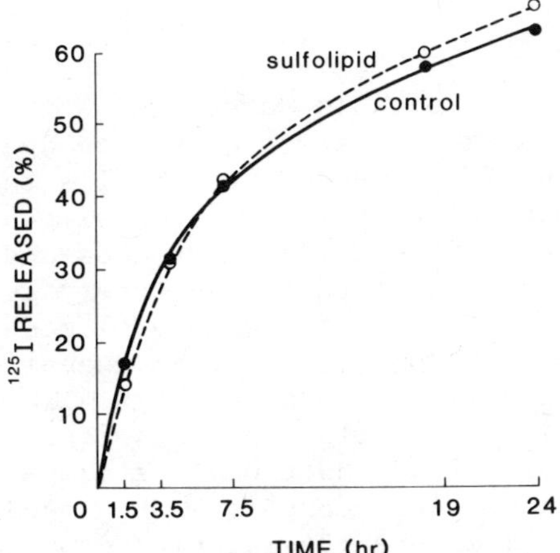

Figure 6.   Release of radiolabel from ingested iodinated zymosan
            by control and sulfolipid-exposed macrophages.

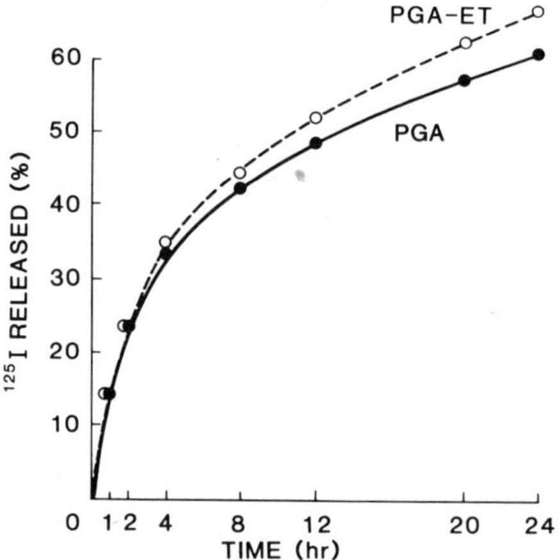

Figure 7.   Release of radiolabel from macrophages exposed 5 days
            to either polyglutamic acid (PGA) or PGA and endotoxin
            (ET). The tracing of the time course for the PGA cells
            is essentially superimposable on that of control cells
            of Fig. 6. ET evidently promotes digestion to some
            extent.

by control and by sulfatide-treated macrophages. The time courses
of radiolabel release are essentially superimposable--and are no
different from that seen for cells exposed to polyglutamic acid
for 5 days with or without endotoxin as a possible stimulant for
relieving any block to digestive processing (Fig. 7). It is clear
that the processing of endocytosed substrates by control and by
polyanion-treated cells is identical.

DISCUSSION

     The recurrent theme of this paper begins to entertain the
conception that secondary lysosomes of macrophages may have a less
important role in the functions of these cells than has been
assumed. The behavior of resident peritoneal cells that have
ingested granulocyte-derived debris, granules and enzymes, which
are presumably incorporated into the secondary lysosomal system
suggests that sequestration of these heterologous enzymes in
macrophage secondary lysosomes makes them unavailable for utilization in host defense against microorganisms. This is inferred
from the very limited utilization of the peroxidase content and
moreover from the more rapid functional as compared with enzymatic
decay that it exhibits. As a corollary, the acquisition of these
potential armaments by mature (resident) macrophages engaged in
primitive scavenging operations in an inflammatory focus may lead
only to a mimicry (thus, a "macrophage in monocyte's clothing")

and not to expression of functions that are actually restricted to
the much less mature cells (monocytes) as they differentiate. It
will be of keen interest to determine--for tumoricidal macrophages
evolved from the resident cell--whether there is indeed mimicry
or not, and how the behavior is influenced by the presence of
granulocyte-derived debris.

With respect to the digestion of endocytosed substrates, in
the past, various investigators have measured this process to
secure evidence concerning how various manipulations affect
PL fusion phenomena: Rachel Goldman (50) and Paul Edelson (51),
respectively (concanavalin A); Pesanti and Axline to study the
effect of colchicine (52); Edith Weiner to study the influence of
cortisone (53); while this paper and earlier studies (38) report
on the influence of polyanionics. I would like to suggest that
the results of such measurements must be interpreted with caution.
If one accepts the evidence that the sequestration of polyanionics
in secondary lysosomes antagonizes their fusion function, then if
the lysosomes are involved in intracellular digestion (as we have
had abundant reason to believe), surely abolition of PL fusion
should have some measurable consequence on this process. Our
evidence suggests that it does not. And this constitutes a two-
fold paradox--for interference with digestion imposed by a block
to fusion should be reinforced by at least some antagonism of
enzymatic activity that may be expected from the presence of
polyanions, even if limited fusion were to occur (vide supra).
Our evidence, however, suggests that whatever activity the polyanions may have in secondary lysosomes, the digestion of radiolabeled (proteins) substrates is unaffected. We are constrained
to ask therefore, are these data evidence of the "shades of gray"
phenomenon alluded to earlier, in which some very low levels of
secondary lysosomal fusion function are still expressed? Is this
sufficient to provide the component of enzymes required to maintain
the digestive process at an undiminished rate--even in the presence
of polyanionics? Or does this instead reflect almost exclusive
contributions from primary lysosomes? This is surely an alternative explanation. This interpretation, which invokes primary
lysosomes as the source for nearly all of the digestive capacity,
is supported by data from Hart, who described ultrastructural
(electron microscopy) evidence that in polyglutamate macrophages
there is essentially undiminished delivery of lysosomal acid
phosphatase to phagosomes (33). Since Hart and colleagues have
consistently confirmed the cut-off of secondary lysosomal fusion
function by polyglutamate, he attributes the source of acid
phosphatase delivery to be macrophage primary lysosomes. Taken
together with our own results, and with the expected antagonism
of enzyme function by polyanionics, the evidence seems to be
mounting that secondary lysosomes, at least in cultured macrophages, may contribute much less to intracellular digestion than
we have believed. The unrestricted digestion of labeled substrates

in polyanion-laden macrophages strengthens the inference that most
of the digestion is dependent therefore on primary lysosomes. I
find this concept somewhat disturbing. It can be tested. It
certainly contradicts well-established dogma, and seems to demand
unequivocal resolution.

ACKNOWLEDGMENT

The participation of Drs. L. Heifets, K. Imai and C.L.
Swendsen, and of Judith Fiscus in various of the studies is
acknowledged with thanks. I thank Nadia de Stackelburg for
illustrations and Betty Woodson for careful preparation of the
manuscript. Original studies reported herein were supported, in
part, by Grant AI 17509 and by Grant AI 08401 from the U.S.-Japan
Cooperative Medical Science Program administered by the National
Institute of Allergy and Infectious Diseases. M.B.G. is Margaret
Regan Investigator in Chemical Pathology, NJHRC.

REFERENCES

1. Nichols, B.A. and Bainton, D.F. (1975). In: R. Van Furth,
         ed., Mononuclear Phagocytes in Immunity, Infection and
         Pathology. Blackwell, Oxford, p. 17.
2. Klebanoff, S.J. (1967). J. Exp. Med. 126:1067.
3. Klebanoff, S.J. (1968). J. Bacteriol. 95:2131.
4. Klebanoff, S.J. (1975). Semin. Hematol. 12:117.
5. Van Furth, R., Hirsch, J.G., and Fedorko, M.E. (1970). J.
         Exp. Med. 132:794.
6. Simmons, S.R. and Karnovsky, M.L. (1973). J. Exp. Med. 138:44.
7. Babior, B.M. (1978). N. Engl. J. Med. 298:659; 721.
8. Lehrer, R.I. (1975). J. Clin. Invest. 55:338.
9. Klebanoff, S.J. and Hamon, C.B. (1975). In: R. Van Furth,
         ed., Mononuclear Phagocytes in Immunity, Infection and
         Pathology. Blackwell, Oxford, p. 507.
10. Goren, M.B. (1977). Ann. Rev. Microbiol. 31:507.
11. Atwal, O.S. (1971). J. Reticuloendothel. Soc. 10:163.
12. Daems, W.Th. and Brederoo, P. (1973). Z. Zellforsch.
         Mikrosk. Anat. 144:247.
13. Daems, W.Th., Wisse, E., Brederoo, P. and Emeis, J. (1975).
         In. R. Van Furth, ed., Mononuclear Phagocytes in
         Immunity, Infection and Pathology. Blackwell,
         Oxford, p. 57.
14. Heifets, L., Imai, K. and Goren, M.B. (1980). J. Reticulo-
         endothel. Soc. 28:391.
15. Klebanoff, S.J. and Clark, R.A. (1977). J. Lab. Clin. Med.
         89:675.
16. Baehner, R.L., Murrmann, S.K., Davis, J. and Johnston, R.B.,
         Jr. (1975). J. Clin. Invest. 56:571.

17. Hooke, A.M., Oeschger, M.P., Zelligs, B.J. and Bellanti, J.A.
         (1978). Infect. Immun. 20:406.
18. Sasada, M. and Johnston, R.B., Jr. (1980). J. Exp. Med. 152:85.
19. Hibbs, J.B., Chapman, H.A. and Weinberg, J.B. (1978). J.
         Reticuloendothel. Soc. 24:549.
20. Ruco, L.P. and Meltzer, M.S. (1977). J. Immunol. 119:889.
21. Ruco, L.P. and Meltzer, M.S. (1978). J. Immunol. 120:1054.
22. Karnovsky, M.L. and Lazdins, J.K. (1978). J. Immunol. 121:809.
23. Alexander, P. and Evans, R. (1971). Nature New Biol. 232:76.
24. Edelson, P.J. and Cohn, Z.A. (1973). J. Exp. Med. 138:318.
25. Klebanoff, S.J., Clark, R.A. and Rosen, H. (1976). In: Schultz,
         J. and Ahmad, F. (eds.) Cancer Enzymology. (Proceedings
         of the Miami Symposia, Jan., 1976). New York, Academic
         Press, p. 267.
26. Selvaraj, R.J., Zyliczynski, J.M., Paul, B.B. and Sbarra, K.
         (1978). J. Infect. Dis. 137:481.
27. Ramsey, P.G. and Klebanoff, S.J. (1980). Clin. Res. 29:389A.
28. Lochsley, R.M., Wilson, C.B. and Klebanoff, S.J. (1981). Clin.
         Res. 29:389A.
29. Wright, C.D., Herron, M.J., Gray, G.R., Holmes, B. and Nelson,
         R.D. (1981). Infect. Immun. 32:731.
30. Sela, M.N., Lahav, M., Ne'eman, N., Duchan, Z. and Ginsburg,
         I. (1975). Inflammation 1:57.
31. Hart, P.D. and Young, M.R. (1975). Nature 256:47.
32. Hart, P.D. and Young, M.R. (1979). Exp. Cell Res. 118:365.
33. Hart, P.D. and Young, M.R. (1980). In: R. Van Furth (Ed.)
         Mononuclear Phagocytes. Functional Aspects II. Martinus-
         Nijhoff, The Hague, p. 1039.
34. Goren, M.B., Hart, P.D., Young, M.R. and Armstrong, J.A.
         (1976). Proc. Natl. Acad. Sci., USA, 73:2510.
35. Kielian, M.C. and Cohn, Z.A. (1980). In: R. Van Furth (Ed.)
         Mononuclear Phagocytes. Functional Aspects II. Martinus-
         Nijhoff, The Hague. p. 1077.
36. Pesanti, E.L. (1978). Infect. Immun. 20:503.
37. Goren, M.B., Swendsen, C.L. and Henson, J. (1980). In: R.
         Van Furth (Ed.) Mononuclear Phagocytes. Functional
         Aspects II. Martinus-Nijhoff, The Hague. p. 999.
38. Goren, M.B., Swendsen, C.L. and Henson, J. (1979). In:
         Proceedings XIV Joint Tuberculosis Conference, U.S.-
         Japan Cooperative Medical Science Program, Denver, CO.
         p. 422.
39. Kielian, M.C. and Cohn, Z.A., (1980). J. Cell Biol. 85:754.
40. deDuve, C., deBarsy, T., Poole, B., Trouet, A., Tulkens, P.
         and Van Hoof, F. (1974). Biochem. Pharmacol. 23:2495.
41. Nairn, R. (1976). Fluorescent Protein Tracing. 4th Ed.,
         Edinburgh:Churchill, Livingston. p. 93.
42. Hart, P.D. and Young, M.R. (1980). Nature 286:79.
43. Seglen, P.O., Grinde, B. and Solheim, A.E. (1979). Eur. J.
         Biochem. 95:215.
44. Amenta, J.S. and Brocher, S.C. (1980). J. Cell Physiol. 102:259.

45. Ohkuma, S. and Poole, B. (1978). Proc. Natl. Acad. Sci.,
        USA 75:3327.
46. Allison, A.C. and Young, M.R. (1969). See specific references
        to P. Jaques and to Lloyd et al. In: J.T. Dingle and
        H.B. Fell (eds.) Lysosomes in Biology and Pathology
        Vol. 2, North Holland, Amsterdam. p. 615.
47. Ginsberg, I. and Sela, M. (1976). Crit. Rev. Microbiol. 4:249.
48. Avila, J.L. and Convit, J. (1975). Biochem, J. 152:57.
49. Calderon, J. and Unanue, E. (1974). J. Immunol. 112:1804.
50. Goldman, R. and Raz, A. (1975). Exp. Cell Res. 96:393.
51. Edelson, P.J. and Cohn, Z.A. (1974). J. Exp. Med. 140:1387.
52. Pesanti, E.L. and Axline, S.G. (1975). J. Exp. Med. 141:1030.
53. Wiener, E., Marmary, Y. and Curelaru, Z. (1972). Lab. Invest.
        26:220.

GENETIC DISORDERS OF GRANULOCYTE FUNCTION: WHAT THEY TELL US

ABOUT NORMAL MECHANISMS

Richard K. Root

Department of Internal Medicine, Yale University
New Haven, Connecticut 06510

INTRODUCTION

Clinically significant genetic defects of phagocyte function
have been most clearly defined for polymorphonuclear leukocytes
(PMN). Several important lessons have been learned from an
analysis of these defects. These include:

1) Defining the role of oxidative metabolism in the
    microbicidal activity of PMNs;

2) determining the existence and activity of "back-up"
    killing mechanisms when deletions in normal pathways
    occur, and;

3) discerning the importance of PMNs relative to other
    cells in host defense against different microorganisms.

For example, the increased incidence of infections with staphylococci, aerobic gram negative rods or opportunistic fungal
infections which occur in patients with neutropenia or impaired
granulocyte killing mechanisms underlines the importance of these
cells in normal defense against these organisms (1-5). The
careful analysis of these PMN defects can provide insights which
may clarify the potential microbicidal mechanisms of monocytes and
macrophages. With these points in mind this discussion will
center on the clinical features of patients with Chediak-Higashi
syndrome, chronic granulomatous disease, or myeloperoxidase
deficiency, and the characteristics of their cellular defects.

THE CHEDIAK-HIGASHI SYNDROME

This rare disorder (approximately one patient per 10 million
population) is inherited as an autosomal recessive (4). A defect
in the formation of single membrane-bound intracellular organelles
involves many cell-types including granulocytes, monocytes, lymphocytes, melanocytes and Schwann cells. As a result these cells
contain giant lysosomes or other single membrane-bound vesicles.
In granulocytes the giant lysosomes contain enzymes and proteins
of both the primary and specific granules of normal PMNs (6,7).
In addition, disordered microtubular assembly has been reported
in CHS granulocytes (8). The functional consequences are impairments in post-phagocytic degranulation (6,9) and in cellular
motility (10). A delay in killing of a wide variety of bacteria
can be demonstrated for CHS granulocytes (9). These same organisms
can be isolated from infections in those patients (4). Monocytes
are also affected (9,11).

Patients with the Chediak-Higashi syndrome have recurrent
infections involving the skin and soft tissues, the respiratory
tract and sinuses, as well as septicemias with a variety of
pyogenic bacteria (4). They also have neutropenia and chemotaxis
of their neutrophils and monocytes is reduced (10,11). The
bactericidal defect of CHS cells is rather mild and similar to
that noted for cells deficient in myeloperoxidase (MPO) activity
(12). Given the relatively infrequent infections seen in the
latter group of patients it seems likely that the severe infections in CHS subjects are a consequence of the multiple functional
deficiencies of their cells and their neutropenia. Of interest
is the fact that the administration of ascorbate or cholinergic
agents to some patients (13) or mice (8) with CHS can correct the
functional disorders of their granulocytes. This correction can
be correlated with improvement in microtubule-dependent activities.
Similar studies in CHS monocytes and macrophages would be of
interest.

CHRONIC GRANULOMATOUS DISEASE

Patients with the chronic granulomatous disease syndrome
suffer from repetitive infections of the lungs, liver and spleen,
bones, skin and soft tissues with staphylococci, aerobic gram
negative rods and opportunistic fungi, in particular aspergillus
and candida species (2,3,14). The majority of affected patients
are male, and a sex-linked mode of inheritance can be demonstrated
(14,15). Some patients are female and have associated glutathione
peroxidase deficiency. They, and affected males, apparently
inherit the disease through an autosomal gene (16). The incidence
of the disorders is rare but more common than CHS (about one
affected patient per million population).

The discovery that the CGD syndrome was due to a defect in neutrophil function was a major milestone in understanding neutrophil physiology (17,18). Not only was it the first demonstration that recurrent infections were a consequence of a qualitative disorder of these cells, but elucidation of the responsible mechanisms revealed the important role of oxidative metabolism in PMN microbicidal activity. Later studies revealed that monocytes from CGD patients were similarly affected (19).

Neutrophils and monocytes from CGD subjects are incapable of generating a burst in respiration during phagocytosis or when stimulated by several soluble activators (14,18). A comparison of the various events in the respiratory burst for each of the granulocyte disorders is listed in Table 1. CGD cells do not consume oxygen, form O_2^- or H_2O_2, iodinate protein, or increase

Table 1. Comparative Features of Three Genetic Defects of Granulocyte Function

	CHS	CGD	MPO Deficiency
INFECTIONS	Pyogenic bacterial	Catalase-positive organisms, bacterial and fungal	Candidiasis with Diabetes mellitus
GRANULOCYTE STRUCTURE	Giant lysosomes	Normal	Peroxidase –deficient azurophilic granules
MICROBICIDAL ACTIVITY	Delayed bacterial killing	Abnormal killing of catalase positive organisms	Mild bactericidal defect Impaired candida-cidal activity
CHEMOTAXIS	Abnormal	Normal	Normal
METABOLISM	Normal	Impairment of O_2 consumption, HMS activity, O_2^- and H_2O_2 formation, MPO-dependent iodination and NBT reduction	Increased O_2 consumption, HMS activity, O_2^- and H_2O_2 formation. Decreased iodination
MONOCYTES AFFECTED	Yes	Yes	Yes

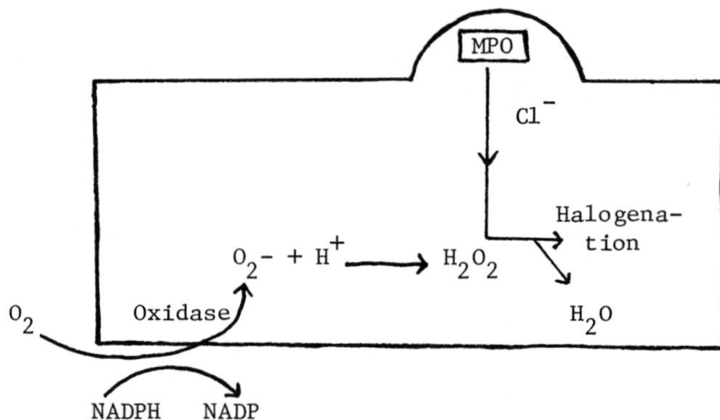

Figure 1.   Major pathways for oxygen utilization in PMN phagosomes.
Oxygen is reduced to O_2^- through the action of an
oxidase which uses pyridine nucleotides (NADPH
preferentially) as electron donors. Through spontaneous
dismutation O_2^- is converted to H_2O_2. H_2O_2 serves as
the substrate for MPO which enters the phagosome by
degranulation of azurophilic lysosomes. Chloride
appears to be the "physiologic" halide cofactor which
is utilized in MPO-H_2O_2 dependent microbicidal activity.
Halogenation of proteins and amino acids can be
measured as a result of MPO activity.

hexose monophosphate shunt activity when stimulated (reviewed in
14). They also do not reduce the yellow dye nitroblue tetrazolium
to formazan, a reaction which forms the basis of a simple test for
this disease (15,20). The cause of this disorder is an impairment
in the function of a critical oxidase which is responsible for
triggering the respiratory burst. This oxidase has a requirement
for flavin, preferentially uses NADPH as a substrate and may
reside in the plasma membrane of neutrophils and (presumably)
monocytes (21). Its nature and function are discussed more fully
by DeChatelet in this series of reports. The relationship of
oxidase activity to the other events in the respiratory burst is
shown in Figure 1.

CGD neutrophils and monocytes phagocytize normally but
exhibit impaired killing of a variety of catalase-positive
organisms, in particular, staphylococci, Enterobacteriaciae,
candida and aspergillus species (14,17). In contrast, catalase-
negative organisms, such as streptococci, are killed normally.
H_2O_2 formed by catalase-negative organisms is not degraded by them,
and thus bypasses the inability of the CGD cells to make this
compound. Similarly, provision of H_2O_2 to CGD cells by an enzymatic system (glucose oxidase) which is delivered into phagocytic

vacuoles (22) or by diffusion of extracellular H_2O_2 into the cells will correct the impaired killing (12). Infusion of H_2O_2 also promotes MPO-mediated protein iodination and stimulates pentose shunt activity (12). This indicates the participation of these pathways in H_2O_2 utilization.

Analysis of these activities of CGD phagocytes has thus shown the critical roles that H_2O_2 and its precursor molecule, O_2^-, play in normal PMN and monocyte antimicrobial activity. Normal mammalian macrophages are capable of O_2^- and H_2O_2 formation (23,24). The importance of these compounds in the microbicidal actions of these cells is currently a subject of intense investigation. Studies of the microbicidal and matabolic activity of CGD macrophages would be of interest although preliminary work has yielded conflicting results (25).

MYELOPEROXIDASE DEFICIENCY

Myeloperoxidase makes up approximately 5% of the wet weight of the neutrophil (26). This azurophilic granule enzyme is also found in monocytes but disappears from these granules as monocytes undergo differentiation into macrophages. Myeloperoxidase together with appropriate anionic cofactors markedly increases the microbicidal and cytocidal activity of H_2O_2 (27). The spectrum of antimicrobial activity of the MPO-halide-H_2O_2 system is broad and includes bacteria, viruses, mycoplasmas and fungi (28); it also can destroy eukaryotic cells (29). When iodide is the halide examined, its fixation to tyrosine in protein can be demonstrated, with MPO playing a major role in this event (30). Iodination, however, is not essential to the microbicidal activity of the MPO-H_2O_2 system. Chloride may in fact be the most important halide in physiologic situations (28,29). The pioneering studies of Klebanoff and his co-workers have demonstrated the function of this system inside intact cells (31).

Until recently MPO deficiency was thought to be a very rare disorder with less than 20 cases described in the world's literature. An autosomal recessive mode of inheritance was demonstrated. The development of an automated system for leukocyte differential counting which uses flow cytochemistry for its technology, has disclosed, however, that MPO deficiency is surprisingly common (32). An analysis of its frequency in over 60,000 subjects examined indicates that partial or complete deficiency may occur in as many as 1/2000 of the general population. Although the majority of MPO deficient subjects are clinically well, the coexistence of diabetes and MPO deficiency can be associated with systemic candidiasis (5,32). Functional analysis of MPO deficient cells indicates that they have relatively mild impairments in killing of bacteria. Killing of <u>Candida albicans</u> is markedly

impaired, a situation similar to that observed for CGD granulocytes.
These findings, together with the clinical manifestations of MPO
deficiency, suggest that the enzyme may normally be involved in
antifungal activity. For normal antibacterial activity O_2^- and
H_2O_2 formation appears essential, but the action of MPO may be
of secondary importance (32).

The relevance of these observations to the function of macrophages, cells which do not contain MPO remains to be determined.
They suggest, however, that the loss of MPO activity during the
differentiation of monocytes into macrophages may have little to
do with their subsequent antimicrobial efficacy.

CONCLUSIONS

In summary, analysis of these three inherited defects of
granulocyte function has highlighted critical events for the antimicrobial function of these cells and placed others in perspective.
Both normal phagosomal fusion and the formation of O_2^- and H_2O_2
during the phagocytic respiratory burst are central to the broad
antimicrobial activity of granulocytes. MPO, on the other hand,
while perhaps normally participating in granulocyte antimicrobial
action, appears to be essential only for the effective killing of
eukaryotic organisms such as fungi. Not described were the non-
oxidative killing mechanisms of neutrophils which appear to be
considerable (28,29) although no specific inherited defects have
been described. While genetic disorders of macrophage effector
function remain to be clearly defined, the lessons learned from
the study of these granulocyte defects have provided both the
technology and approach to the analysis of their antimicrobial
and cytocidal mechanisms.

ACKNOWLEDGMENTS

This work was supported by Public Health Service Grant
#AI 13251 and #AI 107033 from NIAID.

REFERENCES

1. Schimpff, S.C. (1977). Therpay of infection in patients with
          granulocytopenia. Med. Clin. N. Amer. 61:1101.
2. Johnston, R.B., Jr. and Baehner, R.L. (1971). Chronic
          granulomatous disease: correlation between pathgenesis
          and clinical findings. Pediatrics 48:730-379.
3. Cohen, M.S., Isturiz, R.E., Malech, H.L., et al. (1981).
          Fungal infection in chronic granulomatous disease: the
          importance of the phagocyte in defense against fungi.
          Am. J. Med. 71:59-66.

4. Blume, R.S. and Wolff, S.M. (1972). The Chediak-Higashi
        syndrome, studies in four patients and a review of the
        literature. Medicine 51:247-280.
5. Lehrer, R.I. and Cline, M.J. (1969). Leukocyte myeloper-
        oxidase deficiency and disseminated candidiasis: the
        role of myeloperoxidase in resistance to Candida
        infection. J. Clin. Invest. 48:1478-88.
6. Stossel, T.P., Root, R.K. and Vaughan, M. (1972). Phagocytosis
        in chronic granulomatous disease and the Chediak-
        Higashi syndrome. N. Engl. J. Med. 286:120-123.
7. Rausch, P.G., Pryzwansky, K.B. and Spitznagel, J.K (1978).
        Immunocytochemical identification of azurophilic and
        specific granule markers in the giant granules of
        Chediak Higashi neutrophils. N. Engl. J. Med. 298:
        693-698.
8. Oliver, J.M. and Zurier, R.B. (1976). Correction of charac-
        teristic abnormalities of microtubule function and
        granule morphology in Chediak-Higashi syndrome with
        cholinergic agonists: studies in vitro in man and in
        vivo in the beige mouse. J. Clin. Invest. 57:1239-1247.
9. Root, R.K., Rosenthal, A.S. and Balestra, D.J. (1972).
        Abnormal bactericidal, metabolic and lysosomal functions
        of Chediak-Higashi syndrome leukocytes. J. Clin. Invest.
        51:649-665.
10. Clark, R.A. and Kimball, H.R. (1971). Defective granulocyte
        chemotaxis in the Chediak-Higashi syndrome. J. Clin.
        Invest. 50:2645-2652.
11. Gallin, J.I., Klimerman, J.A., Padgett, G.A., et al. (1975).
        Defective mononuclear leukocyte chemotaxis in the
        Chediak-Higashi syndrome of humans, mink and cattle.
        Blood 45:863-870.
12. Root, R.K. (1975). Comparison of other defects of granulocyte
        oxidative killing mechanisms with chronic granulomatous
        disease in J. Bellanti and D.H. Dayton (ed). The
        Phagocytic Cell in Host Resistance. Raven Press,
        New York, p. 201-226.
13. Boxer, L.A., Watanabe, A.M., Rister, M., et al. (1976).
        Correction of leukocyte function in Chediak-Higashi
        syndrome by ascorbate. N. Engl. J. Med. 295:1041-1045.
14. Mills, E.L. and Quie, P.G. (1980). Congenital disorders of
        the functions of polymorphonuclear neutrophils. Rev.
        Inf. Dis. 2:505-517.
15. Windhorst, D.B., Page, A.R., Holmes, B., et al. (1968). The
        pattern of genetic transmission of the leukocyte
        defect in fatal granulomatous disease of childhood.
        J. Clin. Invest. 47:1026-1034.
16. Holmes, B., Park, B.H., Malawista, S.E., et al. (1970).
        Chronic granulomatous disease in females: a deficiency
        of leukocyte glutathione peroxidase. N. Engl. J. Med.
        282:217-221.

17. Quie, P.G., White, J.G., Holmes, B., et al. (1967). In vitro
        bactericidal capacity of human polymorphonuclear
        leukocytes: diminished activity in chronic granulomatous
        disease of childhood. J. Clin. Invest. 46:668-679.
18. Holmes, B., Page, A.R. and Good, R.A. (1967). Studies of
        the metabolic activity of leukocytes from patients with
        a genetic abnormality of phagocytic function. J. Clin.
        Invest. 46:1422-1432.
19. Rodey, G.E., Park. B.H., Windhorst, D.B., et al. (1969).
        Defective bactericidal activity of monocytes in fatal
        granulomatous disease. Blood 33:813-820.
20. Baehner, R.L. and Nathan, D.G. (1968). Quantitative nitro-
        blue tetrazolium test in chronic granulomatous disease.
        N. Engl. J. Med. 278:971-976.
21. Badwey, J.A., Curnutte, J.T. and Kamorsky, M.C. (1979). The
        enzyme of granulocytes that produces superoxide and
        peroxide: an elusive pimpernel. N. Engl. J. Med.
        300:1157-1160.
22. Johnston, R.B., Jr., Keele, B.B., Jr., Misra, H.P., et al.
        (1972). The role of superoxide anion generation in
        phagocytic bactericidal activity: studies with normal
        and chronic granulomatous disease leukocytes. J.
        Clin. Invest. 55:1357-1372.
23. Drath, D.B., Karnovsky, M.L. (1975). Superoxide production
        by phagocytic leukocytes. J. Exp. Med. 141:257-262.
24. Nathan, C.F. and Root, R.K. (1977). Hydrogen peroxide
        release from mouse peritoneal macrophages: depending
        on sequential activation and triggering. J. Exp. Med.
        146:1648-1662.
25. Dilworth, J.A. and Mandell, G.L. (1977). Adults with chronic
        granulomatous disease of childhood. Am. J. Med. 63:233.
26. Schultz, J. and Kaminker, K. (1962). Myeloperoxidase of the
        leucocyte of normal human blood. I. Content and
        localization. Arch. Biochem. Biophys. 96:465.
27. Klebanoff, J.J. (1967). Iodination of bacteria: a bacteri-
        cidal mechanism. J. Exp. Med. 126:1063-1078.
28. Root, R.K. and Cohen, M.S. (1981). The microbicidal mechan-
        isms of human neutrophils and eosinophils. Rev. Inf.
        Dis. 3:565-598.
29. Klebanoff, S.J. and Clark, R.A. (1978). The neutrophil:
        Function and clinical disorders. Amsterdam, North
        Holland Publishing Co. 1978.
30. Klebanoff, S.J. and Clark, R.A. (1977). Iodination by human
        polymorphonuclear leukocytes: a reevaluation. J. Lab.
        Clin. Med. 89:675-686.
31. Klebanoff, S.J. and Hamon, C.B. (1972). Role of myeloper-
        oxidase-mediated antimicrobial systems in intact
        leukocytes. J. Reticuloendothel. Soc. 12:170-196.

32. Parry, M.F., Root, R.K., Metcalf, J.A., et al. (1981).
     Myeloperoxidase deficiency: prevalence and clinical
     significance. Ann. Int. Med. 95:293-301.

MODULATION OF EFFECTOR LYMPHOKINES

Stanley Cohen, Tohru Baba, Matsunobu Suko, Helen
D'Silva and Takeshi Yoshida

Department of Pathology, University of Connecticut
Health Center, Farmington, CT 06032

INTRODUCTION

As is now well known, lymphokines are soluble, non-antibody
mediators produced by appropriately stimulated lymphocytes. They
play important afferent roles with respect to enhancement or
suppression of the induction of immunity. They are important in
efferent reactions as well. These effector lymphokines, initially
described in terms of _in vitro_ bioassays, have been demonstrated
in a variety of _in vivo_ situations and, when exogenously administered, have been shown to exert profound effects _in vivo_ (reviewed
in 1).

As direct agents of host defense, lymphokines are not very
effective. On the other hand, neither are antibodies. Antibodies are very good at neutralizing toxins, or perhaps aggregating
antigen so as to make it more attractive to the reticuloendothelial
system. However, most of the biologically important effects of
antibodies, at least insofar as host defense is concerned, are
indirectly mediated by them. They are based on the ability of the
antibody to activate effector systems such as complement, or
effector cells such as mast cells or basophils. The consequences
of such secondary events include permeability alterations, accumulation and activation of inflammatory cells, cytolytic effects,
etc. Similarly, although there are a few lymphokines such as
lymphotoxin (LT) or Tumor Migration Inhibitory Factor (TMIF) that
may exert direct protective effects, most lymphokines mediate host
defense by indirect means. They do so largely by invoking inflammatory responses. Indeed, although the effector or efferent
lymphokines are usually thought of as the mediators of delayed
hypersensitivity or cellular immunity (which they are), they

should be more correctly regarded as agents by which the immunologic
system is coupled to the inflammatory system.

PROPERTIES OF LYMPHOKINES

For the purpose of this discussion, we will focus attention
on those lymphokines involved in effector functions. In general
these mediators are proteins or glycoproteins, some of which are
known to be composed of subunits (2). The lymphokines of man
range in molecular weight from approximately 10,000-30,000 daltons.
For all species, with rare exception, these factors are less than
about 70,000 daltons. They have been shown to exert effects on
all kinds of inflammatory cells (neutrophils, eosinophils, basophils, monocytes, lymphocytes) as well as fibroblasts and endothelial cells. They also influence vascular permeability and the
clotting system (reviewed in 3). These and other targets of
lymphokine activity are summarized in Table I. There are over
90 defined lymphokine activities and at least half of these represent effector functions. These may be categorized as shown in
Table II. Of these, activation and mobility alterations may play

Table I. Targets for Lymphokine Action

---

1. Inflammatory Cells (granulocytes, monocytes, lymphocytes,
    platelets)

2. Immunologic Cells (T and B Lymphocytes)

3. Reparative Cells (fibroblasts, smooth muscle, endothelium)

4. Stem Cells

5. Specialized Cells of Miscellaneous Origin (bone, thyroid,
    etc.)

6. Pathologic Cells (tumor cells, etc.)

7. Non-Cellular Components (clotting system, vascular
    permeability)

Table II. Effects of Lymphokines

---

1. Cell Motility (inhibition, chemotaxis, etc.)
2. Cell Growth and Proliferation
3. Cell Injury
4. Cell Differentiation
5. Cell Activation
6. Cell Regulation and Repair

---

the major roles in reactions of host defense. With respect to the
usual manifestations of delayed hypersensitivity, it is the cell
of the monocyte/macrophage lineage that appears to be the major
participant, although under certain experimental and clinical
conditions, other inflammatory cells may predominate (3).

CONTROL OF LYMPHOKINE FUNCTION

     One of the most curious things about a cutaneous delayed hyper-
sensitivity reaction is that it goes away. Even though antigen may
persist, the reaction diminishes. This seemingly trivial phenomenon
is interesting because lymphokines are generated locally at delayed
hypersensitivity reaction sites (4). Another interesting observation
is that whenever lymphokines are generated, they tend to occur in
batches; i.e., many lymphokine activities are generated simultaneously. These batches of lymphokines include mitogenic factors. These,
in turn, can induce uncommitted cells to make more lymphokines,
leading to a positive feedback situation that should never end. Yet
it does. These rather commonplace examples point to the existence
of regulatory mechanisms for the control of lymphokine function.
For the most part, studies of such inhibitory mechanisms have utilized the biologic reaction itself, such as contact sensitivity, as
the endpoint, rather than lymphokines (5). This is because lymphokines are rather inaccessible to direct analysis in intact animals
except under special circumstances. One of these circumstances may
be brought into focus by our early observations that macrophage
migration inhibition factor (MIF) could be detected in the sera of
many patients with lymphoproliferative disorders (6) and sarcoidosis
(7). In many such conditions, MIF is detectable in the face of
anergy, that is inability of the patient to express delayed hypersensitivity reactions.

REGULATION OF LYMPHOKINE ACTIVITY IN DESENSITIZATION

     The paradoxical co-existence of a failure of delayed hyper-
sensitivity and the presence of mediators of delayed hypersensitivity may be seen in the experimental models as well. Thus, we and
others (8,9,10) have found that systemic challenge with antigen into
an immunized animal leads to the transient appearance of lymphokines
in the serum. Interestingly, the protocol for generating serum
lymphokine is identical to that used to induce desensitization.
This phenomenon is a transient state of unresponsiveness with respect
to delayed hypersensitivity that can co-exist with continued antibody
production. It is non-specific in the sense that a doubly immunized
animal receiving a (large, systemic) desensitizing dose of one antigen will become unresponsive to both. Finally, the desensitized
state involves an anergic environment; sensitized cells from a non-
desensitized donor cannot function in a desensitized recipient (11).

These observations in association with the previously described
clinical findings suggested that a circulating lymphokine itself
might be responsible for anergy. This prompted experiments in
which we injected exogenous lymphokines systemically into immunized
animals and found that we could achieve desensitization as readily
as when specific antigen was used (10). Of even greater interest,
we could passively transfer the desensitized state with serum,
provided that the serum was taken early in desensitization at a
time when circulating lymphokines were present (12).

What are the explanations for this puzzling phenomenon?
First, the injection of lymphokine-containing supernatants intravenously causes a drop in the monocyte content of peripheral blood
(10) and similarly, the injection of lymphokines intraperitoneally
into animals with pre-existing peritoneal exudates causes a drop
in the macrophage content of those exudates (13). Thus, it is
likely that potential effector cells are pre-empted and/or
sequestered and therefore not available for reaction at the
cutaneous test site. Second, we have found that macrophages
obtained from desensitized animals have a diminished capacity to
respond to a variety of chemotactic stimuli _in vitro_. Finally,
large lymphokines in the systemic circulation can effectively
abolish gradients and thus render chemotactic mechanisms themselves inoperative.

THE SECOND STAGE OF DESENSITIZATION

All the effects we have been discussing occur early within
the first day or two of desensitization. However, anergy may persist for several days thereafter. In this second stage, when
circulating lymphokines are no longer present, it appears as if
lymphokine production itself is blocked (14). In other words, in
the early stages of desensitization, too much lymphokine is
produced and in the later, not enough. It is likely that the
former is the direct cause of the latter, but before we examine
this in detail, a brief digression is in order. If our two-stage
mechanism for desensitization is correct, then administration of
pre-formed mediator in the form of Skin Reacting Factor (SRF),
intracutaneously, should be without effect early in desensitization.
Late in desensitization, when production failure is involved, SRF
should produce a positive reaction, even though the animal still
cannot respond to antigen. In our recent experiments, we have found
this to be the case.

SUPPRESSION OF LYMPHOKINE PRODUCTION

The experiments relating to the inhibition of lymphokine
production in the second stage of desensitization and the irrele-

vance to physiologic regulation are currently underway in our
laboratory, and have appeared only in abstract form. Thus, the
conclusions presented here must be considered preliminary. However,
they fit well with similar findings in other systems by other
investigators, so it is likely that the final picture will be
similar to that sketched in outline form here.

In brief, if lymphocytes from immunized animals are preincubated with preformed lymphokine, they fail to produce new
lymphokine. This apparently simple example of feedback inhibition
is quite complicated. It involves the activation of macrophages
present by one of the lymphokines present in the initial mixture.
This activated macrophage then produces a soluble mediator that
inhibits the subsequent production of lymphokines by the lymphocytes in the mixture.

This is not the only mechanism known for mediator production.
Thus, we have described a T lymphocyte-derived factor (MIFIF) that
can prevent the production of MIF by B cells (15). It is likely,
especially in the light of the suppressor cell studies cited above
that T cells produce similar mediators inhibitory for the production of T cell-derived lymphokines.

DISCUSSION

Perhaps the major theme to emerge from this discussion is
that there are many mechanisms by which lymphokine production and
the expression of lymphokine activity may be regulated. We have
focused on desensitization as a kind of "excessive" regulation,
since this appears to us to represent a simple model that can help
in unravelling these complex events. In brief, it is well known
that the lymphokines themselves, by acting on a variety of target
inflammatory cells, including other lymphocytes, recruit non-immune
T and B cells to produce more lymphokines through various amplify-
ing circuits. For example, mitogenic lymphokines can act on T as
well as B cells (16) and thus, in theory, could serve as agents for
the generation of more lymphokines by uncommitted lymphocytes.
Furthermore, naive T cells may be recruited by macrophage-derived
factors as a consequence of activation of macrophages with lymphokines. B cells could be stimulated by a mitogenic factor generated
by lymphokine-stimulated normal T cells.

In the present discussion, we have described several suppressive circuits. Previously, we have shown B cell production of MIF
can be inhibited by a T cell product (14). As discussed above,
macrophages stimulated by lymphokines may produce factors to
suppress the production of lymphokines and possibly their activities
as well. Some other indirect evidence suggests the role of a subpopulation of T cells and their products in modulating the effector
T cells with respect to production of lymphokines. Moreover, since

many non-lymphoid cells, when appropriately stimulated, can produce
lymphokine-like factors (cytokines) (17), there exists great
potential for regulatory action involving potentially any cell in
the body. Finally, as we have shown, excessive production of
lymphokines may paradoxically suppress local lymphokine-dependent
reactions by pre-emption and target cell inactivation mechanisms
as well as the abolition of chemotactic gradients.

For most of these mechanisms, there is evidence for _in vivo_
activity as well as _in vitro_ effect. Thus, the regulatory circuit
for the lymphokines may be more complex than that involved in
antibody production. Teleologically, this may be necessary because
the lymphokine system, to the extent that it can modulate antibody
formation, effector functions of immunity, and inflammatory and
reparative processes, represents a central locus of host defense.
Since all these reactions, if carried to excess, can be harmful
to the host rather than beneficial, it is obvious that critical
fine-tuning of the system is a biological necessity.

SUMMARY

Lymphokine-dependent reactions form the bulk of what is
commonly thought of as cell-mediated immunity. These reactions
are of importance in biologic responses involving immunologic
activation, inflammation and repair. They are subject to a number
of control mechanisms. These are most readily studied under
conditions in which such reactions are suppressed by systemic
administration of a relatively large dose of antigen. This effect
is known as desensitization of delayed-type hypersensitivity and
mimics of state of clinical anergy seen in various lymphoproliferative and granulomatous disease states. This represents a situation
in which "excessive" regulation appears to be occurring. Studies
of this type of immune unresponsiveness have revealed that the
suppression is due to several mechanisms acting in concert. Thus,
the generation and release of a large amount of circulating lymphokines at the initial stage of desensitization is responsible for
the general suppression of hypersensitive reactions, through both
the loss of mediator gradients from one tissue to another and the
pre-emption of effector cells. In a subsequent stage, factors
inhibitory for further lymphokine production are generated. These
inhibitory mediators may be demonstrated both _in vitro_ and _in vivo_.
At least one of these suppressive molecules is made by lymphokine-
activated macrophages. Thus, there appears to be complex feedback
inhibitory loops for mediator production, as well as for the control
of the expression of lymphokine activity.

ACKNOWLEDGMENTS

This work was supported by National Institutes of Health
Grant #AI-16706. Dr. D'Silva was supported by NIH Postdoctoral
Training Fellowship CA-09205.

REFERENCES

1. Cohen, S. (1977). The role of cell-mediated immunity in the induction of inflammatory responses. Amer. J. Path. 88:502.
2. Possanza, G., Cohen, M.C., Yoshida, T. and Cohen, S. (1979). Human macrophage migration inhibition factor: evidence for subunit structure. Science 205:4403.
3. Adelman, N., Hammond, M.E., Cohen, S. and Dvorak, H.F. (1979). In "Biology of the Lymphokines", S. Cohen, E. Pick and J. J. Oppenheim, eds., Academic Press, NY, p. 13.
4. Cohen, S., Ward, P.A., Yoshida, T. and Burek, C.L. (1973). Biologic activity of extracts of delayed hypersensitivity skin reaction sites. Cellular Immunol. 9:363.
5. Claman, H.N., Miller, S.D., Conlon, P.J. and Moorhead, J.W. (1980). Control of experimental contact hypersensitivity. Adv. Immunol. 30:121.
6. Cohen, S., Fisher, B., Yoshida, T. and Bettigole, R. (1974). Serum MIF activity in patients with lymphoproliferative diseases. New Eng. J. Med. 290:882.
7. Yoshida, T., Siltzbach, L.E., Masih, N. and Cohen, S. (1979). Serum migration inhibitory activity in patients with sarcoidosis. Clin. Immunol. and Immunopath. 13:39.

8. Yamamoto, K. and Takahashi, Y. (1971). Macrophage migration inhibition by serum from desensitized animals previously sensitized with tubercle bacilli. Nature: New Biol. 233:261.
9. Salvin, S.B., Younger, J.S. and Lederer, W.H. (1973). Migration inhibitor factor and interferon in the circulation of mice with delayed hypersensitivity. Infect. Immun. 7:68.
10. Yoshida, T. and Cohen, S. (1974). Lymphokine activity _in vivo_ in relation to circulating monocyte levels and delayed skin reactivity. J. Immunol. 112:1540.
11. Kantor, F.S. (1975). Infection, anergy and cell-mediated immunity. New Eng. J. Med. 292:629.
12. Papermaster, V., Yoshida, T. and Cohen, S. (1978). Desensitization: II. Passive transfer of the desensitized state by serum from desensitized animals. Cell Immunol. 35:378.
13. Sonozaki, H. and Cohen, S. (1971). The macrophage disappearance reaction: Mediation by a soluble lymphocyte-derived factor. Cell Immunol. 2:341.
14. Sonozaki, H., Papermaster, V., Yoshida, T. and Cohen, S. (1975). Desensitization: Effects on cutaneous and peritoneal manifestations of delayed hypersensitivity in relation to lymphokine production. J. Immunol. 115:1657.
15. Cohen, S. and Yoshida, T. (1977). Suppression of B cell MIF production by T cells and soluble T cell-derived factors. J. Immunol. 119:719.
16. Bergenstock, R.W., Cohen, S. and Yoshida, T. (1981). T and B cell activation by mitogenic factor generated from antigen-stimulated lymphocytes. Immunology 42:321.

17. Bigazzi, P.E. (1979). Cytokines: Lymphokine-like mediators
         produced by non-lymphoid cells. In "Biology of the
         Lymphokines", S. Cohen, E. Pick, and J.J. Oppenheim,
         editors. Academic Press, NY, p. 243.

STIMULATION OF HOST RESISTANCE TO METASTATIC TUMORS BY MACROPHAGE

ACTIVATING AGENTS ENCAPSULATED IN LIPOSOMES

Richard Kirsh and George Poste

Department of Tumor Biology, Smith Kline and French
Laboratories, Philadelphia, Pennsylvania 19101

INTRODUCTION

The disappointing results obtained in experimental and clinical
efforts to devise effective, specific, active immunotherapy procedures for the treatment of cancer have stimulated renewed interest
in mechanisms of non-specific "natural" antitumor surveillance
mediated by macrophages and natural killer cells. A significant
effort is now underway in many laboratories to develop effective
biological response modifier (BRM) agents that can stimulate
the antitumor activities of these cells. Liposomes offer a useful
carrier system for delivering BRM agents to macrophages _in vivo_.
When injected i.v. the majority of liposomes are taken up by
phagocytic reticuloendothelial cells in the liver and spleen, and
by circulating monocytes (reviewed in 6). The passive localization of liposomes into mononuclear phagocytes is frustrating to
investigators who wish to target liposomes to other cell types in
the body, including tumor cells, but provides a highly effective
mechanism for "targeting", albeit passively, of liposome-encapsulated materials to macrophages. We have exploited this pathway to
deliver natural and synthetic molecules with macrophage activating
activity to macrophages _in situ_. Data presented here, and in our
previous publications, indicates that systemic administration of
lymphokines or muramyl dipeptide (MDP) encapsulated in liposomes
activates the tumoricidal properties of macrophages _in vivo_ (1-5,
9-14). In this paper we report how this approach can be used to
augment host defense against tumors and eradicate established lung
metastases.

MATERIALS AND METHODS

     Full details of the origin and properties of the tumor systems,
preparation of liposome-encapsulated macrophage activators, assay
of macrophage-mediated tumoricidal activity and methods for evaluating metastatic tumor growth are given in the references cited in
the following sections. Unless stated otherwise all references to
liposomes made in the remainder of this paper refer to multi-
lamellar (MLV) liposomes prepared from phosphatidylserine (PS) and
phosphatidylcholine (PC) (3:7 mole ratio).

RESULTS AND DISCUSSION

     The lung is an important site for metastatic disease. We
have therefore sought to develop methods for efficient activation
of pulmonary macrophages (AM). Detailed comparison of a variety
of liposomes of differing size, surface charge and lipid composition
has established that negatively-charged MLV liposomes (1-2 μm diameter) prepared from PS and PC (3:7 mole ratio) represent the optimal liposome for efficient localization in the pulmonary capillary
bed and activation of AM _in situ_ (2). The activated AM recovered
from animals injected i.v. with these liposomes containing macrophage activators are in fact blood monocytes that engulf liposomes,
are arrested in lung capillaries, and which migrate subsequently
into the alveoli (10). Extravasation of liposomes into the extra-
vascular alveolar compartment does not appear to occur.

     In our initial studies on macrophage activation, we demonstrated
that phagocytic uptake of liposomes containing lymphokines harvested
from mitogen-stimulated lymphocytes produced efficient activation
of rodent macrophages _in vitro_ and _in vivo_ (1-3,9,11,12,14).
Intravenous injection of such preparations into tumor-bearing mice
produced significant destruction of established lung metastases
arising from subcutaneous tumors produced by implantation of the
highly metastatic B16-BL6 melanoma cell line (Table 1). However,
the use of crude, unfractionated lymphokine preparations has a
number of shortcomings. First, the lack of a quantitative assay
for the specific lymphokine(s) responsible for macrophage activation frustrates accurate measurement of dose-response relationships.
Second, many other potent biological mediators are present in such
preparations. This not only complicates interpretation of the
mechanism of the host response but the presence of mediators with
mitogenic, angiogenic and vascular permeabilizing activities could
promote tumor growth under certain conditions. It is therefore
desirable, and probably mandatory for clinical trials in man,
that efforts to augment host resistance to tumors employ agents
of defined composition and purity. For this reason, our current
research is now directed to assessing the efficacy of muramyl

Table 1. Therapy of Pulmonary Metastases by Intravenous Injection
         of Liposome-Encapsulated Lymphokines or Muramyl
         Dipeptide (MDP)[a]

Treatment[b]	Number of pulmonary metastases[c]		Number of mice with metastases/ total mice
	median	range	
untreated control	31	0–94	13/14
lymphokines	27	0–87	13/15
liposomes [HBSS]	33	0–107	13/14
liposomes [lymphokines]	5[d]	0–42	4/14
liposomes [HBSS] + free lymphokines	29	0–78	12/15
MDP (200 µg)	36	0–103	13/14
liposomes [MDP]	0[d]	0–6	3/14
liposomes [HBSS] + MDP (2.5 µg)	32	0–116	12/13

[a] C57BL/6 mice were injected s.c. in the footpad with 5×10^4 viable
B16-BL6 tumor cells suspended in a volume of 0.05 ml of HBSS.
Four to five weeks later, when the tumors reached 10–15 mm in
diameter, the mice were anesthetized by methoxylflurane inhalation
and the tumor-bearing leg, including the popliteal lymph node,
was amputated at the midfemur. Treatments began three days
later as outlined in the footnotes below and were given twice
weekly for four weeks.

[b] Groups of five mice were injected in the tail vein with 0.2 ml
Hanks balanced saline solution containing MLV liposomes (PS/PC,
3:7 mole ratio; 5 µmoles lipid/mouse) containing the encapsulated
materials indicated in the square brackets [lymphokines = 12.5 µl;
MDP = 2.5 µg]. Other test groups of animals received either
lymphokines (200 µl) or MDP (200 µg). Control animals were either
untreated or injected with liposomes containing encapsulated HBSS
and suspended in 0.2 ml HBSS containing lymphokines (12.5 µl) or
MDP (2.5 µg) in equivalent amounts to that encapsulated in the
liposomes administered to test animals.

[c] Mice were killed two weeks after completion of treatment and
autopsied. Lung metastases were determined microscopically and
confirmed histologically.

[d] Significant reduction in the incidence of metastasis ($p < 0.001$,
Chi square test) compared to other treatment protocols and untreated control animals.

dipeptide (MDP) and structurally related compounds as macrophage activating agents.

MDP (N-acetyl-L-alanyl-D-isoglutamine), Fig. 1, is the minimal structural unit (mol wt 492) with immunopotentiator activity that can replace <u>Mycobacteria</u> in Freund's complete adjuvant (FCA) (reviewed in 8). Although MDP has potent effects on macrophage function <u>in vitro</u> (8) its effectiveness <u>in vivo</u> is limited by its extremely rapid clearance from the body (90% of MDP injected i.v. is excreted in the urine within 2 hours; see 8). Even when MDP is injected i.v. at very high doses (500 µg) it fails to stimulate significant macrophage-mediated antitumor activity (5). Intravenous injection of MDP encapsulated in MLV liposomes is highly effective in activating rodent macrophage populations <u>in vitro</u> and <u>in vivo</u> (5,13) and in enhancing destruction of metastases in tumor-bearing animals (Table 1).

In the experiments described in Table 1 the tumor burden in the lung at the start of therapy exceeds 10^7 cells (5). In addition to reducing the number of metastases, parallel experiments revealed that 65% of mice treated with liposome-encapsulated MDP survived at least 120 days after the final treatment (Table 2). This indicates that the tumor burden in these animals was probably reduced to fewer than 10 viable cells because they survived longer than 40-50 days, which is the median life span of mice implanted with 10 viable B16 cells (5,7).

CONCLUSIONS

Although the results reported here are encouraging, we consider it unlikely that liposome-encapsulated macrophage

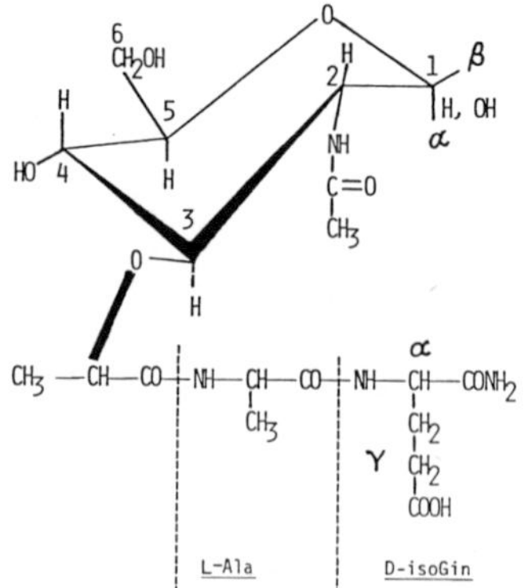

Figure 1. Muramyl dipeptide (Mur NAc)

Table 2. Long Term Survival of Tumor-Bearing Mice After Therapy
          with Liposome-Encapsulated MDP[a]

Treatment	Survival (Days)[b]			
	40	60	80	120
untreated control	19/20	9/20	0/20	0/20
MDP (200 μg)	20/20	12/20	1/20	1/20
liposomes [MDP - 2.5 μg]	20/20	18/20	15/20[c]	13/20[c]
liposomes [HBSS] + MDP (2.5 μg)	20/20	10/20	1/20	0/20

[a] C57BL/6 mice were treated as in footnotes a and b in Table 1
except that survival rather than metastatic burden was measured.

[b] Number of live mice/total number of mice.

[c] Significant increase in median life span compared with other
groups ($p < 0.001$).

activators could serve as a single modality in treating advanced
metastatic disease. As with many other antitumor therapies, optimal
application will probably require its use in combination with other
antitumor agents. Theraputic regimens designed to stimulate the
antitumor properties of macrophages would probably have to follow
cytoreductive treatment which would reduce tumor burden to a level
low enough to allow tumoricidal macrophages to kill the remaining
tumor cells that would otherwise escape destruction. Studies to
evaluate these questions are in progress.

ACKNOWLEDGMENT

     The personal research cited in this paper was supported by
grants CA18260 and CA30192 from the National Cancer Institute.

REFERENCES

1. Fidler, I.J. (1980) Therapy of spontaneous metastases by
         intravenous injection of liposomes containing
         lymphokines. Science 208, 1469-1471.

2. Fidler, I.J., Raz, A., Fogler, W.E., Kirsh, R., Bugelski, P.,
       and Poste, G. (1980). Design of liposomes to improve
       delivery of macrophage-augmenting agents to alveolar
       macrophages. Cancer Research 40, 4460-4466.
3. Fidler, I.J., Hart, I.R., Raz, A., Fogler, W.E., Kirsh, R.
       and Poste, G. (1980). Activation of tumoricidal
       properties in macrophages by liposome-encapsulated
       lymphokines: in vivo studies. In: Liposomes and
       Immunobiology, eds., B.H. Tom & H. Six, pp. 109-118.
       Elsevier/North-Holland, New York.
4. Fidler, I.J., Sone, S., Fogler, W.E. and Barnes, Z. (1981).
       Eradication of spontaneous metastases and activation
       of alveolar macrophages by intravenous injection of
       liposomes containing muramyl dipeptide. PNAS-USA -
       in press.
5. Fidler, I.J., Barnes, Z., Fogler, W.E., Kirsh, R., Bugelski, P.
       and Poste, G. (1982). Evidence for the involvement of
       macrophages in the eradication of established metastases
       produced by intravenous injection of liposomes containing
       macrophage activators. Cancer Research - in press.
6. Gregoriadis, G. and Allison, A.C. (1980). eds. Liposomes in
       Biological Systems. Wiley Interscience, New York.
7. Grisold, D.P., Jr. (1972). Consideration of the subcutaneously
       implanted B16 melanoma as a screening model for
       potential anticancer agents. Cancer Chemotherapy
       Reports, 3, 315-323.
8. Parant, M. (1979). Biological properties of a new synthetic
       adjuvant, muramyl dipeptide (MDP). Semin. Immunopathol.
       2, 101-118.
9. Poste, G., Bucana, C., Raz, A., Bugelski, P., Kirsh, R. and
       Fidler, I.J. (1982). The behavior of intravenously
       inoculated liposomes in the microcirculation: implica-
       tions for liposome targeting and drug design. Cancer
       Research - in press.
10. Poste, G. and Kirsh, R. (1979). Rapid decay of tumoricidal
       activity and loss of responsiveness to lymphokines in
       inflammatory macrophages. Cancer Research 39, 2582-2590.
11. Poste, G., Kirsh, R., Fogler, W.E. and Fidler, I.J. (1979).
       Activation of tumoricidal properties in mouse macrophages
       by lymphokines encapsulated in liposomes. Cancer
       Research 39, 881-892.
12. Poste, G., Kirsh, R., Raz, A., Sone, S., Bucana, C., Fogler,
       W.E. and Fidler, I.J. (1980). Activation of tumoricidal
       properties in macrophages by liposome-encapsulated
       lymphokines: in vitro studies. In: Liposomes and
       Immunobiology, eds. B.H. Tom and H. Six, pp. 93-107.
       Elsevier/North-Holland, New York.

13.   Sone, S. and Fidler, I.J. (1981). In vitro activation of
           tumoricidal properties in rat alveolar macrophages by
           synthetic muramyl dipeptide encapsulated in liposomes.
           Cell. Immunol. 57, 42-50.
14.   Sone, S., Poste, G. and Fidler, I.J. (1980). Rat alveolar
           macrophages are susceptible to activation by free and
           liposome-encapsulated lymphokines. J. Immunol. 124,
           2197-2202.

EFFECT OF PROSTAGLANDINS ON THE PRODUCTION OF INTERLEUKIN-2

R. S. Rappaport and G. R. Dodge

Wyeth Laboratories, Inc.
Philadelphia, Pennsylvania

INTRODUCTION

A substantial body of evidence implicates prostaglandins as
potent, local regulators of the immune response (1). It is well
documented, for example, that prostaglandins, especially of the
E-series, inhibit lectin or antigen-induced lymphocyte proliferation in vitro (2-5). Recently, it has become evident that T-cell
proliferation is dependent on the production and utilization of a
soluble protein known as T-cell growth factor (TCGF) or Interleukin-
2 (IL-2) (6-8). Current evidence suggests that IL-2 is produced
by one subset of T-lymphocytes (producer cells) and that it acts
upon another subset of T-lymphocytes (responder cells via specific
IL-2 receptors (9,10). The precise sequence of events which
regulate the production of IL-2 and the development of specific
receptors are not completely understood.

In the present report, we describe the effect of several
prostaglandins on IL-2 production by mitogen-stimulated normal
human lymphocytes. We also describe the effect of a prostaglandin
synthetase inhibitor in the same system.

MATERIALS AND METHODS

Preparation of Lymphocytes. Peripheral venous blood was drawn
from normal fasting men and women into Vacutainer tubes (Becton-
Dickinson, Rutherford, NJ) containing heparin. The blood was
processed within two hours of collection. To reduce volume and
achieve partial separation of cells, whole blood was centrifuged

at 1200 x g for 4 minutes. After removing the plasma, the white
cell layer was removed and diluted with an equal volume of Hank's
Balanced Salt Solution (HBSS). Mononuclear cells were isolated
by Ficoll-Paque (Pharmacia Fine Chemicals, Piscataway, NJ) density
gradient centrifugation. The resultant lymphocyte-rich bands
were washed 3 times in HBSS and suspended in HEPES-buffered
RPMI-1640 medium (M.A. Bioproducts, Walkersville, MD) containing
streptomycin (100 µg/ml), penicillin (100 units/ml), Fungizone
(2.5 µg/ml) and 10^{-5}M mercaptoethanol (Complete Medium). Cell
viability was routinely assessed by the trypan blue exclusion
method and determined to be >90%.

Production of IL-2. Lymphocytes, at a concentration of 10^6
cells/ml, were cultivated in complete medium supplemented with
2% human AB serum (Grand Island Biological Co. Grand Island, NY)
for 48 hr at 37°C in an atmosphere of 5% CO_2, 95% humidity. In
some experiments, the lymphocytes were first passed through glass
wool (Pyrex wool; Corning Glass Works, Corning, NY) to remove
glass-adherent cells. Prostaglandins (Sigma Chemical Co., St.
Louis, MO) or indomethacin (Merck Sharp and Dohme, West Point,
PA) were added to various cultures containing 1 µg/ml phytohemma-
glutinin (PHA) (Wellcome Reagents, Ltd., Beckenham, England) at
the beginning of the cultivation period. After incubation, culture
supernatants were sterilized by passage through 0.45 µm Acrodisc
filters (Gelman Sciences, Inc., Ann Arbor, MI) and stored frozen
until assayed.

Measurement of IL-2. IL-2 levels were determined by measuring
the ability of IL-2-containing fluids to stimulate proliferation
of sensitized normal human lymphocytes. The assay cells (0.3 x
10^6 per ml) were cultured in Complete Medium with 10% heat-
inactivated fetal calf serum (Δ-FCS) (Grand Island Biological Co.,
Grand Island, NY) and 1 µg/ml PHA for 7 days at 37° in 5% CO_2.
Prior to assay, the cells were washed three times with HBSS,
counted, and suspended (10^6 cells per ml) in Complete Medium containing 40% Δ-FCS. Aliquots of 10^5 cells (100 µl) were distributed
in the wells of Linbro microtiter plates (Flow Laboratories, Inc.,
Hamden, CO) and test samples (100 µl per well) were added in
quadruplicate. The plates were incubated for 48 h at 37° in 5%
CO_2. The cultures were incubated with 1 µCi/well of [^{3}H] thymidine
(New England Nuclear, Boston, MA) and 16-18 hr later, they were
harvested on glass fiber filters using a Titertek Cell Harvester
(Flow Laboratories, Inc.). The filters were counted in a liquid
scintillation counter. Data were evaluated using the following
equation:

$$\frac{\text{c.p.m. control sample} - \text{c.p.m. test sample}}{\text{c.p.m. control sample}} \times 100$$

where the control sample represents the IL-2-containing culture supernatant produced by PHA alone, and test sample represents supernatant produced by PHA and PG or indomethacin. To rule out proliferation due to PHA present in the samples, assay cells were routinely tested for their response to fresh PHA (1 µg/ml) and found to be consistently unresponsive.

RESULTS AND DISCUSSION

The effects of PGA_1, E_1, E_2, $F_{1\alpha}$, and $F_{2\alpha}$ on PHA-induced production of IL-2 by peripheral blood lymphocytes (BPL) are given in Figure 1. The data are from one representative experiment, but similar results were obtained with several other normal donors. The results showed that PGE_1 and PGE_2 inhibited IL-2 production in a dose dependent manner and they were inhibitory at low concentrations ($\sim$3 x 10^{-8}M). In contrast, PGA_1, PGA_2 (not shown), $PGF_{1\alpha}$ and $PGF_{2\alpha}$ did not inhibit IL-2 production, except in the case of PGA, where inhibition was observed only at high concentrations ($\sim$3 x 10^{-6}M). These findings suggested that endogenous prostaglandins regulate IL-2 production, and that a lowering of endogenous PGE levels by a prostaglandin synthetase inhibitor might result in enhancement of IL-2 release. Mitogen-stimulated BPL were, therefore, cultivated in the presence of varying amounts of indomethacin; and the culture supernatants were assayed for IL-2. The results of two representative experiments revealed that IL-2 production increased with increasing concentrations of indomethacin, reaching a maximum at a drug concentration of between 1-5 µg/ml (Fig. 2). Above 5 µg/ml, IL-2 production was severely inhibited, presumably due to the toxicity of the drug.

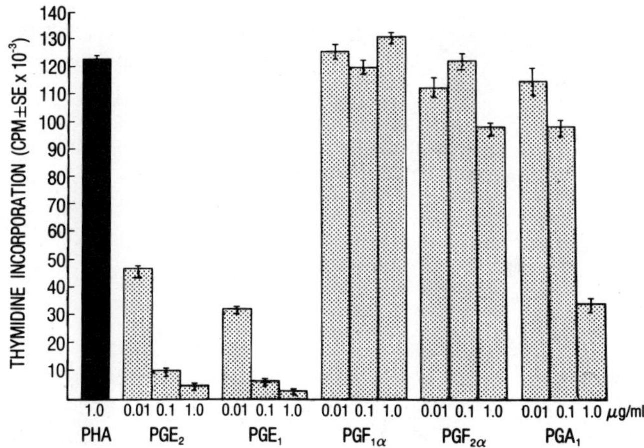

Figure 1.   Effect of Prostaglandins on PHA-Induced Production of IL-2.

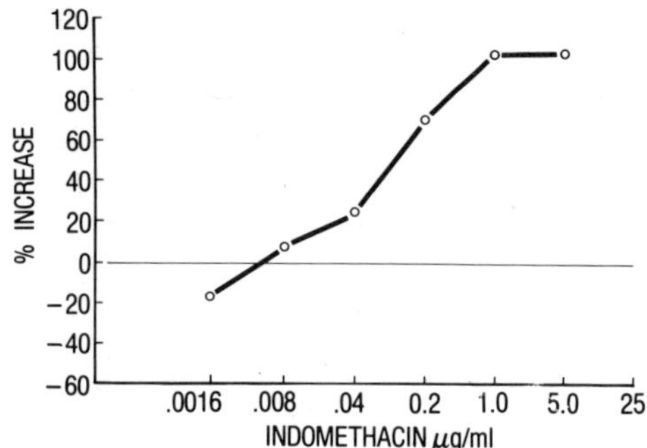

Figure 2.   Percent Increase of PHA (1 μg/ml)-Induced Production
of IL-2 by Indomethacin.

Recently, several investigations have suggested a requirement
for the macrophage factor, Interleukin-1 (IL-1), in the production
of IL-2 by normal murine lymphocytes (11-14). Because of the
importance of these findings we examined IL-2 production by
adherent cell-depleted human lymphocyte cultures. In numerous
comparisons between unfractionated and non-adherent lymphocyte
cultures from the same donors, we observed an increase rather than
a decrease in IL-2 production by the adherent cell-depleted
cultures (Table 1). This observation led to a comparison of the
effects of PG and indomethacin on the same cultures. The results
showed that whereas unfractionated and non-adherent lymphocyte
cultures were equally sensitive to PGE the adherent cell-depleted
culture was comparatively less responsive to indomethacin (Fig. 3).
These findings suggest that there is a population of glass adherent
mononuclear cells that suppress IL-2 production, since their
removal resulted in enhancement of IL-2. Further, the reduced
sensitivity of non-adherent cell cultures to indomethacin indicates
that the glass adherent cell population may be producing PGE.
Suppression of human T-cell mitogenesis by physiological concentrations of PGE and the existence of a prostaglandin-producing
suppressor cell have been previously described by Goodwin et al.
(15). The present findings confirm and extend previous observations by demonstrating a direct effect of PGE on the regulation
of IL-2 production.

SUMMARY

The effect of several prostaglandins on in vitro production
of Interleukin-2 (IL-2) by normal human blood peripheral lymphocytes (BPL) was investigated. Prostaglandins of the E-series
(PGE_1 and PGE_2) inhibited the production of IL-2 in a dose-
dependent manner, and they were inhibitory at concentrations as
low as 10 ng/ml ($\sim$3 x 10^{-8}M). In contrast, prostaglandins of the

Table 1.   Comparison of Interleukin-2 production by unfractionated
           and non-adherent lymphocyte cultures.

Exp. No.	Culture Conditions[1]	Interleukin-2[2] (cpm ± S.E.)	
		unfractionated	non-adherent
1	Control	4,369 ± 280	3,242 ± 266
	PHA	30,085 ± 865	155,308 ±2,389
2	Control	1,104 ± 68	735 ± 113
	PHA	34,062 ±1,159	87,711 ±2,433
3	Control	1,933 ± 241	638 ± 83
	PHA	54,870 ±4,497	78,879 ±3,475
4	Control	1,920 ± 130	1,514 ± 36
	PHA	14,092 ± 541	135,336 ±2,825

[1] Lymphocytes from different donors were cultivated under conditions described in Materials and Methods. Cell-free supernatants
from unstimulated cultures (control) were compared with those
from mitogen-stimulated cultures (PHA, 1 µg/ml).

[2] In each experiment assay cells were tested for their response to
fresh PHA (1 µg/ml). In each case, the response (c.p.m.) was
less than or equal to the response produced by control supernatants.

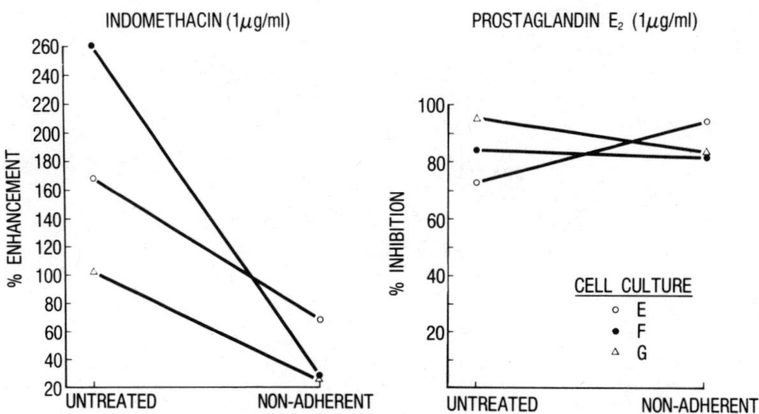

Figure 3.   Effect of Indomethacin and Prostaglandin E_2 on PHA-
            induced IL-2 Production: Untreated vs. Non-Adherent
            Cell Populations.

F and A series ($PGF_{1\alpha}$, $PGF_{2\alpha}$, PGA_1, and PGa_2) did not inhibit the
production of IL-2. Further, indomethacin-treated BPL produced
significantly higher levels of IL-2 than did control cultures,
presumably due to a lowering of endogenous PGE levels. These
effects were also studied in comparisons between unfractionated
and non-adherent lymphocyte cultures. The results suggest that
PGE plays an important role in the regulation of IL-2 production
by normal human lymphocytes.

REFERENCES

1. Goodwin, J.S., Webb, D.R. (1980). Clinical Immunology and
        Immunopathology, 15:106-122.
2. Smith, J.W., Steiner, A.L., Parker, C.W. (1971). J. Clin.
        Invest. 50:442-448.
3. Gordon, D., Bray, M.A., Morley, J. (1976). Nature. 262:3-4.
4. Goodwin, J.S., Bankhurst, A.D., Messner, R.P. (1977). J. Exp.
        Med. 146:1719-1734.
5. Gordon, D., Henderson, D.C., Westwick, J. (1979). Br. J.
        Pharmac. 67:17-22.
6. Morgan, D.A., Ruscetti, F.W., Gallo, R.C., (1976). Science
        193:1007-1008.
7. Smith, K.A., Baher, P.E., Gillis, S., Ruscetti, F.W. (1980).
        Molecular Immunol. 17:579-589.
8. Ruscetti, F.W., Morgan, D.A., Gallo, R.C. (1977). J. Immunol.
        119:131-138.
9. Bonnard, G.D., Yasaka, D., Jacobson, D. (1979). J. Immunol.
        123:2704-2708.
10. Ruscetti, F.W., Gallo, R.C. (1981). Blood. 57:379-394.
11. Smith, K.A., Gillis, S., Baker, P.E. (1979). In The Molecular
        Basis of Immune Cell Function, ed. Kaplan, J.G.
        (Elsevier/North Hollad, Amsterdam), pp. 223-231.
12. Farrar, J.J., Mizel, S.B., Fuller-Farrar, J., Farrar, W.L.,
        Hilliker, M.L. (1980). J. Immunol. 125:793-798.
13. Larsson, E.L., Iscove, N.N., Coutinho, A. (1980). Nature
        (London), 283:664-666.
14. Smith, K.A., Lachman, L.B., Oppenheim, J.J., Favata, M.F.
        (1980). J. Exp. Med. 151:1551-1556.
15. Goodwin, J.S., Bankhurst, A.D., Messner, R.P. (1977). J.
        Exp. Med. 146:1719-1734.

INTERACTION OF MYCOBACTERIA WITH NORMAL AND IMMUNOLOGICALLY

ACTIVATED ALVEOLAR MACROPHAGES

Quentin N. Myrvik, Eva S. Leake and Klara Tenner-Racz*

Department of Microbiology and Immunology, The Bowman
Gray School of Medicine of Wake Forest University,
Winston-Salem, NC 27103 *Bernhard-Nocht Institut für
Schiffs- und Tropenkrankheiten, Dept. of Pathology,
Hamburg, W. Germany

INTRODUCTION

It is well-established that the _Mycobacteria_ differ in their
virulence for various laboratory animals. The mechanisms by which
Mycobacteria express virulence could be the result of a combination
of several factors including a) some form of toxic factor elaborated
by _Mycobacteria_, b) destruction of the phagosomal membrane allowing
Mycobacteria to escape and grow in the cytoplasm of macrophages,
c) inhibition of lysosome fusion and d) a deficient antimicrobial
system in macrophages even in the face of normal lysosome-phagosome
fusion, especially in non-immunologically activated macrophages.

Experiments carried out in our laboratory have demonstrated
that living virulent _Mycobacterium tuberculosis_ (H37Rv) are toxic
for normal rabbit alveolar macrophages in high bacteria/macrophage
ratios (>100:1) as indicated by viability tests (unpublished).
This was further substantiated by the observation that if virulent
Mycobacteria are killed by mild methods such as heating at 65 C for
30 min, this toxicity is lost even in the ratios of 400:1. The
principle of toxicity in this case is associated only with viable
M. tuberculosis and cannot be demonstrated by using supernatants
of culture fluids. Since toxicity in this case is expressed within
a 12-hr period, division of the mycobacteria should be minimal.

The two most likely possibilities to explain the virulence of
intracellular parasites implicate some form of phagosome perturbation. One possibility is that there is inhibition of phagosome-
lysosome fusion and a second possibility would involve phagosome

disruption by the parasite resulting in the parasite growing in the cytoplasm. Both of these observations have been reported in the case of <u>Mycobacteria</u> (1,2).

We have carried out extensive studies on the interaction between <u>Mycobacteria</u> and alveolar macrophages with particular reference toward examining the intracellular events that might take place, including possible destruction of <u>Mycobacteria</u>. In the course of these studies, we have also observed that some species of <u>Mycobacterium</u> upon ingestion are found free in the cytoplasm of normal resident alveolar macrophages. This prompted a study to evaluate on a quantitative basis, the magnitude by which attenuated and virulent <u>Mycobacteria</u> might escape the phagosome and end up free in the cytoplasm of either normal alveolar macrophages or BCG-immune macrophages. This possible mechanism could explain why native resistance to primary infection with virulent tubercle bacilli in normal, susceptible hosts is essentially absent. The early studies of Lurie (3) who measured the growth rates of <u>Myco-bacteria</u> in the lungs of rabbits revealed that both BCG, an attenuated strain, and highly virulent strains multiplied in the lungs until immunity was acquired. If the phagosome destruction mechanism should be of major importance in allowing mycobacteria to to grow in non-immune macrophages, then one would expect that immune macrophages would form a durable phagosome which would remain intact. It should be emphasized that it would not be necessary for all virulent <u>Mycobacteria</u> to escape the phagosome of normal resident pulmonary macrophages to produce initial progressive infection. In this regard, it is possible that pulmonary macrophages at certain stages of maturation might be more vulnerable to this cytopathologic event.

The studies to be described and reviewed compare the interaction of macrophages and <u>Mycobacterium</u> <u>smegmatis</u>, an avirulent microorganism and the BCG strain of <u>Mycobacterium</u> <u>bovis</u>, an attenuated strain that can grow in the lungs of normal rabbits.

MATERIALS AND METHODS

<u>Pulmonary</u> <u>macrophages</u>. Pulmonary macrophages were procured from normal and BCG-vaccinated New Zealand White rabbits by the standard lavage procedure. The vaccinated rabbits received 0.1 mg heat-killed BCG in 0.1 ml mineral oil by the i.v. route. Three weeks after vaccination, the macrophages were harvested.

The lungs from both groups of animals were removed along with the heart while keeping the trachea closed with a hemostat. Without removing the hemostat, 20 ml of sterile physiologic saline were injected through the trachea near its bifurcation. The injected fluid was distributed equally between the two lungs. The lungs

were massaged gently while maintaining the syringe and needle in
place, and the injected fluid withdrawn. This procedure was repeated 5 consecutive times. The cell suspensions were centrifuged,
washed once with sterile saline and suspended in Minimal Essential
Medium with Eagle's salts (MEM, Flow Laboratories, Rockville, MD)
containing 10% heat-inactivated fetal calf serum (FCS), glutamine
(2 mM) and penicillin (100 U/ml). The pH of the medium was adjusted
to 7.2. After determining viability, the cell suspensions were
adjusted to contain 5 x 10^5 cells per ml.

    <u>Mycobacteria</u>. <u>Mycobacterium</u> <u>bovis</u>, BCG strain (ATCC 27289,
Rosenthal strain) and <u>M</u>. <u>smegmatis</u>, NCTC 8159 (ATCC 19420) were
grown in Dubos liquid medium containing 0.04% Tween 80. The BCG
cultures were either 6-days-old or 8-days-old and those of <u>M</u>.
<u>smegmatis</u> were 4-days-old. The cultures were centrifuged, the
bacteria washed once with sterile saline containing 1% FCS and the
sedimented bacteria suspended in saline-FCS. After allowing the
larger clumps of bacteria to sediment, the bacterial suspensions
were standardized to contain approximately 200 million colony-
forming units per ml.

    <u>Phagocytosis</u>. Twenty ml of the macrophage suspensions
(approximately 10 x 10^6 cells) were placed in 50 ml siliconized
Erlenmeyer flasks and the appropriate volumes of the bacterial
suspensions were added to obtain multiplicities of 10, 20, or 50
mycobacteria per macrophage. Control flasks did not receive
bacteria. All the flasks were sealed with non-toxic rubber stoppers
and incubated with gentle oscillation for 18 hr at 37 C in a
regular incubator.

    <u>Electron microscopy</u>. At the end of the incubation period
viability of the cells was determined and the cultures were centrifuged after detaching adherent cells with a rubber policeman. The
cell pellets were fixed for 30 min at room temperature in
Karnovsky's fixative (4). Subsequent fixation was carried out for
2 hr in freshly prepared 2.5% glutaraldehyde in 0.1 M cacodylate
buffer pH 7.2 followed by 1 hr in 2% OsO_4. After fixation, the
samples were dehydrated in graded concentrations of ethanol and
embedded in Spurr rapid cure mixture (5). One and one-half μm
sections were stained with Richardson's azure II-methylene blue
stain. Ultrathin sections were mounted on bare copper grids and
stained with uranyl acetate and lead citrate.

RESULTS

    <u>The problem of detecting phagosomal membranes</u>. In the course of
our studies on the morphology of macrophages and their interaction
with <u>Mycobacteria</u>, we have observed that the phagosomal membrane
may be of the loose type or it may be tightly adherent to the
organism. The problems of visualization of the so-called tight

phagosomal membranes have been discussed in a previous publication
(6). As this study indicated, the outer leaflet of the mycobacterial
cell wall tends to adhere to the inside of the phagosomal membrane.
This creates an artifact in the sense that the phagosomal membrane
appears as a double membrane structure. This potential artifact is
illustrated in Figure 1. It can be noted that a peribacillary space
is clearly visible and that the phagosome which surrounds this
space exhibits a double membrane structure. After examining large
numbers of such phagosomes, we have come to the conclusion that the
double membrane structure is composed of the phagosomal membrane
which is the outer membrane and attached to it is the outer leaflet
of the bacterial cell walls. Although the significance of this
adherence is not known, it could be important in terms of phagosome-
bacterial surface interactions.

Figure 2 illustrates the cell wall and the cytoplasmic membrane
of BCG in an extracellular position. It can be noted that the

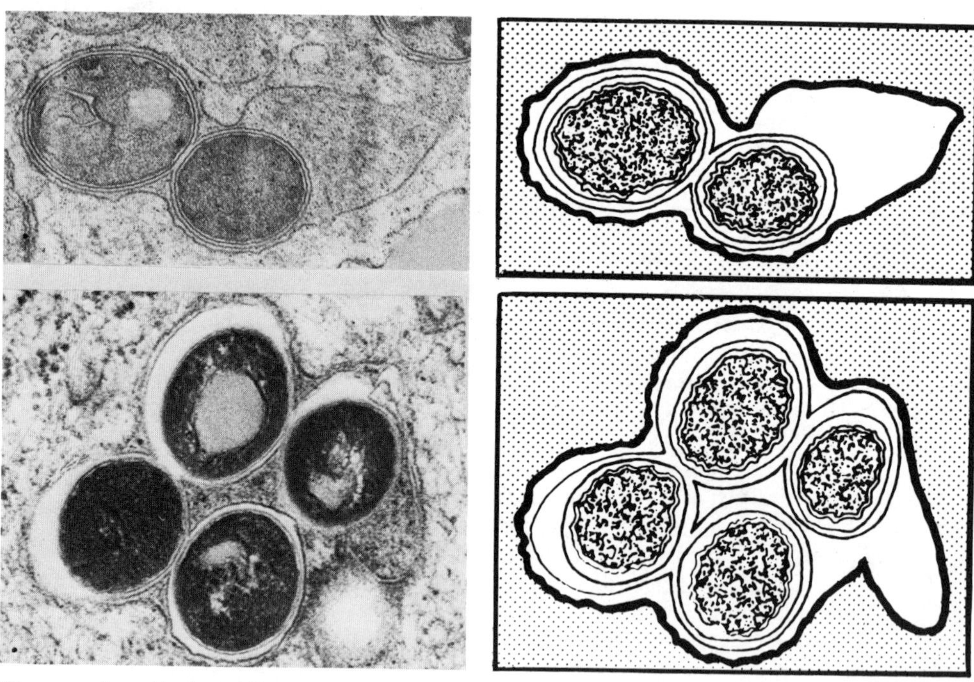

Figure 1. Left: Ultrathin sections of intracellular mycobacteria;
right: diagrammatic representation of the corresponding
micrographs to illustrate how a peribacillary space and
an apparent double phagosomal membrane will result in
relation to the component layers of the bacterial cell
wall. (Permission of J. Reticuloendothel Soc., ref. 6)

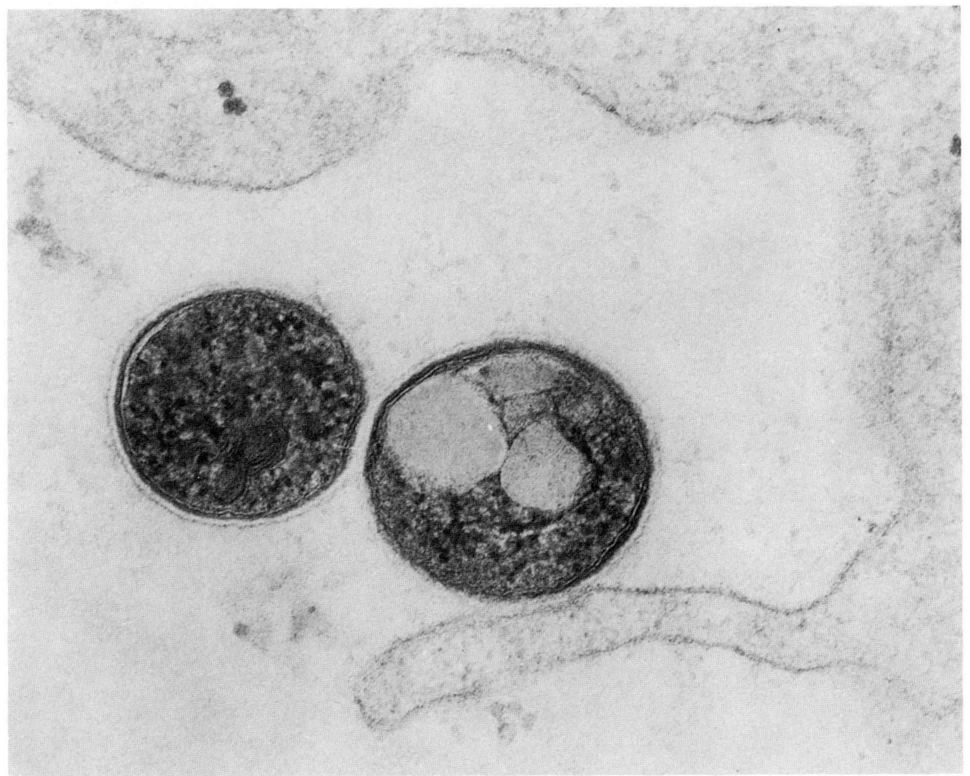

Figure 2. Cross section of two extracellular BCG. Note that the
           outer layer of the cell wall is much thinner than the
           inner layer which is also highly electron dense. X90,500
           (reduced 10% for reproduction)

inner layer of the cell wall is highly electron dense as compared
to the outer layer of the cell wall. It is also striking to note
that when two bacteria are in juxtaposition, the transparent layer
of the cell wall takes on the appearance of fusion. This is a
common observation with Mycobacteria.

   The interaction of Mycobacterium smegmatis and normal alveolar
macrophages. Viable M. smegmatis were incubated with normal resident alveolar macrophages for an 18-hr period as indicated in
Materials and Methods. In examining large numbers of phagocytosed
M. smegmatis, it was observed that essentially all of the bacteria
seemed to be encased in visible phagosomal membranes. It was also
common to see two types of phagosomes, a loose type illustrated in
Figure 3 and the so-called "tight" phagosomal membrane illustrated
in Figure 4. The exact significance of a tight versus a loose
phagosomal membrane is not understood. It is possible that this
could be an artifact in part due to the hypo- or hypertonic

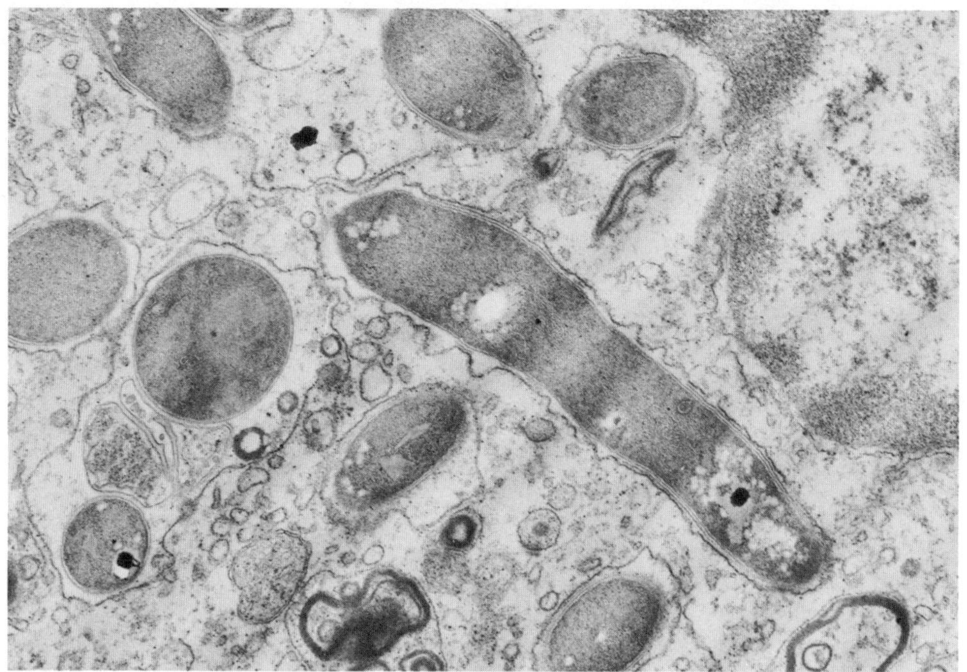

Figure 3.   A "loose type" of phagosomal membrane is seen surrounding
            M. smegmatis in this ultrathin section of an alveolar
            macrophage from a normal rabbit.   X27,700

conditions that might prevail in tissue culture or fixation.
Further studies are needed to establish the significance of this
observation. It should be emphasized that at least 95% of M.
smegmatis was encased in phagosomal membranes in normal alveolar
macrophages.

The interaction of BCG and normal alveolar macrophages. In
an extensive study evaluating this interaction, it was observed
that a significant number of BCG could be found free in the cytoplasm as illustrated in Figure 5. It is particularly noteworthy
in this figure that the exterior layer of the cell wall is
distinct and easily identified. Note that two cells in this figure
have no definable phagosomal membrane. The third cell (arrow)
exhibits a possible phagosome that has fragmented. The BCG which
appeared within phagosomes of normal alveolar macrophages, were
usually encased in tightly adhering phagosomal membranes as
illustrated in Figure 6. It is of interest that in Figure 6,
there is an organism undergoing disintegration which is surrounded
by an identifiable phagosomal membrane. In addition to this, one
can note with one organism (arrow) that the phagosomal membrane is
very difficult to identify on one side of the organism yet it can
be visualized on the opposite side suggesting the possibility that
the phagosomal membrane was in the process of being destroyed.

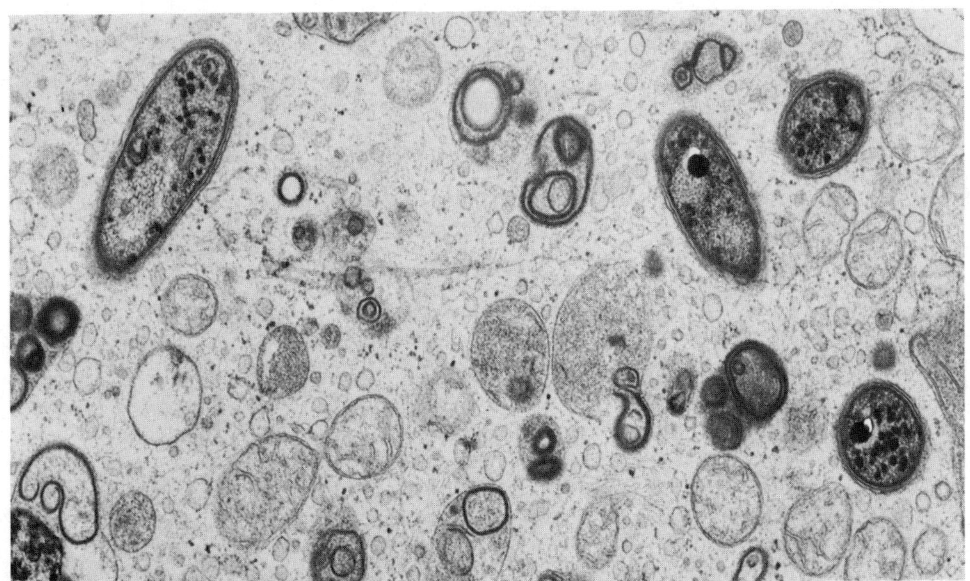

Figure 4. A "tight type" of phagosomal membrane around M. smegmatis is illustrated in this micrograph. X24,800

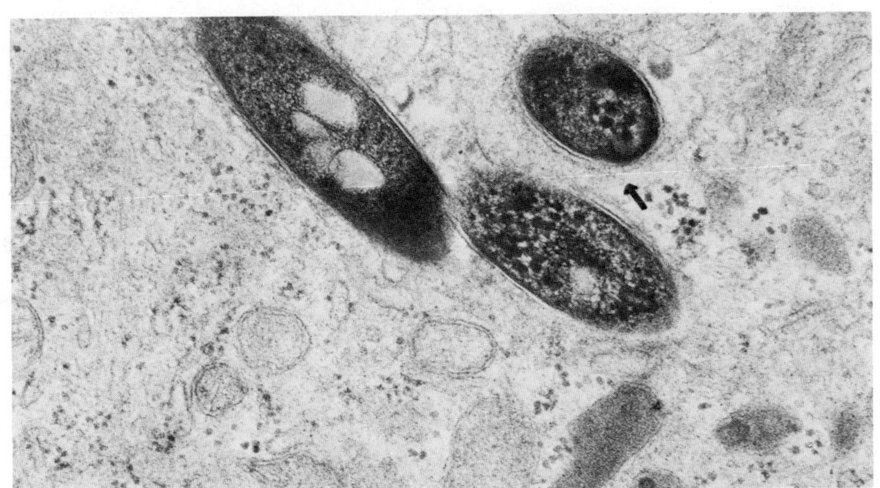

Figure 5. Ultrathin section of an alveolar macrophage from a
normal rabbit where phagocytosed BCG can be seen free
in the cytoplasm. A possible disintegrating phagosomal
membrane is indicated with an arrow. X41,900

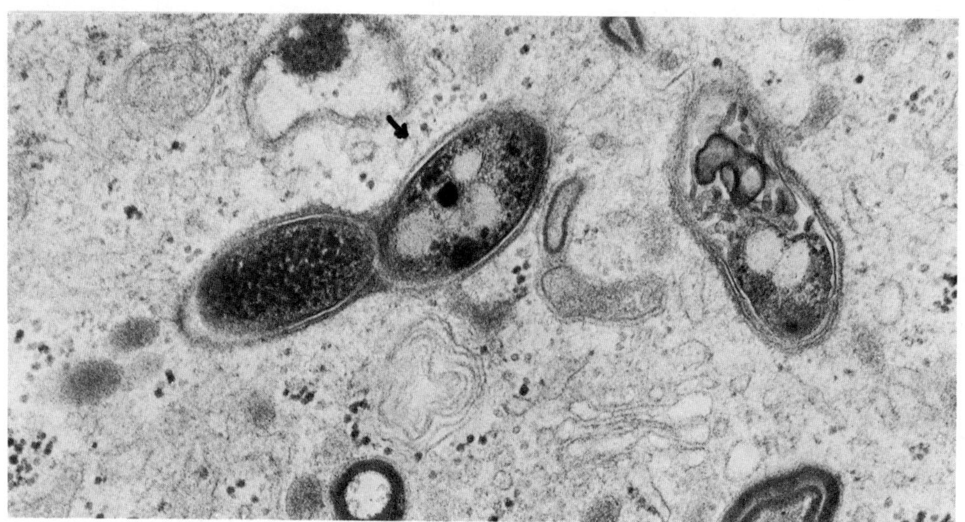

Figure 6.   Intracellular BCG within tightly adherent phagosomal
            membranes in an alveolar macrophage from a normal rabbit.
            A mycobacterium (right) is being digested. The arrow
            points to the side of a mycobacterium where the phago-
            somal membrane appears to be absent. X41,900 (reduced
            10% for reproduction)

The interaction of Mycobacterium smegmatis and BCG-immune
alveolar macrophages. As we indicated previously, M. smegmatis
occurred routinely in identifiable phagosomal membranes in normal
alveolar macrophages. We also observed that M. smegmatis underwent
destruction in normal alveolar macrophages at some possible rate
which is compatible with the loss of viability when the organisms
are injected intratracheally. These studies indicated that when
one introduces M. smegmatis intratracheally into rabbits, the
killing rate is roughly one $\log_{10}$ per day (unpublished). A
striking enhancement of destruction of M. smegmatis was observed
when they were incubated with BCG-immune macrophages. This is
illustrated in Figure 7. In no instance, did we see this level
of destruction of M. smegmatis when they were incubated with
normal alveolar macrophages.

Interaction of BCG with BCG-immune alveolar macrophages. Two
striking observations were made in this aspect of the study. The
phagosomal membranes were commonly present around the microorganisms.
Furthermore, lysosome fusion was evident and where electron-dense
granules appeared, large amounts of electron-dense material was
seen to surround the organisms or clusters of organisms. As can
be noted in Figure 8a, a tight phagosomal membrane is clearly seen
around this organism. In addition, one can note that fusion is
taking place, or about to take place, with this phagosomal

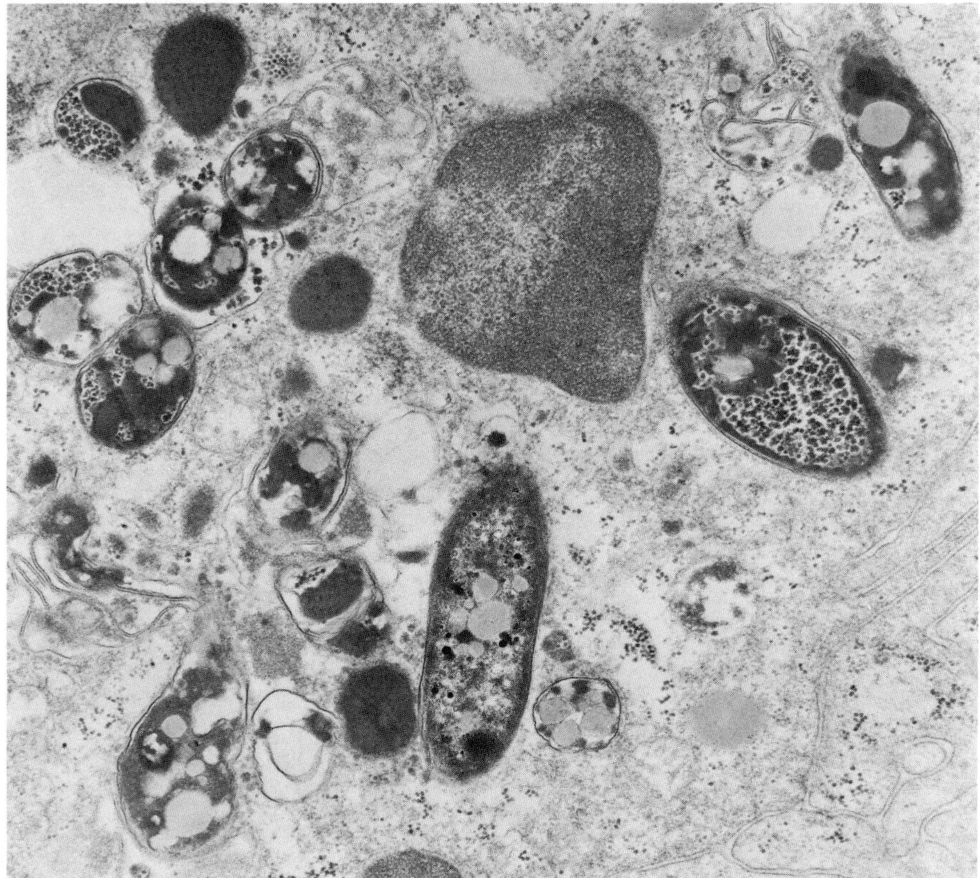

Figure 7.   Severely damaged M. smegmatis are seen in this ultrathin
            section of an immune alveolar macrophage.   X50,100
            (reduced 25% for reproduction)

membrane (arrow). Figure 8b illustrates the common appearance of
phagosomal membranes around ingested BCG. Figure 9 illustrates a
usual appearance when BCG were ingested by highly activated macrophages containing many electron-dense granules. The picture, as
depicted here, shows that the electron-dense material surrounding
the ingested BCG is comparable to that of the material in the
lysosomal-like structures which are very prevalent in BCG-immune
macrophages. In some instances, the phagocytic membrane is
difficult to detect around these segregations of organisms
surrounded by electron-dense material. However, the density of the
electron-dense material tends to obscure the membrane. Figure 10
illustrates two bacterial cells that are in the process of being
engulfed by a macrophage. It is likely that the outermost
organism (B) is still outside the cell. This is a striking

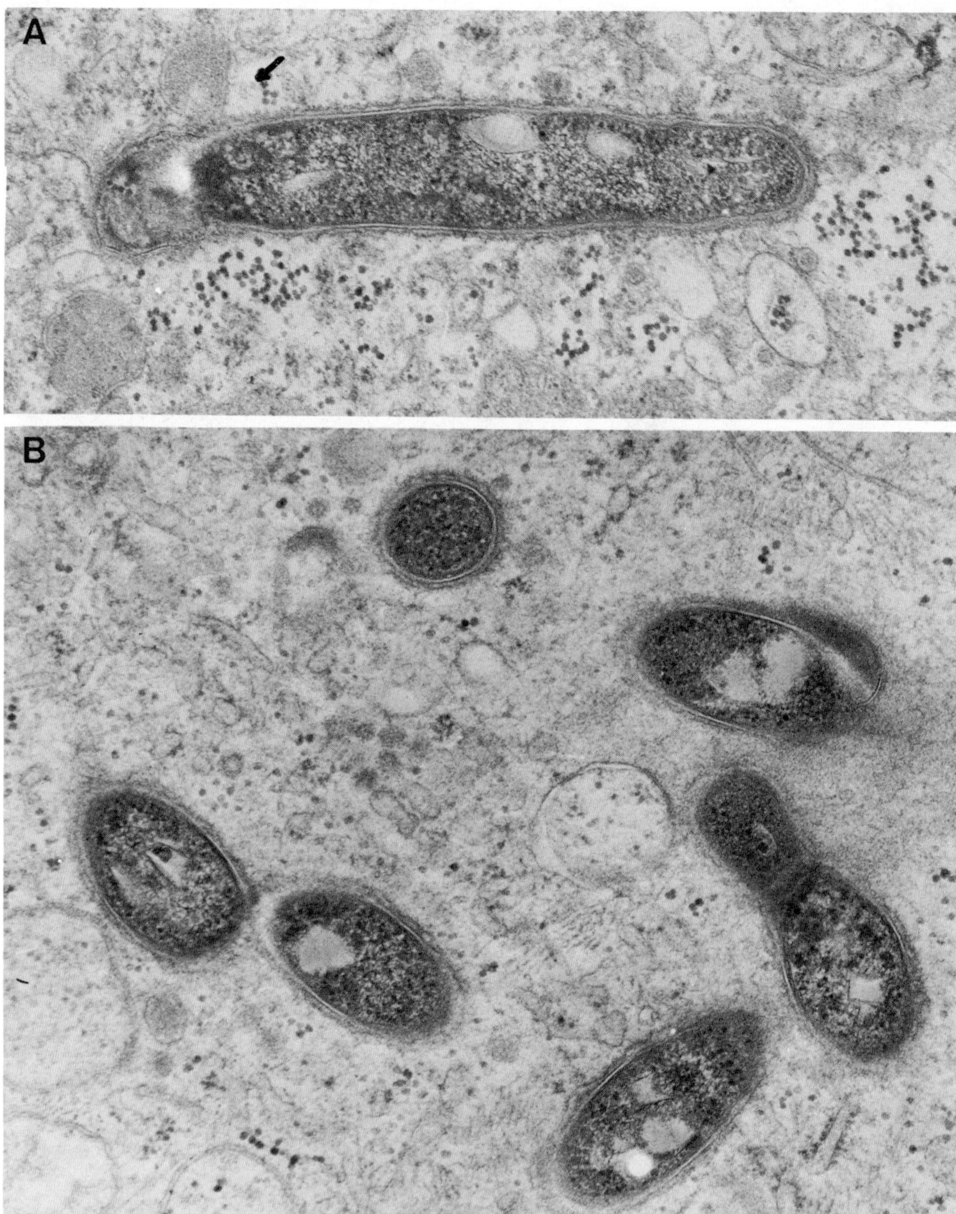

Figure 8.   A)   Longitudinal section of an intracellular BCG in an
immune alveolar macrophage.   Fusion of a lysosome to a
tightly adhered phagosomal membrane is indicated by the
arrow.   B) "Tight type" of phagosomal membranes surrounding intracellular BCG are seen in this ultrathin section
of an immune alveolar macrophage.   X41,000 (reduced 10%
for reproduction)

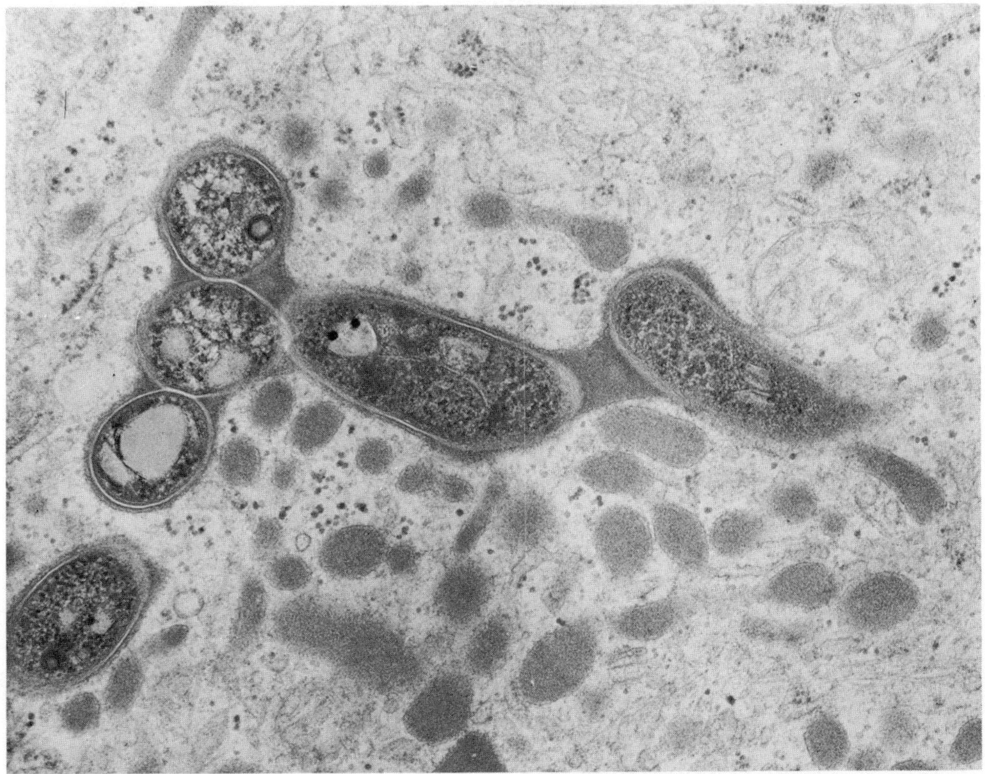

Figure 9. BCG phagocytosed by immune alveolar macrophages contain-
           ing abundant electron dense lysosome-like structures
           commonly appear surrounded by material of similar
           electron density. The phagocytic membrane is difficult
           to detect in these instances. X41,900 (reduced 20%
           for reproduction)

illustration of how the ingested BCG with the apparent tight
phagosome displays a clear-cut delineation of its outer cell wall
layer (A). The bacterium adjacent to A has undergone destruction
(arrow). What is an apparent peeling away of the cell wall
probably reflects the cell wall of the disintegrated organism which
still shows some fusion with the adjacent organism that is in the
process of being ingested.

   An attempt at quantifying the significance of the presence of
intact phagosomal membranes. In the course of examining large
numbers of M. smegmatis and BCG, it is clear that M. smegmatis,
for the most part, is found in intact phagosomes. In a study where
bacterial numeration and phagosome numbers were evaluated, approximately 1-3% of M. smegmatis were found without identifiable phagosomal membranes in normal alveolar macrophages. In the case of

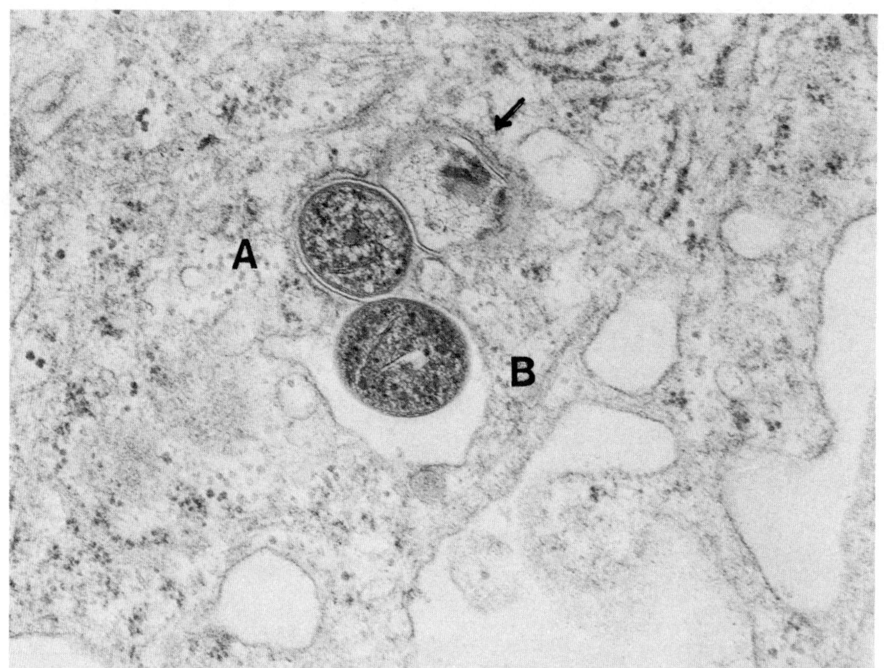

Figure 10.   Cross section of two mycobacteria demonstrating the
             change in electron density observed in the outer leaf-
             let of the cell wall as a bacterium is internalized (A)
             as compared with the adjacent bacterium (B) which is
             extracellular. Note the damaged bacterium adjacent to
             A (arrow). X41,900

BCG, we have observed that between 20-25% of the BCG were free in
the cytoplasm of normal alveolar macrophages. In contrast, studies
involving the interaction of BCG and BCG-immune macrophages indicate
that approximately only 4-6% of the BCG were found free in the
cytoplasm of immune macrophages.

DISCUSSION

     Numerous studies have been directed towards an elucidation of
the mechanisms of virulence of mycobacteria. It is well-established
that virulent bovine and human strains of mycobacteria as well as
the attenuated BCG strain multiply in normal rabbit lung macrophages following initial infection. Lurie (3) clearly showed that
only when immunity was acquired (about 7 to 14 days) did the
pulmonary macrophages inhibit the multiplication of the mycobacteria.
Accordingly, it must be concluded that normal rabbits have essentially no detectable resistance to the intracellular growth of

virulent mycobacteria or even the attenuated BCG strain during the
early intervals after primary infection. This indicates that
normal resident alveolar macrophages collectively are incapable of
inhibiting the intracellular growth of virulent mycobacteria even
though they are capable of killing a variety of other bacteria
including M. smegmatis (unpublished).

One of the first virulence mechanisms proposed for mycobacteria
involved a surface lipid-containing material referred to as cord
factor, largely because virulent M. bovis and M. tuberculosis grow
in serpentine cords, whereas avirulent mutants tend to be non-cord
formers (i.e., grow in random aggregates). The cord factor could
be extracted with petroleum ether. This was a convincing argument
because attenuated strains like BCG tend to be weaker cord formers
than the virulent strains (7). Furthermore, avirulent mutants like
H37Ra are non-cord formers. Whereas originally cord factor was
ill-defined chemically, more recent work suggests that surface
mycosides and certain sulfolipids isolated from mycobacteria may
play a role in virulence (8). The cord factor idea was somewhat
tarnished by the observation that saprophytic mycobacteria also
were cord formers.

A new mechanism of virulence was proposed by Armstrong and
Hart (9) who reported that ferritin-labeled secondary lysosomes did
not fuse with phagosomes containing viable virulent M. tuberculosis
using a mouse peritoneal macrophage system. In contrast, intracellular nonviable M. tuberculosis did not inhibit fusion of the
phagosome with secondary lysosomes based on the observation that
ferritin was detected in mycobacteria-containing phagosomes.

Subsequently, Hart and Young (10) reported that suramin (a
polyanion) if present with yeast in a phagosome blocked fusion of
acridine orange-containing secondary lysosomes. In these experiments, acridine orange was used to label secondary lysosomes. If
fusion of the secondary lysosomes with the yeast-containing phagosome took place the yeast cells would stain with acridine orange.
Conversely, if fusion did not occur, the yeast cells did not stain.
About the same time Armstrong and Hart (1) reported that viable
virulent H37Rv exposed to immune serum before ingestion resulted
in extensive fusion of ferritin-labeled lysosomes with the phagosomes after phagocytosis. However, they reported that in spite
of massive lysosome-phagosome fusion, normal mouse peritoneal
macrophages failed to express antimycobacterial effects on the
ingested virulent mycobacteria. This finding suggested that only
macrophages mobilized and activated by cell-mediated immune mechanisms (specifically immune T cells) could muster significant anti-
mycobacterial effects. Goren et al. (8) reported that when sulfatides from M. tuberculosis, which are anionic trehalose glycolipids
and largely responsible for the neutral red reactivity of virulent
strains, accumulate in lysosomes, fusion with phagosomes did not

occur. In this study, the acridine orange technique was also used.
The authors suggested that sulfatides exerted a selective inhibitory
influence on membrane fusion.

More recently, Goren et al. (11) have concluded that labeling
of lysosomes with acridine orange may lead to artifactual results
based on observations with suramin, dextran sulfate and poly-D-
glutamic acid. These authors have indicated that polyanionic
substances somehow blocked the transfer of acridine orange to the
yeast in phagosomes. Accordingly, yeast cells were not stained
with acridine orange if polyanions were in lysosomes (secondary),
suggesting lack of fusion. However, Goren et al. (11) confirmed
fusions with ferritin-labeled lysosomes under similar conditions.
Somehow added polyanionic substances apparently bind the dye and
prevent the yeast cell from becoming stained.

Tanowitz et al. (12) noted that _Trypanosoma cruzi_ appeared to
survive in host cells mainly outside of vacuoles; however, when
they remained in intact host cell phagosomes they were destroyed.
On the other hand, Berman et al. (13) reported that _Leishmania_
survived and multiplied within phagosomes despite fusion with
secondary lysosomes. Alexander and Vickerman (14) made a similar
observation with _Leishmania_ using a mouse peritoneal macrophage
system.

Jones and Hirsch (15) noted that some phagocytosed _Toxoplasma_
prevented the delivery of lysosomal contents to the phagosome and
considered that such phagosomes become a sheltered microenvironment
for the intracellular growth of toxoplasma.

Kishimoto et al. (16) reported that _Coxiella burnetti_ pretreated with normal serum multiplied within phagosomes with ultimate
destruction of the macrophages. In contrast, _C. burnetti_ incubated
in immune serum were more susceptible to phagocytosis and killing.

Dales (17) and Dales and Kajioka (18) noted that vaccinia virus
disrupted the phagocytic vesicles in L-cells and replicated in the
cytoplasm. However, if specific antibody reacted with the vaccinia
virus, prior to entry, the phagosome remained intact and the viral
genome was ultimately destroyed.

The observations made in the present study suggest that BCG
can disrupt the phagosomal membranes of some normal resident
alveolar macrophages. This immediately raises the question as to
whether immature or mature alveolar macrophages are more or less
vulnerable to this cytopathologic event. If newly mobilized
macrophages are the most susceptible, then such cells could play
an important role in sustaining mycobacterial growth during
primary infection.

Other questions which must be considered relate to the comparative effects of highly virulent mycobacteria such as the Ravenel strain. These studies are in progress. It is of special interest that the phagosomes of BCG-immune alveolar macrophages appear to remain intact, which suggests that these membranes are more resistant to disruption than the membranes of normal resident alveolar macrophages. Based on the observations to date, this mechanism could play an important role in the virulence of mycobacteria, and thus deserves further study.

ACKNOWLEDGMENTS

This grant was supported by U.S. Public Health Service Grant #HL16769.

REFERENCES

1. Armstrong, J.A. and Hart, P. D'Arcy (1975). Phagosome-
       lysosome interactions in cultured macrophages infected
       with virulent tubercle bacilli. J. Exp. Med., 142:1-16.
2. Tenner-Racz, K., Racz, P., Myrvik, Q.N. and Fainter, L. (1977).
       Interaction of normal and activated alveolar macrophages of
       the rabbit with Mycobacterium bovis (BCG) or M. smegmatis
       in vitro and in vivo. (Abstract) J. Reticuloendothel. Soc.
       Suppl., 22:41a.
3. Lurie, M.B. (1964). Resistance to Tuberculosis: Experimental
       Studies in Native and Acquired Defensive Mechanisms.
       Harvard Univ. Press.
4. Karnovsky, M.J. (1965). A formaldehyde-glutaraldehyde fixa-
       tive of high osmolality for use in electron microscopy.
       J. Cell Biol. 27:137A.
5. Spurr, A.R. (1969). A low viscosity epoxy resin embedding
       medium for electron microscopy. J. Ultrastruct. Res. 26:
       31-43.
6. Leake, E.S., Ockers, J.R. and Myrvik, Q.N. (1977). In vitro
       interactions of the BCG and Ravenel strains of Mycobacterium
       bovis with rabbit macrophages: Adherence of the phagosomal
       membrane to the bacterial cell wall and the problem of the
       peribacillary space. J. Reticulendothel. Soc., 22:129-147.
7. Bloch, H., Sorkin, E. and Erlenmeyer, H. (1953). A toxic
       lipid component of the tubercle bacillus ("cord factor").
       Am. Rev. Tuberc., 67:629-643.
8. Goren, M.B., Hart, P. D'Arcy, Young, M.R. and Armstrong, J.A.
       (1976). Prevention of phagosome-lysosome fusion in cultured
       macrophages by sulfatides of Mycobacterium tuberculosis.
       Proc. Nat. Acad. Sci. USA, 73:2510-2514.

9.   Armstrong, J.A. and Hart, P. D'Arcy (1971). Response of
     cultured macrophages to Mycobacterium tuberculosis, with
     observations on fusion of lysosomes with phagosomes.
     J. Exp. Med., 134:713-740.
10. Hart, P. D'Arcy and Young, M.R. (1975). Interference with
     normal phagosome-lysosome fusion in macrophages, using
     ingested yeast cells and suramin. Nature, 256:47-49.
11. Goren, M.B., Swendsen, C.L. and Henson, J. (197). Mycobac-
     terial sulfatides and other polyanionics as inhibitors
     of Phagolysosome formation: Art, fact and artifact.
     Japan Med. Sci. Program, 14th Joint Conference on
     Tuberculosis, p. 422.
12. Tanowitz, H., Wittner, M., Kress, Y. and Bloom, B. (1975).
     Studies of in vitro infection by Trypanosoma cruzi. I.
     Ultrastructural studies on the invasion of macrophages
     and L cells. Amer. J. Trop. Med. Hyg., 24:25-33.
13. Berman, J.D., Dwyer, D.M. and Wyler, D.J. (1979). Multiplica-
     tion of Leishmania in human macrophages in vitro. Infect.
     Immun., 26:375-379.
14. Alexander, J. and Vickerman, K. (1975). Fusion of host cell
     secondary lysosomes with the parasitophorous vacuoles of
     Leishmania mexicana-infected macrophages. J. Protozool.
     22:502-508.
15. Jones, T.C. and Hirsch, J.G. (1972). The interaction between
     Toxoplasma gondii and mammalian cells. II. The absence
     of lysosomal fusion with phagocytic vacuoles containing
     living parasites. J. Exp. Med., 136:1173-1194.
16. Kishimoto, R.A., Veltri, B.J., Canonico, P.G., Shirey, F.G.
     and Walker, J.S. (1976). Electron microscopic study on the
     interaction between normal guinea pig peritoneal macro-
     phages and Coxiella burnetti. Infect. Immun., 14:1087-1096.
17. Dales, S. (1963). The uptake and development of vaccinia
     virus in strain L cells followed with labeled viral
     deoxyribonucleic acid. J. Cell Biol., 18:51-72.
18. Dales, S. and Kajioka, R. (1964). The cycle of multiplication
     of vaccinia virus in Earle's strain L cells. I. Uptake
     and penetration. Virology, 24:278-294.

GUINEA PIG ALVEOLAR MACROPHAGES PROBABLY KILL M. TUBERCULOSIS

H37Rv AND H37Ra IN VIVO BY PRODUCING HYDROGEN PEROXIDE

P.S. Jackett, P.W. Andrew, V.R. Aber and D.B. Lowrie

MRC Unit for Laboratory Studies of Tuberculosis
Royal Postgraduate Medical School
London W12 OHS, England

The major phagocytic cell involved in host resistance to
Mycobacterium tuberculosis infection is probably the alveolar
macrophage. It seems likely that macrophages, at least in the
immune animal, can kill M. tuberculosis (1) and this might involve
the production of hydrogen peroxide (H_2O_2) by the macrophages.
Macrophages release H_2O_2 in vitro when exposed to phagocytic
stimuli or certain soluble agents that perturb the plasma membrane (8,9). H_2O_2 can kill M. tuberculosis and resistance to
H_2O_2 in vitro correlates with high virulence in the guinea
pig (11,5).

To asses the role of macrophage H_2O_2 in killing M. tuberculosis in guinea pig lungs we have compared the fate of parent
M. tuberculosis strains and H_2O_2-susceptible variant strains
during the first 6 days after i.v. infection of normal and BCG-
vaccinated guinea pigs. In parallel we have examined the H_2O_2
releasing capacity of the macrophages recoverable by lavage from
the infected lungs. The studies on bacterial H_2O_2 susceptibility
and viability in the lung formed part of a recent detailed
report (6).

In the basic procedure guinea pigs were vaccinated with
streptomycin-resistant BCG 5 weeks (i.p., 3×10^7 colony-forming
units (c.f.u.)) and 1 week (i.m., 1×10^7 c.f.u.) before challenge
infection. Then normal and vaccinated animals were infected i.v.
with 1×10^7 c.f.u. of one of the bacterial strains. At 3 h,
3 days and 6 days after infection counts of c.f.u. were made from
whole lung homogenates or, in separate experiments, alveolar
macrophages were obtained by pulmonary lavage with medium 199 for
studies of H_2O_2 release. Macrophages that adhered to plastic

petri   dishes after incubation for 1 h were rinsed and assayed for
H_2O_2 release, which was measured fluorimetrically as horse radish
peroxidase-dependent oxidation of p-hydroxyphenylacetic acid (2).
Incubation was for 1 h with opsonised ^{3}H-labelled H37Ra or for
15 min with phorbol myristate acetate (PMA, 1 µg/0.1 ml) as a
stimulus.

The fates of M. tuberculosis strains H37Rv, H37Ra and their
isoniazid-resistant, catalase-negative mutants H37RvHR and H37RaHR
in the lungs of normal and vaccinated guinea pigs are shown in
Fig. 1.   The parent strains were equally resistant and the mutant
strains equally susceptible to H_2O_2 in vitro and in the lungs.
The parent strains fared better than the mutants.   In the normal
animal the effect of the bacterial mutation to H_2O_2 susceptibility
was the same in the first and second 3 day period, suggesting that
H_2O_2 availability was the same in the two periods.   In contrast,
in the vaccinated animal the effect of H_2O_2 susceptibility was
greater in the second period, suggesting an increasing availability
of H_2O_2.   During the first period the effect of H_2O_2 susceptibility
was expressed equally in normal and vaccinated animals, suggesting
equal initial availability of H_2O_2 in normal and vaccinated
animals.

The number of cells obtained by pulmonary lavage of H37Ra-
infected guinea pigs approximately doubled from 3 h to 6 days
after infection, and the proportion of mononuclear cells increased
from 57 to 92% in normal animals, and from 80 to 97% in vaccinated
animals.   The cells obtained contained 0.1% of the viable bacteria
in the lungs at 3 h and this increased to about 2.5% by 6 days.

The capacity of the adherent mononuclear cells (macrophages)
to release H_2O_2 with and without phagocytic stimulation in vitro
is shown in Fig. 2.   Vaccination did not affect the release of
H_2O_2 from macrophages that were removed from the lungs 3 h after
intravenous infection and tested in vitro with or without
phagocytic stimulation.   These observations were consistent with
the evidence that H_2O_2-mediated killing of tubercle bacilli was
the same in normal and vaccinated animals at this early stage of
infection.   Macrophages that were removed from vaccinated animals
on the third and sixth days after infection released progressively
more H_2O_2 than macrophages that were removed at 3 h.   This
increase was not seen with macrophages from the infected normal
animals.   Again, this observation was consistent with the
evidence that H_2O_2-mediated killing of tubercle bacilli in vivo
increased with the onset of acquired immunity.   Phagocytosis of
opsonised H37Ra in vitro enhanced H_2O_2 release.   Figure 2b shows
that the increment per bacterium taken up by the cells increased
with time elapsed after intravenous infection, and that the
increase in responsiveness occurred faster in vaccinated animals
than in normal animals.   The increase in responsiveness was

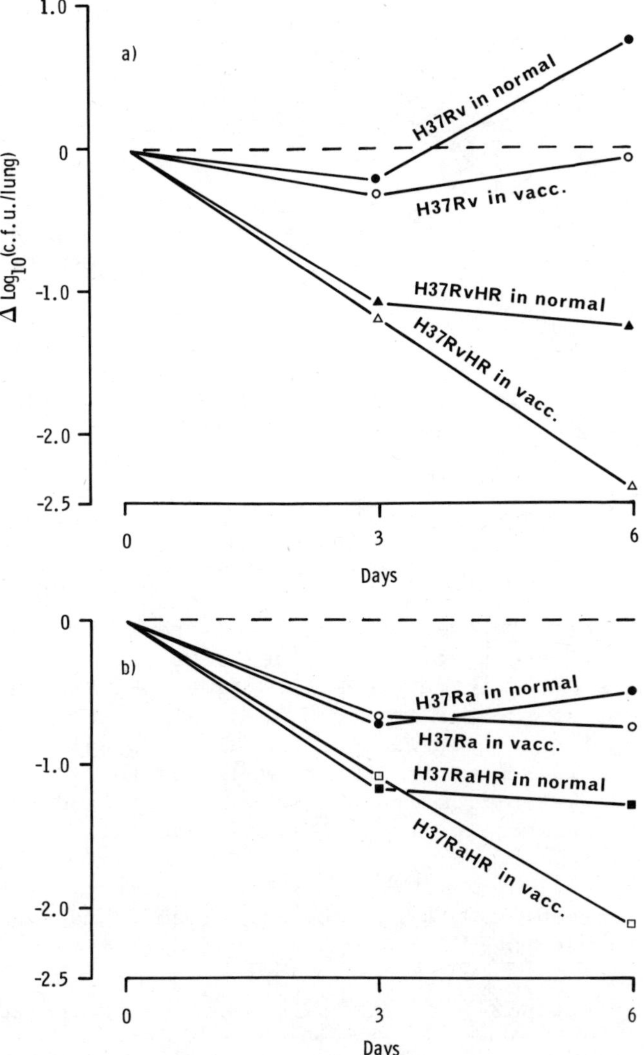

Figure 1. Effect of isoniazid resistance (H_2O_2 susceptibility) of <u>M. tuberculosis</u> on the course of infection in lungs of normal and vaccinated guinea pigs. Each point is the geometric mean of c.f.u. counts from 4 animals minus the mean count from 4 animals killed within 3 h of infection.

selective, in that no differences were seen with time between macrophages from normal and vaccinated animals with respect to H_2O_2 released in response to PMA (Table 1). These observations might be explained by an increase of Fc or C3 receptors for

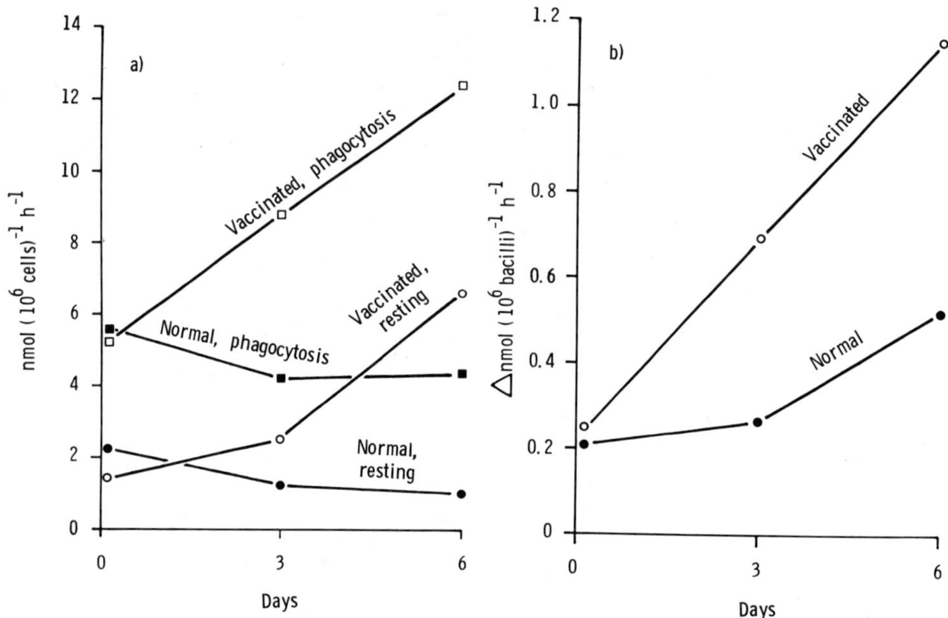

Figure 2.   (a) H_2O_2 release by alveolar macrophages lavaged from normal and vaccinated guinea pigs 3 h, 3 days and 6 days after intravenous infection. Each point is the mean of the results from 5 animals. Cells were incubated for 1 h with or without opsonised H37Ra *in vitro*. (b) Increment in H_2O_2 release due to phagocytosis calculated from the data shown in (a) and the number of cell-associated radio-labelled bacilli.

Table 1.   Release of H_2O_2 by guinea pig alveolar macrophages in response to PMA

Time after intravenous infection	Normal [1]	BCG-vaccinated [1]
3 h	23.0	17.6
3 days	ND[2]	16.1
6 days	17.5	18.9

[1] Macrophage monolayers were established 3 h, 3 days and 6 days after i.v. infection with H37Ra of normal or BCG-vaccinated guinea pigs. H_2O_2 release was assayed in the presence and absence of PMA (1.0 µg/0.1 ml). The increment due to stimulation with PMA is shown as Δ nmol $(10^6$ cells$)^{-1}(15$ min$)^{-1}$.

[2] ND = not determined.

opsonised particles in immune animals (3,4,10,12,14) but not of
receptors for PMA. An increased number of receptors might give
an increased number of receptor-ligand interactions per attached
particle, with a consequent increased release of H_2O_2. The lack
of effect of immunological priming on the response to PMA contrasts
with other reports in the literature (1,7,13). However, our present
findings suggest that the guinea pig alveolar macrophages that
were immunologically primed in vivo differed from the normal
alveolar macrophages, not so much in their potential to respond
to H_2O_2-eliciting stimuli in general (for example as a consequence
of an increased amount of an H_2O_2-generating system or decreased
amounts of H_2O_2-destroying systems) as in the efficiency of
linking of certain stimuli to the response, and perhaps in the
level of basal (unstimulated) activity.

Irrespective of the mechanisms involved, the indirect evidence
that macrophage H_2O_2 might help to kill tubercle bacilli in vivo
(5) is considerably strengthened by the correlation between
availability and effect demonstrated here.

REFERENCES

1. Greening, A.P., Rees, A.D.M. and Lowrie, D.B. (1980). Human
        alveolar macrophage staphylocidal function. Clin. Sci.
        59, 14.
2. Guilbault, G.G., Brignac, P.J. and Juneau, M. (1968). New
        substrates for fluorimetric determination of oxidative
        enzymes. Anal. Chem. 40, 1256-1263.
3. Harmsen, A.G. and Jeska, E.L. (1980). Surface receptors on
        porcine alveolar macrophages and their role in phago-
        cytosis. J. Reticuloendoth. Soc. 27, 631-637.
4. Hunninghake, G.W. and Fauci, A.S. (1976). Immunological
        reactivity of the lung. 1. A guinea-pig model for the
        study of pulmonary mononuclear cell subpopulations.
        Cell. Immunol. 26, 89-97.
5. Jackett, P.S., Aber, V.R. and Lowrie, D.B. (1978). Virulence
        and resistance to superoxide, low pH and hydrogen
        peroxide among strains of Mycobacterium tuberculosis.
        J. Gen. Microbiol, 104, 37-45.
6. Jackett, P.S., Aber, V.R., Mitchison, D.A. and Lowrie, D.B.
        (1981). The contribution of hydrogen peroxide
        resistance to virulence of Mycobacterium tuberculosis
        during the first six days after intravenous infection
        of normal and BCG-vaccinated guinea-pigs. Br. J. Exp.
        Path. 62, 34-40.
7. Johnston, R.B., Godzik, C.A. and Cohn, Z.A. (1978). Increased
        superoxide anion production by immunologically activated
        and chemically elicited macrophages. J. Exp. Med.,
        148, 115-127.

8.   Kaneda, M., Kakinum, K., Yamaguchi, T. and Shimada, K. (1980).
     Comparative studies on alveolar macrophages and poly-
     morphonuclear leukocytes. II. The ability of guinea-
     pig alveolar macrophages to produce H_2O_2. J. Biochem.,
     88, 1159-1165.
9.   Klebanoff, S.J. and Hamon, C.B. (1975). Antimicrobial systems
     of mononuclear phagocytes. In:Mononuclear phagocytes in
     immunity, infection and pathology. pp. 507-529. Edited
     by R. van Furth. Oxford: Blackwell Scientific.
10. Lurie, M.B. (1964). Resistance to tuberculosis; experimental
     studies in native and acquired defensive mechanisms.
     Cambridge, Mass: Harvard University Press.
11. Mitchison, D.A., Selkon, J.B. and Lloyd, J. (1963). Virulence
     in the guinea-pig, susceptibility to hydrogen peroxide,
     and catalase activity of isoniazid-sensitive tubercle
     bacilli from South Indian and British patients. J.
     Path. Bact. 86, 377-386.
12. Montarroso, A.M. and Myrvik, Q.N. (1978). The effect of BCG
     vaccination on the IgG and complement receptors on
     rabbit alveolar macrophages. J. Reticuloendoth. Soc.,
     24, 93-99.
13. Nathan, C.F., Nogueira, N., Juangbhanich, C.W., Ellis, J.
     and Cohn, Z.A. (1979). Activation of macrophages
     in vivo and in vitro. Correlation between hydrogen
     peroxide release and killing of Trypanosoma cruzi.
     J. Exp. Med., 149, 1056-1068.
14. Wolff, L.J., Boxer, L.A., Allen, J.M. and Baehner, R.L.
     (1978). The selective effect of hypoxia on the
     guinea-pig alveolar macrophage membrane. J. Reticulo-
     endoth. Soc., 24, 377-382.

INTERFERON AND HOST DEFENSE SYSTEMS

Howard M. Johnson

Department of Microbiology, The University of Texas
Medical Branch, Galveston, Texas   77550

INTRODUCTION

Interferon (IFN) is defined as a protein(s) which exerts
virus nonspecific, antiviral activity at least in homologous
cells through cellular metabolic processes involving both RNA and
protein synthesis (Reviewed in 1). Interferons are increasingly
being recognized as having pleiotropic effects on cell function.
These effects may be expressed in the form of anticellular,
immunoregulatory, antitumor, and hormonal activities in addition
to the well known antiviral activity. Interferon was discovered
in 1957 by Isaacs and Lindenmann (2), and this discovery was based
on its antiviral properties with a view toward understanding the
nature of host defense against viral infections. In this over-
view we will attempt to convey the present day meaning of interferon in the context of its varied biological activities both at
the cellular and molecular levels. Where appropriate, review
articles will be cited, and these can serve as a source of original
articles.

CLASSIFICATION OF INTERFERONS (IFNs)

Although sometimes presented and discussed in the singular,
there are actually several types of IFNs (Table 1). Human and
murine IFNs may be classifed into three broad groups based on
antigenic properties in a manner analogous to the classification
of immunoglobulin isotypes (3). These are leukocyte (IFNα),
fibroblast (IFNβ), and immune (IFNγ) IFN. IFNα and IFNβ are
classically induced in the appropriate cell by viruses or poly-
ribonucleotides (4-6), while IFNγ is induced in virgin or

Table 1. Classification of IFNs

Interferon	Cellular source	Inducer	Former nomenclature
IFNα	Lymphocytes (B, null, and T)	Viruses, polyribo-nucleotides, tumors, chemicals	Leukocyte, Type I
IFNβ	Fibroblast, epithelial, macrophage	Viruses, polyribo-nucleotides, chemicals	Fibroblast, Type I
IFNγ	T lymphocyte	Antigens, T-cell Mitogens	Immune, Type II

antigen-primed T lymphocytes by T-cell mitogens or specific antigens (7-12). It is of considerable interest that some tumors will induce IFNα in B and/or null cells (13,14), since IFN exerts antitumor activity both _in vitro_ and _in vivo_ (15, reviewed in 16).

Considerable progress has been made in both the purification of the IFN proteins (Reviewed in 17) and cloning of the genes of IFNα (18,19) and IFNβ (20,21). The cloning experiments have revealed eight or more human IFNα genes (22,23), while IFNβ seems to be less complex in gene expression (21,23). Roughly 75 to 80% homology is seen when any two IFNαs are compared based on the inferred amino acid sequences of the various IFNα genes (23). Further classification within the three IFN groups will undoubtedly be required. There is considerable interest in large-scale production and purification of IFNγ and cloning of the gene(s), particularly since IFNγ has been reported to have significantly greater _in vitro_ antitumor activity against certain human transformed cell lines than either IFNα or IFNβ (15).

IFNs, both between and within groups, vary in molecular weights (Reviewed in 17). Human IFNα is predominantly in a 18,500 and 23,000 dalton weight form. IFNβ may vary from roughly 20,000 to 30,000 daltons in molecular weight. IFNγ in humans is approximately 50,000 daltons (24), while IFNγ in the mouse is approximately 40,000 daltons (25-27). Heterogeneity in molecular weight is partly due to the fact that some of the IFNs are glycoproteins (17). With human IFNα, however, the 23,000 dalton form appears to have an adrenocorticotropic (ACTH) molecule associated with it, while the ACTH is absent from the 18,500 dalton form (28). This suggests that there may be factors in addition to glycosylation that are responsible for the variation in the molecular weight of a given IFN type, with possible corresponding differences in function.

ANTIVIRAL PROPERTY OF IFN

The production and action of virus induced IFN at the cellular
level is depicted in Figure 1 (Reviewed in 1 and 29). During early
stages of cellular viral infection, some event (probably the
presence of foreign viral nucleic acid) derepresses cellular genes
located on specific chromosomes which contain the stored genetic
information for the IFN protein. IFN is produced by transcriptional
and translational events as occur in normal protein synthesis. The
produced IFN does not inactivate the virus by direct interaction.
Rather, it leaves the cell and then reacts with a specific receptor
on the cell membrane of the producing and surrounding cells. This
membrane interaction results in the derepression of a gene(s) that
codes for the antiviral proteins. The intracellular antiviral
proteins are thought to be involved in actual inactivation of
virus. We shall examine possible identities and the mechanism of
action of the antiviral proteins.

The IFN system is nonspecific in that it is activated by a
variety of viruses and the IFN produced induces cellular resistance
against a broad range of viruses. IFN is the earliest appearing
of the known host defenses against viral infections and probably
plays an important role in the initial phases of the host protective
response (29). Figure 2 illustrates the early production of IFN
in comparison with antibody during experimental infection of man
with influenza virus. This early appearance of IFN has been shown
to be important in the ultimate outcome of viral infection. Under
experimental conditions of viral infection in mice, where the
produced IFN is neutralized by passively administered anti-IFN
antibody, the viral infections are significantly more severe,
with dramatic increases in animal deaths (30,31). IFN, then, is
probably very important as a natural host defense against viral
infections.

MOLECULAR ASPECTS OF IFN ACTION

Currently, it is not known precisely how IFN inhibits viral
replication. Data have been presented which suggest both a block
of transcription of viral RNA and/or inhibited translation of
viral proteins (Reviewed in 32). Inhibition of protein synthesis
is currently felt to be the primary mode by which IFN exerts its
varied biologic effects.

Three mechanisms by which IFN has been shown to inhibit
protein synthesis are presented in Figure 3 (32). One mechanism
involves induction of a protein kinase, which in turn catalyzes
the phosphorylation of one of the initiation factors (eIF-2α subunit) that is required for protein synthesis (eIF-2α is inactive
in the phosphorylated state). Double-stranded RNA and ATP are

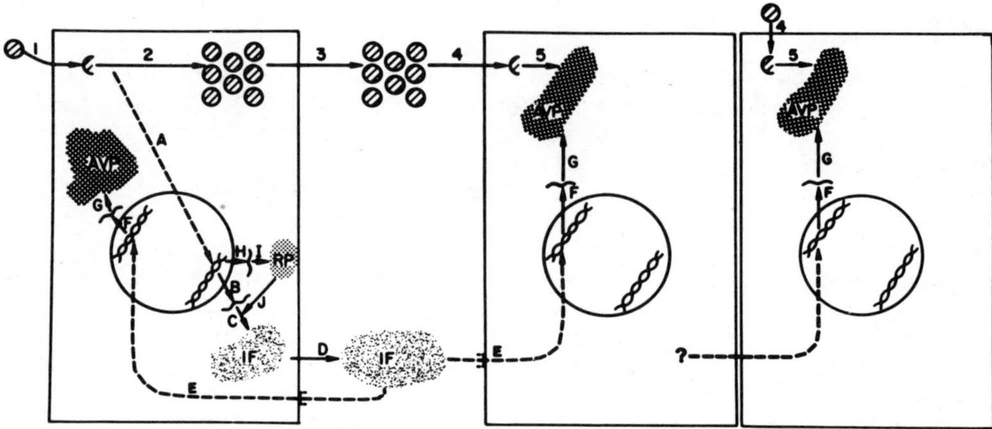

Figure 1. Cellular events of the induction and action of interferon
(IF). Virus comes in contact with the cell and penetrates
the cell membrane. The virus then releases its genetic
material, and replication of the virus occurs. The new
virus leaves the cell, enters the fluid around the first
cell, and some of the replicated virus infects a second
cell, where release of the genetic material again takes
place. During the early stages of infection of the first
cell, some event (viral nucleic acid) stimulates a gene
in the DNA which contains the stored genetic information
for interferon (A). This leads to the production of a
messenger RNA for interferon, which leaves (B) the nucleus
and is translated by the cell's ribosomes (C) into the
interferon protein. Several events now occur more or less
simultaneously. Interferon is secreted by the first infected cell (D), enters the surrounding fluid, where it
comes into contact with and stimulates the second cell
(E) by interacting with a membrane receptor for interferon. The second cell is thereby induced to produce a
new messenger RNA (F) which is translated to a new pro-
tein(s) (G), the antiviral protein (AVP). This in turn
modified the cell's protein-synthesizing machinery, such
that cell mRNA is translated into protein but viral RNA
is poorly bound or translated, or both. In the first cell
processes, E,F, and G may in some instances also operate
to form AVP and thereby reduce the virus yield in the
first cell. Shortly after interferon is synthesized into
the first cell another mRNA (H) is believed to be synthesized from the cell's DNA which is translated (I) into a
regulatory protein (PR) (hypotehsized). This regulatory
protein combines with mRNA for interferon thereby preventing the further synthesis of more interferon (J). There is
recent evidence that the antiviral state may be directly
transferred between adjacent cells (from second to third
cell at right) by the passage of an unknown (?) inducer
of the AVP.

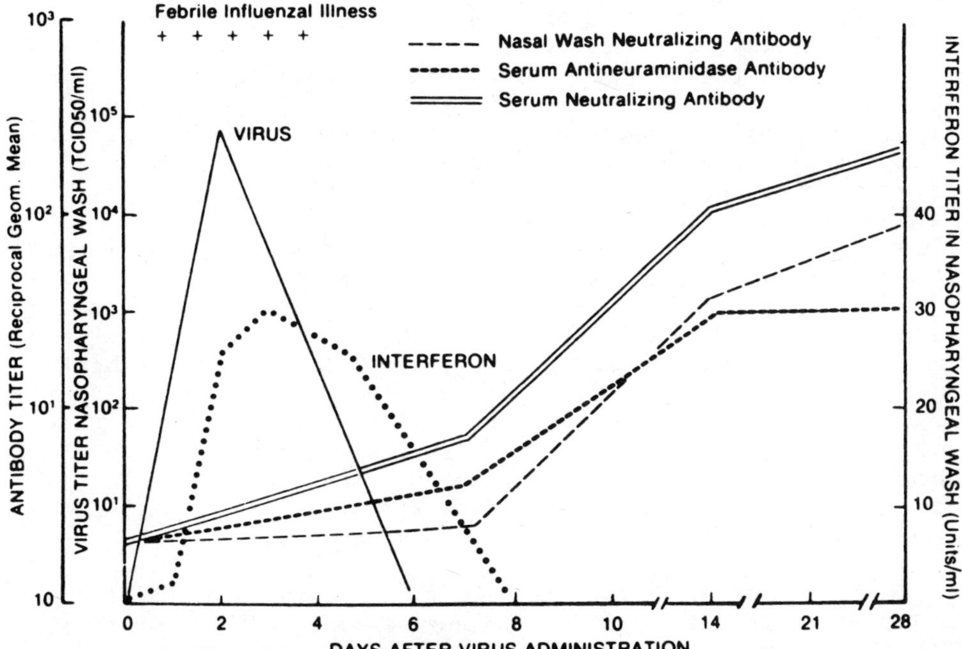

Figure 2.   Production of virus, interferon, and antibody during
experimental infection of man with influenza wild type
virus (H3N2).   (From study by B. Murphy, et al.,
National Institutes of Health).

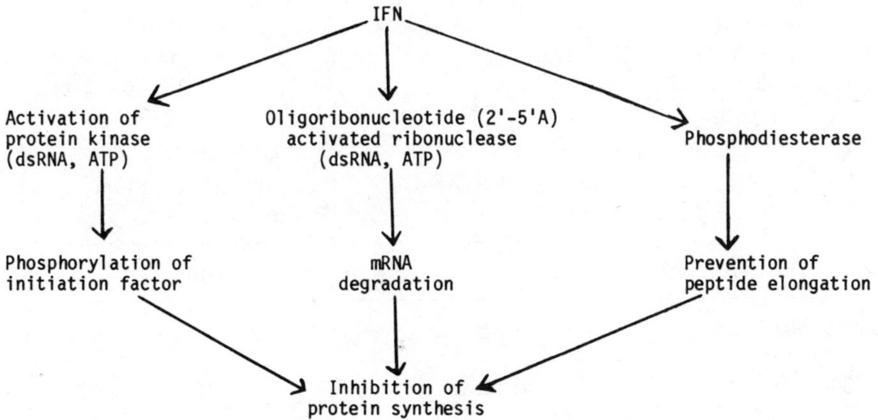

Figure 3.   Possible antiviral and cell regulatory proteins induced
by IFN.

required for activity. The second mechanism involves IFN induction
of a polymerase, called 2'-5'A synthetase. This enzyme, in the
presence of double-stranded RNA, converts ATP to a 2'-5'-linked
oligoadenylate(s) (2'-5'A). 2'-5'A activates an RNA endonuclease,
which catalyzes the degradation of RNA. Destruction of messenger
RNA by this enzyme results in inhibition of cellular protein
synthesis. Single-stranded RNA can substitute for double-stranded
RNA in the kinase and sythetase systems (33). The third mechanism
presented involves IFN induction of a phosphodiesterase which
cleaves the 3'CCA terminal region of transfer RNA (tRNA). 3'CCA
is the site of attachment of amino acids to tRNA, thus the phosphodiesterase prevents peptide elongation by inhibiting the formation
of the aminoacyl-tRNA complex. It is possible that the protein
kinase, 2'-5'A synthetase, and phosphodiesterase systems are the
proposed IFN-induced antiviral proteins. There is no evidence,
however, that any of these systems is selective for inhibition of
virus protein synthesis; they also inhibit eukaryotic protein
synthesis. Thus, these mechanisms for IFN inhibition of protein
synthesis may play a role in a variety of IFN mediated biological
activities, such as inhibition of virus replication, suppression
of the immune response, and inhibition of tumor growth.

We have recently shown that IFN also alters intermediary
metabolism in that it inhibits the hexose monophosphate shunt
(HMP) pathway in endotoxin-stimulated mouse spleen cells while
not affecting the glycolytic or citric acid pathways (Table 2)
(34). The HMP pathway is important in the cell for generation of
reducing power in the form of NADPH, and for conversion of hexoses
into pentoses for nucleic acid synthesis. Reducing agents such as
2-mercaptoethanol block IFN inhibition of the HMP pathway. Oxidized
glutathione, a thio-oxidizing agent, also suppresses the HMP pathway. The data suggest that IFN inhibits the HMP pathway by binding
to, and alteration of, cell membrane sulfhydryl or disulfide
groups, possibly inhibiting membrane-bound glucose 6-phosphate
dehydrogenase activity.

Table 2. IFN inhibition of the hexose monophosphate shunt pathway
          in endotoxin activated mouse spleen cells[a]

Metabolic pathway	Inhibition (%)
Hexose monophosphate shunt	88
Glycolytic and citric acid	20

[a] Data obtained using glucose-1-^{14}C and glucose-6-^{14}C as
described (34).

The generation of NADPH through the HMP pathway is the
principal means by which the cell maintains the reduced state.
Reduced levels of NADPH would result in elevation of the intracellular levels of oxidized glutathione (GSSG). Elevated levels
(as low as 20 μM) of GSSG have been shown to inhibit initiation
of protein synthesis by activation of a protein kinase that
phosphorylates eIF-2 (35). It is possible that the inhibitory
effect of IFN on HMP pathway activity is a mechanism by which it
increases protein kinase activity in the cell for eIF-2a. The
above data with IFN and HMP pathway activity in mouse spleen cells
has recently been confirmed using a mouse fibroblast line (3T3Li)
(36) that converts at confluence or in differentiation media into
adipocytes (R. Saneto and H.M. Johnson, unpublished data). IFN
blocks the differentiation (37,38), and we have shown that there is
a corresponding inhibition by IFN of the HMP pathway in these
cells.

REGULATION OF IFNγ PRODUCTION

The suitability of an IFNγ inducer probably depends on the
temporal events related to its induction of IFNγ and its induction
of suppressor cells that regulate IFNγ production. There is recent
evidence for such suppressor cells (39). Spleen cells obtained
from mice injected 24 to 72 hr previously with staphylococcal
enterotoxin A (SEA) were poor producers of IFNγ when restimulated
with the same mitogen _in vitro_ (Figure 4). Further, co-cultivation
of these spleens with virgin spleen cells inhibited IFNγ production
by as much as 99% by Day 3 of culture. The suppressor cell activity
lacked mitogen specificity in that injection of mice with either
SEA or concanavalin A (Con A) induced potent suppressor cell
activity that inhibited the induction of IFNγ by either mitogen.
The suppressor cell activity could be induced in cultures incubated
with mitogen for 4 days.

The suppressor cell is apparently a T cell, since its activity
was destroyed by treatment with monoclonal antibody to the Thy 1.2
antigen (Table 3). Concentrations of anti-Thy 1.2 that destroyed
suppressor cell activity allowed IFNγ production in the suppressor
cell population. Virgin cells, on the other hand, were blocked
(or destroyed) from IFNγ production by the same concentration of
anti-Thy 1.2. IFNγ-producing cells in the suppressor population
appear to contain a low density of the Thy 1.2 antigen. These
IFNγ-producing cells appear to have either evolved from cells
with a high density of Thy 1.2 or required such cells as helpers
in the production of IFNγ.

The Lyt phenotype of the suppressor cell was studied by
treatment with monoclonal antibodies to the Lyt 1.2 and Lyt 2.2
antigens plus complement (B. Torres and H.M. Johnson, in

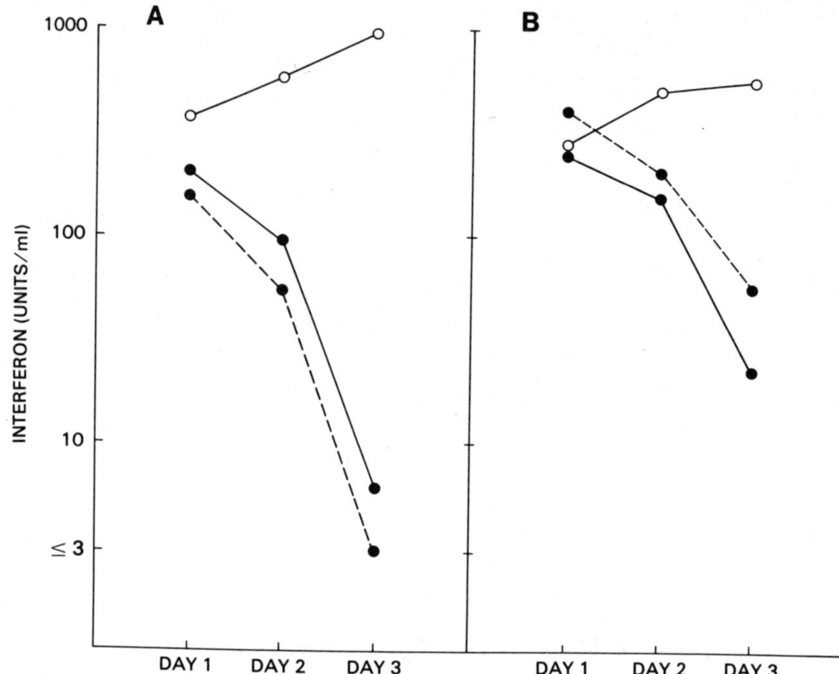

Figure 4.   Kinetics of IFNγ production by cultured spleen cells
obtained from mice injected 24 h earlier with 10µg SEA.
SEA-induced suppressor cells were either cultured alone
or with virgin spleen cells. All cultures were
stimulated with either 0.5 µg SEA (A) or 10 µg Con A
(B). A: o—o , virgin spleen cells and SEA; ●—● ,
SEA-induced suppressor cells and SEA; ●-● , SEA-induced
cells, virgin spleen cells, and SEA. B: o—o , virgin
spleen cells and Con A; ●—● , SEA-induced suppressor
cells and Con A; ●-● , SEA-induced suppressor cells,
virgin spleen cells, and Con A. Coefficients of
variation averaged less than 25% for duplicate
determinations.

preparation). Suppressor cell activity is, or requires, an Lyt 1,
2,3[+] cell with relatively low density of the Lyt 1 antigen for
activity. This cell could possibly be turned on by an Lyt 1[+]
inducer cell and become an Lyt 2,3[+] suppressor cell. The IFNγ-
producing cell(s) required Lyt 1[+] and Lyt 2[+] cells for activity,
but not Lyt 1,2,3[+] cells. Apparently the Lyt 1[+] cell provides
help for IFNγ production by the Lyt 2[+] cell. The role of
suppressor cells in regulation of IFNα and IFNβ production remains
to be determined. It would not be surprising to find such
regulatory cells in IFNα production, since this IFN is a product
of lymphocytes.

Table 3. Effect of anti-Thy 1.2 antibody treatment of suppressor
         cells on regulation of IFNγ production[a]

Cultures	IFN (units/ml)
Virgin spleen cells	112
Suppressor cells	3
Suppressor cells + virgin spleen cells	10
Anti-Thy 1.2-treated (10^{-4} dil) suppressor cells + virgin spleen cells	224
Anti-Thy 1.2-treated (10^{-4} dil) suppressor cells	100
Anti-Thy 1.2-treated (10^{-3} dil) suppressor cells	50
Anti-Thy 1.2-treated (10^{-4} dil) virgin spleen cells	3

[a]All cultures were stimulated with 0.5 µg SEA/ml and IFNγ values
are from Day 3 of culture.

At the molecular level IFNγ production is regulated by adeno-
sine 3':5'-cyclic monophosphate (cyclic AMP). Addition of dibutyryl
cyclic AMP at concentrations of at least 2 x 10^{-5} M to SEA and
Con A stimulated mouse spleen cell cultures suppressed IFNγ production by 85% or more (Figure 5) (40,41). When the same concentrations of dibutyryl cyclic AMP were added to the supernatant
fluids of mouse spleen cell cultures after complete IFNγ production,
no inhibition of antiviral activity of the IFN was observed, so
cyclic AMP blocked the production of IFNγ, and not the established
activity of produced IFN. Substances, such as cholera toxin, that
raise the endogeneous levels of cyclic AMP by stimulating adenylate
cyclase activity also block IFNγ production. Substances, such as
methyl xanthines, that raise the endogeneous level of cyclic AMP
by inhibiting phosphodiesterase (converts cyclic AMP into AMP)
activity similarly block IFNγ production. It would be of interest
to determine the role of cyclic AMP in the mechanism of suppressor
cell activity for IFNγ production. Perhaps the suppressor cell
increases the level of cyclic AMP in the lymphocyte induced to
produce IFNγ.

ANTICELLULAR PROPERTIES OF IFNs

IFNs possess anticellular activity against several cell lines
(Reviewed in 16). Partially purified human IFNγ was compared with
IFNα and IFNβ for their relative abilities to inhibit colony forma-
tion by two transformed human cell lines, WISH and HEp-2 cells (15).

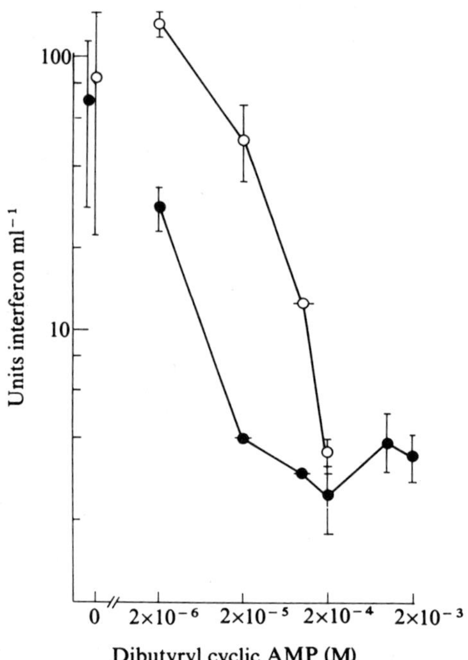

Figure 5.   Dibutyryl cyclic AMP inhibition of the production of IFN
            in Con A (●)   and SEA (o) stimulated C57Bl/6 female
            (8 weeks old) mouse spleen cell cultures. From
            Johnson (40).

Data from a representative experiment are presented in Table 4.
IFNγ at $10^{6.3}$ units/mg protein was 20-100 times more effective
than IFNα and IFNβ in suppression of colony formation. The anticellular activity copurified with the IFNγ, and was characteristic
of IFN in that it showed species preference of action, since it
was inactive on mouse L cells. Similar anticellular activity was
observed with HEp-2 cells treated with IFN.

MODULATION OF THE IMMUNE RESPONSE BY IFN

     Data are quite conclusive that interferon can play a regulatory
role in modulation of the immune response at several levels (Table
5) (Reviewed in 42 and 43). As a matter of fact IFN is probably
the mediator of some forms of suppressor cell activity. Virus-
induced IFN has been shown to suppress the antibody response in
both in vitro and in vivo systems in the mouse.

     The in vitro system for antibody production in the mouse
consists of the use of dissociated spleen cells which are made up

Table 4. Comparison of the antiviral and anticellular activity
          of human IFNs against colony formation by WISH cells[a]

IFN	Titer (U/ml)		Ratio Antiviral/Anticellular Titer
	Antiviral	Anticellular	
α	100	1	100
β	20	1	20
γ	30	20	1.5

[a]From Blalock et al (15).

of B and T lymphocytes and macrophages, all of which are normally
required for an antibody response. Addition of a suitable antigen,
such as sheep red blood cells (SRBC), to the spleen cells in
culture media under defined conditions stimulates the production
of antibody-forming cells. The antibody-forming cells are enumerated by mixing them with a high concentration of SRBC (19%) in a
soft agar medium at 37° for 1 or 2 hours. During this incubation,
the antibody that is produced by the cells binds to surrounding
SRBC. Addition of guinea pig serum, which contains complement,
to this complex results in lysis of the SRBC. Thus a plaque or
zone of hemolysis surrounds a cell that is producing antibody to
SRBC. The antibody-forming cell is frequently referred to as a
plaque-forming cell (PFC), and the antibody response to SRBC is
spoken of as the anti-SRBC PFC response. This system has been of
considerable value in looking at a variety of events related to
regulation of the antibody response. Advantages over _in vivo_

Table 5. Immunoregulatory effects of IFN

---

1. Suppression and enhancement of antibody formation.

2. Suppression of antigen and mitogen-induced lymphocyte
    proliferation.

3. Enhancement of specific cytotoxicity of T lymphocytes.

4. Enhancement of natural killer cell cytotoxicity.

5. Enhancement of antibody-dependent cell-mediated cytotoxicity.

---

experiments are economy of reagents and better control of the
microenvironment of the antibody-forming cell.

To determine the effect of IFN on the in vitro PFC response,
purified IFN is added to the culture at the time of SRBC addition.
Five days later the cells are harvested and examined for the PFC
response to SRBC. Results of a typical experiment are presented
in Figure 6. Virus-induced IFN suppressed the in vitro PFC
response by approximately 99%. IFN is responsible for the suppression, since it is observed with IFN purified to homogeneity and
is specifically blocked by pretreatment of the interferon with
specific antibody. In general, 20-60 units of IFN/ml will suppress
the in vitro PFC response by 90% or more. This represents less
than 0.05 nanograms of IFN/ml. This is a very potent immunosuppression. IFN similarly suppresses the antibody response in

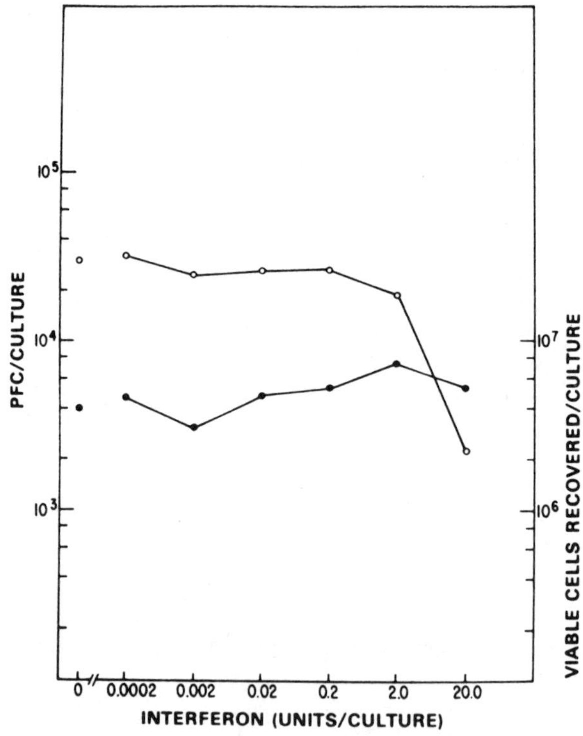

Figure 6. The effect of highly purified mouse ascites tumor
virus-induced IFN on the primary in vitro PFC response.
Direct anti-SRBC PFC/culture (o—o) and viable cells
recovered per culture (●—●) were determined on day
five. From Johnson, Smith and Baron (44).

the mouse, but this requires relatively more IFN (50,000 units or
more) and it is impossible to control the concentrations of IFN in
the microenvironment of the antibody-producing cell.

We may summarize some of the findings to date on the regulatory
effects of IFN on lymphocyte function. IFN is most effective when
added to the system at the time of antigen addition. Secondary
antibody responses are also suppressed. The suppression of the
antibody response could be due to a direct or indirect effect of
IFN on the B cell, the cell actually responsible for antibody
production. We do have data that show the B cell response can
be suppressed indirectly by IFN, through IFN induction of a
suppressor cell, which releases a soluble factor that actually
acts on the B cell (45). This does not preclude a direct effect
of IFN on B cells. There is also data which show that the mechanism
by which IFN suppresses the immune response may be through the
protein kinase and endonuclease systems (46). IFN can suppress the
cellular immune response in a manner similar to suppression of the
antibody response. It can also enhance the killing of T killer
lymphocytes in cooperation with T cell growth factor or interleukin 2 (47). Further, lymphocytes (NK cells) that are neither
B nor T cells can be stimulated by interferon to have increased
killing activity against tumor cells (43). Thus IFN may modulate
the immune system to have increased antitumor activity. These
findings may be related to the reported successful use of interferon as an antitumor agent in certain human cancers.

SUMMARY

Interferons are induced proteins or glycoproteins that have
pleiotropic effects on cell function, which may be expressed in
the form of antiviral, anticellular, immunoregulatory, hormonal,
and antitumor activities. Interferons can exert their varied
biological effects at the molecular level by several mechanisms
which inhibit protein synthesis in the cell. These include the
induction of a protein kinase activity, induction of an RNA
endonuclease activity, and the induction of a phosophodiesterase.
Interferon can also alter intermediary metabolism by inhibition
of the hexose monophosphate shunt pathway, which is important
in providing the cell with reducing power and pentoses for
nucleic acid synthesis. Currently, there are three recognized
interferons based on antigenic properties. Two are typically
induced by viruses and are called leukocyte (IFNα) and fibroblast-
epithelial (IFNβ) interferons to indicate their cellular origin,
and the third is called immune (IFNγ) interferon, since it is
typically produced by T lymphocytes following antigen or T-cell
mitogen stimulation. The interferons are potent suppressors of
both the humoral and cellular immune responses. They also
enhance natural killer and specific T-cell cytotoxicity. The

immunosuppression can occur through interferon induction of
suppressor factor. The immunomodulatory effects of interferon
inducers are regulated by cAMP; increased cAMP activity blocks
induction of the interferons. Additionally, a feedback suppressor
cell regulates the IFNγ induction system. All the interferons
possess anticellular activity against transformed cell lines, and
IFNγ has significantly greater activity against several transformed
cell lines. The complex interactions and varied biological
properties of interferons are apparently significant in host
defense against viral infections, in antitumor properties of
interferons, and in regulation of various cellular functions.

ACKNOWLEDGMENT

This work was supported by grants AI 16428 and CA 27590 from
the Public Health Service, NIH.

REFERENCES

1.   Vilcek, J., Gresser, I., and Merigan, T.C. (eds.) (1980).
         Ann. New York Acad. Sci. 350:1-637.
2.   Isaacs, A. and Lindenmann, J. (1957). Proc. Roy. Soc. London
         Ser. B. 147:258.
3.   Stewart, W.E., et al. (1980). J. Immunol. 125:2353.
4.   Isaacs, A. (1963). Advan. Virus Res. 10:1.
5.   Field, A.K., et al. (1967). Proc. Natl. Acad. Sci. USA 58:
         1004.
6.   Paucker, K. (1977). Tex. Rep. Biol. Med. 35:23.
7.   Wheelock, E.F. (1965). Science 149:310.
8.   Green, J.A., Cooperband, S.R. and Kibrick, S. (1969). Science
         164:1415.
9.   Salvin, S.B., Youngner, J.S. and Lederer, W.H. (1973). Infect.
         Immun. 7:68.
10. Stobo, J., et al. (1974). J. Immunol. 112:1589.
11. Epstein, L.B. (1977). Tex. Rep. Biol. Med. 35:42.
12. Johnson, H.M., Stanton, G.J. and Baron, S. (1977). Proc.
         Soc. Exp. Biol. Med. 154:138.
13. Trinchieri, G., et al. (1978). J. Exp. Med. 147:1299.
14. Weigent, D.A., et al. (1981). Infect. Immun. 32:508.
15. Blalock, J.E., et al. (1980). Cell. Immunol. 49:390.
16. Gresser, I. and Tovey, M.G. (1978). Biochim. Biophys. Acta.
         516:231.
17. Knight, E. (1980). Interferon 2 (I. Gresser, ed.), Academic
         Press 2:1.
18. Nagata, S., et al. (1980). Nature 284:316.
19. Goeddel, D.V., et al (1980). Nature 287:411.
20. Taniguchi, T., et al. (1979). Proc. Japan Acad. Ser. B.
         55:464.

21. Derynck, R., et al. (1980). Nature 285:542.
22. Nagata, S., et al. (1980). Nature 287:401.
23. Goeddel, D.V., et al. (1981). Nature 290:20.
24. Georgiades, J.A., et al. (1979). IRCS Med. Sci. 7:559.
25. Youngner, J.S. and Salvin, S.B. (1973). J. Immunol. 111:914.
26. Sonnenfeld, G., Mandel, A.D. and Merigan, T.C. (1977). Cell.
    Immunol. 34: 193.
27. Osborne, L.C., Georgiades, J.A. and Johnson, H.M. (1979).
    Infect. Immun. 23:80.
28. Blalock, E.J. and Smith, E.M. (1980). Proc. Natl. Acad. Sci.
    USA 77:5972.
29. Stanton, G.J., Johnson, H.M., and Baron, S. (1978). Path.
    Ann. 8:285.
30. Fauconnier, B. (1971). Pathol. Biol. 19:575.
31. Gresser, I., et al. (1976). J. Exp. Med. 144:1305.
32. Revel, M. (1979). Interferon 1 (I. Gresser, ed.), Academic
    Press 1:101.
33. Samuel, C.E. (1979). Proc. Natl. Acad. Sci. USA. 76:600.
34. Saneto, R., Holstun, S., and Johnson, H.M. (1980). IRCS Med.
    Sci. 8:921.
35. Ernst, V., Levin, D.H., and London, I.M. (1978). Proc. Natl.
    Acad. Sci. USA 75:4110.
36. Green, H. and Meuth, M. (1974). Cell 3:127.
37. Cioe, L., O'Brien, T.G. and Diamond, L. (1980). Cell Biol.
    Int. Rep. 4:255.
38. Keay, S. and Grossberg, S.E. (1980). Proc. Natl. Acad. Sci.
    USA 77:4099.
39. Johnson, H.M. (1981). Antiviral Research 1:37.
40. Johnson, H.M. (1977). Nature 265:154.
41. Johnson, H.M., Blalock, J.E., and Baron, S. (1977). Cell.
    Immunol. 33:170.
42. Johnson, H.M. and Baron, S. (1977). Pharmac. Ther. A 1:349.
43. Herberman, R.B. (ed.) (1980). Natural Cell-Mediated Immunity
    Against Tumors, Academic Press, New York.
44. Johnson, H.M., Smith, B.G., and Baron, S. (1975). J.
    Immunol. 114:403.
45. Johnson, H.M. and Blalock, J.E. (1980). Infect. Immun. 29:
    301.
46. Johnson, H.M. and Ohtsuki, K. (1979). Cell. Immunol. 44:125.
47. Farrar, W.L., Johnson, H.M., and Farrar, J.J. (1981).
    J. Immunol. 126:1120.

MEDIATOR INTERACTIONS REGULATING MACROPHAGE SECRETION OF INTER-

LEUKIN 1 AND INTERFERON

Robert N. Moore[1], Patricia S. Steeg[2], and Stephan E. Mergenhagen[2]

[1]Department of Microbiology, University of Tennessee Knoxville, TN 37916; and [2]Laboratory of Microbiology and Immunology, National Institute of Dental Research Bethesda, Maryland 20205

The mononuclear phagocyte system plays a critical role in host resistance to infection. The capacity of mononuclear phagocytes to deal with microorganisms appears, however, to be under regulatory control rather than static. Two mediators which appear to act in the process of macrophage differentiation and activation are interferon and colony-stimulating factor(s) (CSF). A substantial body of evidence indicates that interferon stimulates a variety of macrophage functions including phagocytosis, spreading, killing of microorganisms and tumor cells (1) and, recently, Ia-antigen expression (2). The effects of CSF are, however, less well known. The purpose of the present communication is to discuss recent findings which indicate that CSF as well as interferon can participate in the process of macrophage activation.

CSF is classically considered as a myeloproliferative agent that stimulates _in vitro_ the clonal proliferation and differentiation of granulocyte/macrophage stem cells and more mature monocytes (3). Observations that CSF is produced during early phases of experimental infections, preceding expansion of the granulocyte/macrophage populations, supported the view that CSF might be responsible in part for the proliferative response (4,5). Production of CSF during infections, however, is also associated with enhanced functional activities of the mononuclear phagocyte system. The most obvious changes are enhanced phagocytic activity (5) and elevated secretory responses to parenterally injected LPS (6,7). Based on this temporal association, we have attempted to ascertain if CSF might influence macrophage activities other than proliferation.

Hyperresponsiveness to LPS is considered a criterion for macrophage activation (8). Therefore, our assays for macrophage activation have dealt with macrophage secretory responses to LPS. We have concentrated on two major products, interleukin 1 (IL 1) and interferon. IL 1 is a potent immunostimulant that activates certain subclasses of T lymphocytes (9). Recent evidence indicates that IL 1 and the fever inducing principle, endogenous pyrogen, may also be the same molecule (10,11). Additionally, Mannel et al. (7) have found that hyperresponsive mice (BCG infected) produce substantial levels of this activity upon treatment with LPS.

Initial attempts to determine if CSF can influence LPS-induced IL 1 production were complicated by the finding that CSF itself stimulates macrophages to produce IL 1 (12). Attempts to determine if CSF and LPS acted cooperatively or synergistically often resulted in confusing data. Results obtained using the LPS-unresponsive C3H/HeJ mouse were, however, interesting and are shown in Table 1. As expected C3H/HeJ macrophages did not produce IL 1 in response to

Table 1. Interleukin 1 activity produced by C3H/HeJ exudate macrophages stimulated with partially purified L cell CSF,
LPS or CFS and LPS[a].

Units CSF/ml	LPS[b]	Interleukin 1 Activity[c]	Enhancement[d]
–	–	2,764 ± 556	
–	+	2,721 ± 309	
16,000	–	20,645 ± 2741	
8,000	–	22,442 ± 1369	
4,000	–	29,904 ± 2660	
2,000	–	5,368 ± 474	
1,000	–	2,038 ± 256	
16,000	+	45,073 ± 3790	24,428
8,000	+	41,484 ± 1891	19,042
4,000	+	41,661 ± 2241	11,757
2,000	+	7,348 ± 866	1,980
1,000	+	3,569 ± 329	1,531

[a] 2×10^6 thioglycolate-induced peritoneal exudate macrophages/ 1.0 ml culture in RPMI 1640 incubated 48 h with or without CSF (12) or LPS.

[b] 1 μg/ml Escherichia coli K235 LPS.

[c] ^{3}H-thymidine incorporation by C3H/HeJ thymocytes stimulated by 1/4 dilutions of macrophage supernatants (12).

[d] CPM stimulated by CSF + LPS over corresponding CSF control.

LPS although they did respond to the CSF preparation. CSF and
LPS combined, however, significantly elevated the IL 1 activity
produced over that induced by CSF alone. The explanation for this
enhancement is unknown at present, however, there are three interesting possibilities. First, CSF may activate C3H/HeJ macrophages
to a state where they become reponsive to LPS. Second, CSF may
stimulate these macrophages to produce endogenously a factor(s)
that activates them to respond to LPS. As will be shown later,
this does happen when interferon production is used as a parameter
for measuring macrophage activation. Third, and possibly most
intriguing is the possibility that both CSF and LPS (or other nonspecific agents) fulfill a two signal requirement for IL 1 production. C3H/HeJ macrophages bind LPS but do not secrete CSF or IL 1
in response to it. Exogenous CSF may provide the required signal
to stimulate IL 1 secretion. Similarly, normal LPS-responsive
macrophages stimulated with doses of LPS great enough to induce
CSF production (which peaks within 6 h) endogenously produce their
own second signal (CSF) required for IL 1 induction. Additionally,
direct stimulation of IL 1 production by CSF may actually represent
addition of the second signal to macrophage preactivated either
in vivo or in vitro by medium or serum containing low levels of
contaminating macrophage stimulants such as LPS. Although this
explanation is hypothetical, it is pertinent to note that Kurland
et al. (13) have proposed a similar CSF-dependent mechanism to
explain E prostaglandin production by LPS-stimulated macrophages.

Another macrophage product whose production is subject to regulatory control is interferon. Augmented production of this mediator
in LPS-treated animals is an established correlate of enhanced
nonspecific resistance to infection (6) and is most likely representative of an activation state in which macrophages are highly
capable of neutralizing an infectious challenge. We have
previously reported three methods of inducing this activation
state in vitro: i) propagation of murine macrophages with L cell
CSF (14); ii) 72 h pretreatment of freshly explanted peritoneal
exudate macrophages with L cell CSF (14,15); and iii) overnight
preincubation of peritoneal exudate macrophages with β (type I)
interferon (15). We have also reported on the mechanism by which
CSF induces this activation state. Upon stimulation with CSF,
some portion of the macrophage population is induced to secrete
a small amount of β interferon which acts to prime the culture for
subsequent enhanced interferon production upon secondary challenge
with LPS (15). This work, however, was performed using L cell
CSF. To determine if this is a viable explanation for macrophage
activation in vivo, it was necessary to determine if CSF of T
lymphocyte origin, i.e., lymphokine instead of cytokine CSF,
also had this activity.

Thioglycolate-induced peritoneal exudate macrophages from C3H/
HeN mice were incubated for 24 h with doses of CSF prepared from

Table 2. Inhibitory effect of antiserum to murine fibroblast
interferon on poly rI · poly rC – induced interferon
production by colony-stimulating factor-pretreated
exudate macrophages.

Units CSF/ml[a] (0–24 h)	Anti-Interferon[b] (24–72 h)	Supernatant Interferon Activity[c] (Units/ml) Induced by Poly rI · Poly rC
0	–	40
0	+	42
400	–	110
400	+	52
200	–	100
200	+	30
100	–	80
100	+	45
50	–	55
50	+	38
25	–	46
25	+	42

[a] CSF prepared from supernatant medium of EL 4 cells stimulated
with phorbol myristate acetate (16). The CSF was purified by
chromatography on phenyl sepharose and DEAE-Sephacel as described
by Burgess, et al. (17) (sp. activity > 10^7 units/mg protein).

[b] 1/3000 dilution of antiserum provided by Antiviral Substances
Program, NIAID.

[c] 17 h interferon titer stimulated by 1 h exposure to 10 µg/ml
poly rI · poly rC admixed with 100 µg/ml DEAE-dextran. Interferon was assayed by plaque reduction assay using L_{929} cells
challenged with Vesicular Stomatitis Virus.

the EL-4 murine thymoma cell line. After 24 h the cultures were
washed and exposed to a dilution of highly specific rabbit anti-
mouse β interferon serum sufficient to neutralize 20 U of β
interferon. After an additional 48 h incubation, the macrophage
cultures were again washed and then challenged with poly rI ·
poly rC to induce interferon production (15). Macrophages pretreated with lymphokine CSF produced enhanced levels of interferon in a manner that was dependent on the preincubation dose
of CSF (Table 2). Treatment with anti-interferon totally
neutralized this enhancement but had no effect on interferon
production by control cells that had not been treated with CSF.
Thus, it appears that lymphokine CSF like cytokine CSF can
induce hyperresponsiveness to LPS via an intermediary endogenous
interferon priming stage.

In conclusion, it is apparent that certain types of CSF act as macrophage stimulants as well as proliferative agents. The mechanisms by which CSF acts are, however, complex and appear to involve interplay with other regulatory factors. Based on in vitro findings, it can be inferred that CSF, which is an early product during infections and inflammatory responses (4), plays a critical role at least in the initiation of the activation process that stimulates the mononuclear phagocyte system to deal more efficiently with invading organisms.

REFERENCES

1.   Schultz, R.M., (1980). Lymphokine Rep. 1:63-97.
2.   Steeg, P.S., Moore, R.N. and Oppenheim, J.J.,(1981). Fed.
        Proc. Fed. Am. Soc. Exp. Biol. 40:771.
3.   Burgess, A.W., Metcalf, D. and Watt, S.M. (1978). J. Supramol.
        Struct. 8:489-500.
4.   Trudgett, A., McNeill, T.A. and Killen, M. (1973). Infect.
        Immun. 8:450-455.
5.   Bohme, D.H., Schneider, H.A. and Lee, J.M. (1959). J. Exp.
        Med. 110:9-24.
6.   Youngner, J.S. and Stinebring, W.R. (1965). Nature (London),
        208:456.
7.   Mannel, D.N., Farrar, J.J. and Mergenhagen, S.E. (1980).
        J. Immunol. 124:1106-1110.
8.   Weinberg, J.B., Chapman, H.A. and Hibbs, J.B. (1978). J.
        Immunol. 121:72-80.
9.   Oppenheim, J.J., Moore, R., Gmelig-Meyling, F., Togawa, A.,
        Wahl, S., Mathieson, B.J., Dougherty, S., and Carter, C.
        (1980). in "Macrophage Regulation of Immunity" (E.R.
        Unanue and A.S. Rosenthal, eds.) pp. 379-398, Academic
        Press, New York.
10. Kampschmidt, R.F. (1980). in "Microbiology-1980" (D. Schless-
        inger, ed.) pp. 150-153, ASM, Washington, D.C.
11. Murphy, P.A., Hanson, D.F., Simon, P.L., Willoughby, W.F.
        and Windle, B.E. (1980). in "Microbiology-1980" (D.
        Schlessinger, ed.) pp. 158-161, ASM, Washington, D.C.
12. Moore, R.N., Oppenheim, J.J., Farrar, J.J., Carter, C.S.,
        Waheed, A. and Shadduck, R.K. (1980). J. Immunol. 125:
        1302-1305.
13. Kurland, J.I., Pelus, L.M., Ralph, P., Bockman, R.S. and
        Moore, M.A.S. (1979). Proc. Natl. Acad. Sci. U.S.A.,
        76:2326-2330.
14. Moore, R.N., Vogel, S.N., Wahl, L.M. and Mergenhagen, S.E.
        (1980). in "Microbiology-1980" (D. Schlessinger, ed.),
        pp. 131-134, ASM, Washington, D.C.
15. Moore, R.N., Hoffeld, J.T., Farrar, J.J., Mergenhagen, S.E.,
        Oppenheim, J.J. and Shadduck, R.K. (1981). Lymphokines,
        3:119-148.

16. Farrar, J.J., Fuller-Farrar, J., Simon, P.L., Hilfiker, M.L.,
        Stadler, B.M. and Farrar, W.L. (1980). J. Immunol. 125:
        2555-2558.
17. Burgess, A.W., Metcalf, D., Russell, S.H.M. and Nicola, N.A.
        (1980). Biochem. J. 185:301-309.

MACROPHAGE OXYGEN-DEPENDENT KILLING OF INTRACELLULAR PARASITES:

TOXOPLASMA AND LEISHMANIA

Henry W. Murray

Cornell University Medical College
New York, NY

INTRODUCTION

The importance of cell-mediated immunity in host resistance
to infection caused by intracellular pathogens has been well-
established. Similarly, it has been known for some time that monocytes and macrophages play primary roles in the efferent limb of
the cellular response to such infections (1-9). The mechanisms by
which these phagocytes exert antimicrobial activity, however, have
been largely unexplored. During the past several years, work in
this and other laboratories has indicated that the generation of
toxic oxygen intermediates contributes importantly to the ability
of mononuclear phagocytes to kill or inhibit the growth of ingested
microbial pathogens (10-20). Thus, considerable evidence is now
available to suggest that the well-recognized oxygen-dependent
mechanism present in polymorphonuclear leukocytes, which generates
a number of toxic products including superoxide anion (O_2^-),
hydrogen peroxide (H_2O_2), and hydroxyl radical ($OH\cdot$) (21-23), is
also present and operative in monocytes and macrophages.

This report reviews our recent studies which have focused on
characterizing macrophage oxygen-dependent antiprotozoal activity
(10-15). We have employed a model comprised of a variety of mouse
peritoneal macrophage populations as effector cells and _Toxoplasma
gondii_ trophozoites and _Leishmania_ promastigotes as microbial
targets. Our results indicate that the ability of a parasite to
trigger the macrophage oxidative burst, the magnitude of the phagocyte oxidative response, and intrinsic microbial susceptibility to
H_2O_2 are all important determinants of how macrophages influence
the intracellular fate of pathogenic protozoa.

MACROPHAGE-PROTOZOA INTERACTION

A. <u>Toxoplasma gondii</u>

      After a brief lag period, toxoplasma trophozoites proceed
to multiply within cultivated resident peritoneal macrophages from
normal mice, and 18 h after infection there are 4-5 toxoplasmas per
vacuole (Fig. 1). Replication continues until cells rupture and
liberate parasites which infect other macrophages. By 36-40 h, the
monolayer is destroyed (11). Similar to normal resident cells,
macrophages elicited by inflammatory agents such as thioglycollate
(THIO), proteose peptone (PP), or heart infusion broth (HIB) as
well as phagocytes derived from the J774 macrophage cell line (15)
are also incapable of restricting toxoplasma intracellular growth.
In contrast, peritoneal macrophages obtained from mice chronically
infected with <u>T. gondii</u> display clear toxoplasmastatic activity,

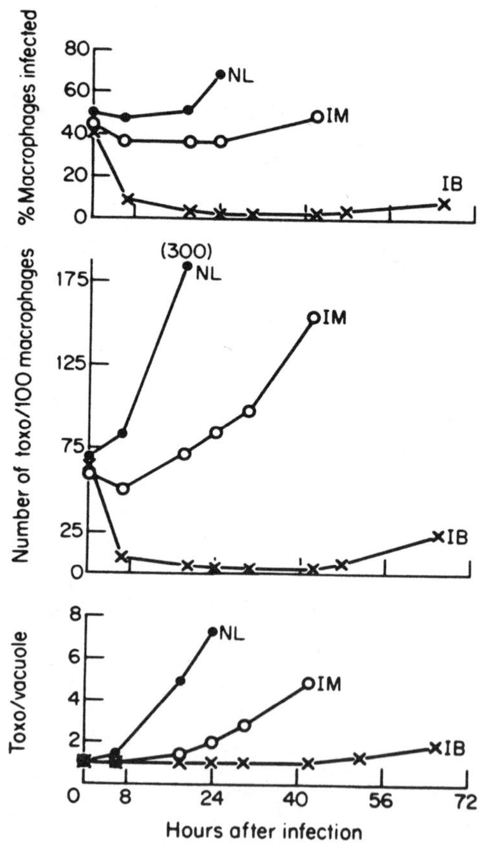

Figure 1. Intracellular fate of <u>T. gondii</u> (TOXO) within normal (NL)
immune (IM), and immune-boosted (IB) macrophages (10).

and 18 h after infection there are 1-1.5 toxoplasmas per vacuole.
After 36 h, however, in vitro replication commences within these
immune (IM) macrophages, and proceeds unchecked unless stimulated
spleen cell products (lymphokines) are added to the culture medium
(11,12). The behavior of macrophages from IM mice boosted intraperitoneally (IP) 3 days before cell harvest with specific antigen
(e.g. heat-killed toxoplasmas), however, is quite different. These
immune-boosted (IB) cells display striking toxoplasmacidal activity,
and kill 75% of ingested T. gondii within 6 h and > 90% by 18 h
(11). Figure 1 summarizes the activities of these three cell types.

Varying degrees of in vitro toxoplasmastatic and toxoplasmacidal activity can also be conferred on peritoneal macrophages if
mice are injected intravenously (iv) or intraperitoneally (ip) with
microbial agents such as viable and dead BCG or killed Coryne-
bacterium parvum (12). Thus, immunologic specificity is not
required for successful activation of macrophages to display antitoxoplasma activity. In addition, resident macrophages from normal
mice can also be induced to inhibit toxoplasma replication by
cultivation with antigen-or mitogen-stimulated lymphokines if a
synergistic inflammatory agent (e.g. THIO, PP, or HIB) is also
present in the medium (12). Since lipopolysaccharide (LPS) can
replace these latter agents in this in vitro activating system (12),
it is likely that they simply serve as sources of endotoxin. By
themselves, however, neither lymphokine nor inflammatory agents
activate normal cells to inhibit toxoplasma replication.

B. Leishmania

In contrast to the inability of normal resident macrophages
to exert intracellular activity against T. gondii, these same cells
readily eradicate 80-95% of ingested L. donovani (LD) and L.
tropica (LT) promastigotes by 24 h (14,15). Killing is prompt, and
over one-half of ingested promastigotes are reduced to intravacuolar
debris within 4 h (Fig. 2). Macrophages activated in vivo by C.
parvum injection or in vitro by lymphokine kill promastigotes at a
faster rate consistent with their enhanced antimicrobial capacity
(12). In contrast, phagocytes derived from a clone (J774G8) of
the J774 cell line exert no leishmanicidal activity (Fig. 2), and
permit virtually all intracellular LD and LT promastigotes to
survive and transform to the amastigote stage (15). These J774G8
cells can be effectively activated by lymphokine, however, to
display considerable leishmanicidal activity (Fig. 2).

We therefore have established models for both T. gondii and
Leishmania comprised of clearly distinguishable macrophage populations which either (a) permit parasite replication or persistence,
(b) inhibit intracellular growth, or (c) promptly kill the majority
of ingested protozoa. These are the cells which have been employed
to investigate the role of oxygen intermediates.

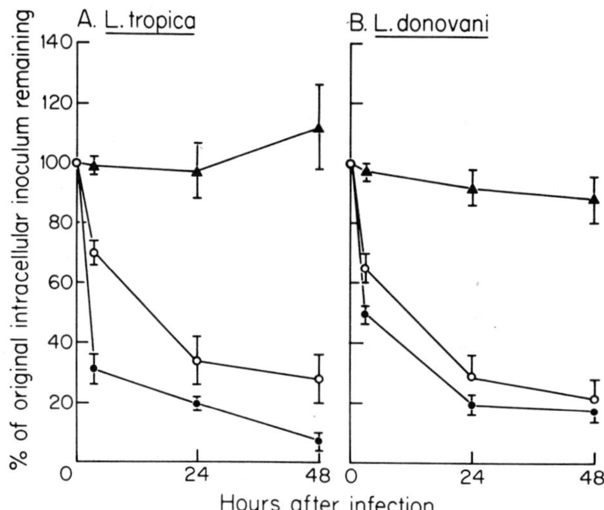

Figure 2. Intracellular fate of _Leishmania_ promastigotes ingested
by normal resident macrophages (●), unstimulated
J774G8 cells (▲), and lymphokine-stimulated J774G8
cells (o) (15).

SUSCEPTIBILITY OF T. GONDII AND LEISHMANIA TO OXYGEN INTERMEDIATES

A. H_2O_2

Before examining macrophage-protozoa oxidative interaction,
we first determined that exposure to oxygen intermediates in the
absence of phagocytes could result in parasite killing. This was
accomplished by subjecting toxoplasmas and promastigotes to a
variety of cell-free oxidative environments (10-14). In the
presence of glucose-glucose oxidase (GO), a reaction which generates
no intermediates other than H_2O_2, _Leishmania_ were readily immobilized
and lysed by remarkably low fluxes of H_2O_2 (Fig. 3A). In contrast,
toxoplasmas were unaffected by fluxes of H_2O_2 which were uniformly
leishmanicidal, and were resistant to > 15 nmol/minute. Furthermore, a 1 h exposure to 10^{-3}M reagent H_2O_2 also failed to kill
toxoplasmas (10).

The striking resistance of _T_. _gondii_ to H_2O_2 suggested that
this parasite, but not _Leishmania_, possessed particularly effective
defense mechanisms against the toxicity of exogenous H_2O_2. As
shown in Table 1, toxoplasmas contain abundant catalase and
glutathione peroxidase (GPO), enzymes which scavenge H_2O_2, at
levels over 100-fold more than _Leishmania_ (14). To illustrate the
potential importance of exogenous catalase, toxoplasmas were pretreated with 50mM aminotriazole which inhibited > 80% of their

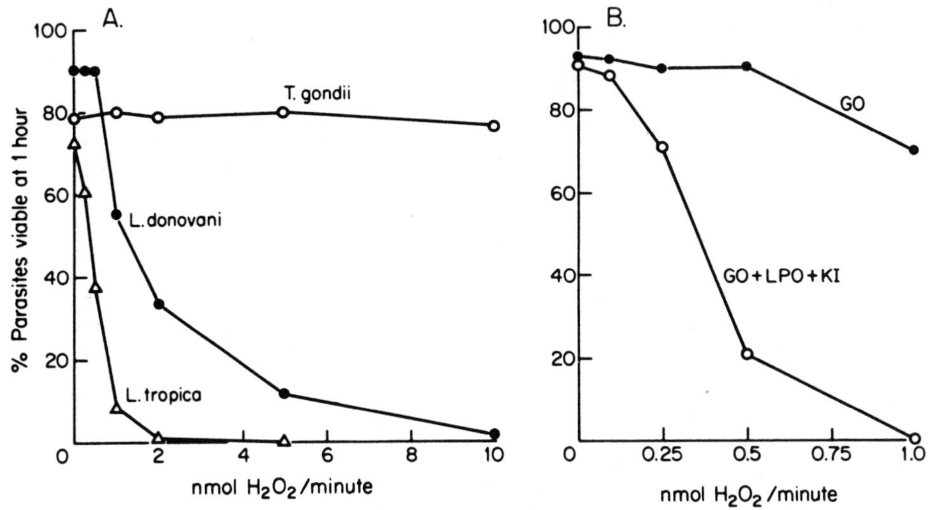

Figure 3.   (A) Susceptibility of T. gondii and Leishmania promastigotes to H_2O_2 generated enzymatically by glucose
oxidase (GO) (14).   (B) Enhancement of L. donovani
killing by H_2O_2 in the presence of lactoperoxidase
(LPO) and KI (14).

Table 1.   Endogenous Scavengers of O_2^- and H_2O_2 (14)

	SOD	Catalase	Glutathione Peroxidase
	(U/mg)	$(BU/mgx10^{-2})$	(nmol/min/mg)
T. gondii	6.1	4.8	117
L. donovani	4.1	0.05	0.5
L. tropica	6.4	0.03	0.2

catalase activity without altering GPO levels. The 50% lethal
dose (LD_{50}) of reagent H_2O_2 for T. gondii was decreased 100-fold
from 5×10^{-3}M to 5×10^{-5}M by aminotrizole treatment (13).

B. H_2O_2 - Peroxidase-Halide

Although macrophages lack granular peroxidase, monocytes do
contain myeloperoxidase and phagocytize Leishmania (24) and T.
gondii (17); thus, we also investigated whether protozoan susceptibility to H_2O_2 could be enhanced by the addition of a peroxidase

Table 2. Susceptibility to Enzymatically Generated and Reagent
          H_2O_2 (10,14)

	LD_{50} of H_2O_2 after 1 h			
	GO[*]	GO+ LPO–KI	Reagent H_2O_2	Reagent H_2O_2 +LPO–KI
T. gondii	> 15[†]	1.5	5×10^{-3}M	5×10^{-6}M
L. donovani	1.5	0.3	——	——
L. tropica	0.5	< 0.1	——	——

[*]Glucose oxidase (GO), lactoperoxidase (LPO)

[†]nmol of H_2O_2 per minute

and an oxidizable halide cofactor. The potent H_2O_2-myeloperoxidase-
halide oxidizing system has been shown to be highly microbicidal
against virtually all classes of micro-organisms (21). For both
toxoplasmas and <u>Leishmania</u>, the presence of lactoperoxidase (LPO)
and KI considerably enhanced susceptibility to H_2O_2 (Fig. 3B,
Table 2) (10,14).

C. Other Oxygen Intermediates

     To assess the effects of other oxygen intermediates, <u>T</u>. <u>gondii</u>
and promastigotes were also exposed in a cell-free system to
xanthine-xanthine oxidase (XO). This reaction generates a full
complement of intermediates including O_2^-, H_2O_2, OH·, and perhaps
singlet oxygen ($^1O_2$) as well (25). O_2^-, which is formed from the
univalent reduction of molecular oxygen, dismutates to H_2O_2, and in
the presence of trace metal (iron) ions, O_2^- and H_2O_2 appear to
interact to generate OH· (23,25). Thus, the xanthine-XO reaction
seems to closely mimic the sequence of biochemical events which
are believed to occur during the phagocyte respiratory burst.
After exposure to xanthine (1.5×10^{-4}M) and XO (50 μg/ml), a
comparable proportion (75-80%) of toxoplasmas and <u>Leishmania</u> were
killed (Table 3). However, the intermediates responsible for the
killing activity of XO appeared to be quite different. Thus, for
promastigotes only catalase inhibited the leishmanicidal activity
of XO, and scavengers and quenchers of O_2^- (superoxide dismutase
(SOD)), OH· (mannitol, benzoate), and $^1O_2$ (histidine, DABCO) (25)
had no protective effect. This suggested that H_2O_2 alone was both
necessary and sufficient to lyse promastigotes, and supported the
findings of the GO system. In contrast, both SOD and catalase

Table 3. Effect of Oxygen Intermediate Scavengers on Parasite
Killing by Xanthine-Xanthine Oxidase System[*]

| | % Parasites Viable at 1 h[†] | | |
Scavenger added	T. gondii	L. donovani	L. tropica
None	26	18	21
Catalase, 200 µg/ml	68	62	72
SOD, 100 µg/ml	63	25	25
Mannitol, 50mM	65	18	23
Benzoate, 10mM	63	15	24
DABCO, 1mM	66	17	–
Histidine, 10mM	67	19	–

[*] Xanthine, 1.5×10^{-4}M, plus xanthine oxidase, 50 µg/ml

[†] Viability of control preparations incubated with xanthine or
xanthine oxidase alone was 75-80% for T. gondii (10) and
74-87% for Leishmania (14).

inhibited the toxoplasmacidal effects of XO (Table 3). Since
catalase does not affect O_2^- production and SOD does not diminish
H_2O_2 formation (25), these findings indicated that although neither
O_2^- nor H_2O_2 were toxoplasmacidal, products of their interaction
(e.g. OH· and/or $^1O_2$ (25)) were required and appeared responsible
for killing. The observation that proposed scavengers of OH· and
quenchers of $^1O_2$ (25) also effectively inhibited the activity of
the XO system supports a toxoplasmacidal role for these latter two
more distal products of oxygen reduction. It should also be
pointed out that both T. gondii and Leishmania contain comparable
SOD activity (Table 1), and each parasite was resistant to O_2^-
alone (10,14).

TRIGGERING OF THE MACROPHAGE OXIDATIVE BURST

     In order for macrophages to deliver toxic oxygen intermediates
to phagocytic vacuoles housing ingested protozoa, an obvious prerequisite is effective triggering of the O_2^--generating mechanism.
It has been amply demonstrated by others (26,27) that plasma
membrane perturbation, achieved by exposing macrophages to soluble
agents such as phorbol myristate acetate (PMA) or to particulate
phagocytic stimuli, promptly triggers respiratory burst activity.
Thus, we utilized both qualitative and quantitative assays to
characterize the oxidative response to phagocytized toxoplasmas
and Leishmania.

Table 4. Macrophage Reduction of Nitroblue Tetrazolium (NBT)[*]

Macrophage	% of Cells with Precipitated Formazan 1 h after Injestion of:			
	Zymosan	T. gondii	L. donovani	L. tropica
Normal	85	16	80	84
Normal + LK[†]	79	65[††]	82	86
Inflammatory	87	18	–	–
J774G8	8	9	10	12
J774G8 + LK[†]	60	–	65	70
In vivo activated[**]	88	76	87	–

[*] Qualitative (microscopic) assay (14,15)

[†] Lymphokine (LK)-activated

[**]Macrophages from mice immunized with C. parvum or T. gondii

[††]HIB included during lymphokine activating period

    Normal resident macrophages and those elicited by inflammatory agents readily respond to the ingestion of opsonized zymosan with nitroblue tetrazolium (NBT) dye reduction, a reaction which is O_2^--dependent (28). These same macrophages fail, however, to respond to toxoplasma ingestion (Table 4). The mechanism by which T. gondii avoids triggering the oxidative burst of cells it readily parasitizes is unknown (17). In vivo and in vitro activated macrophages, cells capable of displaying antitoxoplasma activity, behave differently, however, and promptly reduce NBT upon toxoplasma ingestion. In contrast, Leishmania promastigotes readily trigger the oxidative burst of normal resident as well as activated macrophages (Table 4), and as noted, all of these cells exert leishmanicidal activity. Oxidatively deficient J774G8 macrophages, however, fail to reduce NBT during Leishmania ingestion, and these cells were the only ones to permit promastigotes to persist intact within their cytoplasm (15).

    Since the results of the cell-free experiments indicated that H_2O_2 alone was sufficient for Leishmania killing, we also employed the fluorometric scopoletin assay to determine if promastigote ingestion stimulated H_2O_2 generation by normal macrophages (Fig. 4). L. tropica and L. donovani were equally effective as triggers for H_2O_2 production which was dependent upon both the size and the duration of exposure to the promastigote inoculum. Ninety minutes after incubation with 5×10^6 promastigotes per ml, normal macrophages released 76–110 nmol of H_2O_2 per mg of adherent cell protein; comparable to that released after triggering with zymosan or PMA (14). J774G8 cells, however, generated scant amounts of H_2O_2 (16–18 nmol/

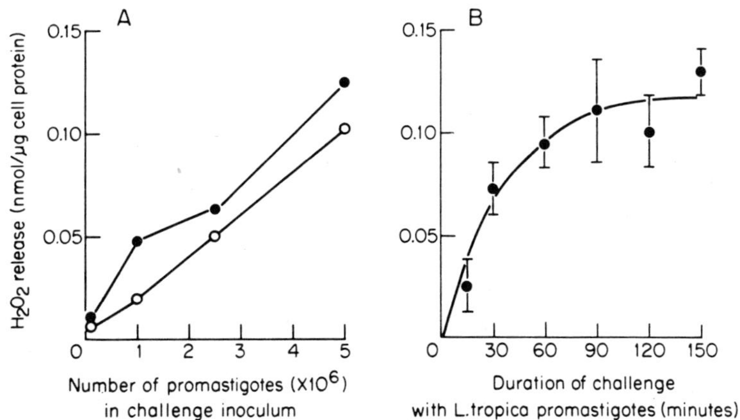

Figure 4. Capacity of promastigote ingestion to stimulate H_2O_2 release by normal resident macrophages (14).

mg protein) in response to promastigote ingestion consistent with their inherent inability to mount an effective oxidative response (previously indicated by the qualitative NBT assay (Table 4)) (15). On the other hand, the O_2^- – H_2O_2 generating capacity of J774G8 cells could be enhanced three to five-fold by 24 h of exposure to lymphokine (15). In parallel, this treatment induced intracellular leishmanicidal activity (Fig. 2).

Thus, using both quantitative and qualitative assays for oxidative burst activity, it appeared that the ability of various macrophage populations to respond to toxoplasma or <u>Leishmania</u> ingestion correlated closely with the intracellular fate of these two protozoa.

THE MAGNITUDE OF THE OXIDATIVE BURST

It has been previously demonstrated that immunologically activated macrophages display an enhanced capacity to generate O_2^- and H_2O_2 (26,27); thus, we also investigated whether there was a quantitative relationship between oxygen intermediate production and the ability to carry out an intracellular antiprotozoal act. As an index of oxidative prowess, the PMA-triggered H_2O_2 release of a variety of macrophage populations was measured, and compared to their capacity to kill <u>T</u>. <u>gondii</u> or inhibit its replication (11,12). As indicated in Figure 5, there were striking differences between toxoplasmacidal (IB), toxoplasmastatic (IM), and normal (NL) macrophages in terms of the rate and magnitude of extracellular H_2O_2 release. Macrophages capable of killing <u>T</u>. <u>gondii</u> generated 20-25 times more H_2O_2 than normal cells and 6-7 times more than cells which inhibited toxoplasma replication (11). We also

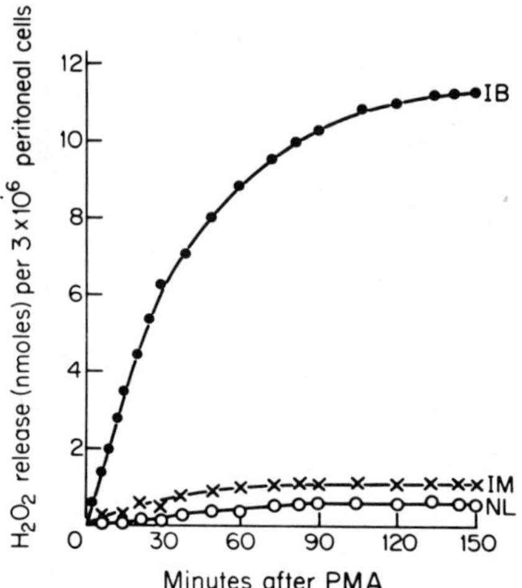

Figure 5. Capacity of normal (NL), toxoplasma immune (IM), and
immune-boosted (IB) macrophages to release H_2O_2 after
triggering with PMA (100 ng/ml) (11).

extended this analysis beyond macrophages from mice specifically
immunized with T. gondii, and examined in parallel the antitoxoplasma and oxidative activities of peritoneal cells from mice
stimulated in vivo by a variety of inflammatory and unrelated
immunologic agents. As judged by the release of H_2O_2 (Fig. 6),
there was a close correlation between macrophage oxidative activity
and the capacity to inhibit intracellular toxoplasma replication.

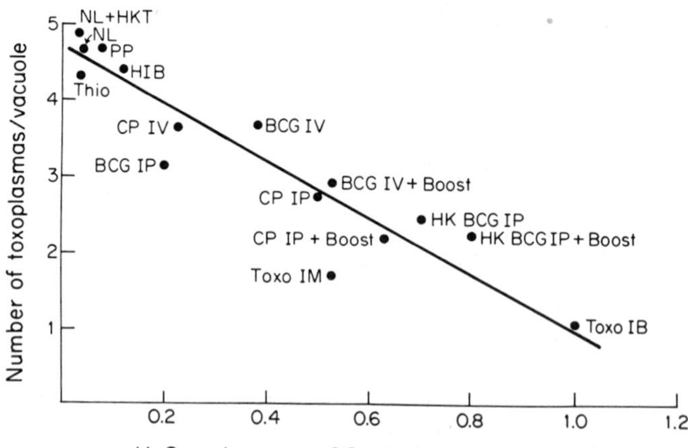

Figure 6. Correlation between macrophage capacity to release H_2O_2
and ability to inhibit toxoplasma replication. CP = C.
parvum-immunized cells (12).

Macrophages elicited by inflammatory agents (THIO, PP, HIB) behaved similar to normal resident cells, and released low amounts of H_2O_2 and failed to display antitoxoplasma activity. In contrast, most macrophage populations activated _in vivo_ by iv or ip delivered immunologic microbial stimuli (viable and heat-killed BCG, killed _C. parvum_, viable _T. gondii_) demonstrated both enhanced H_2O_2 release and either toxoplasmacidal or toxoplasmastatic activity (12). Moreover, ip boosting with respective microbial antigen before cell harvest heightened both activities in parallel.

A similar correlation between the magnitude of oxidative burst activity and protozoal intracellular fate was also observed for _Leishmania_ (Table 5). Normal macrophages, which readily kill both LD and LT promastigotes, generate up to 10 times more O_2^- and H_2O_2 after PMA triggering than J774G8 cells, and the latter exert no leishmanicidal activity. If, however, J774G8 cells are first exposed to lymphokine, their capacity to produce O_2^- and H_2O_2 is enhanced (Table 5), and these activated J774G8 cells kill appreciable numbers of ingested _Leishmania_ (Fig. 2) (15).

EVIDENCE THAT OXYGEN INTERMDIATES PARTICIPATE IN MACROPHAGE ANTI-PROTOZOAL ACTIVITY

A. Effect of Impairing O_2^- and H_2O_2 Production

Taken together, the prior observations suggested that an oxygen-dependent mechanism, which generated O_2^-, H_2O_2, or products

Table 5. Oxidative Activity of J774G8 Cells vs. Normal Macrophages

Macrophage	Response to Triggering with PMA (100ng/ml)		
	% Cells NBT-Positive[*]	O_2^- Release[†]	H_2O_2 Release[†]
Normal Resident	97	178	112
J774G8	3	46	13
J774G8 + LK[**]	34	114	28

[*] As judged by qualitative reduction of NBT (15).

[†] nmol/mg protein per 90 min as determined by the reduction of ferricytochrome C for O_2^- (27) and the scopoletin technique for H_2O_2 (26).

**Cells first activated by lymphokine (LK) exposure (15).

of their interaction, could be implicated in macrophage activity against pathogenic protozoa. To provide more firm evidence for the participation of oxygen intermediates, we assessed the intracellular survival of T. gondii and Leishmania within microbicidal macrophages whose oxidative activity had been impaired. This was achieved by three diverse techniques; glucose deprivation, pretreatment with PMA, and the administration of exogenous soluble scavengers. Although their mechanisms are quite different, these three manipulations depress macrophage O_2^- and H_2O_2 generation, and render cells oxidatively unresponsive to subsequent phagocytic stimuli (11-14,29,30).

Depriving macrophages of exogenous glucose limits substrate for the hexose monophosphate shunt (HMPS), presumably decreases NADPH availability, and thus retards the reduction of oxygen to O_2^- and the formation of H_2O_2 (29). Polymorphonuclear leukocytes from patients with severe glucose-6-phosphate dehydrogenase deficiency and impaired HMPS activity, for example, fail to elaborate normal amounts of H_2O_2 (31). Since PMA is an effective triggering stimulus for the generation and release of O_2^- and H_2O_2, we also employed this agent to deplete macrophages of the capacity to produce oxygen intermediates by pretreating cells with 200 ng/ml for 90 minutes. This treatment inhibited by > 90% the capacity of macrophages to generate O_2^- or H_2O_2 in response to a subsequent phagocytic challenge (30). In addition, macrophage oxidative activity was also impaired by preincubation with high concentrations of soluble scavengers of oxygen intermediates. These included SOD (for O_2^-), catalase (for H_2O_2), mannitol and benzoate (for OH·), and DABCO and histidine for ($^1O_2$) (25). We have, for instance, previously demonstrated that macrophages readily interiorize exogenous catalase by fluid-phase pinocytosis (10), and that this treatment readily decreases H_2O_2 release upon subsequent stimulation (13). To maximize the efficacy of these scavengers and quenchers, they were also included in the extracellular medium during parasite ingestion (11, 12,14).

The results of these manipulations yielded consistent evidence indicating that an intact ability to respond oxidatively to microbial ingestion and an intact capacity to generate O_2^- and H_2O_2 were required for full expression of macrophage antiprotozoal activity (Table 6; Figs. 7, 8 and 9). Thus, the toxoplasmacidal and toxoplasmastatic activity of in vitro and in vivo activated macrophages and the leishmanicidal capacity of normal macrophages were all reversed to a comparable extent by glucose deprivation, PMA pretreatment, and the administration of exogenous scavengers (11,12,14). In addition, studies with the latter agents suggested that while the generation of H_2O_2 alone was sufficient to account for the killing of LD and LT promastigotes by normal cells (14), macrophage antitoxoplasma activity required products of O_2^- - H_2O_2 interaction, presumably OH· and/or $^1O_2$ (11,12) (Table 6). Thus, catalase but

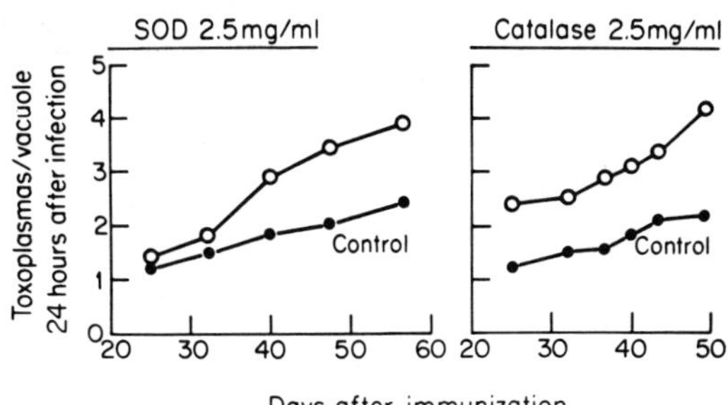

Figure 7.   Inhibition of immune (IM) macrophage toxoplasmastatic
activity by exogenous superoxide dismutase (SOD) and
catalase (11).

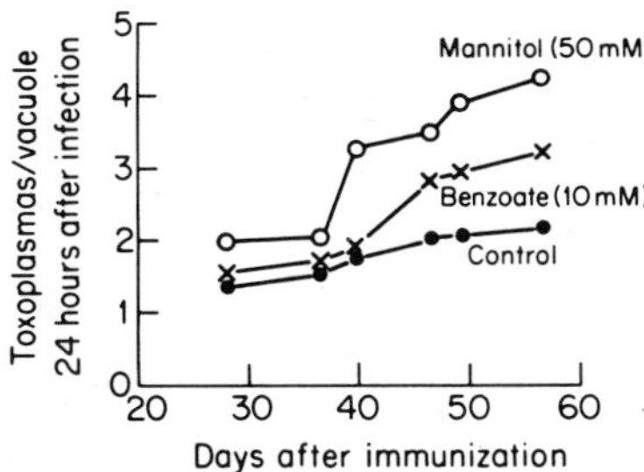

Figure 8.   Inhibition of immune (IM) macrophage toxoplasmastatic
activity by exogenous OH· scavengers, mannitol and
benzoate (11).

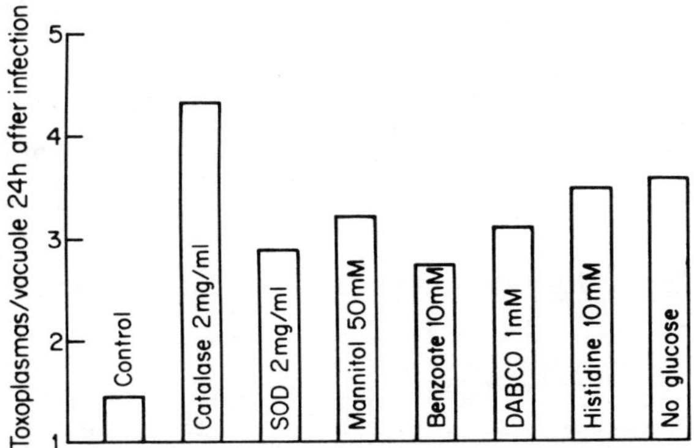

Figure 9.   Inhibition of lymphokine-induced toxoplasmastatic
activity of normal macrophages by exogenous soluble
scavengers and glucose deprivation (12).

Table 6. Effect of Exogenous Scavengers, Glucose Deprivation and
          PMA Pretreatment on Macrophage Protozoal Killing (11,14,30)

Additions	% T. gondii killed*		% L. donovani killed*	
	4 h	18 h	4 h	18 h
None	81	92	47	85
SOD, 2.5 mg/ml	50	53	45	78
Catalase, 2.5 mg/ml	48	52	28	41
Mannitol, 50mM	50	49	48	83
DABCO, 1mM	60	60	51	76
No Glucose	25	65	33	49
PMA Pretreatment	0	0[†]	21	36

* Immune-boosted (IB) macrophages challenged with T. gondii; normal
  resident macrophages challenged with L. donovani promastigotes.

[†] Toxoplasma replicated in PMA pretreated immune-boosted (IB)
  macrophages.

not SOD or OH· scavengers inhibited Leishmania killing, indicating
a primary role for H_2O_2. In contrast, both SOD and catalase were
effective in reversing toxoplasma killing and inhibition by
activated macrophages, similar to the results observed in the XO
cell-free system. The requirement for both O_2^- and H_2O_2 for expression of antitoxoplasma activity as well as the inhibitory effect of
OH· scavengers and $^1O_2$ quenchers suggested key roles for OH· and
$^1O_2$ in this macrophage model. Because the specificity of the
inhibition produced by mannitol, benzoate, DABCO, and histidine has
not been fully established (25), the nature of the active inter-
mediate(s) remains in question. Given the specificity of SOD and
catalase, however, O_2^- and H_2O_2 appear to function as precursors of
the macrophage toxoplasmacidal agents in the systems we employed.

B. Parasite Fate Within Phagocytes Inherently Deficient in O_2^- and
    H_2O_2 Production

     To provide further evidence for an oxygen-dependent mechanism
in macrophage antiprotozoal activity, we also examine the intracellular fate of T. gondii and Leishmania promastigotes within the
oxidatively deficient J774G8 macrophage-like cell line (15). As
previously noted, these cells inherently generate scant amounts of
O_2^- and H_2O_2 and fail to respond with NBT reduction to parasite
ingestion. Not unexpectedly, J774G8 cells both support toxoplasma
replication and permit promastigotes to survive and transform to
the amastigote stage (15) (Fig. 2).

C. <u>Intracellular Effects of O_2^- and H_2O_2 Generated from Extracellular Sources</u>

Finally, we also explored whether providing macrophages with
an extracellular source of oxygen intermediates could augment the
destruction of intracellular parasites. Adding glucose-glucose
oxidase (GO), which generated up to 15 nmol of H_2O_2 per minute, to
cultures of normal macrophages after <u>T</u>. <u>gondii</u> infection had no
effect, and toxoplasma replicated freely. The addition of xanthine-
XO to the extracellular medium as a source of O_2^-, H_2O_2 and OH·,
however, resulted in the killing of a small portion of intracellular
toxoplasmas and consistent inhibition of their replication for 18 h
(Fig. 10). In contrast, for J774G8 cells infected with the H_2O_2-
susceptible <u>Leishmania</u>, a 1 h exposure to an exogenous source of
enzymatically generated H_2O_2 was quite effective, and resulted in
the killing of > 60% of intracellular promastigotes (15).

SUMMARY

These studies provide several lines of evidence which together
suggest that macrophages exert antimicrobial activity against <u>T</u>.
<u>gondii</u> and <u>Leishmania</u> by utilizing an oxygen-dependent mechanism.
It appears that the same basic mechanisms, the generation of O_2^- and
H_2O_2, contribute to the killing of both toxoplasmas and promastigotes. At the same time, however, this work also indicates clear

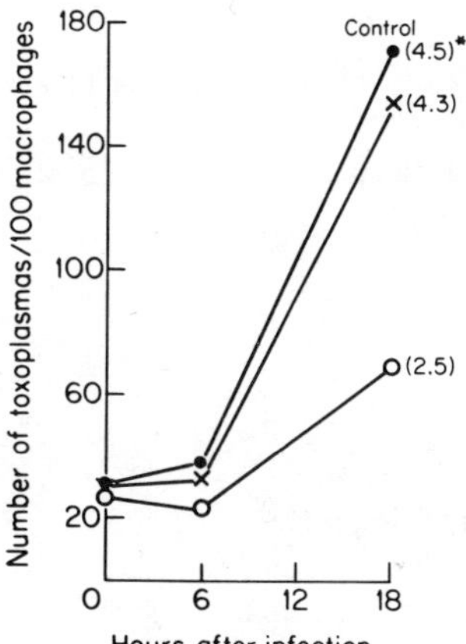

Figure 10. Inhibition of intracellular toxoplasma replication by
            normal macrophages provided after infection with an
            extracellular source of oxygen intermediates by adding
            xanthine plus xanthine oxidase (o) to the culture medium.
            Adding xanthine or xanthine oxidase alone (X) had no
            effect. (*) indicates the number of toxoplasmas per
            vacuole 18 h after infection (11).

differences between these two protozoans in terms of susceptibility
to oxygen intermediates, triggering of the macrophage oxidative
burst, and intracellular fate within the cytoplasm of normal macrophages.

    T. gondii appears to stand at one end of the protozoal spectrum.
This organism contains abundant levels of catalase and GPO, is
highly resistant to H_2O_2, and during ingestion avoids stimulating
the respiratory burst activity of normal macrophages which under
the best conditions generate only small amounts of O_2^- and H_2O_2.
At the opposite end of this spectrum are Leishmania promastigotes.
These hemoflagellates are low in catalase and GPO activity, highly
susceptible to H_2O_2 and readily trigger normal macrophages to
display oxidative burst activity including the generation of H_2O_2.
Thus, in view of these differences and given that oxidative
mechanisms are pertinent to intracellular protozoal killing, toxoplasmas appear to be particularly well-suited to parasitize normal
macrophages whereas Leishmania promastigotes are not. Once activated either in vitro or in vivo, however, macrophages acquire the
ability to generate considerably higher levels of O_2^- and H_2O_2 and
to respond oxidatively to T. gondii ingestion. In parallel,
these cells develop the capacity to kill toxoplasmas or inhibit
their intracellular replication. These immunologically induced
biochemical changes appear to contribute in large measure to the
enhanced antiprotozoal activity associated with macrophage
activation.

ACKNOWLEDGMENTS

    Figures 1-10 were reprinted with permission from the Journal
of Experimental Medicine (10-15). Grants from the USPHS (AI 16963-
01) and the Rockefeller Foundation (RF 78021) supported proteins
of this work.

REFERENCES

1. Mackaness, G.B. (1962). J. Exp. Med. 116:381.
2. North, R.J. (1970). J. Exp. Med. 132:535.
3. McLoed, R. and Remington, J.S. (1977). Cell Immunol. 34:156.
4. North, R.J. (1978). J. Immunol. 121:806.
5. Jones, T.C., Len, L. and Hirsch, J.G. (1975). J. Exp. Med.
        141:466.
6. Mauel, J., Buchmuller, Y. and Beelin, R. (1978). J. Exp. Med.
        148:393.
7. Nogueira, N. and Cohn, Z.A. (1978). J. Exp. Med. 148:288.
8. Nacy, C.A. and Meltzer, M.S. (1979). J. Immunol. 123:2544.
9. Nathan, C.F., Murray, H.W. and Cohn, Z.A. (1980). N. Engl. J.
        Med. 303:622.

10. Murray, H.W. and Cohn, Z.A. (1979). J. Exp. Med. 150:938.
11. Murray, H.W., Juangbhanich, C.W., Nathan, C.F. and Cohn, Z.A.
        (1979). J. Exp. Med. 150:950.
12. Murray, H.W. and Cohn, Z.A. (1980). J. Exp. Med. 152:1596.
13. Murray, H.W., Nathan, C.F. and Cohn, Z.A. (1980). J. Exp.
        Med. 152:1601.
14. Murray, H.W. (1981). J. Exp. Med. 153:1302.
15. Murray, H.W. (1981). J. Exp. Med. 153:1690.
16. Nathan, C.F., Nogueira, N., Juangbhanich, C., Ellis, J., and
        Cohn, Z.A. (1979). J. Exp. Med. 149:1056.
17. Wilson, C.B., Tsai, V., and Remington, J.S. (1980). J. Exp.
        Med. 151:328.
18. Sasada, M. and Johnston, R.B. (1982). J. Exp. Med. 152:85.
19. Sagone, A.L., King. G.W. and Metz, E.N. (1976). J. Clin.
        Invest. 57:1352.
20. Klebanoff, S.J. and Hamon, C.B. (1975). In Mononuclear
        Phagocytes in Immunity, Infection, and Pathology, Blackwell
        Scientific Publications, Oxford, p. 507.
21. Klebanoff, S.J. (1975). Semin. Hematol. 12:117.
22. Babior, B.M. (1978). N. Engl. J. Med. 298:659.
23. Badwey, J.A. and Karnovsky, M.L. (1980). Ann. Rev. Biochem.
        49:695.
24. Berman, J.D., Dwyer, D.M. and Wyler, D.J. (1979). Infect.
        Immun. 26:375.
25. Rosen, H. and Klebanoff, S.J. (1979). J. Exp. Med. 149:27.
26. Nathan, C.F. and Root, R.K. (1977). J. Exp. Med. 146:1648.
27. Johnston, R.B., Godzik, C.A. and Cohn, Z.A. (1978). J. Exp.
        Med. 148:115.
28. Baehner, R.L., Boxer, L.A. and Davis, J. (1976). Blood. 48:309.
29. Nathan, C.F., Silverstein, S.C., Brukner, L.H. and Cohn, Z.A.
        (1979) J. Exp. Med. 149:100.
30. Murray, H.W. (1981). Manuscript submitted.
31. Baehner, R.L., Johnston, R.B. and Nathan, D.G. (1972). J.
        Reticulendothel Soc. 12:150.

IMMUNOLOGIC LESIONS DURING TOXOPLASMA GONDII INFECTION

Edmond A. Goidl, Urban Ramstedt and Thomas C. Jones

Divisions of Allergy and Immunology and International
Medicine, Department of Medicine, Cornell University
Medical College, 1300 York Ave., New York, NY 10021
Department of Immunology, Biomedicum Centrum
University of Uppsala, Uppsala, Sweden

INTRODUCTION

One mechanism of successful parasitism may be the induction
of relatively localized, but non-specific, lesions in the mammalian
host's immune response. To investigate this, we have initiated
studies of the responses in mice to various antigens during
Toxoplasma gondii infection. Experiments reported here indicate
that different responses to thymic-independent (TI) and thymic-
dependent (TD) antigenic challenges are seen which are consistent
with an immunologic lesion or lesions in thymus regulatory pathways. The possibility is raised that the immunologic injury may
be an important correlate of microbial infectivity.

MATERIALS AND METHODS

Brain cysts of the avirulent Toxoplasma strain, Pe, were
inoculated from chronically infected CF_1 mice into the thigh of
LAF_1 male mice (6 to 8 weeks of age, purchased from Jackson
Laboratories, Bar Harbor, Maine). Brains of mice infected 3 to 6
months previously were removed using sterile procedures, placed in
1 ml of minimal essential medium, diced, and passed several times
through a #18 gauge needle to produce a suspension. The number of
toxoplasma cysts/ml was enumerated, and 0.1 ml containing 2 to 4
cysts was inoculated into each LAF_1 mouse. By previous experience,
this dose of cysts is known to produce a chronic toxoplasma infection, which has been evaluated in other experiments with regard

145

to lymph node and thymus histology (1), protection against challenge
in vivo and in vitro with virulent organisms (2), lymphokine production and macrophage function (3), and T-suppressor cell activity
(P. Erb and T.C. Jones, unpublished observations). One month
after infection, the animals appeared outwardly healthy when compared with recipients of normal brain, but on autopsy they displayed
signs of infection, such as enlarged spleen and Toxoplasma brain
cysts.

Methods concerning immunization schedules and Jerne plaque
forming cell assays have been extensively described previously (4–7).

RESULTS

The magnitude of the direct anti-TNP splenic PFC response to
TNP-F (trinitrophenyl-lisyl-Ficoll) immunization, in Toxoplasma
gondii infected LAF_1 mice is presented in Table 1. There is a
dramatic increase in the primary splenic anti-TNP PFC response in
the toxoplasma-infected animals. Increases range from 54 to 198%
over the control anti-TNP-PFC response ($p < 0.05$). In contrast
to the augmentation seen during the primary response to the TI
antigen TNP-F, the secondary immune response in toxoplasma-infected
animals, although of higher magnitude, is not statistically
different from control animals. The increase in magnitude of the
direct anti-TNP PFC in the primary response to TNP-F, may be
indicative of deregulation of thymic suppressor cells, which have
been shown to down-regulate the immune response to TI antigens in
euthymic mice. Such increased responsiveness has been shown in
athymic nude (Nu/Nu) mice by Baker et al. (8). Some direct
effect upon B cells, however, cannot be ruled out.

The characteristics of the immune response in toxoplasma-
infected mice following primary immunization with a TD antigen
(TNP-BGG), when compared to that of control animals, further

Table 1. Comparison of the anti-TNP PFC resonse following primary
          and secondary TNP-F immunization of Toxoplasmosis
          gondii infected LAF_1 mice.

ANTIGEN CHALLENGE	CONTROL MICE	TOXOPLASMA INFECTED MICE
Primary	Normal	Increased
Secondary	Normal	Normal

Table 2. Comparison of the anti-TNP PFC response following
          primary and secondary TNP-BGG immunization of
          Toxoplasmosis gondii infected LAF$_1$ mice.

	CONTROL MICE		TOXOPLASMA INFECTED MICE	
ANTIGEN CHALLENGE	DIRECT PFC/SPLEEN	INDIRECT PFC/SPLEEN	DIRECT PFC/SPLEEN	INDIRECT PFC/SPLEEN
Primary	Normal		↓	↓↓↓
Secondary	Normal		Normal	

implicate alterations in T-cell function. The primary and
secondary responses to TNP-BGG are presented in Table 2. The
toxoplasma infected group shows a 74% depression in the magnitude
of the indirect PFC response, and a 50% depression in the direct
anti-TNP PFC response. Further alterations are demonstrated by
the change in the IgG/IgM ratio of the infected animals (<2) when
compared to that of the control group (IgG/IgM = 3.5). The
secondary immune response of the toxoplasma infected animals,
although slightly elevated, is not statistically different from
the secondary immune response of the control group.

DISCUSSION

     Immune suppression during the primary immune response has
been described for a number of protozoal diseases including
malaria (9), trypanosomiasis (10,11), Chagas' (12) and leishmaniasis
(13). In toxoplasmosis, immune suppression of response to sheep
RBCs (14,15) and viruses (16) has been demonstrated. Hibbs, et al.
has recently shown prolonged allograft rejection during toxoplasma
infection (17). Though these changes are clear, and appear to
correlate with a delay in the appearance of a delayed-type hypersensitivity response (1,18), they do not cause marked changes in
response to toxoplasma or unrelated antigens (19) or an increase
in susceptibility to other infections (20). When the antibody
response to toxoplasma antigens was evaluated comparing T-deprived
mice with normal animals, a persistence of IgM antibody during
infection of T-deprived mice was observed (21). Correlation of
changes in the immune response as a function of the virulence of
the microbe strain used has been suggested for trypanosomes (10),
but for toxoplasma strains of differing mouse virulence, no
significant alterations have been found (22). Studies of virulence
may not be relevant to the issues raised here because factors
determining virulence may be quite different than those determining

successful microbial infectivity. When infectivity or response to
complex antigens has been examined in other microbial systems, such
as Leishmania (23,24) or Mycobacterium bovis (25), abnormalities
of macrophage-lymphocyte interactions have been recorded among
susceptible (or low-responder) and resistant (high-responder)
strains of mice. Mouse strain variation in response to toxoplasma
has been recorded (26), but it is not as impressive as that
observed with Leishmania, and immunologic evaluation of different
mouse strains has not been done. Toxoplasma is certainly one of
the most successful microbes for infecting mammals across all
strains and genetic backgrounds.

The lesions in immune responsiveness displayed in the primary
immune response to both TI and TD antigens in Toxoplasma gondii
infected animals point to several possible alterations in T-cell
function. First, the increase in the response to TNP-F may
indicate a decrease in suppressor T-cell activity which normally
depress antibody responses to TI antigens. It may also point to
a suppression of the T cell responsible for the production of auto-
anti-idiotypic antibody which has recently been shown to participate in the down-regulation of the antibody response to TI antigens (5). Although a direct stimulatory nonspecific polyclonal
effect upon B cells cannot at this time be ruled out, this
possibility is unlikely since no such effect was observed in the
PFC responses to TNP-BGG. Second, the decrease in the magnitude
of the primary anti-TNP PFC responses to TNP-BGG may be due to an
increase in suppressor activity following Toxoplasma gondii
infection. This has recently been suggested by Howard, et al.
during Leishmania tropica infection (27) and suppressor cells have
been documented during toxoplasma infection (P. Erb and T.C. Jones,
unpublished observations). In addition, as indicated by the marked
decrease in the IgG/IgM antibody response in the infected animals,
the TD function of the IgM to IgG switch in the immune response
has been profoundly affected.   The generation of immune memory
during the primary response is normal in the Toxoplasma gondii
infected animals (as shown by the normal secondary response
following TD antigenic challenge).

The non-specific immune suppression consequent to toxoplasmosis
involves a degree of complexity which has not been apparent in previous studies. The possible lesions in the regulatory cellular
circuitry remain to be identified, but we can point to decreases
in the following T-cell functions: a) the down-regulation of the
antibody response to TI antigens, b) a decrease in the magnitude
of the primary response to TD antigens and c) a depression in the
"switch" from IgM to IgG antibody during the primary immune response
while presuming a normal secondary immune response.

These studies demonstrate relatively localized, but nonspecific immunologic lesions in T-cell regulation of the immune

response. Careful dissection of mechanisms of such special lesions may be helpful in understanding microbial pathogenicity.

ACKNOWLEDGMENTS

This research was supported in part by NIH Grants AI 12146 and AI 16282.

REFERENCES

1. Jones, T.C. (1980). Springer Semin. Immunopathol. 2:387.
2. Jones, T.C., Len, L. and Hirsch, J.G. (1975). J. Exp. Med. 141:466.
3. Jones, T.C., Masur, H., Len, L. and Fu, T.L.T. (1977). Am. J. Trop. Med. Hyg. 26:187.
4. Mitchell, G.F., Humphrey, J.H., and Williamson, A.R. (1972). Eur. J. Immunol. 2:460.
5. Schrater, A.F., Goidl, E.A., Thorbecke, G.J. and Siskind, G.W. (1979). J. Exp. Med. 150:138.
6. Jerne, N.K. and Nordin, A.A. (1963). Science (Washington, D.C.) 140:405.
7. Dresser, D.W. and Greaves, M.F. (1973). In Handbook of Experimental Immunology. D.M. Weir, ed., Blackwell Scientific Publications, Oxford, 271.
8. Baker, P.J., Reed, N.D., Stashak, P.W., Ansbaugh, D.F. and Prescott, B.J. (1973). J. Exp. Med. 137:1431.
9. Greenwood, B.M.J., Playfair, J.H.L., Torrigiani, G. (1971). Clin. Exp. Immunol. 8:467.
10. Sacks, D.L., Selkirk, M., Ogilvie, B.M. and Askonas, B.A. (1980). Nature (London) 283:476.
11. Jayawardena, A.N. and Waksman, B.H. (1977). Nature (London) 265:539.
12. Reed, S., Larsen, C.L., and Speer, C.A. (1977). Z. Parsitenk, 52:11.
13. Preston, P.M. (1978). J. Clin. Lab. Immunol. 1:207.
14. Huldt, G., Gard, S. and Olovson, S.G. (1973). Nature (London) 244:301.
15. Strickland, G.T., Petitt, L.E. and Voller, A. (1973). Am. J. Trop. Med. Hyg. 22:452.
16. Strickland, G.T. and Sayles, P.C. (1977). Infect. Immun. 15:184.
17. Hibbs, J.B., Jr., Rem, J.S. and Steward, C.C. (1980). Pharmacol. Ther. 8:37.
18. Anderson, S.E., Jr., Krahenbuhl, J.L. and Remington, J.S. (1979). J. Clin. Lab. Immunol. 2:293.
19. Handman, E., Chester, P.M. and Remington, J.S. (1980). Infect. Immun. 28:524.

20. Ruskin, J. and Remington, J.S. (1968). Science (Washington,
         D.C.) 160:72.
21. Aryanpour, J., Hafizi, A. and Modabber, F. (1980). Infect.
         Immun. 27:1038.
22. Handman, E. and Remington, J.S. (1980). Infect. Immun. 29:215.
23. Bradley, D.J., Taylor, B.A., Blackwell, J., Evans, E.P. and
         Freeman, J. (1979). Clin. Exp. Immunol. 27:7.
24. Handman, E., Ceredig, R. and Mitchell, F.G. (1979). Aust. J.
         Exp. Biol. Med. Sci., 57:9.
25. Nakamura, R.M., Tokunaga, T. and Yamamoto, S. (1980). Infect.
         Immun. 27:268.
26. Araujo, F., Williams, D., Grumet, F. and Remington, J. (1976).
         Infect. Immun. 13:1528.
27. Howard, J.G., Hale, C. and Liew, F.Y. (1980). J. Exp. Med.
         152:594.

IMMUNODEPRESSION IN BALB/c MICE INFECTED WITH <u>LEISHMANIA</u> <u>TROPICA</u>

Phillip A. Scott and Jay P. Farrell

Department of Pathobiology, University of Pennsylvania
Philadelphia, Pennsylvania

<u>Leishmania</u> <u>tropica</u> is an obligate intracellular parasite of
macrophages which causes cutaneous leishmaniasis in man. Infections
in BALB/c mice with this parasite are characterized by the development of large primary ulcers and multiple non-healing metastatic
cutaneous lesions (1,2,3). To date, the immunological mechanisms
responsible for the inability of BALB/c mice to resolve leishmanial
infections are unknown; however, specific suppressor T cells
responsible for regulating skin test responses to leishmanial
antigens have been reported (3,4). In addition, BALB/c mice
infected with <u>L</u>. <u>tropica</u> develop a non-specific immunodepression,
as described in this report. This immunodepression is mediated
by an adherent suppressor cell which suppresses mitogen and
leishmanial antigen responses in a lymphocyte transformation assay.

Spleen cells from BALB/c mice infected for 15 weeks with <u>L</u>.
<u>tropica</u> were stimulated <u>in vitro</u> with concanavalin A (Con A), phy-
tohemagglutinin (PHA) or lipopolysaccharide (LPS) in a lymphocyte
transformation assay. As seen in Fig. 1, proliferative mitogenic
responses by spleen cells from infected mice were significantly
less than the responses of spleen cells from normal age-matched
mice.

In order to determine if a suppressor cell was responsible
for the depressed mitogen responses, mitomycin C treated spleen
cells from 15 week infected mice were co-cultured with spleen cells
from normal mice and stimulated with Con A. It was found that
cells from infected mice could suppress the Con A responses of
normal cells, and that optimal suppression occurred with 2.5×10^5
cells from infected mice were co-cultured with an equal number of
normal cells (Fig. 2). This suppressor cell was first demonstrable

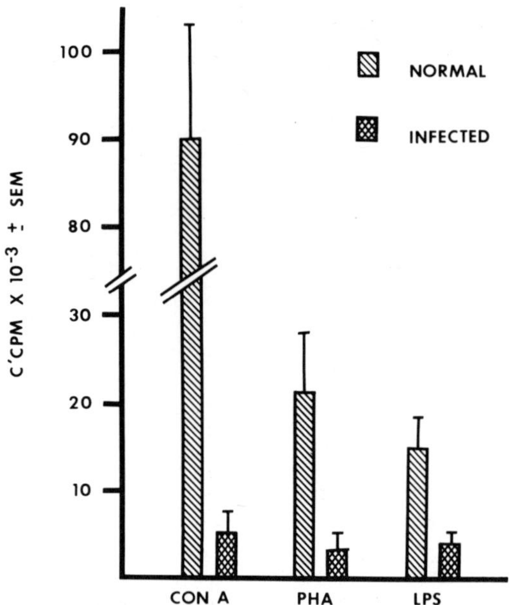

Figure 1.   In vitro response to mitogens of spleen cells from mice after 15 weeks of infection. C' CPM = corrected counts per minute and represents cpm of stimulated cultures with the background cpm subtracted.

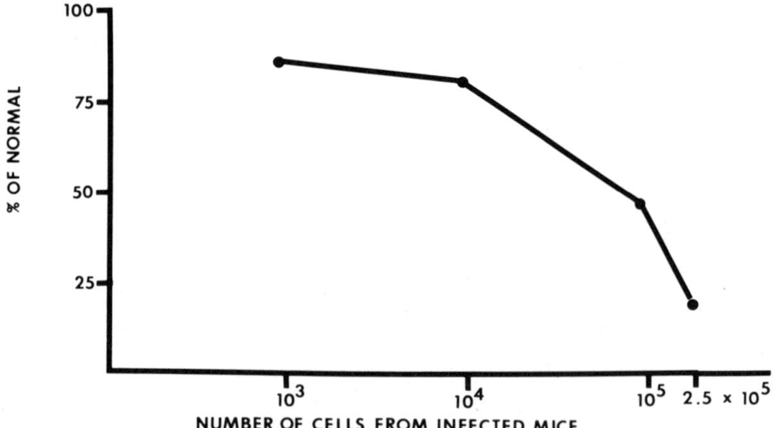

Figure 2.   Effect of the addition of varying numbers of mitomycin C treated spleen cells from 15 week infected BALB/c mice on Con A induced proliferation by 2.5×10^5 normal spleen cells. Results are compared to responses obtained when the same number of mitomycin C treated normal cells are added to 2.5×10^5 normal cells and are expressed as % of normal.

by six weeks of infection, and its presence was closely correlated
with the level of immunodepression seen in these mice (Fig. 3).

By fractionating spleen cells into adherent and non-adherent
populations by Sephadex G-10 passage (5), it was found that the
suppressor cell was an adherent cell. Thus, when Sephadex G-10
adherent and non-adherent fractions were added to normal spleen
cells, only the adherent cell population suppressed normal spleen
cell responses to Con A (Fig. 4).

To further investigate the mechanisms of action of these
suppressor cells, indomethacin, and prostaglandin synthetase inhibitor, was added to co-cultures of spleen cells from normal or infected
mice. Indomethacin (2µg/ml) effectively reversed the suppression
to Con A of spleen cells from mice infected for 12 weeks, suggesting
that prostaglandin production may be involved in the suppression
seen in these mice (Fig. 5). In addition, the removal of Sephadex
G-10 adherent cells or the addition of indomethacin enhanced the
ability of spleen cells from infected mice to respond to leishmanial
antigen (Fig. 6).

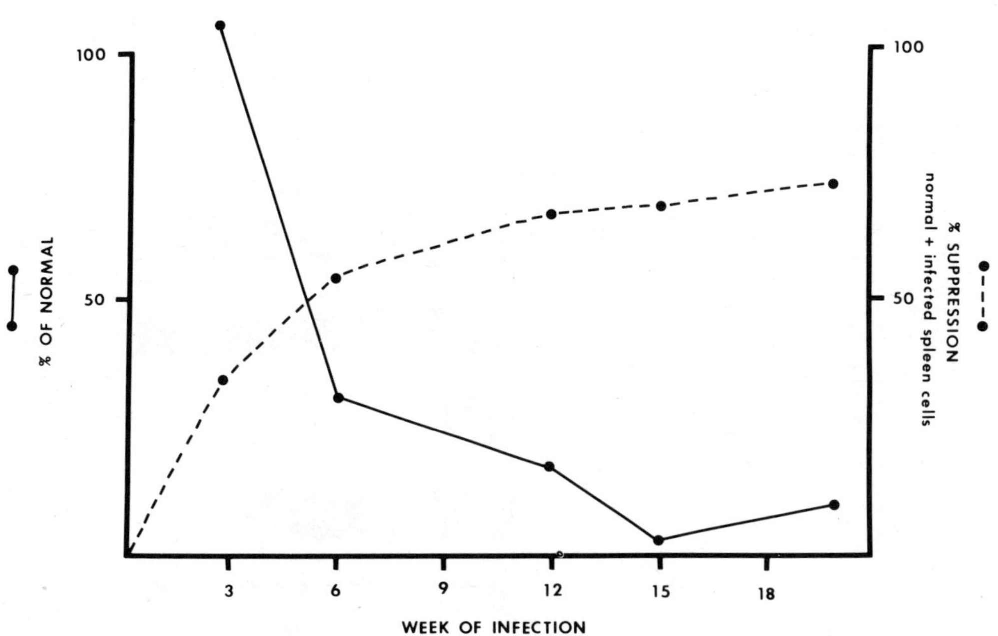

Figure 3.   Comparison of depressed Con A responses and development
of suppressor cells during the course of infection in
BALB/c mice.

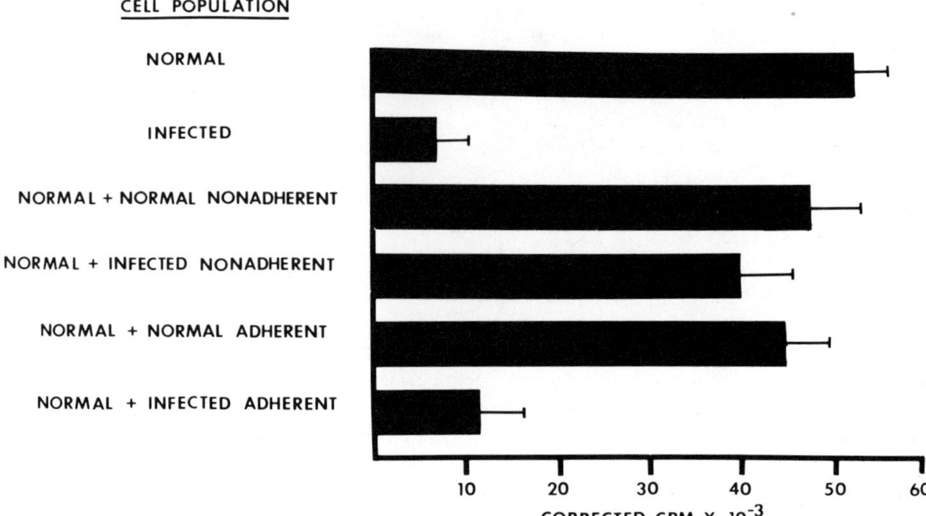

Figure 4.   Comparison of nonadherent and adherent cell suppressive activity upon normal spleen cell responses to Con A. Sephadex G-10 fractionated cell populations (2.5 x 10^5/ culture) were mitomycin C treated prior to addition to normal spleen cells (2.5 x 10^5/culture). Infected spleen cells were obtained from BALB/c mice infected with L. tropica for 12 weeks.

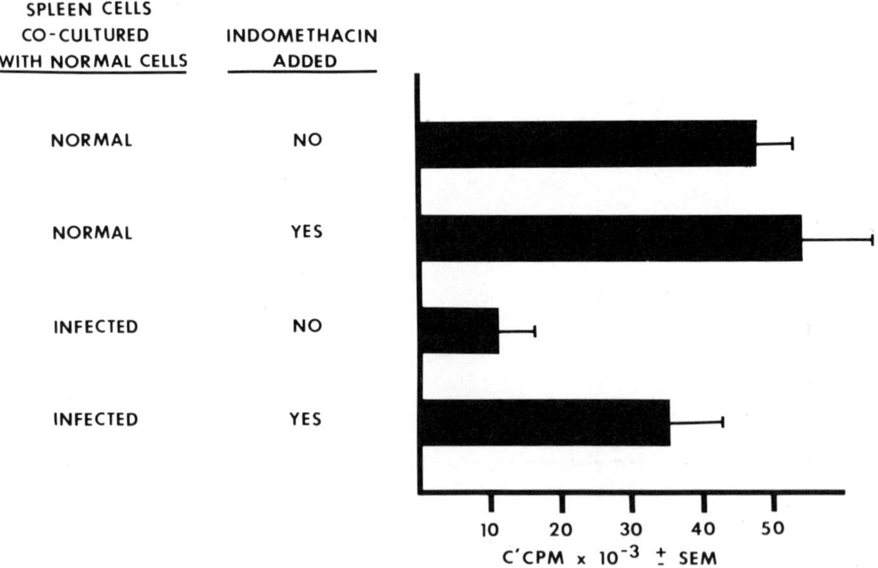

Figure 5.   Ability of indomethacin (2.0 µg/ml) to reverse the suppressive effect of spleen cells from L. tropica infected mice.

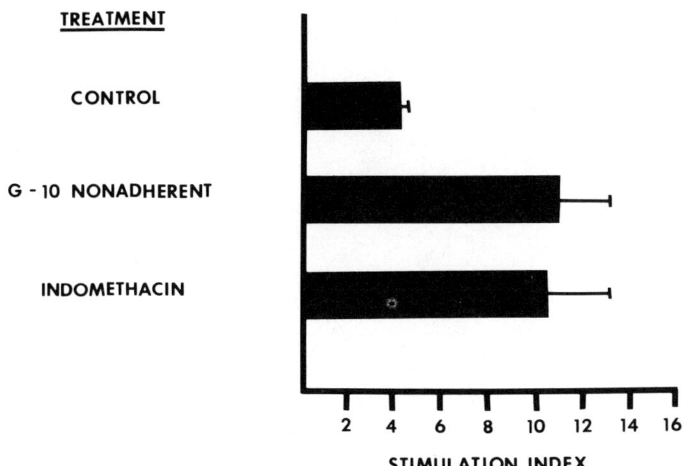

Figure 6. Effect of Sephadex G-10 passage or addition of indomethacin on the responses to Leishmania antigen of
spleen cells from L. tropica infected BALB/c mice (12
weeks of infection). Stimulation index = c'cpm with
antigen ÷ c'cpm without antigen. Control responses
represent the response to antigen of spleen cells from
infected mice without Sephadex G-10 passage or indomethacin addition.

Thus, both antigen-specific and mitogen responses were modulated by an adherent suppressor cell which appeared to mediate
suppression through the production of prostaglandins. Further
work is in progress to directly determine the role that prostaglandins may play in leishmanial infections.

REFERENCES

1. Bjorvatn, B., and Neva, F.A. (1979). Am. J. Trop. Med. Hyg.
        28:472-479.
2. Howard, J.G., Hale, C. and Liew, F.Y. (1980). J. Exp. Med.
        152:594-607.
3. DeTolla, L.J., Scott, P.A. and Farrell, J.P., Immunogenetics
        (in press).
4. Howard, J.G., Hale, C. and Liew, F.Y. (1981). J. Exp. Med.
        153-568.
5. Ly, I.A. and Mishell, R.I. (1974). J. Immunol. Meth. 5:239-247.

CELLULAR MECHANISMS OF ANTI-MYCOBACTERIAL IMMUNITY

Frank M. Collins

Trudeau Institute, Inc.
Saranac Lake, NY

INTRODUCTION

Tuberculosis is an ancient human disease (1), which continues
to be an important public health problem, despite the development
of effective vaccines and chemotherapeutic agents. Widespread use
of chemotherapy has dramatically reduced worldwide mortality rates
for this disease, but has not yet been able to eliminate it (2).
Approximately 28,000 new cases of pulmonary tuberculosis were
reported in the United States last year, with nearly 3,000 deaths
directly attributable to this disease (3,4). This continued high
rate of incidence some 30 years after the development of potentially
effective control measures is both surprising and disturbing.
However, it is only when the problem of controlling this disease
in Asia, Africa, and Central America is examined, where over-
crowding, malnutrition and inadequate health delivery services
contribute still further to its continued persistence, that the
true dimensions of this world public health problem begin to
emerge (5).

Mortality due to pulmonary tuberculosis has certainly decreased
dramatically in this country over the past 30 years or so (6). As
a result, this disease no longer carries the frightening health
implications of a half century ago. However, this relief has been
tempered lately by an increasing concern over the emergence of
drug-resistant strains of M. tuberculosis, especially in alcohol
and drug-abuse patients, as well as concern for the increasing
incidence of life-threatening pulmonary disease caused by the so-
called "atypical" or nontuberculous mycobacteria (7). The latter
organisms are resistant to most first-line antituberculous drugs
(8) and treatment usually involves both surgery and aggressive

157

chemotherapy with more than one hepatotoxic, second line drug (9).
Many of these nontuberculous mycobacteria were originally thought
to be harmless members of the normal soil and water microflora,
but it is now clear that they can cause life-threatening infections
under both natural and iatrogenically-induced circumstances (7).
As a result, many of these organisms are now considered to be
"opportunistic" human pathogens (10). Many such infections are
characterized by a persisting type of immunosuppression affecting
the host's ability to respond to a variety of antigenic stimuli
(11). The resulting suppressor cells may be responsible for a
permanent state of immune tolerance to the inducing microbial
antigens and discussion of the role of these cells in the development of a number of chronic systemic human and experimental
diseases will make up a substantial aspect of the present review.

CHARACTERISTICS OF THE ANTITUBERCULOSIS RESPONSE

     Tuberculosis is usually spread by direct person-to-person
contact involving inhalation of infectious droplets from an index
case (6). Although mortality rates for pulmonary tuberculosis
have steadily declined, the corresponding morbidity figures have,
for some reason, failed to keep pace, and may have even begun to
climb again (4). Virulent M. tuberculosis is highly infectious for
most normal individuals who undergo tuberculin conversion a few
weeks after exposure (12). However, only a small portion (2%) of
these converters go on to develop clinical disease (13). Apparently,
most people develop sufficient antituberculous immunity during the
primary infection to limit it to subclinical proportions. Prior
vaccination with an attenuated strain (BCG) statistically increases
the chances of such a response (14) by presensitizing the host's
cellular defenses so that they will react to the virulent challenge
infection with an accelerated anamnestic response (15). The protective value of living attenuated vaccines of this sort has been
established by means of field trials carried out in many parts of
the world (16). However, not all of these trials have provided
clear-cut evidence of enhanced protection (17,18,19). The reason(s)
for this lack of protection remains controversial (16,20) but it is
clear that a number of apparently unrelated environmental, nutritional and immunological factors can affect the level of protection
achieved in different vaccine trials (21). In particular, the
existence of inapparent infections caused by a number of nontuberculous mycobacteria can raise the background level of resistance
to an M. tuberculosis challenge, which BCG vaccination will be
unable to further enhance (16,20). In addition to such cross-
reactive interference, extensive differences can be shown to exist
in the immunogenicity of various commercially available BCG preparations (22,23,24). These differences have arisen during extended
periods of cultivation by different commercial manufacturers. This
effect has been amply confirmed experimentally (25,26,27,28). Even

when a single BCG vaccine preparation is used, considerable variation in the resulting immune response may still occur depending upon the test protocol and the host strain (29,30). In any human population, a few BCG vaccinated individuals consistently fail to develop detectable levels of tuberculin hypersensitivity, even after several inoculations (31). The reason for this lack of responsiveness is not immediately clear since such individuals will usually respond normally to other immunogens. The immunological significance of nonresponsiveness seems clear cut, however, since there is good epidemiological evidence that such tuberculin negative individuals are more susceptible to subsequent attacks of tuberculosis compared to the normal responders. Apparently this immunological defect is responsible for an inability on the part of the normal host defenses to respond to the triggering antigens released by the actively growing mycobacteria within the inoculation site (32). A similar type of unresponsiveness can be seen in lepromatous leprosy patients and it has been suggested that a genetically-mediated defect may be responsible for their inability to recognize the M. leprae antigens in vivo (33). Genetically determined susceptibility or resistance to mycobacterial infection was first demonstrated 30 years ago by Lurie using inbred rabbits (34). Further studies of this nature have also been carried out using inbred rats and mice of differing susceptibility to this microbial parasite (35,36). Such resistance appears to be controlled by a single gene not associated with the H-2 complex (37,38).

The level of immunity achieved with any given vaccine will also vary extensively depending upon the size and route of vaccination and challenge, the nutritional status of the host, the use of concomitant immunosuppressive chemotherapy and the existence of other predisposing diseases such as emphysema and silicosis (6,39, 40). In the past, most experimental vaccine studies have concentrated on the survival or death of BCG-vaccinated mice or guinea pigs challenged after some arbitrary time interval with a lethal dose of virulent M. tuberculosis (41). The precise experimental details usually reflect the individual preferences of the investigator and have varied extensively from one laboratory to another (42). Many of the protocols used can be criticized on the grounds that they are not biologically relevant to the human disease (41), and several alternative challenge procedures have been recommended (29). In this review, most of the protection studies were carried out using intravenously challenged, specific pathogen-free B6D2 F_1 hybrid mice. These animals were immunized with approximately 10^6 CFU of BCG Pasteur some 1 to 3 months prior to their sublethal challenge. The mycobacteria were grown in modified Sauton's$^{®}$ liquid medium (MSTA) containing 0.05% Tween ^{R} 80 and 0.5% albumin, incubated in 1 liter roller bottles at 37°C for 10-14 days (43). Viable yields of mycobacteria as high as 10^9 CFU per ml can be routinely obtained in this way, with excellent viable recoveries after storage at -70°C for as long as 24 months (44). Suspensions

of known viability are inoculated directly into experimental animals
after the thawed suspension has been homogenized briefly to break
up any clumps of bacilli (45).

Protection may be assessed as the relative 30 or 90 day survival rate, or mean survival time, of vaccinated compared to unvaccinated controls (41). A more quantitative procedure involves
the demonstration of an anti-bacterial immune response within the
lungs and spleen of the vaccinated host (46). Protection can be
deduced from a significant reduction in the growth of the virulent
M. tuberculosis challenge population (47) within the lungs and
spleens of the vaccinated mice (Fig. 1). Using this test method
the challenge period can be limited to 2 to 3 weeks since, after
that time, the infection will have produced its own de novo immune
response in the unvaccinated controls (42,48).

A second parameter of the immune response is the development
of tuberculin hypersensitivity by the BCG infected host, some 14–21
days into the vaccinating infection (Fig. 1). Some caution must be
exercised in the interpretation of this response, however, since
some animals may not express significant tuberculin hypersensitivity
although they may possess a high level of protective cellular
immunity (48). This type of anomaly is well illustrated in the case
of the mice vaccinated intravenously with 10^8 CFU of BCG Pasteur
(Fig. 1). None of the mice developed significant levels of tuberculin hypersensitivity although they did restrict the growth of the
virulent M. tuberculosis. In fact, the immune response seen in the
anergic host is quantitatively little different from that occuring
in the tuberculin hypersensitive mice receiving only 10^6 CFU of BCG.
When the size of the vaccinating inoculum was reduced further to
10^4 CFU, the mice again failed to develop significant levels of
tuberculin hypersensitivity although they were able to mount an
accelerated immune response against the subsequent challenge (Fig.
1). Thus, it can be seen that the level of peripheral tuberculin
hypersensitivity does not provide a reliable index of the level of
cell-mediated immunity expressed within the lungs, liver, and
spleen (42). In this regard, it should be noted that the only
relevant parameter of the host response to a given vaccinating
regimen will be its ability to prevent the development of clinically
significant disease in naturally infected individuals many months
or even years after immunization (15,21). In order to meet this
kind of assay need, Wiegeshaus and Smith (29,41) recommended that
BCG vaccines be assessed in terms of the ability of the immunized
host to limit the in vivo growth of a minimal aerogenic challenge
inoculum of virulent M. tuberculosis (49). In addition to limiting
the primary growth of the inoculum within the lung itself, the
cell-mediated immune response should be able to prevent the hema-
togenous spread of the infection to other uninvolved tissues and
organs of the body (50,51). This latter effect may be the most
important parameter of the immune response, so that this proposed

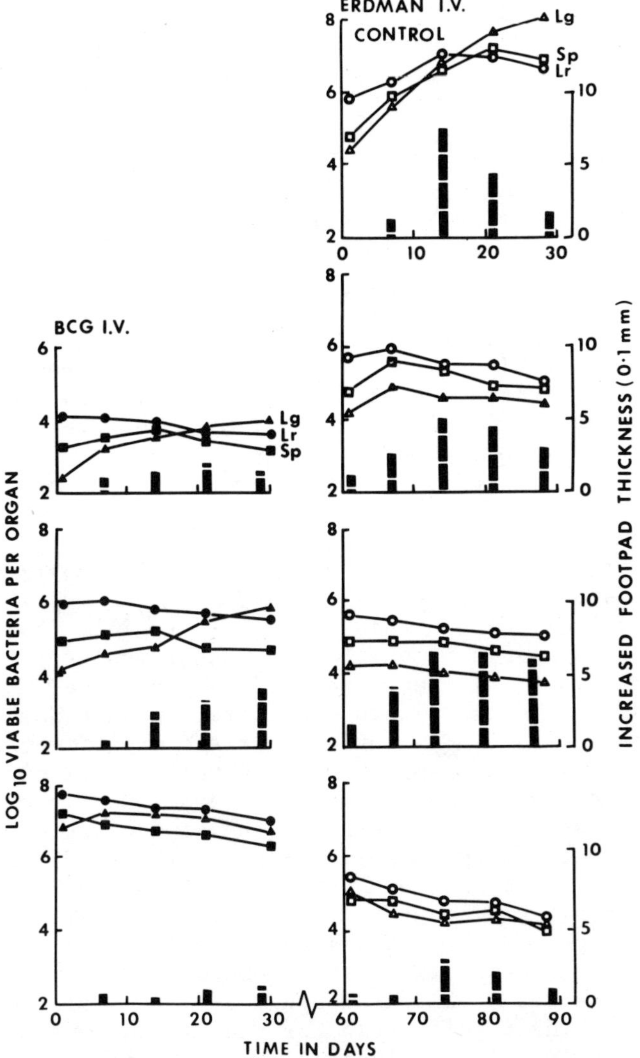

Figure 1. Growth curves for 10^4, 10^6 or 10^8 viable BCG Pasteur in B6D2 mice following intravenous infection. Lg = lungs; Sp = spleen; Lr = liver. The three groups of vaccinated mice and a group of age-matched controls were challenged after 60 days with 8×10^5 viable units of <u>M</u>. <u>tuberculosis</u> Erdman. The histograms represent the 24 hr foot swelling response following the injection of 2.5 μg PPD (increases of 2 U or more significant at the 1% level).

type of experimental protocol probably represents the most relevant type of assay procedure since the ensuing host-parasite interactions are likely to closely resemble those seen in the naturally infected individual (30).

The third important parameter of any vaccination response will
be the induction of a residual memory T-cell population able to
induce an accelerated recall of the earlier cell-mediated response
whenever the host is reexposed to the same infectious agent (51).
This anamnestic response should be able to limit the growth of the
virulent tubercle bacillus _in vivo_ before it can establish a
clinically significant infection (42). Most virulent tubercle
bacilli are extraordinarily resistant to the anti-microbial effects
of even specifically activated macrophages (52) so that the most
that the immune defenses can hope to achieve against a highly
virulent challenge inoculum will be the limitation of further
growth by the organism within the lung and its draining lymph nodes.
Even when an initially effective antimicrobial response is seen,
the rate of inactivation usually slows with time and the host
enters the so-called 'carrier state' (53). This results in a
latent tuberculous infection which may persist for many years and
which can be reactivated at any time by the use of appropriate
immunosuppressive chemotherapy (54). Less virulent strains of
mycobacteria may be totally eliminated from the tissues of normal
animals over a relatively short period of time (25). Challenge
experiments using $\underline{M}$. $\underline{tuberculosis}$ R_1Rv, or $H_{37}Ra$ or BCG INH^R will
be eliminated from the controls relatively rapidly (55), which
indicates higher levels of antituberculous immunity than would be
the case if a highly virulent challenge organism was used. This
difference should be taken into consideration when assessing such
protection data (25,47,56). By appropriate experimental manipulation, it is possible to demonstrate the paradoxical persistence of
a primary infecting population within a host readily capable of
eliminating a secondary challenge by the same organism (42). Such
carrier effects can be seen in other infections (53) as well as
with parasitic agents, and even tumors (51). Clearly we need to
know a great deal more about the nature of the cellular interactions which occur during the establishment of a persistent infection _in vivo_. It seems likely that some microbial parasites can
manipulate the normal cellular responses of the host to their own
advantage by making it difficult for the immunocompetent host to
recognize the triggering antigens of the invading parasite and thus
block the development of an effective cell-mediated immune response
(11,32,33). We will consider some of these host-parasite interactions again in a later section.

T-LYMPHOCYTES AND THE EXPRESSION OF ANTITUBERCULOUS IMMUNITY

Acquired antituberculous resistance is mediated by a population
of immunocompetent T-cells present in the thymus-dependent areas of
the spleen and lymph node (51). Immunity to this disease is
mediated by sensitized T-cells, since transfer studies make it clear
that the specific anti-mycobacterial antibodies have little or no
protective value for the host (57). Hyperimmune serum does not

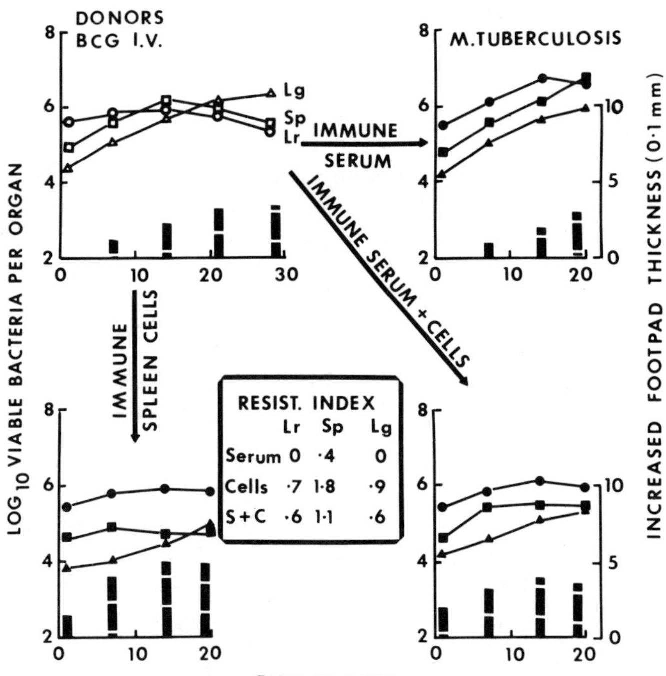

Figure 2.   Transfer of tuberculin hypersensitivity and antituberculous immunity by splenic T-cells but not by immune serum
obtained from day 28 BCG vaccinated B6D2 mice. Lg =
lungs; Sp = spleen; Lr = liver. Normal recipients were
infused intravenously with 5x10^7 viable T-lymphocytes
and/or 0.5 ml hyperimmune serum 24 hours prior to
intravenous challenge with 5x10^5 M. tuberculosis Erdman.
The histograms represent the 24 hr foot swelling
reactions to 2.5 µg PPD. The Resistance Index represents
the difference (log $_{10}$) between the number of viable
Erdman in the test and control organ. An R.I. difference
of 1.0 or more was significant at the 1% level.

even enhance the response to splenic T-cells harvested from BCG
immunized mice (Fig. 2), at least in terms of the growth behavior
of the challenge M. tuberculosis population within the lungs and
spleens of the recipients. Adoptive transfer of immunity could be
abrogated by prior treatment of the immune spleen cells with anti-
Thy-1.2 antiserum and complement, but not by anti-Ig antisera (58).

Depletion of the immunocompetent T-cell population by neonatal
thymectomy or by adult thymectomy, followed by lethal whole-body
irradiation and bone marrow reconstitution (THXB) largely ablates
this ability. Even when the T-cell depleted host was infected with
normally immunogenic doses of live BCG, most of the animals developed

a progressive mycobacteriosis with eventual death (59). This was
not due to the effects of unrestricted mycobacterial growth in vivo.
Viable counts carried out on lung and spleen homogenates prepared
from the BCG-infected THXB mice showed a plateau in the viable
counts once a maximum of about 10^7 CFU per organ was observed
(Fig. 3). The injection of as many as 10^{10} heat-killed M. tuberculosis failed to reveal any evidence of tissue toxicity in normal
mice (Collins and Auclair, unpublished data). Death of the BCG-
infected, T-cell depleted mice appeared to be due to a progressive
consolidation and destruction of lung tissue by the continuing
mononuclear infiltrate in the T-cell deficient lung (60). In the
absence of immunocompetent T-cells, the ongoing mononuclear response
occurs with no evidence of evolving cell-mediated immune response.
At the same time, the blood, liver, spleen and bone marrow samples
contained increasing numbers of tubercle bacilli while the
corresponding lesions were rapidly resolving within the immunocompetent controls. Limitation of the secondary spread by the
infection to other lymphoreticular organs within the immunocompetent
host (50) constitutes one of the most important characteristics of
the normal cell-mediated immune response to this infectious agent
(51).

     Histologically, the granulomas which developed within the T-
cell depleted lung were characterized by the presence of large
numbers of foamy macrophages, many of which contained acid-fast
bacilli (60). Furthermore, because many of the cells within these
lesions were actively dividing, the T-cell depleted lung showed an
increased uptake of tritiated thymidine (^{3}H-TdR) throughout the
infection (Fig. 3). The corresponding incorporation curve for the
control lungs peaked about day 28, followed by a sharp decline to
preinfection levels. The late peak in ^{3}H-TdR incorporation by the
T-cell depleted lung could be reversed by means of thymus grafts
or by the repeated injection of a thymic hormone such as 'thymosin'
(Fig. 4). The T-cell depleted (THXB) mice were injected daily with
20 mg of calf thymosin per kg of body weight for up to 14 days.
As shown in Fig. 4, a substantial restoration in anti-tuberculous
immunity was seen in these mice. There was also a restoration
in their ability to express significant levels of tuberculin
hypersensitivity (61). Cell transfer studies indicated that the
thymic hormone injections brought about the maturation of a
population of immunocompetent T-cells from bone marrow precursor
cells which were responsible for the restoration of both DTH and
acquired antituberculous immunity (Fig. 5). The time course of
the thymosin treatment protocol was compatible with the estimated
maturation time for T-cell precursors present in the spleen and
bone marrow (62).

     Restoration of antituberculous immunity in THXB mice may also
be achieved using spleen cells harvested from thymosin-treated
THXB mice (Fig. 5). This restorative effect (measured in terms of

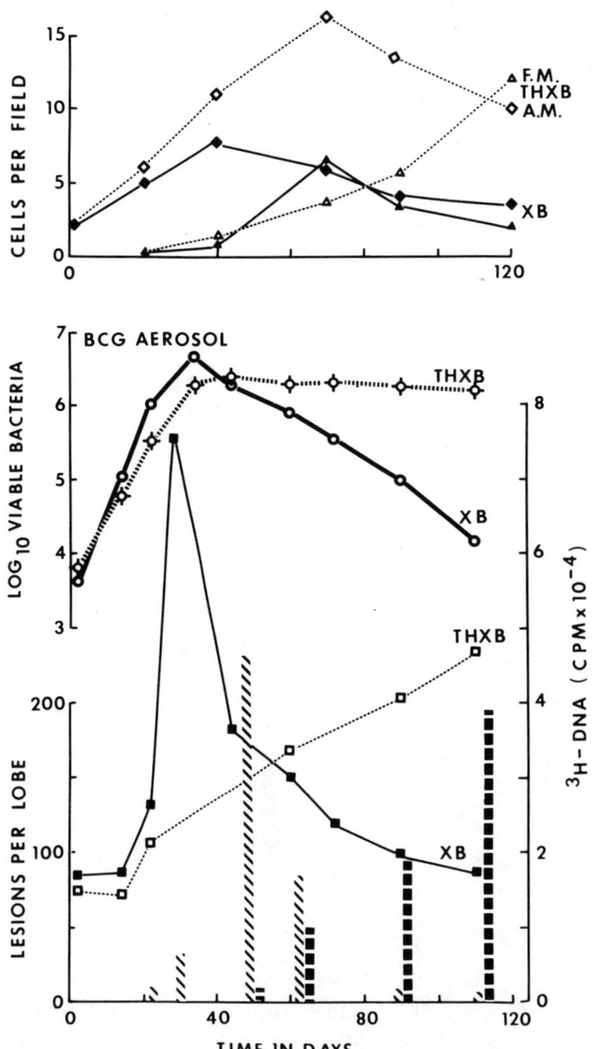

Figure 3. Growth of BCG Pasteur in adult thymectomized, lethally
irradiated, bone marrow reconstituted (THXB) or sham
thymectomized (XB) B6D2 mice following aerogenic
infection with 6,000 viable units. Tritiated thymidine
uptake by lung cells in the THXB (□) and the XB (■)
mice and the number of lung tubercles in the THXB (solid)
or XB (hatched bars) mice respectively. The average
number of alveolar macrophages per high power field
is represented by the diamonds (top) while the number
of foamy macrophages are shown as triangles.

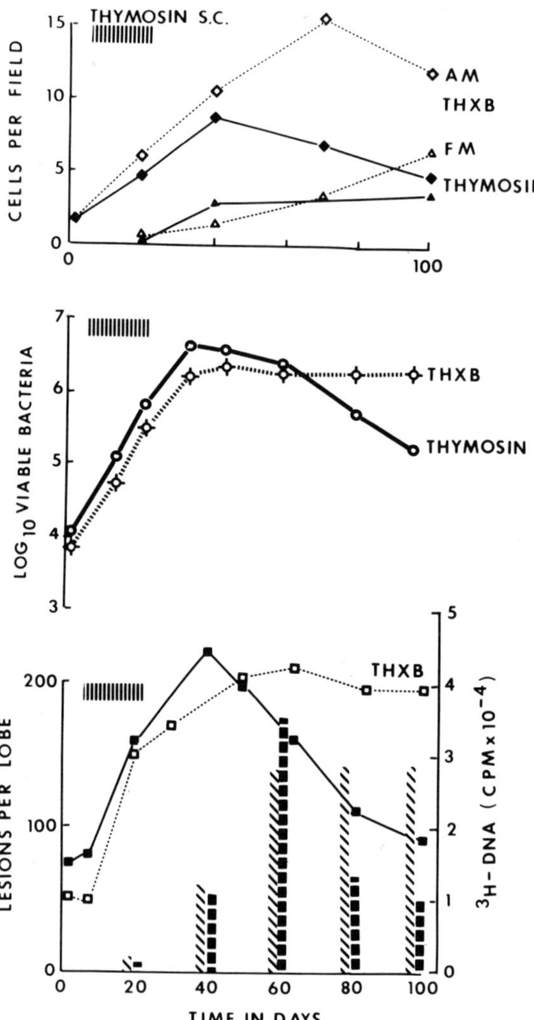

Figure 4.   The effect of 15 daily injections of 20 mg thymosin per
            kg body weight given to THXB and normal B6D2 mice
            following aerogenic infection with 5000 viable BCG
            Pasteur.   See legend to Fig. 3 for further details.

logs of protection) was similar to that achieved with immunocompetent T-cells harvested from normal donors. Protection was
mediated by a population of anti-Thy-1.2 antiserum sensitive,
non-adherent lymphocytes which could express their activity in
recipient mice in the absence of thymosin (63). This indicates
that the response seen in the THXB mice was not due to contaminants
(such as endotoxin) in the calf thymosin which might have nonspecifically activated the treated host macrophages.

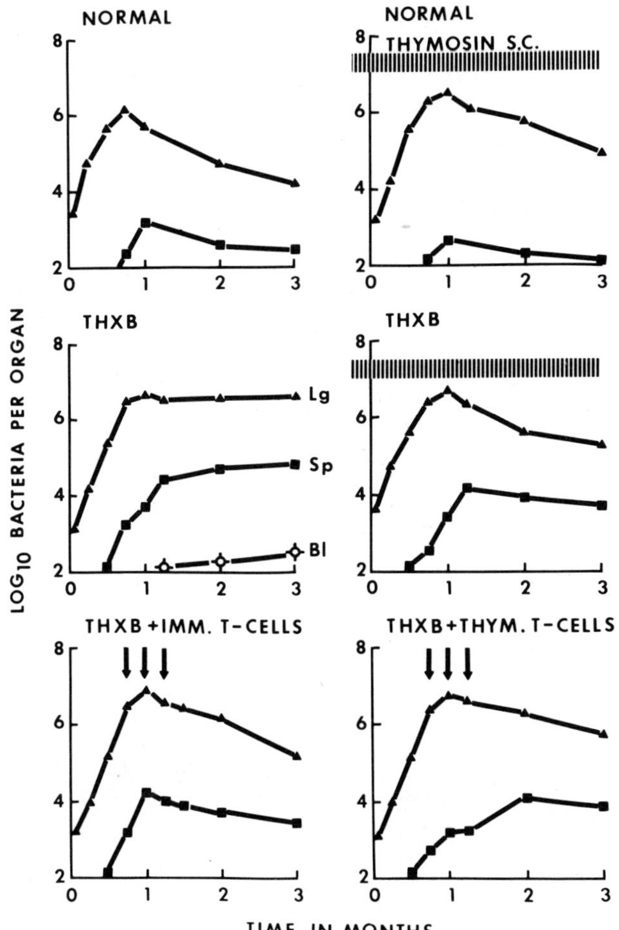

Figure 5. Growth curves of BCG Pasteur in the lungs (triangles) and spleens (squares) of aerogenically infected THXB mice adoptively immunized with 3 doses (arrows) of 5×10^6 splenic T-cells harvested from immune control mice (bottom left) or thymosin-treated THXB (bottom right) donors. Untreated THXB mice (middle left) or thymosin-treated (hatched bars) THXB mice (middle right) should be compared with normal control mice with or without thymosin treatment (top).

PERSISTENT MYCOBACTERIAL INFECTIONS, ANERGY AND IMMUNOSUPPRESSION

Live M. tuberculosis induces an extensive proliferative response within the thymus-dependent areas of the draining lymph

node(s) and the spleen, some 10-14 days after infection (46,52).
The sensitized T-cells migrate into the developing lung granulomas
where they are exposed to specific sensitin(s) released by the
infected tissue macrophages (64). The resulting lymphokines activate the blood-derived monocytes entering the lesion and these
"angry" or activated macrophages are able to kill the mycobacteria
(62,65). The corresponding cellular response which develops within
the draining lymph node may also help to limit the further spread
of the infection from the lung (66). However, the effectiveness
of this secondary line of defense will largely depend upon the
size and route of the challenge infection (66).

Most individuals infected with M. tuberculosis will develop
tuberculin hypersensitivity which may then be followed by a persistent anergic state lasting for the remainder of the lifetime
of the untreated host (13). Such a persistent loss of tuberculin
skin hypersensitivity usually has poor prognostic implications in
pulmonary tuberculosis (67). However, the decline also represents
a natural progression in the cellular responsiveness by the infected
host and occurs even in BCG-vaccinated individuals (64). Normal
B6D2 mice infected with 10^7 viable BCG will be highly responsive
to tuberculin for only a few weeks (42). The subsequent suppressive
effect can be shown to be mediated by a population of splenic T-
cells which are sensitive to treatment with anti-Thy-1.2 antiserum
and complement (Table 1). Mice infected with similar numbers of
such nontuberculous mycobacteria as M. simiae or M. avium remain
unresponsive to the specific footpad test antigen virtually in-
definitely (Fig. 6), and the heavily infected mice seem unable to
express an effective cell-mediated immunity against these organisms
(45). The mice develop little splenomegaly or obvious lung
consolidation, despite the continued presence of large numbers of
acid-fast bacilli within the tissues (Collins, unpublished data).
On the other hand, even closely related species of mycobacteria
(such as M. intracellulare in Fig. 6) can be eliminated from the
tissues quite rapidly, apparently by a non-immunologically mediated
killing mechanism (68). As yet there is no known antigenic,
cultural or structural differences which can adequately explain
this striking difference in in vivo behavior (45).

When splenic T-cells were harvested from BCG infected mice,
an enhanced blastogenic response (day 7-14) was observed (Fig. 7)
following their in vitro exposure to such nonspecific T-cell mitogens as PHA and Con A, as well as the specific mitogen PPD (69).
In fact, all 3 of the heavily infected spleen cell preparations
showed an early peak followed by a sharp drop in thymidine incorporation (immunosuppression) 1-2 months into the inducing infection.
Specific suppression of the M. simiae and M. avium infected spleen
cells continued for the remainder of the study suggesting that the
unresponsive state represents some form of infectious tolerance or
immune paralysis (70). This correlates with the persistent

Table 1. Effect of Anti-Thy 1.2 Antiserum Treatment on Spleen
          Cells Obtained from Mice Infected with Mycobacteria

Cell Source	Response (CPM±SEM)	
	PHA	PPD
Before Anti-Thy 1 Antiserum		
A. Normal	39,440±1,960	480±80
B. 10^6 BCG D-14	268,940±4,050	48,870±2,060
C. 10^8 BCG D-60	6,500±780	670±100
After Anti-Thy 1 Treatment		
D. Normal	2,500±480	220±40
E. 10^6 BCG D-14	8,090±1,990	2,700±340
F. 10^8 BCG D-60	900±85	380±70
Cell Mixing Experiment		
A+C	6,690±850	550±110
A+F	35,490±1,200	420±50
B+C	62,440±2,990	3,950±870
B+F	149,090±2,200	23,480±1,900
Half #A Cells	22,510±920	320±50
Half #B Cells	150,050±3,800	22,660±3,100

infections seen in these mice (Fig. 7). On the other hand, the
BCG infected mice showed a gradual restoration of their blastogenic responses to PPD, which coincided with the decline in the
viable mycobacterial population within the spleen to below 10^5 CFU
per spleen (71). Cell-mixing experiments using splenic T-cells
harvested from the M. simiae or M. avium anergic mice confirmed the
persistence of the specific suppressor T-cell population within
the spleen (Table 2).

Induction of suppressor T-cells within the spleen appeared to
require the presence of viable mycobacteria (72), since injection
of equivalent numbers of heat-killed mycobacteria into normal B6D2
mice failed to induce either delayed hypersensitivity or suppressor
T-cell production (46,48). However, once suppressor cells had been
induced in vivo, they seemed to persist for some time, possibly
because of the presence of excess antigen or antigen-antibody
complexes within the spleen (73).

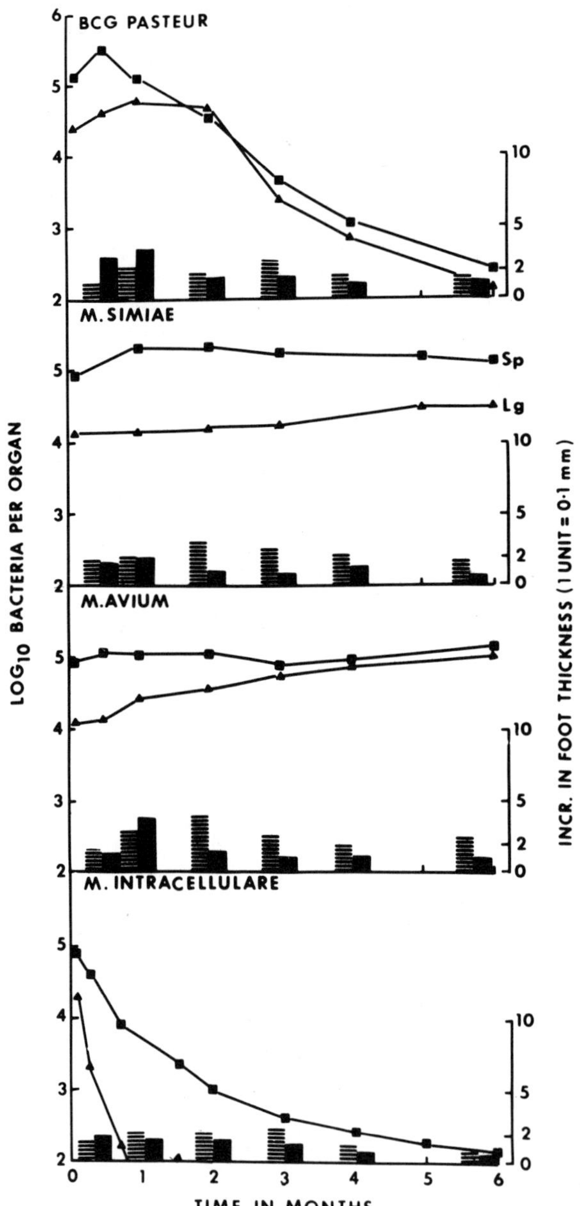

Figure 6. Growth curves for BCG Pasteur (top), M. simiae 1226, M. avium 702 and M. intracellulare 1403 (bottom) following intravenous injections of about 10^6 CFU. Lung = Lg; Spleen = Sp; Liver = Lr. The hatched bars represent the 3 hr foot swelling response and the solid bars represent the 24 hr swelling responses to 5 μg CPA injected into a hind footpad. Increases of 2 units or more significant at the 1% level.

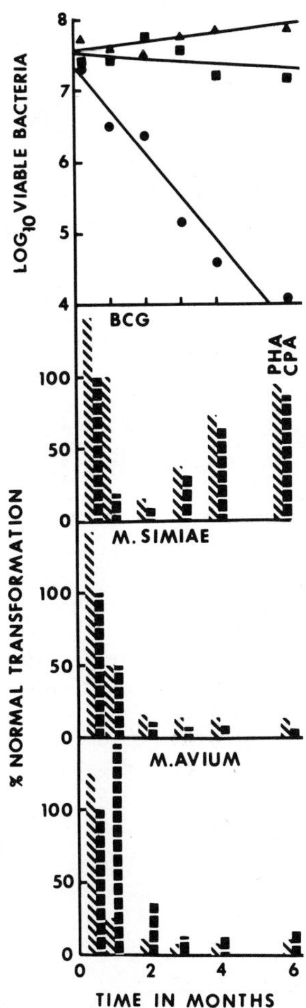

Figure 7.   Regression lines for BCG TMC #1011 (●), M. simiae
TMC #1226 (■) and M. avium TMC #702 (▲) in intravenously infected B6D2 mice. The hatched bars represent the percent of normal transformation by T-cell
enriched spleen cells harvested from the infected mice
at increasing time intervals and exposed to 0.1 µg PHA.
The solid bars represent transformation following
exposure to 0.1 µg of the specific cytoplasmic protein
antigen (CPA).

Tuberculin anergy may be accompanied by reduced reactivity to
other unrelated skin-test antigens in tuberculous patients (11,74).
However, the nonspecific immunosuppression tends to disappear
following successful chemotherapy (13). One exception to this
rule appears to be the persistent lepromin anergy seen in dapsone

Table 2.   Cell Mixing Experiments with Splenic T-Cells Harvested
           from Mice Infected with Increasing Numbers of Mycobacteria

Cell Source	Response (CPM±SEM)	
	PHA	CPA
A. 10^6 BCG D-14	149,060±4,050	30,470±1,500
B. 10^8 BCG D-60	3,470±980	350±60
C. 10^6 _M. Simiae_ D-14	149,860±2,140	5,690±320
D. 10^8 _M. Simiae_ D-60	3,110±180	300±20
E. 10^6 _M. Avium_ D-14	48,940±2,900	3,964±500
F. 10^8 _M. Avium_ D-60	1,950±100	900±14
Cell Mixing		
A+B	62,440±3,000	3,990±70
C+D	3,500±200	490±70
E+F	6,500±510	460±25
Half #A Cells	74,000±1,020	18,100±2,940
Half #C Cells	69,200±2,100	4,100±1,900
Half #E Cells	21,060±2,000	2,000±110

treated lepromatous leprosy patients (33,74). Normal individuals
develop a local skin reaction 4 to 6 weeks after injection of heat-
killed _M. leprae_ (lepromin) into the forearm. Lepromatous leprosy
patients suffer a loss of such reactivity which persists even
after all acid-fast bacilli have apparently been eliminated from
the tissues (76). This type of persistent unresponsiveness may
be the result of an antigenic "overload" phenomenon (77), since it
is known that untreated lepromatous leprosy patient may carry as
many as 10^{10} acid-fast bacilli per gram of tissue (32). Chemotherapy may reduce the number of acid-fast bacilli seen in tissue
biopsies substantially (78). However, tissue sections from such
patients will still contain large numbers of silver staining non-
acid fast bacillary carcasses (79), suggesting that substantial
amounts of antigenic material may persist within the spleens and
draining lymph nodes of heavily infected mice for many months, and
this antigenic load may be responsible for the ongoing state of
high-dose immune tolerance seen in the _M. simiae_ and _M. avium_
infected mice (80).

EFFECT OF CHEMOTHERAPY ON THE PERSISTENCE OF SUPPRESSOR CELLS
IN VIVO

M. bovis (BCG) has been widely used as an immunopotentiator
of various antimicrobial and anti-tumor responses in man and experimental animals (81,82,84,85). Attempts to potentiate these
responses still further has led to the use of repeated inocula of
an ever-increasing size (83). However, experimental studies carried
out in mice, guinea pigs and monkeys indicate that, far from prolonging or increasing the desired cell-mediated immune response,
such massive doses of live BCG may lead to an increasingly severe
state of immunosuppression (84). This suppression seems to be
mediated by both T-cells (71,73,85) and by adherent cells (86).
Such suppression is probably expressed through a series of feedback control loops affecting the expressor T-cells present within
the infected tissue (87). Under some experimental conditions, the
suppressor cells may disappear relatively quickly once the systemic
bacterial load drops below a threshold level, either as a result of
the naturally expressed immunity (70) or by appropriate chemotherapy (88). When drug treatment (isoniazid and rifampin) is
initiated immediately after BCG vaccination (Fig. 8, top) none
of the mice develop either tuberculin hypersensitivity or
suppressor T-cell activity (Table 3). This lack of cellular
responsiveness correlates with the very rapid reduction in the
number of viable BCG present in the spleens of the treated mice
(Fig. 8). When the drug treatment was delayed for 30 days, the
BCG-infected mice had already developed substantial levels of
suppressor T-cell activity (Fig. 9) and the sharp drop in the
number of viable BCG (Fig. 8) was accompanied by a slow recovery
in blastogenic responsiveness to PPD, beginning some 10 days after
the commencement of chemotherapy (day 30). This suggests that the
half-life of the suppressor T-cell population in these mice was
relatively short (2 to 3 weeks at the most), a conclusion which
seems to be consistent with the loss of suppression observed in
the cell-mixing experiments shown in Table 3.

CONCLUSIONS

Tuberculosis remains a severe, life threatening human disease
with important public health implications in many parts of the
world (5,6). The disease persists despite the ready availability
of effective prophylactic and chemotherapeutic measures which
could be used against it. The increasing use of immunosuppressive
therapy for both transplant and cancer patients makes it likely
that many nontuberculous mycobacterial infections will also
continue to be important human pathogens for many years to come.

Acquired anti-tuberculosis immunity is mediated by a population of immunocompetent T-lymphocytes which enter the lesion and

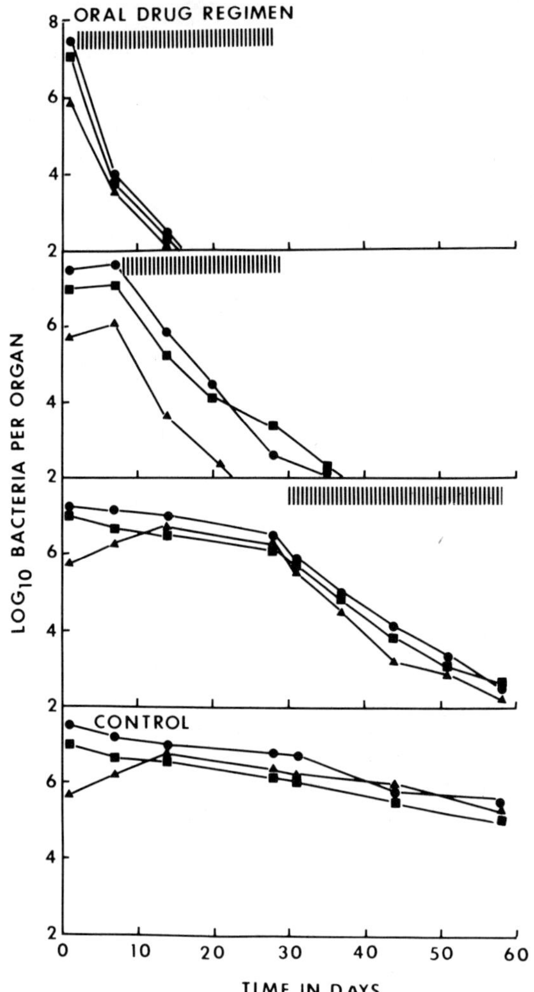

Figure 8.   Growth curves for BCG Pasteur in intravenously infected
            B6D2 mice given drinking water containing 100 mg of INH+
            rifampin per liter, beginning on day 1 (top), 7 days
            or 30 days after infection. Untreated controls are
            shown at the bottom. Liver (●), spleen (■) and
            lungs (▲).

activate the blood-derived monocytes accumulating within the
developing tissue granuloma. Little or no humorally-mediated
immune protection can be demonstrated for this disease. Despite
dramatic advances in our understanding of the immunology of
tuberculosis, we still need to know a great deal more about those
host-parasite interactions which occur during the establishment
of the parasite within the tissues. These early cellular responses

Table 3. Cell-Mixing Experiments to Demonstrate Suppressor Cell
          Activity in Drug-Treated BCG Infected Spleens

Cell Mixtures	No Drug	Drug	Response (CPM±SEM) to PPD Exp. 1.	Drug	Exp. 2.
Indicator D-14 10^6 BCG	45,500±1,500	–	–	–	–
Indicator Cells Plus D-1 10^8 BCG	24,150±1,520	+	24,150±1,520	–	24,150±1,520
D-7 10^8 BCG	15,420±1,850	+	25,500±2,100	–	28,760±2,100
D-14 10^8 BCG	7,860±120	+	18,750±3,200	–	7,375±1,090
D-28 10^8 BCG	2,880±110	+	23,600±1,500	+	3,440±220
D-42 10^8 BCG	2,140±200	–	19,450±2,000	+	4,320±560
D-56 10^8 BCG	3,100±620	–	18,400±1,100	+	7,900±1,000
Half # Indicator Cells	27,340-±4,560		–		–

appear to determine the type of disease which ultimately develops
(89). Recent experimental studies using chronically infected
animals suggest that suppressor cells may play a pivotal role in
the limitation to the various cellular responses to the mycobacterial antigens, both _in vivo_ and _in vitro_ (70). It is tempting
to speculate that the same suppressor cells may be responsible
for the inability of the host defenses to recognize the continuing
presence of the sensitizing antigens of the microbial parasite in
the tissue and thus contribute to the persistence of the infection
within the apparently normal, immunocompetent host. Mice infected
with non-tuberculous mycobacteria provide a useful experimental
model with which to study the immunological interactions responsible
for the establishment of these facultative intracellular parasites
within the tissues.

ACKNOWLEDGMENTS

Technical assistance was provided by William Woodruff, Vincent
Montalbine, Linda Auclair and Cindy Tremblay. This work was
supported by Grant No. AI-14065 administered by the U.S.-Japan
Cooperative Medical Sciences Program; by Grant No. HL-19774 from
the Heart, Lung and Blood Institute and by the Biomedical Research

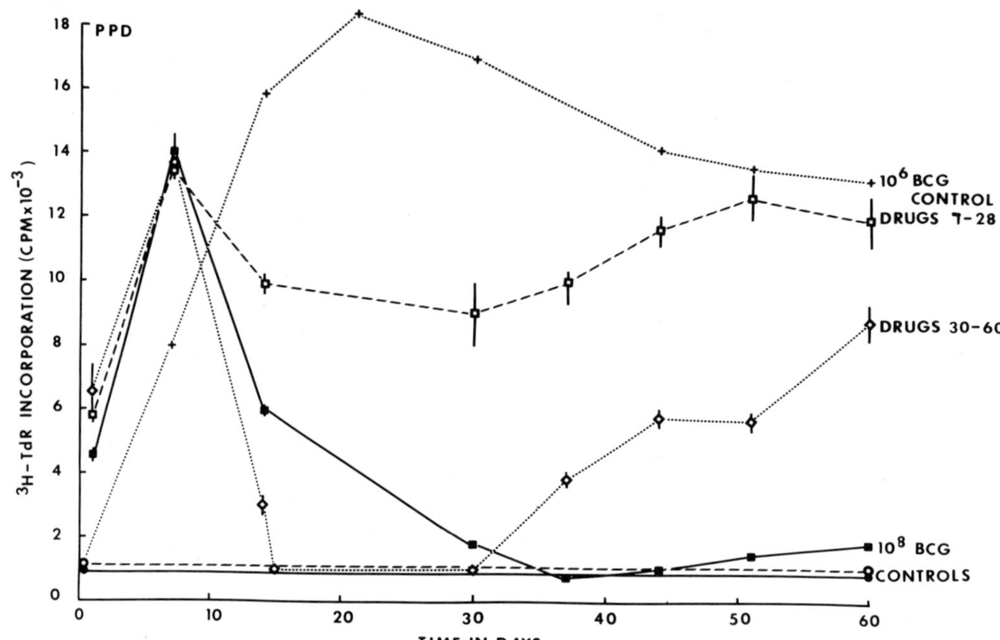

Figure 9. Tritiated thymidine incorporation by T-cell enriched
spleen cells exposed in vitro to 0.1 µg of PPD. Spleen
cells from mice treated with the drug regimen over the
7-28 day period (□) or from days 30 to 60 (◇) were
compared with drug-free control mice which were infected
with 10^6 (+) or 10^8 (■) viable units of BCG Pasteur.
Incorporation levels for normal control (○) mice and
uninfected mice receiving the drug treatment (●) were
included.

Grant RR-05705 from the General Research Support Branch, Division
of Research Resources, NIH.

REFERENCES

1.   Zimmerman, M. R. (1979). Pulmonary and osseous tuberculosis
        in an Egyptian mummy. Bull. N.Y. Acad. Med. 55:604-698.
2.   Rouillon, A., Perdizet, S. and Parrott, R. (1976). Trans-
        mission of tubercle bacilli: the effect of chemotherapy.
        Tubercle. 57:275-299.
3.   Anon. (1979). "Annual Summary of Reported Morbidity and
        Mortality in the U.S." Morb. Mort. Weekly Rep. 27:67-69.
4.   Anon. (1981). Tuberculosis-United States. 1980. Morb. Mort.
        Weekly Rep. 30:55-56.

5. Holm, J. (1969). Tuberculosis in the world today. Bull.
       Intern. Union Tuberc. 43:26-33.
6. Lowell, A.M., Edwards, L.B. and Palmer, C.E. (1969). Tubercu-
       losis. Harvard Press, Mass.
7. Wolinsky, E. (1979). Nontuberculous mycobacteria and associated
       diseases. Amer. Rev. Resp. Dis. 119:107-159.
8. Burjanova, B. and Urbancik, R. (1971). Experimental chemo-
       therapy of mycobacteriosis provoked by atypical myco-
       bacteria. Adv. Tuberc. Res. 17:154-188.
9. Corpe, R.F. (1964). Clinical aspects, medical and surgical,
       in the management of Battey-type pulmonary disease. Dis.
       Chest. 45:380-382.
10. Marks, J. and Jenkins, P.A. (1971). The opportunistic myco-
       bacteria: a 20 year retrospect. Postgrad. Med. 47:705-711.
11. Kantor, F.S. (1975). Infection, anergy, and cell-mediated
       immunity. New Engl. J. Med. 292:629-634.
12. Stead, W.W. (1978). Undetected tuberculosis in prison.
       J. Amer. Med. Assoc. 240:2544-2547.
13. Lester, C.F. and Arvell, R.J. (1958). The tuberculin skin
       reaction in active pulmonary tuberculosis. Amer. Rev.
       Tuberc. 78:399-402.
14. von Magnus, P., Eugbank, C.H. and Jespersen, A. (1967).
       Tuberculosis and BCG. Med. Clinics N. Amer. 51:753-764.
15. ten Dam, N.G., Toman, K., Hitze, K.L. and Guld, J. (1976).
       Present knowledge of immunization against tuberculosis.
       Bull. World Health Organ. 54:255-269.
16. Hart, D'A.P. (1967). The place of vaccination in tuberculosis
       control. Amer. Rev. Resp. Dis. 96:1-6.
17. Anon. (1980). Trial of BCG vaccines in South India for
       tuberculosis prevention: first report. Bull. Int.
       Union Tuberc. 55:14-20.
18. Comstock, G.W. and Edwards. P.Q. (1972). An American view
       of BCG vaccination illustrated by results of a controlled
       trial in Puerto Rico. Scand. J. Resp. Dis. 53:207-217.
19. Comstock, G.W. and Palmer, C.E. (1966). Long-term results of
       BCG vaccination in the Southern United States. Amer.
       Rev. Resp. Dis. 93:171-183.
20. Palmer, C.E. and Long, M.W. (1966). Effects of infection with
       atypical mycobacteria on BCG vaccination and tuberculosis.
       Amer. Rev. Resp. Dis. 94:554-568.
21. Eickhoff, T.C. (1977). The current status of BCG immunization
       against tuberculosis. Annual Rev. Med. 28:411-428.
22. Dubos, R.J. and Pierce, C.H. (1956). Differential character-
       istics _in vitro_ and _in vivo_ of several substrains of
       BCG. IV. Immunizing effectiveness. Amer. Rev. Tuberc.
       74:699-717.
23. Vallishayee, R.S., Shashidhara, A.N., Bunch-Christiansen, K.
       and Guld, J. (1974). Tuberculin sensitivity and skin
       lesions in children after vaccination with 11 different
       BCG strains. Bull. World Health Organ. 51:489-494.

24. Sher, M.A., Chaparas, S.D., Pearson, J. and Chirigos, M. (1973).
        Virulence of six strains of Mycobacterium bovis (BCG) in
        mice. Inf. Immun. 8:736-742.
25. Collins, F.M. and Miller, T.E. (1969). Growth of a drug-
        resistant strain of Mycobacterium bovis (BCG) in normal
        and immunized mice. J. Infec. Dis. 120:517-533.
26. Mackaness, G.B., Auclair, D.J. and Lagrange, P.H. (1973).
        Immunopotentiation with BCG. 1. Immune response to
        different strains and populations. J. Nat. Cancer Inst.
        1655-1667.
27. Freudenstein, H., Pranter, W. and Schweinsberg, H. (1979).
        Assessment of several BCG vaccines in different animal
        test systems. J. Biol. Stand. 7:203-212.
28. Ladefoged, A., Bunch-Christensen, K. and Guld, J. (1976).
        Tuberculin hypersensitivity in guinea pigs after vaccina-
        tion with varying doses of BCG of 12 different strains.
        Bull. World Health Organ. 53:435-533.
29. Smith, D.W. and Harding, G.E. (1977). Approaches to the
        validation of animal test systems for assay of protective
        potency of BCG vaccines. J. Biol. Standard. 5:131-318.
30. Wiegeshaus, E.H., Harding, G., McMurray, D., Grover, A.A.
        and Smith, D.W. (1971). A cooperative evaluation of
        test systems used to assay tuberculosis vaccines. Bull.
        World Health Organ. 45:543-550.
31. Anon. (1955). Experimental studies of vaccination, allergy,
        and immunity in tuberculosis. V. Relation between degree
        of tuberculin sensitivity and resistance in the vaccinated
        individual. Amer. J. Hyg. 62:185-189.
32. Bullock, W.E. (1975). Anergy and infection. Adv. Internal
        Med. 21:149-173.
33. Godal, T. (1978). Immunological aspects of leprosy-present
        status. Prog. Allergy. 25:211-242.
34. Lurie, M.G., Abramson, S. and Heppleton, A.G. (1952). On the
        response of genetically resistant and susceptible rabbits
        to the qualitative inhalation of human type tubercle
        bacilli and the nature of resistance to tuberculosis.
        J. Exp. Med. 95:119-134.
35. Gray, D.F. (1961). The relative natural resistance of rats
        and mice to experimental pulmonary tuberculosis. J. Hyg.
        59:471-477.
36. Youmans, G.P., Youmans, A.S. and Kanai, K. (1959). The
        difference in response of four strains of mice to
        immunization against tuberculous infection. Amer. Rev.
        Tuberc. 80:753-756.
37. Forget, A., Skamene, E., Gros, P.K., Miailhe, A.C. and
        Turcotte, R. (1981). Differences in response among
        inbred mouse strains to infection with small doses of
        Mycobacterium bovis BCG. Inf. Immun. 32:42-47.

38. Allen, E.M., Morre, V.L. and Stevens, J.O. (1977). Strain
         variation in BCG-induced chronic pulmonary inflammation
         in mice. 1. Basic model and possible genetic control
         by non H-2 genes. J. Immunol. 196:343-347.
39. Daniel, R.M. (1980). The immunology of tuberculosis. Clinics
         Chest Med. 1:189-201.
40. Lurie, M.B. (1964). Resistance to tuberculosis: experimental
         studies in native and acquired defense mechanisms.
         Harvard Univ. Press, Cambridge, Mass.
41. Weigeshaus, E.H. and Smith, D.W. (1968). Experimental models
         for study of immunity in tuberculosis. Ann. N.Y. Acad.
         Sci. 154:174-199.
42. Collins, F.M. (1972). Acquired resistance to mycobacterial
         infections. Adv. Tuberc. Res. 18:1-30.
43. Collins, F.M., Wayne, L.G. and Montalbine, V. (1974). The
         effect of cultural conditions on the distribution of
         Mycobacterium tuberculosis in the spleens of specific
         pathogen-free mice. Amer. Rev. Resp. Dis. 110:147-156.
44. Kim, T.H. and Kubica, G.P. (1972). Long term preservation
         and storage of mycobacteria. Appl. Microbiol. 24:311-317.
45. Collins, F.M., Morrison, N.E. and Montalbine, V. (1978).
         Immune response to persistent mycobacterial infection
         in mice. Infec. Immun. 20:430-438.
46. Blanden, R.V., Lefford, M.J. and Mackaness, G.B. (1969). The
         host response to Calmette-Guerin bacillus infection in
         mice. J. Exp. Med. 129:1079-1107.
47. Lefford, M.J. (1975). Transfer of adoptive immunity to tuber-
         culosis in mice. Immunol. 21:369-381.
48. Collins, F.M. and Mackaness, G.B. (1970). The relationship
         of delayed hypersensitivity to acquired antituberculous
         immunity. 1. Tuberculin sensitivity and resistance to
         reinfection in BCG-vaccinated mice. Cell. Immunol. 1:
         253-265.
49. Wiegeshaus, E.H., McMurray, D.N., Grover, A.A., Harding, G.E.
         and Smith, D.W. (1970). Host-parasite relationships in
         experimental airborne tuberculosis. III. Relevance of
         microbial enumeration to acquired resistance in guinea
         pigs. Amer. Rev. Resp. Dis. 102:422-429.
50. Fok, J.S., Ho, R.S., Arora, K., Harding, G.E. and Smith, D.W.
         (1976). Host-parasite relationships in experimental
         tuberculosis. V. Lack of hematogenous dissemination of
         M. tuberculosis to the lungs in animals vaccinated with
         Bacille Calmette-Guerin. J. Inf. Dis. 133:137-144.
51. Collins, F.M. (1979). Cellular antimicrobial immunity.
         Crit. Rev. Microbiol. 7:27-91.
52. Mackaness, G.B. (1971). The induction and expression of cell-
         mediated hypersensitivity in the lung. Amer. Rev. Resp.
         Dis. 104:813-828.
53. Collins, F.M. (1971). Mechanisms in antimicrobial immunity.
         J. Reticuloendothel. Soc. 10:58-99.

54. McCune, R.M., Feldman, F.M. and McDermott, W. (1966). Microbial
     persistence II. Characteristics of the sterile state of
     tubercle bacilli. J. Exp. Med. 123:469-486.
55. Collins, F.M. (1973). The relative immunogenic of virulent and
     attenuated strains of tubercle bacilli. Amer. Rev. Resp.
     Dis. 107:1030-1040.
56. Lefford, M.J., McGregor, D.D. and Mackaness, G.B. (1973).
     Immune response to M. tuberculosis in rats. Inf. Immun.
     8:182-189.
57. Reggiardo, Z. and Middlebrook, G. (1974). Failure of passive
     serum transfer of immunity against aerogenic tuberculosis
     in guinea pigs. Proc. Soc. Exp. Biol. Med. 145:173-175.
58. North, R.J. (1973). Importance of thymus derived lymphocytes
     in cell-mediated immunity to infection. Cell.Immun. 7:
     166-176.
59. Collins, F.M., Congdon, C.C. and Morrison, N.E. (1975).
     Growth of Mycobacterium bovis (BCG) in T-lymphocyte
     depleted mice. Infec. Immun. 11:57-64.
60. Morrison, N.E. and Collins, F.M. (1975). Immunogenicity of
     an aerogenic BCG vaccine in T-cell depleted and normal
     mice. Inf. Immun. 11:1110-1121.
61. Collins, F.M. and Auclair, L.K. (1979). Effect of thymosin
     treatment on anti-tuberculous immunity in immunosuppressed
     mice. J. Reticulendothelial Soc. 26:143-153.
62. Bach, J.F. (1976). The mode of action of thymic hormones
     and its relevance to T-cell differentiation. Transplant.
     Proc. 8:243-248.
63. Collins, F.M. and Morrison, N.E. (1979). Restoration of T-
     cell responsiveness to thymosin: expression of anti-
     tuberculous immunity in mouse lung. Inf. Immun. 23:
     330-335.
64. Mackaness, G.B. (1974). Delayed hypersensitivity in lung
     disease. Ann. N.Y. Acad. Sci. 221:312-316.
65. North, R.J. (1974). T-cell dependence of macrophage activa-
     tion and mobilization during infection with Mycobacterium
     tuberculosis. Infect. Immun. 10:66-71.
66. Collins, F.M., Auclair, L.K. and Mackaness, G.B. (1977).
     Mycobacterium bovis (BCG) infection of the lymph nodes
     of normal, immune and cortisone-treated guinea pigs.
     J. Natl. Cancer Inst. 59:1527-1535.
67. Rooney, J.J., Croceo, J.A. and Kramer, S. (1976). Further
     observations on tuberculin reactions in active tubercu-
     losis. Amer. J. Med. 60:517-522.
68. Collins, F.M. (1971). Immunogenicity of various mycobacteria
     and the corresponding levels of cross-protection
     developed between species. Infec. Immun. 4:688-696.
69. Watson, S.R., Morrison, N.E. and Collins, F.M. (1979).
     Delayed hypersensitivity responses to Mycobacterium leprae
     M. vaccae and M. nonchromogenicum cytoplasmic proteins.
     Inf. Immun. 25:229-236.

70. Watson, S.R. and Collins, F.M. (1980). Development of
         suppressor T cells in mice heavily infected with myco-
         bacteria. Immunol. 39:367-373.

71. Collins, F.M. and Watson, S.R. (1979). Suppressor T-cells in
         BCG-infected mice. Infec. Immun. 25:491-496.

72. Watson, S.R. and Collins, F.M. (1979). The development of
         suppressor T-cells in _Mycobacterium_ _habana_-infected
         mice. Inf. Immun. 25:497-506.

73. Parish, C.R. (1972). The relationship between humoral and
         cell-mediated immunity. Transplant. Rev. 13:35-66.

74. Uberoi, S., Malairga, A.N., Chatopadhyay, C., Kamar, R. and
         Shrivias, X. (1975). Secondary immunodeficiency in mil-
         iary tuberculosis. Clin. Exp. Immunol. 22:404-408.

75. Bullock, W.E. (1978). Leprosy: a model of immunological pertur-
         bation in chronic infection. J. Inf. Dis. 137:341-354.

76. Godal, T., Myrvang, B., Forhland, S.S., Shao, J. and Melaku,
         G. (1972). Evidence that the mechanism of immunological
         tolerance (central failure) is operative in the lack of
         host resistance in lepromatous leprosy. Scand. J.
         Immunol. 4:311-321.

77. Rook, G.A.W. (1975). The immunological consequences of anti-
         gen overload in experimental mycobacterial infections
         mice. Clin. Exp. Immunol. 19:167-178.

78. Nyka, W. (1963). Studies on M. _tuberculosis_ in lesions in
         human lung. Amer. Rev. Resp. Dis. 88:670-679.

79. Sutter, E. and Roulet, F.C. (1968). Staining _M._ _leprae_ in
         paraffin sections by the Gomori methenamine-silver
         method. Stain Technol. 40:49-51.

80. Watson, S.R. and Collins, F.M. (1981). The specificity of
         suppressor T-cells induced by _Mycobacterium_ _avium_
         infections in mice. Clin. Exp. Immunol. 43:10-19.

81. Bast, R.C., Bast, B.S. and Raff, H.J. (1976). Critical
         review of previously reported animal studies of tumor
         immunotherapy with non-specific immunostimulants.
         Annals N.Y. Acad. Sci. 277:60-92.

82. Mastrangelo, M.J., Berd, D. and Beller, R.E. (1976). Critical
         review of previously reported clinical trials of cancer
         immunotherapy with non-specific immunostimulants.
         Annals N.Y. Acad. Sci. 277:94-123.

83. Mathé, G. (1976). Surviving in company of BCG. Cancer
         Immunol. Immunother. 1:3-6.

84. Portelance, V., Quevillan, M. and Page, Y. (1976). Toxic
         effects of massive doses of BCG in mice, guinea pigs
         and monkeys. Cancer Immunol. Immunother. 1:63-68.

85. Gefford, M. and Orbach-Arbouys, S. (1976). Enhancement of
         suppressor activity in mice by high doses of BCG.
         Cancer Immunol. Immunother. 1:41-44.

86. Klimpel, G.R. and Henny, C.S. (1978). BCG-induced suppressor
        cells. 1. Demonstration of a macrophage-like suppressor
        cell that inhibits cytoxic T-cell generation. J.
        Immunol. 12:563-569.
87. Cantor, H. (1979). Control of the immune system by inhibitor
        and inducer T-lymphocytes. Ann Rev. Med. 30:269-278.
88. Collins, F.M. and Watson, S.R. (1980). Effect of chemotherapy
        on suppressor T-cells in BCG-infected mice. Immunol.
        40:529-537.
89. Lenzini, L., Rottoli, P. and Rottoli, L. (1977). The
        spectrum of human tuberculosis. Clin. Exp. Immunol.
        27:230-237.

HOST RESPONSE TO INFECTION WITH MYCOBACTERIUM BOVIS (BCG) IN MICE:

GENETIC STUDY OF NATURAL RESISTANCE

Philippe Gros[1], Emil Skamene[1], Adrien Forget[2] and
Ben Taylor[3]

[1]Montreal General Hospital Research Institute, Montreal
Quebec, Canada; [2]University of Montreal, Montreal
Quebec, Canada; [3]Jackson Laboratories, Bar Harbor, Maine

Resistance of mice to several infectious agents has recently
been shown to be under simple genetic control (reviewed in 1).
Genetic differences in the response of outbred and inbred mice to
infection with mycobacteria such as Mycobacterium tuberculosis,
Mycobacterium bovis and Mycobacterium lepraemurium have been noted
for many years (2,3). We have recently reported that clear segregation of resistance and susceptibility to Mycobacterium bovis (BCG),
exists among mice of inbred strains infected with a single intravenous dose of this bacillus (4). The present study deals with
detailed genetic analysis of the trait of BCG resistance.

Briefly, groups of mice were inoculated intravenously with
dispersed monocellular suspension of approximately 2.5×10^4
colony-forming units (CFU) of BCG. Mycobacterial burden in the
spleens of infected animals was determined 3 weeks later: serial
dilutions of spleen homogenates were plated on Dubos solid medium
and the number of viable bacilli was established by CFU count 18
days later. Results are presented either as the mean of spleen
CFU (log) or as a BCG multiplication index (M.I.) which is defined
as the mean number of BCG recovered from the spleen divided by the
number of viable BCG injected at the time of infection.

Clear differences in host response to BCG characterized by
either absence of BCG multiplication (resistance) or by a log-
arithmic BCG growth (susceptibility) was observed on inbred strain
survey: mice of CBA, C57Br, C3H, DBA/2, AKR, and A/J strains were
resistant (R) while mice of C57BL/6-derived strains,(B6, B10, B10.A,
B10.D2), BALB/c, CE/J and DBA/1 were sensitive (Fig. 1). The fact
that B10.A and B10.D2 strain mice were as susceptible as the B10

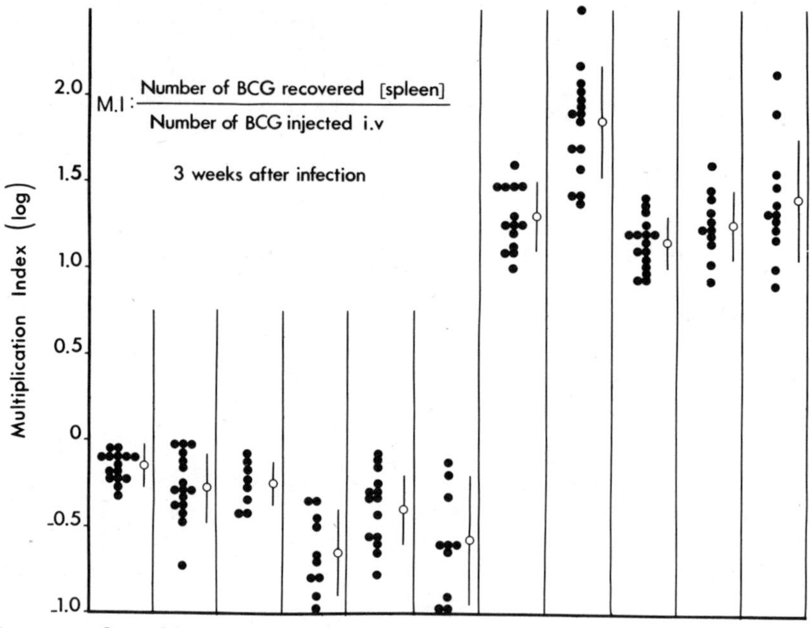

Figure 1.   Strain distribution of resistance or sensitivity to BCG.
Multiplication Index (log) of BCG 3 weeks after intravenous infection.

mice, in contrast to the resistance of A/J and DBA/2 mice,
clearly indicates that the trait of BCG-resistance is not linked to
H-2 genes. To estimate the number of genes involved in regulation
of the trait of BCG resistance, a classical Medelian analysis was
performed. Individual mice of the hybrid (F_1,F_2) and backcross
progeny derived from BCG resistant A/J progenitors and BCG sensitive B10.A progenitors were examined for their level of resistance
to BCG, 3 weeks after infection (Fig. 2). 26 out of 26 animals
(100%) of the F_1 hybrid generation were BCG resistant, as were
53 out of 53 animals of the backcross to resistant progenitor
(F_1xA). Among the offspring of F_1 hybrids backcrossed to sensitive
progenitor (F_1xB10.A) 12 out of 26 animals (46%) were resistant
while 14 (54%) were sensitive. Out of 82 animals of F_2 hybrid
generation, 72% were BCG resistant. No sex differences were
observed when the level of BCG resistance was compared in males
and females. The ratios of resistant to susceptible individuals
of the F_2 and backcross populations are compatible with the
hypothesis that the trait of resistance to BCG infection is controlled by a single, dominant, autosomal gene which was given a
provisional designation Bcg.

In order to map the Bcg gene BXD recombinant inbred strains
were typed for resistance to BCG. Recombinant inbred strains

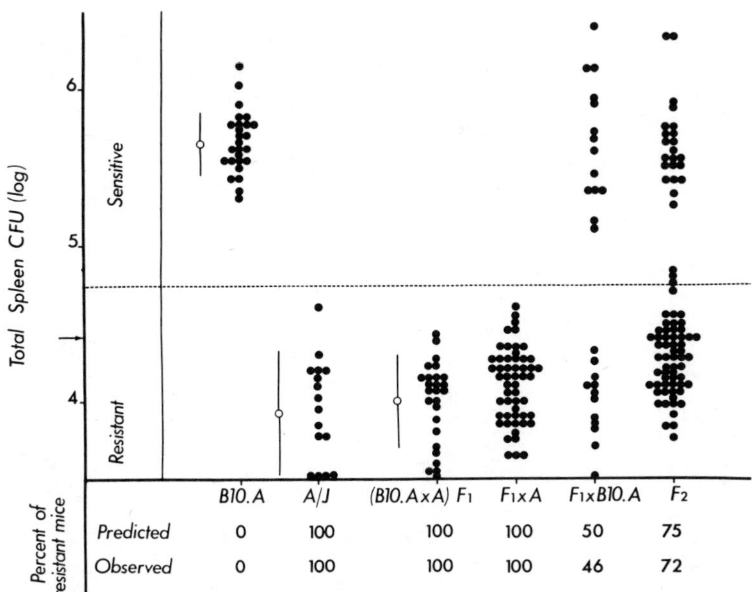

Figure 2. Segregation analysis of response to BCG. Viable bacilli
in spleens of individual animals 3 weeks after intravenous injection. The horizontal line represents 99%
confidence limit for typing an animal as resistant
(below the line) or sensitive (above the line).

represent a stable population of segregants derived from one original
cross between mice of two inbred strains different at many loci.
Many separate sublines so derived have been characterized for an
extensive library of marker genes. Gene mapping can be performed
by establishing a concordance of the strain distribution pattern
(SDP) of a new trait with that of a marker gene, without the need
for exhausting tactics of classic linkage analysis of genes in the
eukaryotic genome. 25 out of the 27 available BXD strains which were
derived from BCG-sensitive C57BL/6 and BCG-resistant DBA/2 progenitors were examined for their level of resistance, 3 weeks after i.v.
inoculation of BCG (Table 1). Clear segregation into resistant and
susceptible strains was again observed thus supporting the notion
of a single gene difference responsible for this genetic variability.
Furthermore, the SDP of BCG resistance among these BXD strains
showed near-perfect concordance with SDP of Lsh (a gene regulating
natural resistance to Leishmania donovani) and of Ity (a gene
regulating natural resistance to Salmonella typhimurium). Since
these two genes have already been mapped to the 1st chromosome (5,
6) this finding thus represents either an identity or a close
linkage of Bcg, Lsh and Ity on the first chromosome, approximately
10 centimorgans distally from the centromere.

Table 1. Strain distribution pattern SDP) of the <u>Bcg</u>, <u>Lsh</u> and <u>Ity</u> genes in BXD recombinant inbred strains.

Strain	Bcg type	(Bradley,1977) Lsh type	(O'Brien,1980) Ity type
BXD-1	R	R	R
BXD-2	S	S	S
BXD-5	S	S	S
BXD-6	R	R	R
BXD-8	R	R	R
BXD-9	S	S	S
BXD-11	S	S	S
BXD-12	S	S	S
BXD-13	R	R	R
BXD-14	S	S	S
BXD-15	R	R	R
BXD-16	S	S	S
BXD-19	S	S	S
BXD-20	S	R	S
BXD-21	S	S	S
BXD-22	S	S	S
BXD-23	R	R	R
BXD-24	S	S	S
BXD-25	R	R	R
BXD-27	R	R	R
BXD-28	R	R	?
BXD-29	S	S	R
BXD-30	R	R	R
BXD-31	S	S	?
BXD-32	R	R	R
C57BL/6	S	S	S
DBA/2	R	R	R

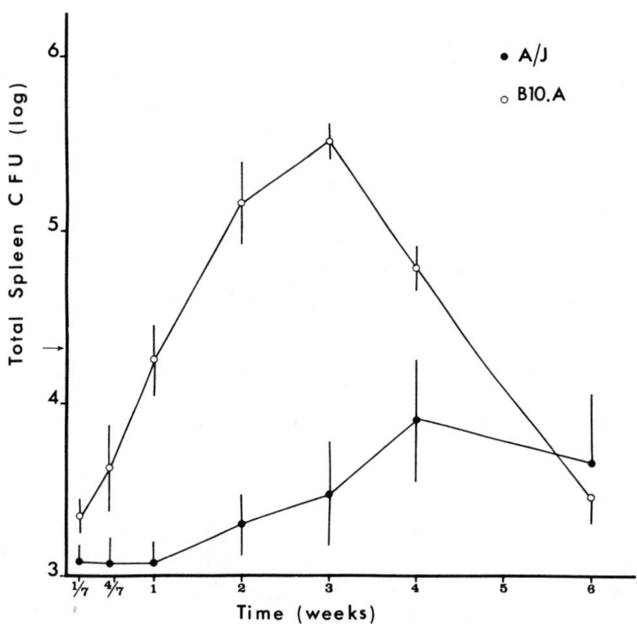

Figure 3.   Kinetics of bacterial growth of BCG. Viable bacilli
             in spleens of resistant A/J and susceptible B10.A
             strain mice after infection with $2x10^4$ CFU. Data are
             expressed as geometric mean ± S.D.

     We have further confirmed this observation by the linkage
analysis between Lsh and Bcg as performed by successive infections
of individual animals from segregating populations and by linkage
analysis between Ity and Bcg performed by progeny testing. In both
cases, 100% correlation was obtained in the phenotypic expression
of those 3 genes among individual backcross animals. Our present
data are compatible with hypothesis that a single gene regulates
resistance to (at least) three intracellular pathogens (BCG,
S. typhimurium, L. donovani) although a possibility of three genes
linked in a chromosomal locus being responsible for these findings
cannot be excluded.

     To initiate studies on phenotypic expression of the Bcg gene,
we have follwed the time-course of BCG infection in representative
resistant (A/J) and sensitive (B10.A) hosts (Fig. 3). The rate of
BCG multiplication in A/J animals was very low during the 6 week
experiment; in the sensitive B10.A animals, BCG multiplied freely
up to the third week of infection. At this point mice of the
sensitive strain had 100 times more viable BCG in their spleens
than their resistant counterparts. After the third week, bacterial
counts started to decline in the spleens of sensitive animals.
Phenotypic expression of the Bcg gene is first detectable very

early in the course of the infection (within the first few days) and it is thus likely that the natural or non-induced resistance is under its regulatory influence. It is noteworthy that similar phenotypic expression have been observed for the _Lsh_ (6) and _Ity_ (7) genes; in both cases, bacteriostatic intracellular milieu of the macrophage has been proposed to account for the observed differences. Animals possessing the sensitive allele of _Bcg_ gene depend heavily on the specific immune response for their protection against BCG infection. The downturn in splenic BCG counts observed in susceptible animals from 3rd to 6th week of infection is likely an immune phenomenon as indicated by the presence of numerous granuloma-like structures in the spleens and livers of these animals.

This study clearly demonstrates that a superior resistance of certain inbred mouse strains to BCG infection is controlled by a single gene, _Bcg_, on the 1st chromosome. The gene product enables the host to resist bacterial multiplication at the phase of natural resistance. Animals which possess the sensitive allele of _Bcg_ gene resist the infection only when specific immune response develops later on in the course of host-parasite interaction.

ACKNOWLEDGMENT

This research was supported by MRC grants 5389 and 6431.

REFERENCES

1.   Genetic Control of Natural Resistance to Infection and Malignancy. Edited by Skamene, E., Kongshavn, P.A.L. and Landy, M. Academic Press, New York, 1980.
2.   Gray, D.F., Graham-Smith, P. and Noble, J.L. (1960). J. Hyg. Camb. 58:215.
3.   Lefford, M.J., Patel, P.J., Poulter, L.N. and Mackaness, G.B. (1977). Infect. Immun. 18 (3):654.
4.   Forget, A., Skamene, E., Gros, P., Miailhe, A.C. and Turcotte, R. (1981). Infect. Immun. 32:42.
5.   O'Brien, A.D., Rosenstreich, D.L. and Taylor, B.A. (1980). Nature. 287:440.
6.   Bradley, D.J. and Kirkley, J. (1977). Clin. Exp. Immunol. 30:119.
7.   Hormaeche, C.E. (1979). Immunology 37:111.
8.   Bradley, D.J. (1979). Acta. Tropica. 36:171.
9.   Hormaeche, C.E. (1980). Immunology 41:973.

CELL MEDIATED LYSIS OF LYMPHOCYTES EXPRESSING BACTERIAL ANTIGENS

J. A. Hank and P. M. Sondel

University of Wisconsin
Madison, WI

INTRODUCTION

Two T cell subpopulations are involved in the primary _in vitro_ response to allogeneic cells. The cytotoxic T lymphocyte (T_c) recognizes HLA-A, B, and C antigens. However, the T_c is activated to kill only after collaboration with a second subpopulation, the helper T cell (T_h), which responds proliferatively to HLA-D antigens (1). We have asked whether purified protein derivative (PPD) and soluble candida antigens (CAN) can substitute for foreign HLA-D in providing a helper interaction to alloreactive T_c.

The _in vitro_ proliferative response to PPD requires antigen presenting cells that bear at least one HLA-DR antigen of the responding individual (2). Comparable _in vitro_ proliferative responses to viral and hapten antigens require HLA-D bearing autologous cells. Furthermore, these _in vitro_ cultures can generate T_c that kill viral or hapten modified autologous cells. We have thus examined the ability of sensitization with PPD alone to activate T_c against autologous cells labeled with PPD.

MATERIALS AND METHODS

Sensitization of Responding Cells

Ten x 10^6 Ficoll purified PBL from healthy donors (A, B, etc.) were cultured with varying combinations of the following stimuli: soluble antigens PPD and CAN, the mitogen PHA, PBL irradiated with 2500R (B_x), or PBL heated to 45°C for 1 hr (B_H).

Note that heat-treated, stimulating cells neither stimulate T_h nor T_c, but can induce allospecific T_c if irradiated cells bearing a foreign HLA-D antigen are used to provide a helper stimulus. The PPD, CAN, and PHA were used at dilutions previously shown to induce optimal _in vitro_ proliferation.

Proliferative and Cytotoxic Assays

On day 6 of culture, sensitized cells were resuspended, plated in quadruplicate, and pulsed with ^{3}H-thymidine (^{3}HTdR) for 8 hrs. Remaining sensitized cells were washed and added to 5 x 10^3 ^{51}Cr labelled target cells at three different effector to target (E/T) ratios, in quadruplicate u bottom microtiter wells for 4 hrs. at 37°.

After incubation the plates were spun at 500 X G for 10 min. at 4° and the supernatant harvested. Percent of cytotoxicity was calculated using the formula,

$$\% \text{ Cytotoxicity} \quad \frac{\text{Exp cpm} - \text{spon cpm}}{\text{Max cpm} - \text{spon cpm}} \text{ X } 100$$

In some experiments, PPD labelled target cells (A^{PPD}) were produced by simultaneous culturing for 3 hours with 10 µg of PPD in addition to 250 µCi of ^{51}Cr.

RESULTS

Antigen Specific Proliferative Response Activates T_c Help

Table 1 demonstrates the ability of PPD and CAN to activate an _in vitro_ proliferative response and to induce a helper collaboration in the development of T_c to heat treated cells. Responders A, B and C give proliferative and cytotoxic responses after stimulation with D_x, indicating antigenic differences that activate both T_h and T_c. After heat treatment, D_H cells do not effectively express their HLA-D antigen, and thus activate neither T_h nor T_c by A, B, and C. Individuals A and B respond to both PPD and CAN while C responds proliferatively to only CAN. In parallel to these proliferative responses, A and B are helped to generate T_c to the heat-treated cells by both soluble antigens, while C is helped only by CAN.

Mitogens Provide a Proliferative Response but no Help

In an attempt to determine whether any proliferative response could initiate a helper collaboration, PHA was tested for its ability to help in the generation of T_c by heat-treated cells. Table 2 indicates that addition of PHA at an optimal mitogenic

Table 1. Antigen specific proliferative response activates T_c help

Responder		E/T	Stimulus					
			Dx	D_H	D_HPPD	PPD	D_HCan	Can
A	% Cytotoxicity	3	19.7	0.7	10.7	2.8	7.8	0.4
		10	34.6	−0.2	25.2	2.7	20.0	0.8
		30	49.9	0.6	40.7	2.4	31.8	1.1
	^{3}H−TdR c.p.m.		21,110	178	26,672	23,362	18,814	19,446
B	% Cytotoxicity	3	23.0	−0.9	4.1	2.1	6.6	2.7
		10	39.2	−0.3	9.3	3.3	13.9	4.7
		30	55.7	−1.5	17.5	5.5	23.3	7.4
	^{3}H−TdR c.p.m.		18,666	133	15,486	12,779	43,492	45,095
C	% Cytotoxicity	3	11.5	−1.3	−2.0	−0.7	2.3	−0.2
		10	22.7	−1.6	−0.2	−1.2	9.0	1.3
		30	40.6	−1.9	−0.2	−1.1	16.8	1.0
	^{3}H−TdR c.p.m.		31,865	692	4,193	4,949	37,127	36,742

Lymphocytes from donors A, B and C were cultured 6 days with irradiated (Dx) or heat−treated (D_H) cells from unrelated donor D, and/or soluble PPD or Candida antigens. Cytotoxicity was measured on ^{51}Cr labelled lymphocytes from D at 3 different effector to target (E/T) ratios. Proliferation was measured by 5 hr ^{3}H−Tdr incorporation.

Table 2. Mitogens provide a proliferative response but no help

	E/T Ratio	AB_x	AB_H	$AB_H PPD$	APPD	$AB_H PHA_o$ [1]	$APHA_o$	$AB_H PHA_3$ [2]	$APHA_3$	AC_x
^{3}H-thymidine incorporation (CPM)		16,821	117	27,757	22,742	8,672	12,127	24,591	26,701	11,947
% Cytotoxicity on B targets	3:1	15	1	10	3	1	0	2	1	1
	10:1	27	1	22	2	1	1	5	6	3
	30:1	50	0	42	1	0	1	12	14	6
% Cytotoxicity on C targets	3:1	1	-1	0	1	1	0	2	-1	4
	10:1	3	-1	1	0	0	0	3	3	10
	30:1	8	-1	4	0	0	0	10	7	21

[1] PHA added on day 0 of culture

[2] PHA added on day 3 of culture

dose on day 0 does not provide an effective helper interaction.
Addition of PHA on day 3 induces lectin dependent, non-specific
cytotoxicity, but does not help generate specific T_c to the heat-
treated cells.

Cells Primed In Vitro to PPD Kill PPD Labelled Cells

Responder cells primed to PPD alone kill autologous and
allogeneic cells labelled with PPD (Table 3). Responders A and B
proliferate in response to PPD, and mediate PPD-specific cytotoxicity. Individual C has a weak proliferative response and
does not kill PPD labelled target cells. However, cells of
individual C bind PPD and are recognized by PPD primed T_c from A
and B. Addition of excess soluble PPD does not inhibit the
killing of cells surface labelled with PPD. However, nonchromated
PPD-labelled cells competitively inhibit this PPD specific killing
(data not shown).

DISCUSSION

These experiments demonstrate that soluble bacterial and
mycotic antigens can induce an effective helper stimulus that
collaborates in the generation of antigen specific allo-reactive
T_c. Strong proliferative responses induced by PHA do not help
generate alloreactive T_c. This difference might reflect activation of different cell subpopulations by these distinct stimuli
(ie: suppressor versus helper cells), or that these 2 types of
signals activate the same subpopulation along separate differentiation pathways.

The proliferative response to PPD requires antigen presenting
(HLA-D bearing) cells that share at least one DR antigen with the
responding individual. Thus the response to PPD might really be
an immune response to "altered self" in which both PPD and HLA-D
together are recognized by a mechanism comparable to foreign
HLA-D recognition in MLC.

"Altered-self" T_c have been generated in response to viral
and hapten antigens (3). We have demonstrated that T_c able to
kill PPD-labelled target cells are induced following culture with
soluble PPD. These T_c recognize autologous PPD-labelled targets
better than they recognize allogeneic PPD targets, however there
is extensive cross-reactivity. All allogeneic PPD-labelled cells
tested so far have been lysed by PPD primed T_c. This cell
mediated killing by autologous PPD activated cells is not blocked
by excess soluble PPD, but is competitively inhibited by nonchromated PPD-labelled target cells. These results parallel those
observed for responses to viral and hapten modified cells where
simultaneous recognition of H-2 or HLA antigens is required.

Table 3. Stimulation with PPD alone activates T_c able to kill autologous targets labelled with PPD

| Effectors[2] | ^{3}H-TdR Incorporation CPM | % Cytotoxicity[1] | | | | | |
| | | Targets | | | | | |
		A	A^{PPD}	B	B^{PPD}	C	C^{PPD}
APPD	41,097	0	15	-1	9	2	22
BPPD	11,807	3	8	-1	9	8	10
CPPD	4,676	0	-1	-1	-1	1	1
ABx	38,498	-2	-1	16	19	3	7
BCx	48,608	6	6	0	-1	46	42
CAx	54,075	22	17	9	10	1	2

[1] Effector to target ratio of 30:1

[2] PBL were primed with either soluble PPD (5 µg/ml) or irradiated allogeneic cells. After 6 days, cultures were pulsed with ^{3}H-TdR or tested for cytotoxic ability in a 4 hr ^{51}Cr release assay. A^{PPD}, B^{PPD} and C^{PPD} target cells were cultured in 0.2 ml H5-RPMI for 3 hr with 10 µl soluble PPD (1 mg/ml) and 25 µl ^{51}Cr (10 m Ci/ml). Control targets A, B, and C were incubated with ^{51}Cr only. Background CPM for ^{3}H-TdR incorporation were: A-250, B-380, C-984.

We anticipate that simulatneous recognition of HLA and these soluble bacterial antigens will parallel that required for viral and hapten antigens.

Is there any evidence that T_c to bacterial antigens are important _in vivo_? First, we have not in any way altered the target cells or the PPD antigen in this system. PPD passively adheres to the target cells. Only individuals who have been primed _in vivo_ can mount an _in vitro_ response. This suggests that _in vivo_ priming with either BCG or M.tuberculosis induces the expansion of PPD specific T_c and T_h by activating them with the equivalent of "PPD-altered self". Secondly, immune cytolysis mediated by antibody and complement or by antibody dependent cellular killing (ADCC) occurs by osmotic lysis (4). In contrast, cytotoxic T cells induce cellular destruction by first causing an intracellular nuclear disintegration prior to any loss of cell surface membrane integrity. In this latter mechanism the target cell surface membrane remains intact for nearly an hour, while the nucleus undergoes rapid lysis. If this nuclear disintegration reflects the activation of intracellular, toxic products, these products might have the opportunity to destroy intracellular pathogens before they exit the cell via membrane destruction.

ACKNOWLEDGMENT

This work was supported by CA 14520-08 and a general research grant to U.W. Medical School from NIH. J. A. H. is a Fellow of the Cancer Research Institute. P. M. S. is a scholar of the Leukemia Society of America and a JA Hartford Foundation Fellow.

REFERENCES

1. Bach, F. H., Bach, M.L. and Sondel, P.M. (1976). Nature
         259:273.
2. Bergholtz, B.O. and Thorsby, E. (1978). Scand. J. Immunol.
         8:63.
3. Zinkernagel, R.M. and Doherty, P.C. (1975). J. Exp. Med.
         141:1427.
4. Russel, J.H. and Dubos, C.B. (1980). J. Immunol. 125:3, 1256.

IMPROVEMENT OF ABNORMAL LYMPHOCYTE RESPONSES IN "ATYPICAL"

MYCOBACTERIOSIS WITH INDOMETHACIN

U. G. Mason III, L. E. Greenberg, S. S. Yen, and
C. H. Kirkpatrick

Department of Medicine, National Jewish Hospital and
Research Center, Denver, CO   80206

INTRODUCTION

Although chemotherapy of tuberculosis has resulted in a
dramatic decrease in the number of new and active cases in the
United States, relatively little improvement has been made in the
treatment of diseases caused by nontuberculous mycobacteria.
Depending on the type of hospital and population served by the
hospital, the incidence of "atypical" mycobacteria may be as high
as 30% of all mycobacterial infections (1). Aside from the
clinical importance and economic consequences of these infections,
these chronically ill patients provide a valuable source of new
information concerning cellular defense mechanisms and resistance
to chronic infectious diseases. Unlike patients with chronic
fungal infections who commonly develop disease during early childhood, patients with "atypical" mycobacterial diseases are usually
well during early childhood and adolescence (2). Thus, to explain
the predisposition of individuals to infections with these
organisms, one would predict an abnormality of immunoregulatory
mechanisms rather than a failure to develop immunocompetent
lymphoid cells.

Our laboratory has been investigating cell-mediated immune
responses in patients with chronic "atypical" mycobacterial
infections. The frequency with which these patients demonstrated
negative responses to skin tests with tuberculin and other antigens suggested that the patients were immunologically impaired.
Subsequent experiments which examined in vitro T-cell proliferation in response to antigens and a mitogen revealed that although
these responses were subnormal, marked enhancement and, in some
cases, normalization was observed when the T-cells were cultured

197

in media containing indomethacin. This finding prompted a systematic study of the relationship of arachidonic acid metabolites to T-cell proliferation in response to antigens and a mitogen.

MATERIALS AND METHODS

Nine patients with culturally proven, clinically active "atypical" mycobacterioses were studied. The controls, consisting of four healthy males and two healthy females, three of whom had delayed hypersensitivity to PPD and others were negative, were assayed over the 8 month period during which the patients were studied.

Cultures of Ficoll-Hypaque prepared peripheral blood mononuclear cells received one of the following inhibitors of arachidonic acid metabolism: indomethacin (final concentration 10^{-7}M and 10^{-6}M), a preferential inhibitor of cyclooxygenase; nordihydoguaiaretic acid (NDGA) ($3x10^{-7}$, $3x10^{-6}$, $3x10^{-5}$M), a preferential inhibitor of lipoxygenase, and phenidone ($3x10^{-7}$, $3x10^{-6}$, $3x10^{-5}$M), an inhibitor of both enzymes. Control cultures received no inhibitor. Dose response relationships were obtained for the mitogen (phytohemagglutinin 0.33 μg/ml; PHA) and antigens (PPD, 10 μg/ml and Candida, 10 μg/ml; CAN) which were used to stimulate T-cell proliferation, which in turn was monitored by thymidine incorporation.

RESULTS

The studies with healthy control subjects confirmed the findings of others; tuberculin-sensitive subjects responded to PPD with marked T-cell proliferation while tuberculin-insensitive subjects did not. These findings were in contrast to those in patients with "atypical" mycobacterial infections. Even though they had known exposures to mycobacterial antigens, all 9 patients were hyporesponsive to PPD both by skin testing and by T-cell proliferation. The extent of the hyporesponsive state was illustrated by the finding that 5 patients were also hyporesponsive to CAN and 3 of 3 patients were poorly responsive to PHA.

In most instances, addition of indomethacin to the cells of the hyporesponsive patients enhanced and, in some cases, even normalized the T-cell proliferative responses to the microbial antigens and to PHA. Lymphocytes from 6 of 9 patients demonstrated significant ($p < .05$) improvement of thymidine incorporation in response to PPD (Figure 1); 3 of 9 patients had a significant ($p < 0.01$) improvement of proliferative responses to CAN, and 1 of 3 patients had significant ($p < 0.01$) improvements in responses to PHA. In contrast, inhibition of the lipoxygenase

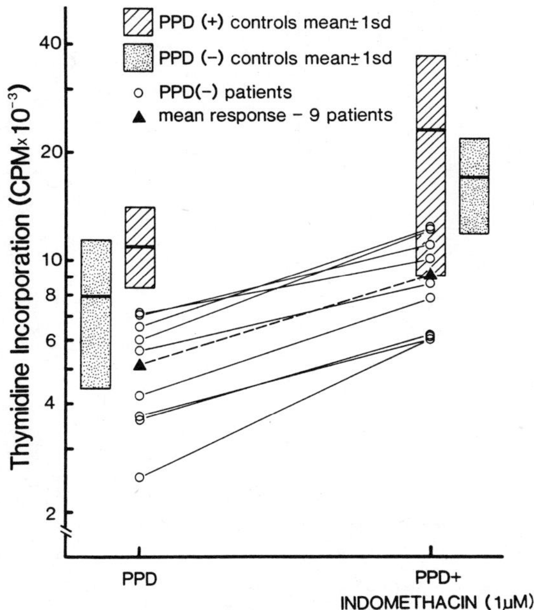

Figure 1. Effect of indomethacin on PPD-induced DNA synthesis in
T-lymphocytes as measured by thymidine incorporation.
Note that the responses by all patients improved and
4 patients had normal responses when their cells were
cultures with PPD (10 μg/ml) in the presence of
indomethacin (10^{-6}M).

pathway with NDGA (30 μM) produced a significant (p < 0.05)
reduction of lymphocyte proliferation in response to PPD in 5 of
9 patients and to CAN in 4 of 9 patients.   Moreover, this
lymphoproliferative response was even lower than that of the
antigen-stimulated cultures without any drug.   Studies with
phenidone (30.0 μM) produced proliferative responses similar to
those cultures without any inhibitors. In fact, significantly
different responses were noted in only 2 of 9 experiments with PPD
and 1 of 9 with CAN. Figure 2 summarizes the effects of the
blockers of arachidonic acid metabolism on PPD-induced T-cell
proliferation.

CONCLUSIONS

     Our data indicate that in vitro proliferative malfunctions
are characteristic of lymphocytes from PPD skin test-negative
patients with nontuberculous mycobacterial disease. This is
demonstrated by the subnormal responses to antigens and a
mitogen.

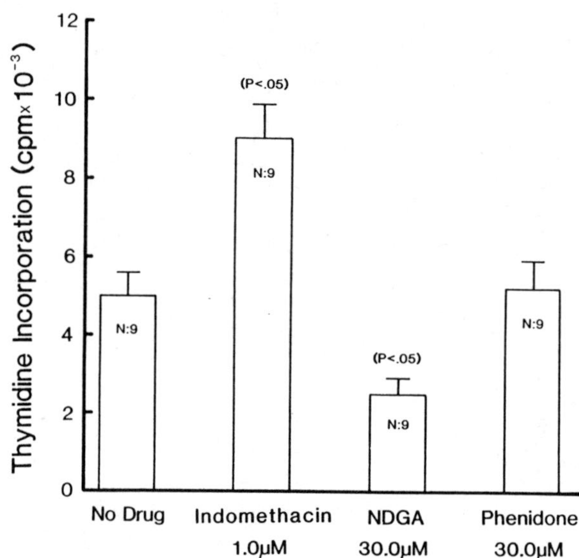

Figure 2.   Effects of inhibitors of arachidonic acid metabolism
            on T-cell responses to PPD in patients with chronic
            "atypical" mycobacterial infections. Note that
            significant enhancement of responses occurred when
            indomethacin was added to the cell cultures. NDGA
            produced significant depression in the responses, but
            phenidone had no effect.

    The finding of an enhancement of lymphocyte response when the
cells are cultured with indomethacin and even greater suppression
of the response with NDGA suggests that subnormal DNA synthesis
may result from abnormal immune regulation, and that the regulatory mechanism is related to arachidonic acid metabolism. By
preventing formation of the cyclooxygenase products, indomethacin
enhances antigen and mitogen-dependent T-cell proliferation.
Since this pathway leads to production of prostaglandins, prosta-
cyclins and thromboxanes (3), and because prostaglandins are known
to have immunosuppressive activities (4), it is likely that the
suppressive effects in our system are mediated by prostaglandins
or other cyclooxygenase products. An alternative mechanism,
which is also compatible with the data, is that a lipoxygenase
derivative(s) is important for optimal proliferative responses by
T-cells. Compatible with either possibility is the effect of NDGA,
the inhibitor of lipoxygenase, which would make available more
substrate for the production of the inhibitors via the cyclooxygenase pathway or block production of the enhancing substance.
Phenidone, by inhibiting both arachidonic acid pathways (5) re-
stores the milieu found in the "drugless" cultures which originally
demonstrated poor lymphocyte responses.

REFERENCES

1. Youmans, G.P. (1979). Diseases due to mycobacteria other
        than mycobacterium tuberculosis. In: Tuberculosis,
        W. B. Saunders Co., Philadelphia.
2. Wolinsky, E. (1979). Nontuberculous mycobacteria and
        associated diseases. Am. Rev. Resp. Dis. 119:107.
3. Hamber, M. and Samuelsson, B. (1974). Prostaglandin endo-
        perioxides. Novel transformations of arachidonic
        acid in human platelets. Proc. Nat. Acad. Sci.
        USA 71:3400.
4. Goodwin, J.S., Bankhurst, A.D. and Messner, R.P. (1977).
        Suppression of human T-cell mitogenesis by prosta-
        glandin. Existence of a prostaglandin-producing
        suppressor cell. J. Exp. Med. 146:1719.
5. Blackwell, G.J. and Flower, R.J. (1978). 1-phenyl-3-pyrazoli-
        done: An inhibitor of cyclooxygenase and lipoxygenase
        pathways in lung and platelets. Prostaglandins 16:417.

IMMUNOREGULATORY DEFECTS IN LEPROSY

Susan R. Watson and Ward E. Bullock

The Division of Infectious Diseases, Department of
Medicine, University of Cincinnati College of Medicine
Cincinnati, Ohio 45267

INTRODUCTION

It was over one hundred years ago that the causative organism
of human leprosy, Mycobacterium leprae was recognized by the
Norwegian investigator, Armauer Hansen. Since that time our
knowledge of this disease has advanced but little. M. leprae is
an obligate intracellular parasite growing in the cells of the
mononuclear phagocyte system with a special affinity for those
tissue macrophages found in association with the skin, nerves,
and lymphoreticular system. The organism has a very long generation time, estimated to be 10 to 15 days, which stands in comparison
to M. tuberculosis that divides approximately every 20 hours, and
coliform bacteria which divide every 20 to 30 minutes. To date
the cultivation of M. leprae in vitro has proved elusive, thus
hampering studies of the basic biology of the organism as well as
the development of a potential vaccine against leprosy. For twenty
years the principal method of studying the growth kinetics of the
organism, the pharmacology of anti-leprous drugs, and the local
pathology and immunity to the bacillus has been the mouse footpad
model (1). However, in 1971, it was discovered that armadillos
were susceptible to M. leprae and will support its multiplication
in enormous numbers in vivo (2). As a consequence, the quantity
of organisms available to the research community has increased
greatly.

Leprosy presents with a variety of clinical manifestations
ranging from a single, often self-healing lesion, as seen in polar
tuberculoid leprosy, to a progressive and disseminated disease,
as manifested in lepromatous leprosy. Man appears to be the only
significant infectious reservoir of leprosy but definite

information concerning the mode of transmission is lacking. Nasal
discharge from a lepromatous patient is considered an important
source of infectious M. leprae, but the site of primary infection
remains unknown. In countries endemic for leprosy, the leprosy
bacillus appears to cause clinically detectable disease in only a
small percentage of those who acquire infection; in the great
majority of people, an effective cellular immune response prevents
extensive multiplication of M. leprae. However, in those individuals who fail to limit the growth of M. leprae, as is the case in
lepromatous leprosy, 10^9 bacilli may be present per gram of
infected tissue, without overt clinical symptoms.

THE LEPROSY SPECTRUM

The dermatologic signs in leprosy range from a single localized
lesion to a diffuse generalized infiltration of the dermis. The
most widely employed and best known sytematic classification of
leprosy is that of Ridley and Jopling (3). This classification
consists of five groups: TT = polar tuberculoid; BT = borderline
tuberculoid; BB = borderline; BL = borderline lepromatous; LL =
lepromatous leprosy. Ridley has described, in great detail, the
histopathological features of these groups (4).

Tuberculoid (TT) Leprosy

The lesion of TT leprosy is characterized by a well-developed
granuloma formation composed of epithelioid cells with a uniform
appearance and giant cells of the Langhans' or foreign body type.
Lymphocytes are abundant at the periphery of the granuloma and
acid-fast bacilli usually cannot be identified. Dermal nerves
involved in the granulomatous inflammation are destroyed.

Borderline Tuberculoid (BT) Leprosy

The pathology of TT and BT is similar, except that acid-fast
bacilli are more readily seen in the latter, especially in the
dermal nerves, although the number is small.

Borderline (BB) Leprosy

In BB leprosy, the histologic picture may vary considerably
from lesion to lesion, or even within the same lesion. Typically,
the granuloma formation is less well developed. Epithelioid cells
are spread more diffusely throughout the granuloma, giant cells
are not present, and there are fewer lymphocytes within the infiltrate. The dermal nerves are less damaged and therefore more
easily visible. Acid-fast bacilli are numerous.

Borderline Lepromatous (BL) Leprosy

In this form, histiocytes are the predominant cell type with relatively few lymphocytes scattered among them, sometimes in aggregates. Epithelioid cells are absent and large numbers of acid-fast bacilli are present.

Lepromatous Leprosy (LL)

The inflammatory infiltrate is composed, almost exclusively of histiocytic cells that have a foamy appearance. Masses of bacilli are present intracellularly, many in large clumps called globi. Lymphocytes are very sparse. Characteristically, there is a "clear zone" beneath the epidermal basement membrane in which there is no inflammatory infiltrate, as contrasted with the tuberculoid forms of leprosy in which the granulomas extend to the dermal-epidermal junction.

SUSCEPTIBILITY TO LEPROSY

The host factors that determine susceptibility to disease once an individual has been infected with $\underline{M}$. $\underline{leprae}$ are poorly understood. Leprosy is diagnosed more frequently in males than in females but this may be a specious finding, as census figures in many regions are uncontrolled for sex selection. Various workers have suggested that genetical factors may play an important role in the development of a particular form of leprosy. Many groups have studied a wide variety of population samples for an association between leprosy or leprosy types and HLA antigens (5,6) however the results have not been definitive. Recent work has suggested a possible HLA-linked control of susceptibility to tuberculoid leprosy only and suggests a recessive inheritance of this trait for which HLA-DRw2 appears to be a genetic marker (7).

IMMUNE FUNCTION AND THE LEPROSY SPECTRUM

Studies of the immune responses of leprosy patients have documented a progressive decrease in cell mediated immunity (CMI) from the tuberculoid to the lepromatous end of the disease spectrum. Intracutaneous injection of patients with TT or BT leprosy with lepromin (a crude presentation of heat-killed $\underline{M}$. $\underline{leprae}$) will induce local granuloma formation within a 3-4 week period whereas LL patients are completely anergic to lepromin. Thus, the lepromin test is used as a clinical test of the patient's capacity to mount a CMI response against $\underline{M}$. $\underline{leprae}$. Using more purified preparations of $\underline{M}$. $\underline{leprae}$ antigens in both $\underline{in}$ $\underline{vivo}$ skin testing and $\underline{in}$ $\underline{vitro}$ lymphoproliferative responses, the decrease

in CMI towards the end LL end of the spectrum is again demonstrated.
Moreover, a large percentage of LL patients manifest a generalized
impairment of delayed-type hypersensitivity responsiveness to a
variety of "recall" antigens both _in vivo_ and _in vitro_.

Serum immunoglublin levels are usually within the normal
range in tuberculoid patients, whereas polyclonal hypergammaglobulinaemia is a common feature of lepromatous leprosy (8). Most
serum samples from LL patients contain antibodies to _M. leprae_ that
cross-react with other _Mycobacteria_. Other serological abnormalities
include false positive VDRL tests, cryoglobulinaemia, the presence
of circulating immune complexes, elevated levels of amyloid-related
serum protein component, increased levels of C-reactive protein
and, less frequently, the presence of autoantibodies such as anti-
nuclear and anti-thyroglobulin antibodies (8). Thus, the picture
emerges of a LL patient who is compromised with respect to CMI but
has an intact and highly active humoral immune system. As the
early work of Mackaness and his colleagues (9) has shown, immunity
to intracellular bacteria is dependent on cell mediated responses
rather than humoral aspects of the immune system. Thus, the proliferation of antibody responses in lepromatous patients appear
to serve no recognizable protective function.

IMMUNE DEFECTS AND LEPROMATOUS LEPROSY

As attempts to establish a link between the type of disease
and the patient's HLA type have not been conclusive, many groups
have tried to establish immunological mechanisms that would explain
the lack of CMI in LL patients. This work has employed human
subjects and animal models. It is possible to infect mice (and
rats) with _M. lepraemurium_ and, depending on the mouse strain,
induce a disease that resembles either the lepromatous or tuberculoid forms of leprosy (10). A wide variety of immune functions
have been studied in lepromatous leprosy patients with results
that often vary from group to group. This is not surprising when
one considers the differences between patient populations with
respect to treatment status and socioeconomic conditions that may
contribute to the variations in scientific findings.

As the causative organism of leprosy grows in macrophages,
one of the obvious questions that has been asked is, do macrophages
from LL patients retain their antibacterial function. Beiquelman
(11) described a failure of the lepromatous macrophage to lyse
autoclaved _M. leprae in vitro_. He claimed that this defect was
highly specific as the macrophages could lyse other _Mycobacteria_.
Other studies (12,13) failed to confirm this finding. Cultivation
of _M. leprae_ in human macrophages _in vitro_ has been attempted (14),
and interestingly the results suggested that the persistence of
the bacillus within these cells was similar regardless of whether

the macrophages were derived from LL or TT patients or healthy
individuals. Convit and his co-workers (15) have shown that by
injecting large amounts of killed M. leprae with viable BCG organisms
into lepromatous skin, epithelioid cell granulomas are induced by
the BCG that apparently will eliminate the killed M. leprae. Thus,
they suggest that if the lepromatous macrophage can be activated
non-specifically, it may be capable of digesting M. leprae.

Using the mouse model, two groups have found defects in a
variety of cell mediated immune responses in vivo (16,17). In
addition, suppression of the in vitro antibody responses to both
T-dependent and, at later states of infection, to T-independent
antigens has been demonstrated unequivocally (18,19). To explain
the immunosuppression, Watson postulated a defect in macrophage
function due to overloading with excessively large numbers of
Mycobacteria which could prevent processing of antigens (18),
whereas Bullock obtained evidence for both macrophage and T-
lymphocyte mediated suppressor cell activity (19).

SUPPRESSOR CELLS AND LEPROSY

Immunoregulation involving a subset of T cells (suppressor T
cells), acting by means of negative feedback mechanisms, has been
shown to be of great importance in controlling immune responses to
various antigens in both mice (20) and man (21). Other studies
(35) have demonstrated the suppressive effect of macrophages.
Therefore several groups have made attempts to define a suppressor
cell population in LL patients that could be invoked to explain
the lack of CMI responsiveness in this group. It has been shown
that macrophages, alone or in concert with T cells, may be immunosuppressive in human tuberculosis (22) and in murine leprosy (19).
Hirschberg (23) has demonstrated a defect in T cell-macrophage
interactions in LL patients, but these results are difficult to
evaluate as he employed allogeneic cells in his culture system.

More recent evidence suggesting the presence of suppressor
cell activity in LL patients comes from the work of Mehra, et al.
(24). These workers observed that when peripheral blood mononuclear (PBM) cells from patients with lepromatous leprosy were
cultured with the non-specific T cell mitogen Concanavalin A (Con
A) in the presence of Dharmendra lepromin, the blastogenic response
of these cells was suppressed as compared to LL cells that were
cultured in the presence of Con A alone. This non-specific
suppressor activity in the presence of Dharmendra lepromin was
not observed in cultures of mononuclear cells obtained from normal
or tuberculoid donors. Cell fractionation studies indicated at
least two cell populations involved in this in vitro suppression -
adherent cells and T lymphocytes. Further work by this group
identified a subset of T cells bearing a surface antigen recognized

by anti-TH$_2$ hetero antiserum as being the cells that exerted the lepromin-induced suppression (25).

Although this work is of interest, several problems remain to be addressed. One is the data from other laboratories demonstrating that the addition of M. leprae bacilli to mononuclear cell cultures from healthy individuals can also cause a suppression of in vitro lymphocyte transformation responses (26,27). Moreover these laboratories have found that cells from either normal donors or TT patients also showed suppressed in vitro blastogenic responses to both antigens and mitogens when cultured with M. leprae antigens. Whether these discrepancies can be accounted for by the fact that Mehra used Dharmendra lepromin (autoclaved, chloroform - ether treated bacilli) and the other workers used the cruder preparation of standard lepromin (autoclaved bacilli) is as yet unknown.

CON A INDUCIBLE SUPPRESSOR CELLS AND LEPROSY

Exposure of normal human peripheral blood lymphocytes to Con A induces a non-specific suppressor cell that modulates immune responses (28). This fact led our laboratory to hypothesize that if the connection between lepromatous patients and their inability to mount CMI responses was indeed an excess of suppressor T cell activity, then the PBM cell population from these patients should contain increased numbers of Con A-inducible suppressor cells. As the work unfolded, it became apparent that this was not the case. In fact, the study very clearly demonstrated that patients with lepromatous leprosy and other disseminated granulomatous infections had much lower levels of demonstrable Con A-induced suppressor cell activity as compared to normal individuals or patients with localized granulomatous infections (29).

The methodology has been described elsewhere (29), but briefly, PBM cells from patients and normal age matched control were cultured in the presence or absence of Con A for 72 hours. These cells were then assayed for their capacity to suppress both mitogen and antigen-stimulated responses by allogeneic PBM cells in vitro.

In support of the findings of others, pre-incubation of PBM cells from normal donors with Con A clearly suppressed the mitogenic response of allogeneic PBM cells to Con A (Fig. 1). When this experiment was performed using Con-A pretreated PBM cells from patients with disseminated granulomatous disease, the suppressive effect was significantly less than that observed with pretreated cells from normal donors. Similar results were obtained when Con A-pretreated cells from patients and normal controls were assayed for their effects upon the response to other antigens (major histocompatability complex antigens or histoplasmin) by

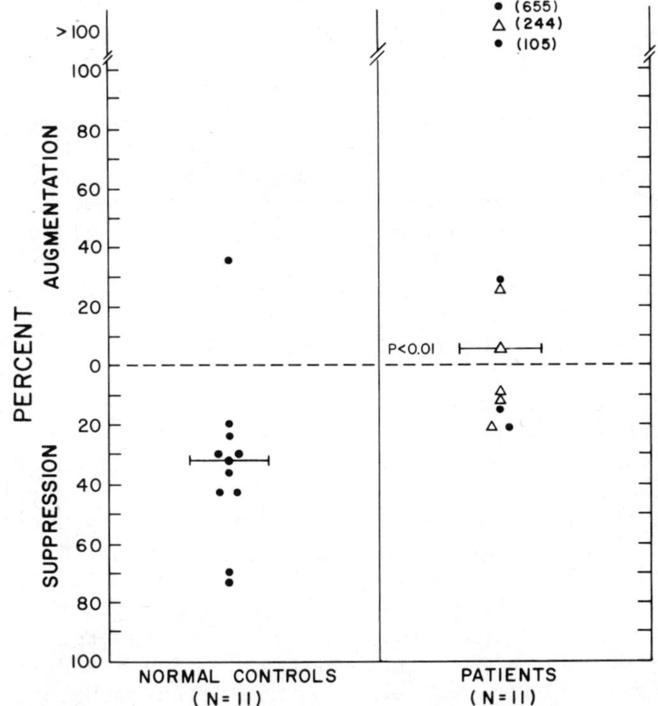

Figure 1.   Effect of Con A-pretreated cells from patients or normal
controls on responses to Con A by allogeneic cells from
normal donors. Median values are indicated by horizontal bars; (Δ) leprosy patients. [From Artz, et al.(29).]

allogeneic cells from normal donors. Thus, these findings
demonstrated a reduction in the Con A-inducible suppressor
activity of PBM cells from patients with lepromatous leprosy and
other disseminated granulomatous infections. Very similar data
have been reported by Nath, et al. (30) in studies of leprosy
patients.

IN VITRO IMMUNOGLOBULIN (Ig) SYNTHESIS IN LEPROSY PATIENTS

     As previously stated, serum immunoglobulin (Ig) levels in
LL patients usually are elevated considerably above normal levels.
This elevation of serum Ig is, in part, due to antibodies reactive
with M. leprae. However, LL patients also have other serological
abnormalities such as cryoglobulinaemia, circulating immune
complexes and autoantibodies in varying degree. These facts
coupled with the previous finding concerning the levels of Con A-
inducible suppressor cells, lead us to embark on a study of the

in vitro responsiveness of PBM cells from normal controls and
leprosy patients to pokeweed mitogen (PWM). PWM mitogen induces
proliferation in vitro of both T and B cell populations such that
at the end of a 6-7 day culture period, the B cells are secreting
Ig. The number of plasmacytoid cells that are non-specifically
secreting Ig can then be quantitated by means of a reverse haemolytic plaque assay (31). This technique allowed us to assess the
ability of PBM cells from normal donors and leprosy patients to
secrete non-specific Ig in response to PWM stimulation. In addition
we were able to study the mechanisms by which this response is
controlled in both normal individuals and patients.

PWM-STIMULATED PLAQUE FORMING CELL RESPONSES

        In these experiments, PBM cells from normal donors, LL and
TT leprosy patients were cultured for 7 days with PWM. At the
end of this time, the number of Ig secreting cells was determined
by the reverse haemolytic plaque assay. The PWM-induced plaque
forming cell (PFC) responses of PBM cells from LL patients was
significantly greater (median 33,000 PFC per 10^6 cells in culture)
than those mounted by cells from healthy controls (median 6,625
PFC/10^6 cells) and by cells from four patients with tuberculoid
leprosy (P < 0.01). The time course of this elevated response was
studied and, at each time point studied (3, 5 and 7 days), the
mean PFC response by cells from LL patients was substantially
higher than the responses of normal healthy controls.

        The high PFC responses to PWM by PBM cells from LL patients
prompted further investigation to determine if the blood of these
individuals contained large numbers of B lymphocytes spontaneously
secreting Ig on day 0. This was found not to be the case as the
number of spontaneously Ig secreting cells in LL patients, at day
0, was not significantly higher than the spontaneous PFC in
healthy controls.

        Another question raised by these findings was the possibility
that the LL patients had a greater number of B cells in their
peripheral blood than did the normal controls, thus, explaining
the increased number of PFC in the LL group. However, quantitation
of the B lymphocytes in the blood of LL patients revealed that the
absolute count of B cells per cubic millimeter was not significantly
higher than in the blood of normal controls. On the other hand,
the mean number of peripheral blood T cells in LL patients was
significantly lower than normal controls as has been demonstrated
previously (32). Evaluation of the numbers of the T4[+] (helper)
and T8[+] (suppressor) cell subsets (see below) revealed that they
also were correspondingly reduced in LL patients as compared to
normal controls. Of note was the absence of a preferential loss
of one T-cell subset relative to the other.

STUDIES ON IMMUNOREGULATORY DEFECTS IN THE PFC RESPONSES OF
LEPROMATOUS PATIENTS

Two of the more likely explanations for our finding that PBM
cells from LL patients gave an elevated PFC response _in vitro_ were
(1) hyperactive function by helper T cells or (2) loss of functional
suppressor cell activity. Some foundation for the latter hypothesis
was provided by our previous studies that demonstrated loss of
Con A-inducible suppressor T-cell activity in these patients.
Therefore, we conducted experiments to study the function of helper
and suppressor T-cell subpopulations from LL patients. These
experiments have been facilitated by the recent production of
monoclonal antibodies that specifically recognize helper T cells
($T4^+$) and suppressor T cells ($T8^+$) in humans and are also cytotoxic for these same subpopulations in the presence of complement (33).

Helper ($T4^+$) and suppressor ($T8^+$) cell subsets from normal or
LL patients were co-cultured with normal B-enriched cells to
differentiate between a possible deficiency of $T8^+$-suppressor
function versus hyperfunction of $T4^+$ cells. Cultures of B-enriched
cell populations alone did not produce plaques when stimulated with
PWM. Co-culture of $T4^+$ cells from either normal controls or LL
patients with normal allogenic B cells resulted in a vigorous PFC
response. Moreover, the LL $T4^+$ cells did not appear to drive the
response of B cells to supranormal levels under the experimental
conditions employed. When normal $T8^+$ cells were added to cultures
containing B cells and $T4^+$ cells, the PFC response was suppressed.
Conversely, when LL $T8^+$ cells were added to similar cultures of
normal B cells and $T4^+$ cells, they failed to suppress the response
(34).

DISCUSSION

Previously it has been suggested that the polyclonal hypergammaglobulinaemia observed in lepromatous patients is induced by
an "adjuvant" effect of _M. leprae_ present in the host tissues.
This view seems overly simplistic since we were unable to detect
elevated numbers of PBM cells that spontaneously secreted Ig
when assayed by the reverse haemolytic plaque assay on day 0. Nor
is it likely that the B cell hyperactivity can be explained on the
basis of increased numbers of B cells in the blood of LL patients,
since the absolute numbers of B cells were not significantly
elevated above those of healthy donors.

The reason for the failure of LL $T8^+$ cells to suppress the
in vitro PFC response is unclear at the present time. It is
possible that LL patients lack an auxiliary cell population _in vivo_

that may be necessary for the development of a functional $T8^+$ cell
population. One candidate for such a cell is the macrophage that
is known to provide an accessory cell function, and can potentiate
the activity of suppressor T cells (35). Furthermore, defective
T cell-macrophage interactions have been previously demonstrated
in human leprosy (23,36).

As previously stated, autoantibodies can be detected in the
serum of some lepromatous patients. Therefore, it could be
postulated that some of these antibodies may react specifically
with suppressor T cells to ablate their effect. There was, in
fact, an absolute decrease in the number of T cells in the peripheral blood of LL patients. Nevertheless, there was no evidence
that the number of $T8^+$ cells was reduced selectively, relative to
the number of $T4^+$ cells. Thus, if such an autoantibody to $T8^+$
suppressor cells were present, it must mediate its effect by non-
cytotoxic means. Additional studies are in progress to determine
the validity of this assumption. Of interest with reference to
our data is a recent report which describes an imbalance between
the $T4^+$ and $T8^+$ subsets in the blood of LL patients undergoing
reactions of erythema nodosum (37).

During the past decade, numerous theories have been generated
concerning the mode of cell-cell communication, the most comprehensive of which is that of Jerne, who postulated that cells "talk"
to each other by a series of reactions against their own antigen
receptors on the cell surface (38). These cellular antigen
receptors are considered comparable to the antigen binding sites
of the immunoglobulin molecules and are called, in consequence,
idiotypes. It has been demonstrated experimentally that responses
against idiotypes can alter the course of an immune response by
enhancing and/or suppressing responses (39), and that an animal
can generate responses against its own idiotypes (40). Therefore
it could be postulated that LL patients may develop autoanti-
idiotypic responses which lead to an inability of the $T8^+$ cell
population to function. Of interest in this regard are recent
reports (41,42) demonstrating that the serum or IgG fraction of
some patients with systemic lupus erythematosus can inhibit the
development of Con A-induced suppressor T cell activity (a function
of $T8^+$ cells) by means of a non-cytotoxic mechanism.

At present, the question remains whether the inability of the
$T8^+$ cell population to function normally is a primary or secondary
defect in LL leprosy. Notwithstanding the fact that the nature of
the primary defect in leprosy remains to be elucidated, experiments
such as reported here, hold considerable promise for improving our
understanding of the immunopathogenesis of hypergammaglobulinaemia
and autoantibody formation in leprosy.

ACKNOWLEDGMENT

The studies discussed in this communication were supported by NIH grants AI 10094 and AI 16308.

REFERENCES

1. Shepard, C.C. (1960). The experimental disease that follows the injection of human leprosy bacilli into footpads of mice. J. Exp. Med. 112:445.

2. Kirchheimer, W.F. and Storrs, E.E. (1971). Attempts to establish the armadillo as a model for the study of leprosy. Int. J. Lepr. 39:693.

3. Ridley, D.S. and Jopling, W.H. (1966). Classification of leprosy according to immunity - A five group system. Int. J. Lepr. 34:255-273.

4. Ridley, D.S. (1974). Histological classification and the immunological spectrum of leprosy. Bull. Wld. Hlth. Org. 51:451.

5. Escobar-Gutierrez, A., et al. (1973). Distribution of the HLA-A system lymphocyte antigens in Mexicans II studies in atopics and lepers. Vox Sang. 25:151-155.

6. Massoud, A., et al. (1978). A study of cell mediated immunity and histocompatibility antigens in leprosy patients in Iran. Int. J. Lepr. 46:149.

7. deVries, R.R.P., et al. (1980). HLA-linked control of susceptibility to tuberculoid leprosy and association with HLA-DR types. Tissue Antigens. 16:294.

8. Bullock, W.E. (1978). Leprosy: A model of immunological perturbation in chronic infection. J. Inf. Diseases. 137:341.

9. Mackaness, G.B. (1971). Delayed hypersensitivity and the mechanism of cellular resistance to infection. Prog. Immunol. 1:413.

10. Closs, O. and Haugen, O.A. (1975). Experimental murine leprosy III early reaction to _Mycobacterium lepraemurium_ in C3H and C57BL/6 mice. Acta. Path. Microbiol. Scand. Section A 83:51-58.

11. Beiquelman, B. (1967). Leprosy and genetics. Bull. Wld. Hlth. Org. 37:461.

12. Drutz, D.J. and Cline, M.J. (1974). Polymorphonuclear leukocyte and macrophage function in patients with leprosy. J. Clin. Invest. 53:380.

13. Godal, T. and Rees, R.J.W. (1970). Fate of _M. leprae_ in macrophages of patients with lepromatous or tuberculoid leprosy. Int. J. Lepr. 38:439.

14. Samuel, D.R., et al. (1973). Behavior of _M. leprae_ in human macrophages _in vitro_. Infect. Immun. 8:446.

15. Convit, J., et al. (1974). Elimination of M. leprae subsequent
     to local in vivo activation of macrophages in lepromatous
     leprosy by other mycobacteria. Clin. Exp. Immunol. 17:
     261.
16. Ptak, W., et al. (1970). Immune responses in mice with murine
     leprosy. Clin. Exp. Immunol. 6:117.
17. Bullock, W.E., et al. (1977). Impairment of cell mediated
     immune responses by infection with Mycobacterium leprae-
     murium. Infection and Immunity. 18:157.
18. Watson, S.R., et al. (1975). Defect of macrophage function in
     the antibody response to SRBC in systemic M. lepraemurium
     infection. Nature 256:206.
19. Bullock, W.E., et al. (1978). The evolution of immunosuppres-
     sive cell populations in experimental mycobacterial
     infections. J. Immunol. 120:1709.
20. Gershon, R.K. (1974). T cell control of antibody production.
     Contemp. Top. Immunobiol. 3:1.
21. Thomas, Y., et al. (1980). Functional analysis of human T
     cell subsets defined by monoclonal antibodies. I col-
     laborative T-T interactions in the immunoregulation of
     B cell differentiation. J. Immunol. 125:2402.
22. Ellner, J.J. (1978). Suppressor adherent cells in human
     tuberculosis. J. Immunol. 121:2573.
23. Hirschberg, H. (1978). The role of macrophages in the lympho-
     proliferative response to M. leprae in vitro. Clin.
     Exp. Immunol. 34:46.
24. Mehra, V., et al. (1979). Lepromin-induced suppressor cells
     in patients with leprosy. J. Immunol. 123:1813.
25. Mehra, V., et al. (1980). Delineation of a human T cell sub-
     set responsible for lepromin-induced suppression in
     leprosy patients. J. Immunol. 125:1183.
26. Nath, I. and Singh, R. (1980). The suppressive effect of M.
     leprae on the in vitro proliferative responses of
     lymphocytes from patients with leprosy. Clin. Exp.
     Immunol. 41:406.
27. Touw, J., et al. (1980). Effect of M. leprae on lymphocyte
     proliferation: suppression of mitogen and antigen
     responses of human peripheral blood mononuclear cells.
     Clin. Exp. Immunol. 41:397.
28. Shou, et al. (1976). Suppressor activity after concanavalin
     A treatment of lymphocytes from normal donors. J. Exp.
     Med. 143:1100.
29. Artz, R.R., et al. (1980). Decreased suppressor cell activity
     in disseminated granulomatous infections. Clin. Exp.
     Immunol. 41:343.
30. Nath, I., et al. (1979). Con A induced suppressor activity in
     human leprosy. J. Clin. Lab. Immunol. 2(4).

31. Fauci, A.S., et al. (1980). Activation of human B lympho-
        cytes. XIV Cellular requirements, interactions and
        immunoregulation of pokeweed mitogen total-immuno-
        globulin producing plaque forming cells in peripheral
        blood. Cell. Immunol. 54:230.
32. Dwyer, J.M., et al. (1973). Disturbance of the blood T-B
        lymphocyte ratio in lepromatous leprosy. N. Eng. J.
        Med. 288:1036.
33. Reinherz, E.L. and Schlossman, S.F. (1980). The differentia-
        tion and function of human T lymphocytes. Cell 19:821.
34. Bullock, W.E., et al. (1981). Hyperactivity of B lymphocyte
        function in lepromatous leprosy. Abst. 5066, Federa-
        tion Proceedings. 40:1122.
35. Stobo, J.P. (1977). Immunosuppression in man: Suppression
        by macrophages can be mediated by interactions with
        regulatory T cells. J. Immunol. 119:918.
36. Nath, I., et al. (1980). Natural suppressor cells in human
        leprosy: The role of HLA-D identical peripheral lympho-
        cytes and macrophages in the in vitro modulation of
        lymphoproliferative responses. Clin. Exp. Immunol.
        42:203.
37. Bach, M.A., et al. (1981). Studies of T cell subsets and
        functions in leprosy. Clin. Exp. Immunol. 44:491.
38. Jerne, N.K. (1974). Towards a network theory of the immune
        system. Ann. Immunol. 125c:373.
39. Ramsier, H. and Lindenmann, J. (1972). Allotypic antibodies.
        Transplant. Rev. 10:57.
40. Rodkey, L.S. (1974). Studies of idiotypic antibodies: Pro-
        duction and characterization of autoanti-idiotypic
        antisera. J. Exp. Med. 139:712.
41. Twomey, J.J., et al. (1978). A serum inhibitor of immune
        regulation in patients with systemic lupus erythematosus.
        J. Clin. Invest. 62:713.
42. Sagawa, A. and Abdou, N.I. (1979). Suppressor-cell antibody
        in systemic lupus erythematous: possible mechanism
        for suppressor cell dysfunction. J. Clin. Invest.
        63:536.

CELLULAR MECHANISMS OF RESISTANCE TO LISTERIA MONOCYTOGENES

Emil Skamene and Patricia A. L. Kongshavn

Montreal General Hospital Research Institute and
Department of Physiology, McGill University, Montreal
Quebec, Canada

INTRODUCTION

Listeria monocytogenes is a facultative intracellular bacterial
parasite that can survive and multiply in mammalian macrophages.
The infected host has at its disposal several lines of defense,
each of which is mobilized in successive steps in the course of
listeriosis. It has recently been demonstrated that the key
mechanisms underlying successful anti-listerial resistance are
genetically regulated. The availability of well-defined inbred
mouse strains facilitated greatly the study of host-parasite interaction in listeriosis and it has become obvious that genetic
control of host defense processes is simpler than originally
anticipated. This review will deal mainly with the mechanisms of
natural resistance to Listeria monocytogenes as uncovered in our
laboratory using the genetic probe. The newer aspects of acquired
resistance to this infection, a process which is now recognized
as a classical example of cell-mediated immunity, will also be
summarized briefly.

MODEL OF EXPERIMENTAL LISTERIOSIS IN MICE

Murine listeriosis (induced by intravenous or intraperitoneal
infection) can manifest itself, depending on the dose of infectious
inoculum and on the genetic make-up of the host, as either destructive infection resulting in death within a few days, as an acute
disease with subsequent recovery, or inapparent infection. When
mice are infected with an intravenous sublethal dose of Listeria,
the course of infection can be followed by examining the bacterial
burden in the liver and spleen. In these organs, a well-recognized

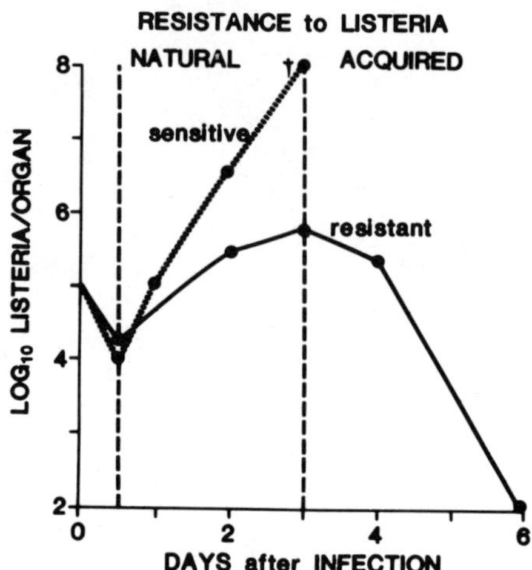

Figure 1.   Model of natural and acquired resistance to <u>Listeria</u>
<u>monocytogenes</u>.   Based on Ref. 5 and 6.

pattern is seen. The liver captures 90-95% of the intravenous
<u>Listeria</u> inoculum and destroys 50-75% of this load within 6-12
hours. Surviving organisms start to multiply and the bacterial
load increases in the liver and spleen for 2-3 days. After
reaching peak numbers, the bacteria are progressively eliminated
from the organs in the course of the next 5-6 days (Fig. 1). The
downturn in bacterial numbers, with subsequent rapid and complete
elimination of <u>Listeria</u> from the organs, is undoubtedly the result
of T cell-driven macrophage activation. However, one can easily
appreciate, by following a course of infection in different mouse
strains, that a significant degree of anti-bacterial protection is
demonstrable in several strains as early as 24-48 hours after
infection, before any detectable signs of substantial T cell
immunity develop. This phase of the host response can therefore be
termed a natural (or non-immunologically induced) resistance. It
would appear that mechanisms of anti-bacterial defense which
operate at this point serve to hold bacterial multiplication in
check until such time as specific acquired immunity develops.

The phenomenon of natural resistance is quantitative and a
relative one. Death will occur in mice of any strain, even in
those with a high level of natural resistance when they are
infected with high doses of infectious inoculum (>10^6 bacteria).
On the other hand, all mice, even those of naturally susceptible
strains, will survive infection with a low dose of <u>Listeria</u>
(<10^3). However, when an intermediate range of doses (10^3-10^6
<u>Listeria</u>) is used to infect mice of various inbred strains, the
latter can be separated into two distinct, non-overlapping groups
in terms of their resistance to listeriosis, namely, resistant and

susceptible. It is within this dose range that the biological
significance of natural resistance to <u>Listeria</u> is most convincing:
susceptible mice, deficient in this facet of the host response,
fail to prevent bacterial numbers from reaching a lethal level
of approximately 10^8 organisms/organ (spleen and liver). It can be
demonstrated that mice of susceptible strains possess the potential
to develop effective acquired immunity, but they succumb to
infection before this potential is realized.

GENETIC CONTROL OF NATURAL RESISTANCE TO <u>LISTERIA</u>

The median lethal dose of infection (LD_{50}) can be used as a
reliable parameter of natural resistance to <u>Listeria</u>. Mice which
are classified as resistant (C57BL-derived strains, NZB,SJL) can
survive an infectious dose of close to 5×10^5 organisms while a
hundredfold lower dose is lethal to susceptible mice (A/J, BALB/c,
DBA/2, CBA and others).

The pattern of resistance and sensitivity, as examined in
hybrid and backcross populations which were bred from the representative resistant and susceptible progenitors, is compatible
with the distribution of a trait which is regulated by a single,
autosomal gene. This gene locus has been designated <u>Lr</u> (for
<u>Listeria</u> resistance) and is recognized as having two alleles,
resistant ($\underline{Lr}^r$) and sensitive ($\underline{Lr}^s$) (1), with resistance being
dominant (2). Linkage studies with known chromosomal markers
decisively mapped the <u>Lr</u> gene away from the MHC locus but they
failed, as yet, to localize the gene with certainty (1,3).

PHENOTYPIC EXPRESSION OF Lr GENE

In order to analyze the mechanisms underlying the genetically-
determined, <u>Lr</u> gene-controlled natural resistance to <u>Listeria</u>, the
course of infection was followed in the resistant (C57BL/6 or B10.A)
and susceptible (A/J) mice. No difference in the numbers of viable
<u>Listeria</u> in the livers and spleens of these two strains was observed
in the first few hours of infection. This period of infection is
characterized by trapping of <u>Listeria</u> by fixed tissue macrophages,
mainly Kupffer cells in the liver, as demonstrated by light or
electron microscopy (4). A large proportion (50-75%) of this
trapped inoculum is destroyed within the next 12 hours, killing
being identical in both the resistant and susceptible mice. Almost
certainly, this antibacterial activity is expressed by the infected
resident tissue macrophages, since it is known that this mechanism
of protection is preserved by 900 R irradiation (10,19,25). The
actual infecting dose of <u>Listeria</u> that gives rise to subsequent
bacterial growth is thus identical in the animals of resistant and
susceptible strains. From this point on, the genetic advantage of

the resistant mice quickly becomes apparent. Within 24 hours
following infection, the slope of the bacterial proliferation of
Listeria starts to follow a significantly steeper course in the
susceptible A/J mice and, by 48 hours, a difference of up to a
100-fold in bacterial liver and spleen burden between the resistant
and susceptible mice can be detected (5). One could assume
intuitively that some facet of the macrophage response is responsible
for this superior antibacterial activity of genetically resistant
mice, although most of the evidence in support of this notion is,
at this point, indirect. Thus, this phase of resistance is intact
in T cell deficient mice (7,8,9,10), in B cell deficient mice prepared by treatment with rabbit anti-mouse immunoglobulin M serum
from birth (28), and in NK cell-deficient (beige) mice (Skamene,
E., and Roder, J., unpublished observation). Although polymorphonuclear cell infiltration is the prominent histologic feature in
the livers of infected mice in the early phase of resistance, this
picture has an identical character in resistant and susceptible
strains, thus making it unlikely that this cell type contributes
to natural anti-Listeria resistance. On the other hand, maneuvers
that interfere with macrophage function (silica, dextran sulphate
DS-500, carrageenan, colloidal carbon) or macrophage production
(irradiation, treatment with the marrow-seeking isotope Sr^{89})
effectively abrogate this stage of resistance (11,25,26). More
importantly, these treatments have a differential effect when applied
to either resistant or susceptible animals (11). The genetic
advantage of resistant strains can be eliminated totally by whole-
body-irradiation, at a dose as low as 200R. Such radiation-induced
susceptibility to Listeria of genetically resistant mice can be
fully corrected by the transfer of normal bone marrow cells.
These observations lead to the conclusion that the high level of
anti-Listeria activity of genetically-resistant mice is mediated
by a radiosensitive cell of the mononuclear phagocyte system,
which is almost certainly derived from blood monocytes that
originate from rapidly-dividing bone marrow precursors (12,13).
This population of effector cells was, in fact, first observed by
North (14) who established that an effective host response to
Listeria, at this stage of infection, involves the progressive
influx, to the foci of infection, of macrophages derived from
rapidly-dividing extra-hepatic precursors.

     Mechanisms of natural resistance to Listeria are quite
different in genetically susceptible strains of mice as compared
to genetically resistant strains. Irradiation, which so drastically
abrogated host defense in the resistant strain, had no effect for
at least the first 2 days of infection, on the anti-Listeria
response of susceptible animals. However, treatment of genetically
susceptible hosts with dextran sulphate resulted in acceleration
of a fulminant course of infection. It can be concluded that the
genetically susceptible host relies heavily, if not exclusively,
on the activity of the fixed macrophage system for its protection

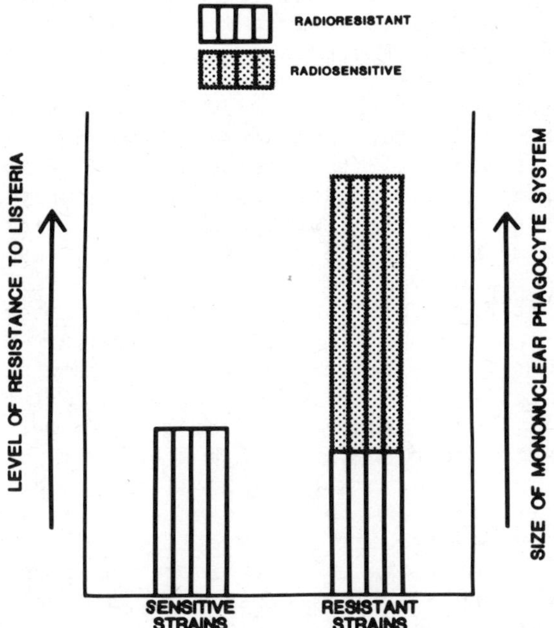

Figure 2.   Contribution of two macrophage populations to natural
resistance aganst Listeria monocytogenes.   Based on
Ref. 11.

in the phase of natural resistance.   The genetically resistant
host, on the other hand, is able to mobilize promptly a new macrophage population derived from radiosensitive precursors, thus
greatly augmenting its defenses (Fig. 2).   The emphasis here is
definitely on the promptness of the monocyte-macrophage response.
Mice of susceptible strains must eventually also recruit newly-
formed macrophages to the infectious sites, but the host may pay
a high price for this sluggish response in being unable to prevent,
in the short time available, bacterial multiplication from reaching
lethal levels.

Given the significance of the Lr gene product in protecting
mice against uncontrolled proliferation of Listeria, it was of
importance to determine how this gene might augment the macrophage
response in a manner that resulted in a rapid influx of new cells
to the sites of infection.   Does the macrophage precursor of the
resistant strain possess intrinsic genetic characteristics which
are responsible for its prompt response to an infectious stimulus,
or does the gene product express its regulatory influence in the
hemopoietic microenvironment in which such macrophage response
takes place?   This question was answered by testing the natural
resistance to Listeria of radiation bone marrow chimeras created

between Lr^r and Lr^s animals. Chimeras carrying macrophages of the
Lr^s genotype that have differentiated in the Lr^r microenvironment
proved to be resistant to Listeria while the chimeras carrying the
Lr^r macrophages differentiating in the Lr^s host were susceptible
(15). The Lr gene product thus clearly influences the hemopoietic
microenvironment and has the functional characteristics of an
inducible monocytopoietin. Its precise mode of action is not fully
defined at present but it appears to act by promptly decreasing
monocyte generation time (T_G), within hours following Listeria
infection of the animals of the resistant strain (29). Several
other components of the mononuclear phagocyte system were shown to
express an augmented response in mice of Listeria-resistant strains.
Thus, prompt peripheral blood monocytosis is observed in the
resistant, but not in the susceptible, mice following intravenous
injection of Listeria (or of a soluble extract derived from Listeria
cell walls) (11). Young mononuclear cells derived from rapidly
dividing precursors under the influence of the Lr gene product have
a superior response to chemotactic stimuli (27) and subsequently
the monocyte traffic between the blood and tissues is accelerated
as demonstrated by a decreased monocyte T1/2 in the mice of
genetically-resistant strains. Furthermore, an augmented peritoneal
macrophage response to a variety of stimuli is also a characteristic
feature of Lr^r mice. When challenged with a sterile irritant
injected into the peritoneal cavity, these animals respond by the
accumulation of large numbers of macrophages which far exceed the
numbers induced in Lr^s mice. This trait is also genetically
regulated in a manner similar to the genetic control of resistance
to Listeria. It could be demonstrated, by linkage analysis, that
both events (macrophage inflammatory response to a sterile irritant
and natural resistance to Listeria) are regulated by an identical
chromosomal locus (Stevenson et al., this volume). The Lr gene
product is, therefore, augmenting the monocyte response to any
appropriate stimulus, be it a peritoneal irritation (with resulting
accumulation of macrophages in the peritoneal cavity) or Listeria
organisms multiplying in the liver (with the resulting division
and migration of macrophage precursors to the foci of infection).
Such augmented monocytopoiesis seems to lead to shifts in the pool
size within the compartments of the macrophage differentiation
pathway, resulting in an increased pool of promonocytes and monocytes, and in a relatively contracted pool of macrophage stem
cells. Evidence for the latter statement stems from the observation that the bone marrow of genetically resistant animals, although
hypercellular and presumably containing an expanded pool of promonocytes, has a very low plating efficiency for macrophage colonies
in vitro when compared with bone marrow derived from genetically
susceptible hosts (16).

The Lr gene, therefore, although originally uncovered by its
regulation of natural resistance to Listeria, is a perfect example
of a gene with pleiotropic effects. Its action would be

demonstrable in any number of biologic situations which call for
prompt monocytopoiesis. One would envisage that the Lr gene
product, when isolated and identified, would be useful not only
in converting the Listeria-susceptible animal into a resistant
one (as the chimera experiment demonstrated) but in many other
circumstances where an enhancement of monocytopoiesis is desirable.

ACQUIRED RESISTANCE TO LISTERIA

    The course of infection in the later phase of the host response
(from the 2nd or 3rd day on) is characterized by adaptive changes
in macrophages which are induced by specifically-sensitized
mediator T lymphocytes (17,18). It should be noted that the Lr
gene action is also demonstrable in the phase of acquired resistance,
presumably reflecting the bone marrow monocytopoietic response to
lymphokines (5,6). Similarly, both the early secondary response
and the adoptive transfer of resistance to naive recipients, is
restricted by the Lr gene, in that the resistant strain macrophages respond much better to the T cell stimuli than do the
sensitive strain macrophages. With an ongoing secondary immune
response to Listeria, however, the macrophage system of the
Listeria-susceptible strain eventually reaches effective levels of
production with resulting abrogation of Lr-controlled genetic
differences in response to Listeria (5,6).

    The characteristics of T cell-induced macrophage activation have
been comprehensively reviewed elsewhere (20). With the recent
appreciation of the heterogeneity of T cell subsets and their roles
in the regulatory network of immunity, it was of interest to
identify the T cell subpopulations serving a mediator role in macrophage activation for Listeria resistance. Experiments using Lyt
antisera to eliminate defined T cell subsets from adoptively
transferred inoculum in the Listeria system indicate that cells of
the Lyt 1,2,3+ type are involved (21). The cooperation of Lyt 1+
cells with an Lyt 2,3+ population in this function is, however,
not excluded. It is further evident that the T cell subset which
regulates macrophage activation can do so only in the situation in
which both the mediator T cell and the effector macrophage share
I-A region coded characteristics. This was demonstrated in the
models of macrophage activation for anti-listerial protection in
vivo,macrophage activation for tumor cytotoxicity in vitro and the
production of monokines in vitro (22-24). Furthermore, anti-Ia
antibody treatment of macrophages abrogates such T cell-macrophage
interaction in vitro. It is apparent that the same kind of MHC
restrictions, that in many models have been demonstrated convinc-
ingly to be operating in the afferent arc of the immune response,
apply also in the effector phase, at least in the case of listerial
immunity.

It can be concluded that macrophages, at several stages of their cellular history, contribute to the outcome of host-parasite interaction in infection with <u>Listeria</u>. At least two levels of the genetically-controlled macrophage response can be identified, (1) the <u>natural resistance</u> which is regulated by the <u>Lr</u> gene operating at the level of monocytopoiesis and, (2) <u>acquired resistance</u> (specific immunity) where the T cell-macrophage interaction is restricted by the I-A locus of MHC.

ACKNOWLEDGMENT

This research was supported by MRC grants 5389 and 6431.

REFERENCES

1. Cheers, C. and McKenzie, I.F.C. (1978). Infect. Immunity
        19:755.
2. Cheers, C., McKenzie, I.F.C., Mandel, T.E. and Chan, Y.Y.
        (1980). In Genetic Control of Natural Resistance to
        Infection and Malignancy. Edited by Skamene, E.,
        Kongshavn, P.A.L., and Landy, M., Academic Press, New
        York, p. 141.
3. Skamene, E., Kongshavn, P.A.L. and Sachs, D. (1979). J.
        Infect. Dis. 139:228.
4. Cheers, C., Mandel, T.E., and McKenzie, I.F.C. (1978). In
        Function and Structure of the Immune System. Edited by
        Muller-Rucholtz, W., and Muller-Hernelink, H.K., Plenum
        Press, New York, p. 703.
5. Skamene, E. and Kongshavn, P.A.L. (1979). Infect. Immun.
        25:345.
6. Cheers, C., McKenzie, I.F.C., Pavlov, H., Waid, C. and York,
        J. (1978). Infect. Immunity 19:763.
7. Emmerling, P., Finger, H. and Bockemuhl, J. (1975). Infect.
        Immun. 12:437.
8. Cheers, C. and Waller, R. (1975). J. Immunology 115:844.
9. Chan, C., Kongshavn, P.A.L. and Skamene, E. (1977). Immunol-
        ogy 32:529.
10. Newborg, M.F. and North, R.J. (1980). J. Immunology 124:571.
11. Sadarangani, C., Skamene, E. and Kongshavn, P.A.L. (1980).
        Infect. Immunity 28:381.
12. Volkman, A. and Gowans, J.L. (1965). Brit. J. Exp. Path.
        44:62.
13. van Furth, R., and Cohn, Z.A. (1968). J. Exp. Med. 128:415.
14. North, R.J. (1970). J. Exp. Med. 132:521.
15. Kongshavn, P.A.L., Sadarangani, C., and Skamene, E. (1980).
        Cell. Immunology 53:341.

16. Stewart, C.C., Skamene, E. and Kongshavn, P.A.L. (1980). In
         Genetic Control of Natural Resistance to Infection and
         Malignancy. Edited by Skamene, E., Kongshavn, P.A.L.
         and Landy, M., Academic Press, New York. p. 499.
17. Mackaness, G.B. (1969). J. Exp. Med. 129:973.
18. North, R.J. (1973). J. Exp. Med. 138:342.
19. Takeya, K., Shimotori, S., Taniguchi, T. and Nomoto, K. (1977).
         J. Gen. Micro. 100:373.
20. North, R.J. (1974). In Activation of Macrophages. Edited by
         Wagner, W.H. and Hahn, H. Excerpta Medica, Amsterdam.
         p. 210.
21. Kaufmann, S.H.E., Simon, M.M. and Hahn, H. (1979). J. Exp.
         Med. 150:1033.
22. Zinkernagel, R.M., Althage, A., Adler, B., Blanden, R.V.,
         Davidson, W.F., Kees, U., Dunlop, M.B.C. and Shreffler,
         D.C. (1977). J. Exp. Med. 145:1353.
23. Farr, A.G., Kiely, J-M. and Unanue, E.R. (1979). J. Immunol-
         ogy. 122:2395.
24. Farr, A.G., Wechter, W.J., Kiely, J-M. and Unanue, E.R. (1979).
         J. Immunology 122:2405.
25. Mitsuyama, M., Takeya, K., Nomoto, K. and Shimotori, S. (1978).
         J. Gen. Micro. 106:165.
26. Bennett, M. and Baker, E.E. (1977). Cell. Immunology 33:203.
27. Stevenson, M.M., Kongshavn, P.A.L. and Skamene, E. (1981).
         J. Immunology 127:in press.
28. Gordon, J., Brodt, P., Vargas, F. and Kongshavn, P.A.L. (1981).
         Infect. Immun. in press.
29. Sadarangani, C., Kongshavn, P.A.L. and Galsworthy, S. (1981).
         J. RES. in press.

EFFECT OF INTERFERON INDUCERS AND PURIFIED MOUSE INTERFERON ON

THE SUSCEPTIBILITY OF MICE TO INFECTION WITH <u>LISTERIA</u> <u>MONOCYTOGENES</u>

Sarah F. Grappel, Michael J. Polansky, and Paul Actor

Smith Kline and French Laboratories
Philadelphia, PA

INTRODUCTION

Depression of cell-mediated immune reactions during viral
infection has been a repeatedly confirmed phenomenon since the
original observation of von Pirquet (20). Reports have included
depression of tuberculin activity in man following viral infection
or vaccination (2,3,12,16,19) as well as prolongation of allograft
survival in virus-infected mice (14). Numerous investigations
have indicated that virus-induced Type I interferons can modulate
cell-mediated immune responses. These also include inhibition of
the development of delayed-type hypersensitivity to tuberculin
and sheep red blood cells (6,7) and prolongation of skin allograft
survival in mice (13).

Tilorone, a potent inducer of Type I interferon has also been
shown to be capable of suppressing delayed-type hypersensitivity
reactions in mice (15). In addition, pretreatment of mice or
rats with a single dose of Tilorone, as well as several other
inducers of Type I interferon, has been shown to enhance virus-
induced oncogenesis (9,10). Collins (4) showed that treatment of
CD-1 mice with tilorone suppressed their cell-mediated immune
response to the bacterial pathogens, <u>Listeria</u> <u>monocytogenes</u>,
<u>Mycobacterium</u> <u>tuberculosis</u> and <u>Salmonella</u> <u>enteritidis</u>. The
susceptibility of the pretreated mice to these intracellular
pathogens was increased significantly. Gruenwald and Levine (11)
confirmed the increase in susceptibility of CD-1 mice to <u>L</u>. <u>mono-
cytogenes</u> following treatment with tilorone. They also showed
that susceptibility to an extracellular pathogen, <u>Staphylococcus
aureus</u> was not increased by such treatment.

This investigation was carried out to determine whether other inducers of Type I or Type II interferon, as well as crude and partially purified mouse interferon preparations are also capable of increasing susceptibility of mice to intracellular infection with L. monocytogenes.

METHODS AND RESULTS

The EGD strain of Listeria monocytogenes was used to establish lethal infections in male albino Webster derived CD-1 mice weighing 18-20 g. Log. phase trypticase soy broth cultures of L. monocytogenes were diluted in sterile saline and injected intraperitoneally, 0.5 ml/mouse. The interferon inducers, tilorone, poly A-U, poly I-C, statolon, and levamisole, were administered in a single dose prior to infection as described in the tables. Partially purified and crude mouse interferons were obtained from Cal. Biochem. These were also administered in a single dose prior to infection. The final survival percentages for groups of 10 mice at each level of inoculum after 7 days were used to determine the LD_{50} for L. monocytogenes before and after treatment with the interferon inducers. The statistical significance of the effect of the interferon inducers was also determined (1,8).

Pretreatment of mice with tilorone, 100 mg/kg, p.o., 3 hours prior to infection, resulted in a significant decrease in the LD_{50} for L. monocytogenes i.e., a decrease of approximately 2 Logs. (Table 1). This increase in virulence was significant within 95% confidence limits.

Treatment of mice 3 hours prior to infection with poly I-C, poly A-U or statolon, other compounds which also induce Type I interferon, resulted in a significant increase in susceptibility to L. monocytogenes (Tables 2 and 3). The decreases in LD_{50}'s for L. monocytogenes following treatment with poly I-C and poly A-U were significant within 95% confidence limits. The decrease in the LD_{50} for L. monocytogenes following treatment with statolon was significant within 90% confidence limits. In each of these tests, tilorone pretreatment increased the susceptibility of the mice significantly.

Pretreatment with levamisole, an inducer of Type II interferon, also increased the susceptibility of mice to infection. The increase in susceptibility was significant when mice were infected 24 hours post-treatment (Table 4) and corresponded to the time at which the serum level of Type II interferon is optimal (18). No significant change in LD_{50} was obtained in mice treated with levamisole three hours prior to infection. These time intervals were in sharp contrast to those seen with tilorone administration.

Table 1. Effect of Tilorone on the Susceptibility of CD-1 Mice to L. monocytogenes[a]

L. monocytogenes EGD	Mice	
	Control	Tilorone Pre-Treatment[b]
Virulence (LD_{50}, cfu/mouse)	5.2×10^6	4.8×10^4
95% Confidence limits	$(1.2 \times 10^6 - 3.7 \times 10^7)$	$(1.3 \times 10^4 - 1.8 \times 10^5)$
Relative virulence in tilorone treated		108.5[*]

[a] Test duration = 7 days. L. monocytogenes injected i.p. in Webster-derived CD-1 mice.

[b] Tilorone, 100 mg/kg, p.o., three hours prior to infection.

[*] Significant increase in virulence within 95% confidence limits.

Table 2. Effect of Interferon Inducers Poly I-C and Poly A-U on the Susceptibility of CD-1 Mice to L. monocytogenes EGD[a]

Compound[b]	Regimen	L. monocytogenes Infection		
		LD_{50} cfu/mouse	95% Confidence Limits	Virulence Relative to Control[b]
Control	–	8.6×10^5	$2.1 \times 10^5 - 3.9 \times 10^6$	
Tilorone	100 mg/kg p.o.	1.4×10^4	$4.3 \times 10^3 - 4.6 \times 10^4$	62.3[*]
Poly I-C	0.6 mg/kg i.p.	9.4×10^3	$2.6 \times 10^3 - 3.2 \times 10^4$	91.5[*]
Poly A-U	1.2 mg/kg i.p.	1.3×10^5	$3.4 \times 10^4 - 4.5 \times 10^5$	6.9[*]

[a] L. monocytogenes injected i.p. Test duration = 7 days.

[b] Compounds administered 3 hours prior to infection.

[*] Significant within 95% confidence limits.

Table 3.   Effect of Statalon on the Susceptibility of CD-1 Mice to
           $\underline{L}$. monocytogenes EGD[a]

		L. monocytogenes Infection		
Compound[b]	Regimen	LD_{50} cfu/mouse	95% Confidence Limits	Virulence Relative to Control[b]
Control		1.8×10^6	2.9×10^5–1.0×10^7	
Tilorone	100 mg/kg p.o.	5.2×10^4	1.7×10^4–1.5×10^5	35.4[*]
Statalon	100 mg/kg p.o.	2.9×10^5	7.9×10^4–9.9×10^5	6.4[**]

[a] $\underline{L}$. monocytogenes EGD injected i.p.   Test duration = 7 days.

[b] Compounds administered 3 hours prior to infection.

[*] Significant within 95% confidence limits.

[**] Significant within 90% confidence limits.

Table 4.   Effect of Levamisole on the Susceptibility of CD-1 Mice
           to $\underline{L}$. monocytogenes EGD[a]

		L. monocytogenes Infection		
Compound[b]	Regimen	LD_{50} cfu/mouse	95% Confidence Limits	Virulence Relative to Control[b]
Control		4.8×10^4	1.4×10^4–1.8×10^5	
Tilorone	100 mg/kg p.o. 3 hrs pre	1.2×10^3	2.0×10^2–5.6×10^3	41.6[*]
Tilorone	100 mg/kg p.o. 24 hrs pre	1.2×10^4	4.0×10^3–3.6×10^4	3.97 NS
Levami-sole	10 mg/kg p.o. 24 hrs pre	4.2×10^3	5.5×10^2–2.4×10^4	11.6[*]
Levami-sole	10 mg/kg p.o. 3 hrs pre	3.9×10^4	2.7×10^3–7.7×10^5	1.24 NS

[a] $\underline{L}$. monocytogenes EGD injected i.p.   Test duration = 7 days.
Compounds administered prior to infection.

[b*] Significant, N.S. = not significant within 95% confidence limits.

Table 5. Effect of Mouse Interferon Preparations on the Susceptibility of CD-1 Mice to _L. monocytogenes_ EGD[a]

Compound	Regimen[b]	_L. monocytogenes_ Infections		
		LD_{50}	95% Confidence Limits	Virulence Relative to Control[c]
Control		8.0×10^4	1.3×10^4 – 5.3×10^5	
Tilorone	100 mg/kg p.o.	3.8×10^3	6.6×10^2 – 1.7×10^4	21.1[*]
Crude Mouse Interferon	800 units i.v.	6.5×10^4	6.4×10^3 – 5.2×10^5	1.23 NS
Partially Purified Mouse Interferon	800 units i.v.	9.3×10^4	1.6×10^4 – 6.2×10^5	0.86 NS
Mock Crude Mouse Interferon	Equiv. weight 800 units i.v.	2.2×10^5	3.4×10^4 – 1.5×10^6	0.36 NS
Mock Partially Purified Mouse Interferon	Equiv. weight 800 units i.v.	8.1×10^4	1.9×10^4 – 3.9×10^5	0.99 NS

[a] Mouse interferon preparations from Cal. Biochem. _L. monocytogenes_ injected i.p. in Webster-derived, CD-1 mice. Test duration = 7 days.

[b] All compounds given in a single dose prior to infection; tilorone given 3 hours pre; others 1 hour pre.

[c] * Significant, NS = Not significant within 95% confidence limits.

In an attempt to determine whether the increased susceptibility
to L. monocytogenes of mice treated with these agents was due to
interferon, mice were pretreated with crude and partially purified
interferons, as well as with the corresonding mock preparations.
With the regimen used, these preparations had no significant effect
on the susceptibility of these mice to infection (Table 5).
Tilorone continued to increase susceptibility significantly in the
same experiment.

DISCUSSION

Tilorone has been reported to cause a depletion of the thymus-
dependent areas of lymphoid tissue (17) as well as to affect re-
circulation of lymphocytes (15). In addition, Collins (5) has
shown that following tilorone treatment, there is a decrease in
peripheral blood leukocytes resembling that obtained following
sublethal x-irradiation. Such effects may account for the increase
in susceptibility of CD-1 mice to intracellular infection with
L. monocytogenes following tilorone treatment.

All of the interferon inducers, regardless of whether they
induced Type I or Type II, significantly increased the susceptibility of mice to L. monocytogenes. Further investigations would
be required to determine whether similar changes in the lymphoid
tissue and peripheral blood occur with the other interferon
inducers and with mouse interferon itself.

REFERENCES

1. Berkson, J. (1953). Statistically precise and relatively
        simple method of estimating the bioassay with quantal
        response based on the logistic function. J. Amer. Stat.
        Assoc. 48:565-599.
2. Brody, J.A. and MacAlister, R. (1964). Depression of
        tuberculin sensitivity following measles vaccination.
        Amer. Rev. Resp. Dis. 90:607-611.
3. Brody, J.A., Overfield, T. and Hammes, L.M. (1964). Depression
        of the tuberculin reaction by viral vaccines. N. Eng.
        J. Med. 271:1294-1296.
4. Collins, F.M. (1975). Effect of tilorone treatment on intra-
        cellular microbial infections in specific-pathogen-free
        mice. Antimicrob. Agents Chemother. 7:447-452.
5. Collins, F.M. (1980). Mechanism of cellular suppression
        induced by oral tilorone treatment of mice. Infect.
        Immun. 30:289-296.
6. DeMaeyer, E. (1976). Interferon and delayed-type hypersensi-
        tivity to a viral antigen. J. Infect. Dis. 133, suppl:
        A63-A65.

7.   DeMaeyer, E., DeMaeyer-Guignard, J. and Vandeputte, M. (1975).
     Inhibition by interferon of delayed-type hypersensitivity
     in the mouse. Proc. Nat. Acad. Sci. 72:1753-1757.

8.   Finney, D.J. (1971). Probit Analysis. Cambridge Univ. Press
     3rd ed., London.

9.   Gazdar, A.F. (1972). Enhancement of tumor growth rate by
     interferon inducers. J. Nat. Cancer Inst. 72:1435-1438.

10. Gazdar, A.F., Steinberg, A.D., Spahn, G.F. and Baron, S.
     (1972). Interferon inducers: Enhancement of viral
     oncogenesis in mice and rats. Proc. Soc. Exp. Biol.
     Med. 139:1132-1137.

11. Gruenwald, R. and Levine, S. (1976). Effect of tilorone on
     susceptibility of mice to primary of secondary infection
     with L. monocytogenes. Infect. Immun. 13:1613-1618.

12. Hall, C.B. and Kantor, F.S. (1972). Depression of established
     delayed hypersensitivity by mumps virus. J. Immunol.
     108:81-85.

13. Hirsch, M.S., Ellis, D.A., Black, P.H., Monaco, A.P. and Wood,
     M.L. (1974). Immunosuppressive effects of an interferon
     preparation in vivo. Transplantation 17:234-236.

14. Howard, R.J., Mergenhagen, S.E., Notkins, A.L. and Dougherty,
     S.F. (1969). Inhibition of cellular immunity and
     enhancement of humoral antibody formation in mice
     infected with lactic dehydrogenase virus. Transplant
     Proc. 1:586-588.

15. Kettman, J.R. (1978). Modulation of the acquisition and
     expression of immunity by tilorone: I. Delayed-type
     hypersensitivity responses.   Immunopharmacology 1:21-28.

16. Lamb, G.A. (1969). Effect of HPV-80 rubella vaccine on the
     tuberculin reaction. Amer. J. Dis. Child. 118:261.

17. Levine, S., Gibson, J.P. and Megel, H. (1974). Selective
     depletion of thymus dependent areas in lymphoid tissue
     by tilorone. Proc. Soc. Exp. Biol. Med. 146:245-248.

18. Matsubara, S., Suzuki, F. and Ishida, N. (1979). The
     induction of interferon by levamisole in mice. Cell
     Immunol. 43:214-219.

19. Starr, S. and Berkovich, S. (1964). Effects of measles,
     gamma-globulin-modified measles and vaccine measles
     on the tuberculin test. New Eng. J. Med. 270:386-391.

20. von Pirquet, C. (1908). Das Verhalten der Kutanen tuberkulin-
     reaktion während der Masern. Deut. Med. Wochenschr.
     34:1297-1300.

NATURAL RESISTANCE TO <u>LISTERIA MONOCYTOGENES</u> AS A FUNCTION OF
MACROPHAGE INFLAMMATORY RESPONSE

Mary M. Stevenson, Patricia A. L. Kongshavn and Emil
Skamene

The Montreal General Hospital Research Institute
Montreal, Quebec

The accumulation and mobilization of mononuclear phagocytes
to foci of infection provide a vital source of effector cells for
host defense against intracellular pathogens. For example, during
the course of infection in mice with <u>Listeria monocytogenes</u>, the
accumulation of inflammatory macrophages has been shown to be
critical in preventing fulminant bacterial growth in the infected
organs before the appearance of sensitized T-cells and subsequent
activation of macrophages (1,2). The recruitment of adequate
numbers of inflammatory macrophages may, therefore, be related to
the level of genetically-determined host resistance to infection
with Listeria. The level of resistance in inbred mice is genetically controlled by a single, dominant, non H-2 linked autosomal
gene which has been named the <u>Lr</u> gene (3,4). In surveying various
inbred strains of mice which can be distinguished on the basis of
the presence or absence of the resistant allele at the <u>Lr</u> gene
locus, we have found differences in the number of macrophages
which accumulate in the peritoneal cavity following intraperitoneal
(ip) treatment with thioglycollate. Mouse strains characterized as
Listeria-resistant (LD_{50} > 10^5 CFU: C57BL/6J, B10.A, SJL, B6.C-H-
2^{ba}/ByJ) exhibited a high macrophage inflammatory response while
strains characterized as sensitive (LD_{50} < 10^4:A/J, DBA/1J, DBA/2J,
BALB/cJ) had low responses (Table 1).

The correlation between resistance to Listeria and a high
macrophage inflammatory response was examined in greater detail
using the strain combination B10.A and A/J. B10.A mice infected
with a sub-lethal dose of Listeria show the characteristic pattern
of bacterial growth described by Mackaness (5). In contrast,
Listeria-sensitive A/J mice infected with a comparable dose are
unable to control bacterial proliferation in their livers and

Table 1. Strain Survey - Macrophage Inflammatory Response in
          Inbred Mice

Strain	H-2 type	Resistance to Listeria[a]	Macrophage Inflammatory Response[b,c]
C57BL/6J	b	+	22.4 ± 1.1
B10.A	a	+	21.6 ± 1.2
B6.C-H-2^{ba}/ByJ	ba	+	21.8 ± 2.7
SJL	s	+	16.3 ± 1.5
DBA/2J	d	-	6.5 ± 0.5
DBA/1J	q	-	9.5 ± 1.0
A/J	a	-	7.0 ± 0.5
C3H/HeJ	k	±	20.5 ± 2.9
Balb/c	d	-	6.8 ± 3.5

[a] + LD_{50} 10^5 - 5×10^5 cfu

   - LD_{50} 5×10^3 - 10^4 cfu

   ± LD_{50} 3×10^4 - 5×10^4 cfu

[b] Numbers of macrophages were calculated from total and differential counts performed on samples of peritoneal exudate cells
from 5 to 40 individual mice treated 3 days previously with
thioglycollate.

[c] High >15×10^6 macrophages, Low <10×10^6 macrophages.

spleens (6). Furthermore, studies in our laboratory of the
phenotypic expression of the Lr gene have shown that the biological
advantage of genetically resistant strains is dependent upon the
early appearance of a radiosensitive precursor cell of the mononuclear phagocyte system (7).

    Since the size of the mononuclear phagocyte system may be
the basis for the superior anti-listerial ability of resistant
mice, we obtained an approximation of the size of one of its
compartments by determining the number of macrophages in the unstimulated peritoneal cavity. There was a significant difference
between the two strains in the yield of resident peritoneal
macrophages (p < 0.05; Figure 1). Analysis of the ability to
produce an inflammatory exudate to a variety of non-specific
sterile stimuli magnified the quantitative difference between the
sizes of the mononuclear phagocyte systems of Listeria sensitive
A/J and resistant B10.A mice. In addition to 2-3 fold differences

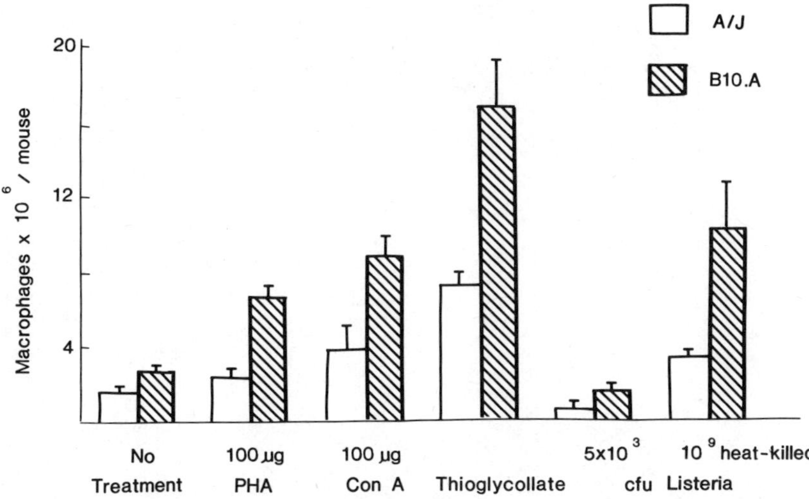

Figure 1. Macrophage inflammatory response: specific and non-specific stimuli. Mice were untreated or were injected ip with 1.0 ml of the agents indicated 3 days before determining macrophage inflammatory responses except for live Listeria (24 hr) and heat-killed Listeria (48 hr). Numbers of macrophages were determined from total cell and differential counts of peritoneal washouts from individual mice. Results represent mean ± SEM of 3 to 4 mice per group.

following ip treatment with thioglycollate, PHA or Con A, there was an enhanced accumulation of peritoneal macrophages in resistant B10.A mice following ip infection with live or heat-killed Listeria. The differences between the resistant and sensitive strains in macrophage inflammatory response during infection and to non-specific stimuli suggest that a high level of anti-listerial resistance may be related to host ability to produce and/or recruit large numbers of inflammatory macrophages to the site of infection.

To determine if the enhanced macrophage inflammatory response evident in B10.A mice was due only to quantitative differences between the mononuclear phagocyte systems of the two strains, we examined the course of accumulation of inflammatory cells after ip injection with thioglycollate. B10.A mice produced a 2-3 fold greater accumulation of inflammatory cells in comparison to A mice through 4 days after injection of thioglycollate (Figure 2). The enhanced inflammatory response to B10.A mice was evidenced by the recruitment of both more total peritoneal cells and more macrophages. In addition to quantitative differences, B10.A mice produced an inflammatory exudate sooner than A/J mice. The number of

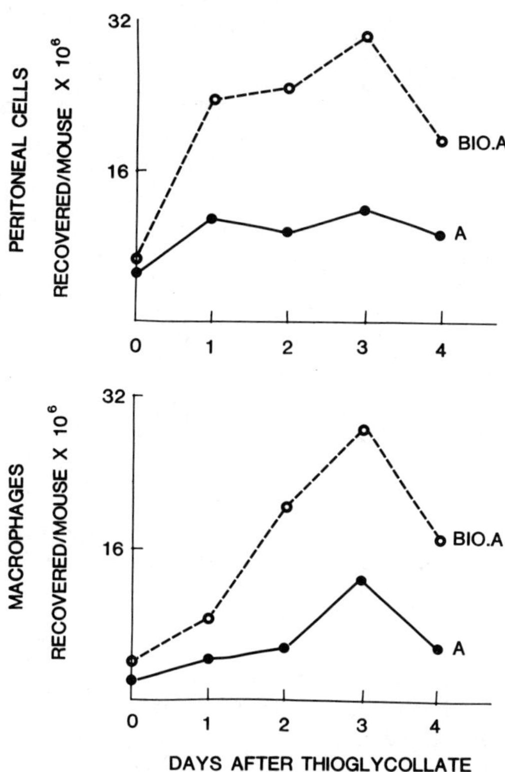

Figure 2.   Accumulation of inflammatory cells in B10.A and A/J
            mice. Mice were injected ip with 1.0 ml of thiogly-
            collate at various times. The numbers of total cells
            and macrophages were determined on samples of perito-
            neal washouts pooled from 3 mice per group. Data are
            presented as the number of cells recovered per mouse.

inflammatory macrophages harvested from B10.A mice was doubled by
day 1, reached approximately a 7-fold peak at day 3 and was
returning to normal levels by day 4. Although there was a significant increase by day 3 in A mice, the accumulation of inflammatory macrophages in these mice was not as dramatic as that
evident in B10.A strain hosts.

    We have previously shown that the quantitative differences in
inflammatory responsiveness _in vivo_ were paralleled by qualitative
differences in macrophage chemotactic responsiveness _in vitro_ (8).
The differences in chemotactic responsiveness to the serum derived
chemotactic factor C5a between thioglycollate-induced macrophages
from B10.A and A mice were not dependent upon the dose of C5a,
the time of incubation or the time following treatment of the
mice with thioglycollate. However, the differences have been

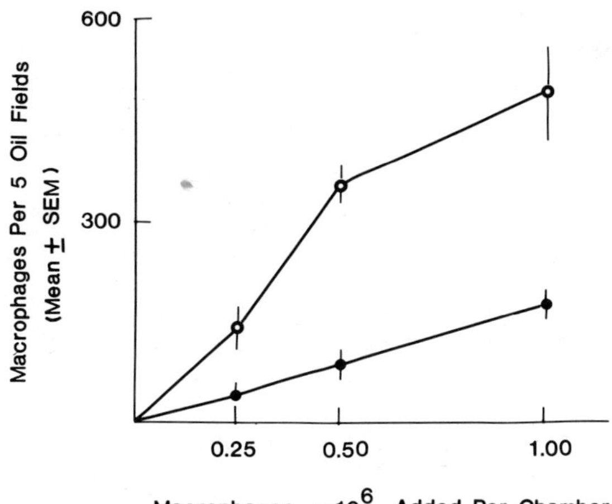

Figure 3.   Chemotactic responses of various numbers of thiogly-
            collate-induced macrophages from B10.A and A/J mice.
            Chemotaxis to 1/500 dilution of endotoxin activated
            mouse serum containing C5a(9) was measured after 4 hr
            of incubation. Results are expressed as mean number
            of macrophages per 5 oil fields ± SEM for triplicate
            filters.

found to be due to differences between the strains in the number
of responsive macrophages within the population of thioglycollate-
induced cells. When 1×10^6 macrophages from A mice were added to
the top of the modified Boyden chamber, the number of migrated
macrophages was equal to the B10.A response when one-fourth
(0.25×10^6) the number of macrophages was used (Figure 3). Thus,
the differences in the _in vitro_ chemotactic responsiveness of
macrophages from the two strains appears to be due to quantitatively fewer responsive cells in A/J strain mice. From the
in vivo and _in vitro_ results, it can be concluded that Listeria
resistant B10.A mice can promptly mobilize large numbers of mononuclear phagocytes to local sites of inflammation or infection.

     The results of the strain survey suggested to us that the
level of macrophage responsiveness was influenced by the genetic
background of the host. To determine if the macrophage inflammatory response is under genetic control, we compared the inflammatory responses of A and B10.A parental strains with those of
(B10.A x A)F_1, F_2 and backcross animals. The cumulative results
of five experiments are summarized in Table 2. The number of
macrophages recovered from individual mice following injection
with thioglycollate was expressed as the percent of the mean
response of all B10.A mice (n = 29; mean ± SEM = 100 ± 5%). The
mean macrophage inflammatory response of all A mice (n = 26) was

Table 2.   Segregation of High and Low Responders to Macrophage
           Inflammatory Stimulus

Segregating Population[a]	High Responders[b]		Low Responders[c]	
	% Obtained	% Expected[d]	% Obtained	% Expected[d]
B10.A (29)	100	100	0	0
A/J (26)	0	0	100	100
(B10.AxA)F_1 (34)	100	100	0	0
F_1xB10.A (25)	92	100	8	0
F_1xA (34)	55	50	45	50
F_2 (31)	72	75	28	25

[a] number of mice per group

[b] > mean ± 2Sd of A/J response

[c] < mean ± 2Sd of A/J response

[d] percentage expected for a trait controlled by a single dominant
gene.

37 ± 2%. The upper limit of low macrophage inflammatory responses
was chosen as 63% (mean ± 2 standard deviations of the A strain
response; 95% confidence limits). Using these criteria to
distinguish high and low responders, 34 of 34 F_1 mice had high
intermediate macrophage inflammatory responses. Of the backcross
animals, 92% (23 of 25) of F_1xB10.A mice and 55% (19 of 34) F_1xA
mice exhibited high macrophage inflammatory responses. Analysis
of responses of F_2 mice showed that 15 of 31 (48.4%) fell within
an intermediate level resembling the inflammatory responses of
F_1 mice. The responses of the remaining 16 F_2 animals were
equally distributed as either high (B10.A-like) or low (A-like)
responders. The χ^2 values are not significant (p > 0.25) and are
consistent with the hypothesis that the trait of macrophage
inflammatory responsiveness is genetically controlled by a single,
dominant gene, in the genome of B10.A mice, which is being expressed
as incompletely dominant in the heterozygotes. The trait of
resistance to Listeria is under similar genetic control (5).

     The finding of genetic control of the macrophage inflammatory
response gave us the opportunity to use genetic analysis to more
closely examine the relationship between the mobilization of inflammatory cells and anti-listerial resistance. Direct association
between the macrophage inflammatory response and anti-listerial
resistance was demonstrated using two different experimental

Table 3. Linkage of Mø Yield and Resistance to Listeria

Mice[a]	Number of peritoneal macrophages $\times 10^6$ (Mean + SEM)[b]	Resistance to Listeria $(\log_{10}CFU/liver)$[c] (Mean) [Range]
A(6)	0.61 ± 0.08	sensitive (8.26) [8.00-8.5]
B10.A (6)	2.33 ± 0.11	resistant (5.05) [3.62-7.27]
F_1(AxB10.A) (6)	2.54 ± 0.42	resistant (5.12) [3.60-7.17]
Backcross F_1 x A		
BC1	1.33	resistant (6.11)
BC2	1.15	resistant (7.17)
BC3	0.74	sensitive (8.17)
BC4	0.32	sensitive (8.30)
BC5	0.74	sensitive (8.27)
BC6	1.84	resistant (6.86)
BC7	0.96	sensitive (8.30)
BC8	1.38	resistant (7.20)
BC9	0.43	sensitive (8.30)
Backcross F_1 x B10.A		
BC10	2.34	resistant (4.50)
BC11	2.01	resistant (5.63)
BC12	2.91	resistant (4.83)
BC13	2.48	resistant (7.01)
BC14	2.66	resistant (6.22)
BC15	1.54	resistant (4.35)
BC16	3.82	resistant (6.17)
BC17	2.77	resistant (7.00)
BC18	2.87	resistant (3.90)
BC19	1.57	resistant (5.23)

[a] ( ) number of mice per group.

[b] Number of macrophages were calculated from total and differential counts performed on samples of peritoneal cavity washout from individual mice. Total macrophage count was based on volume of peritoneal washout recovered.

[c] Numbers of viable organisms in the liver were established by plating serial 10-fold dilutions of organ homogenates in saline on trypticase agar. The colony counts were performed 18 to 24 hrs later.

approaches. First, individual A and B10.A progenitor mice, F_1 hybrid and backcross mice were tested simultaneously for resistance to infection with Listeria and total number of peritoneal macrophages recoverable during the infection. All mice which had low macrophage numbers like the sensitive parental A mouse were typed as sensitive to Listeria (> $\log_{10}$ 8.0 CFU Listeria recovered from the liver) and all animals exhibiting high macrophage numbers like resistant B10.A mice were typed as resistant to Listeria (Table 3).

For the second approach, (B10.AxA)F_1xA backcross mice were first selected for high anti-listerial resistance with a typing dose that discriminates between the presence or absence of the resistant allele at the _Lr_ locus. Following recovery from infection, the surviving animals were tested for recruitment of inflammatory macrophages. Of the surviving mice, 92% (13 of 14) exhibited high macrophage inflammatory responses (Table 4). This experiment has been repeated 3 times with 28 of 31 F_1xA mice (90%) surviving

Table 4. Macrophage Inflammatory Responses of Listeria-Resistant
          Backcross Segregants

Strain[a]	Listeria iv[b]	Total Macrophages x 10^6 [c]
A(3)	–	8.96 ± 0.87
(B10.AxA)F_1 (3)	–	19.60 ± 1.5
(B10.AxA)F_1 (5)	+	16.70 ± 3.2
(F_1xA)BC		
BC1	+	18.90
BC2	+	11.60
BC3	+	13.60
BC4	+	27.25
BC5	+	14.97
BC6	+	14.09
BC7	+	12.15
BC8	+	16.50
BC9	+	15.01
BC10	+	15.46
BC11	+	17.18
BC12	+	13.80
BC13	+	14.05
BC14	+	2.70

[a] ( ) number of mice per group.

[b] 2×10^4 cfu _Listeria_ _monocytogenes_ 4 weeks previously.

[c] Numbers of macrophages were calculated from total and differential counts performed on samples of peritoneal cavity washouts from individual mice. Results for A/J and F_1 mice are mean ± SEM. Total macrophage count was based on volume of peritoneal washout recovered.

infection with Listeria, that is, having the resistant allele at the $\underline{Lr}$ gene, also showing high $\underline{in\ vivo}$ macrophage inflammatory responses. The data are consistent with the hypothesis that a high macrophage inflammatory response and resistance to Listeria are genetically linked (χ^2 = 0.31, p > 0.50). Thus, these experiments demonstrate linkage between genes which control the quantitative response of the mononuclear phagocyte system to inflammation and the level of natural resistance to an intracellular pathogen. The quantitative response evident as a superior recruitment of inflammatory macrophage provides an augmented population of effector cells which can control bacterial growth before the phase of sensitized T-cell macrophage cooperation in the anti-listerial response.

ACKNOWLEDGMENT

This work was supported by MRC grant 6431.

REFERENCES

1. North, R.J. (1970). The relative importance of blood monocytes and fixed macrophages to the expression of cell-mediated immunity to infection. J. Exp. Med. 132:521.

2. Mitsuyama, M., Takeya, K., Nomoto, K. and Shimotori, S. (1978). Three phases of phagocyte contribution to resistance against $\underline{Listeria\ monocytogenes}$. J. Gen. Microbiol. 106:165.

3. Skamene, E., Kongshavn, P.A.L. and Sachs, D.H. (1979). Resistance to $\underline{Listeria\ monocytogenes}$ in mice is genetically controlled by genes which are not linked to the H-2 complex. J. Infect. Dis. 139:228.

4. Cheers, C. and McKenzie, I.F.C. (1978). Resistance and susceptibility to bacterial infections: genetics of listeriosis. Infect. Immun. 19:755.

5. Mackaness, G.B. (1962). Cellular resistance to infection. J. Exp. Med. 116:381.

6. Skamene, E. and Kongshavn, P.A.L. (1979). Phenotypic expression of genetically controlled host resistance to $\underline{Listeria\ monocytogenes}$. Infect. Immun. 25:345.

7. Sadarangani, C., Skamene, E. and Kongshavn, P.A.L. (1980). Cellular basis for genetically determined enhanced resistance of certain mouse strains to listeriosis. Infect. Immun. 28:381.

8. Stevenson, M.M., Kongshavn, P.A.L. and Skamene, E. (1981). Genetic linkage of resistance to $\underline{Listeria\ monocytogenes}$ with macrophage inflammatory responses. J. Immunol. In press.

9. Snyderman, R., Pike, M.C., McCurley, D. and Lang, L. (1975).
        Quantification of mouse macrophage chemotaxis _in vitro_:
        role of C5 for production of chemotactic activity.
        Infect. Immun. 11:488.

EFFECT OF ACUTE NUTRITIONAL DEPRIVATION ON HOST DEFENSES AGAINST

LISTERIA MONOCYTOGENES -- MACROPHAGE FUNCTION

Edward J. Wing

University of Pittsburgh, School of Medicine
Montefiore Hospital, Pittsburgh, PA   15213

INTRODUCTION

Acute nutritional deprivation or starvation occurs frequently
in the United States during such conditions as chemotherapy for
malignancy and surgery. Despite this, the effect of starvation on
host defense mechanisms has largely been ignored. In a previous
study (1), we established a model of acute starvation in mice and
used Listeria monocytogenes as an immunologic probe to investigate
host defenses. Paradoxically, starved mice were resistant to
normally lethal doses of Listeria. The protective mechanism in
starved mice was not defined, although the tumoristatic capacity
of macrophages from starved mice was increased. In the experiments
described here, the effect of starvation on macrophage function
was studied in detail as a possible mechanism for the increased
resistance to Listeria.

MATERIALS AND METHODS

The model, Listeria infection, harvesting of macrophages, and
macrophage assays have been described in detail previously (1).
In brief, Swiss-Webster mice were starved for 48 hours or starved
for 48 hours and refed for 24 hours. Control mice were fed normal
rodent laboratory chow. The number of Listeria in spleens and
livers were determined by dilution and quantitative cultures.
Peritoneal exudate cells (PEC) were used as a source of macrophages. The capacity of macrophages (adherent PEC) to kill Listeria
in vitro was measured by the method of Cole (2). The tumoristatic
assay and the capacity of macrophages to inhibit Toxoplasma gondii
in vitro have been described by us previously (3).

245

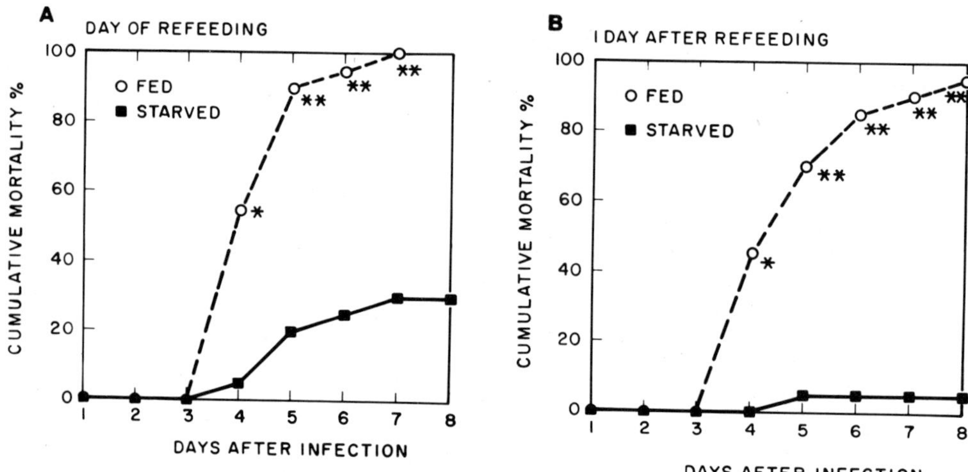

Figure 1.   <u>Effect of starvation on mortality due to L. monocyto-
<u>genes</u></u>.   Swiss-Webster female mice were either fed,
starved for 48 hours (A) or starved for 48 hours and
refed for 24 hours (B).   Groups of 20 mice were then
injected IV with 2 x LD$_{50}$ of <u>L. monocytogenes</u> and
cumulative mortality was recorded over 10 days.
Starved mice were significantly protected against
infection compared to fed mice.

RESULTS

     Experiments were performed to measure the resistance of fed
and starved mice to <u>Listeria</u> (Fig. 1).   Mice starved for 48 hours
had a lower mortality after injection of 2 x 10^5 (2 x LD$_{50}$)
<u>Listeria</u> compared to fed mice.   In additional studies, the number
of <u>Listeria</u> in spleens and livers of starved mice was found to be
significantly less than in those of fed mice 48 hours after
inoculation of <u>Listeria</u> (Fig. 2) (p < .05).   These results
demonstrated that starved mice were markedly more resistant to
<u>Listeria</u> infection than fed mice and that multiplication of
<u>Listeria</u> within the reticuloendothelial system during the first
48 hours after injection was decreased in starved mice.   This
suggests that the activity of resident macrophages in starved
mice is enhanced compared to fed mice.

     Initial <u>in vitro</u> experiments were carried out to determine
whether starvation altered the numbers and inflammatory responses
of PEC in starved mice.   Starved mice had fewer mean PEC than
fed mice, but the differences were not consistently significant.
In contrast, the exudative response after intraperitoneal
thioglycollate was slightly greater in starved mice (15.8 ± 4.1 x
10^6) than in fed mice (13.7 ± 3.2 x 10^6).   The percentage of cells

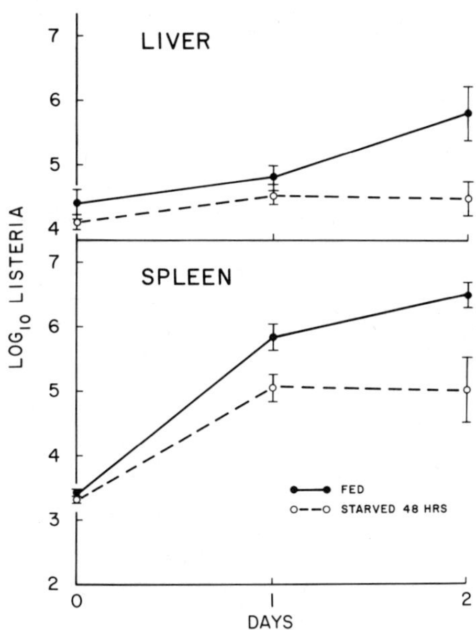

Figure 2.   Effect of starvation on the number of Listeria in
spleen and liver. Multiplication of Listeria in spleen
and liver was used as an indirect measure of resident
macrophage function. After starvation, mice were
injected with 1 x LD$_{50}$ of L. monocytogenes and bacterial
counts were determined at 0, 24 and 48 hours after
infection. Starved mice had less Listeria/spleen after
24 and 48 hours than did fed mice (p < .05) and had
less Listeria/liver after 48 hours (p < .05).

that were adherent and phagocytic was similar in fed and starved
mice.

The in vitro listeriacidal activity of macrophages from
starved-refed mice was measured directly. Results are shown in
Table 1. Macrophage monolayers that had been challenged with
Listeria were allowed to incubate for 24 hours, and the number of
Listeria was determined by quantitative plating. Counts immediately
after challenge were equal in all groups. After 24 hours macrophages from starved and starved-refed mice had significantly less
Listeria than fed mice. In addition, low doses (10-100 ng/ml) of
lipopolysaccharide (LPS) significantly increased the capacity of
macrophages from starved mice to inhibit Listeria. Neither
macrophages from starved nor starved-refed mice inhibited Listeria
to the same degree as macrophages from BCG-infected mice.

Tumoristatic activity was used as a nonspecific measure of
macrophage function. Macrophages from starved mice inhibited

Table 1. Capacity of Macrophages to kill <u>L</u>. <u>monocytogenes</u> <u>in</u> <u>vitro</u>

	No. Listeria/plate[c] mean 1SD x 10^3	
Addition to Culture	Fed	Starved 48 hrs.
None	2,298 ± 688	513 ± 133[a]
Endotoxin 10 ng/ml	1,027 ± 354	172 ± 47[a]
Endotoxin 100 ng/ml	1,143 ± 354	147 ± 25[b]

[a] p < .05 compared to fed mice

[b] p < .01 compared to fed mice

[c] Macrophages were challenged with Listeria, washed and incubated
for 24 hours. LPS was added as noted. The macrophages were
then lysed in distilled water, and the number of Listeria was
determined by plating on tryptic soy agar.

tumor cell ^{3}H–TdR incorporation more than normal macrophages.
Addition of small concentrations of LPS increased the inhibitory
activity of macrophages.

Macrophages from starved mice were studied for their ability
to inhibit the multiplication of <u>T</u>. <u>gondii</u> over 18 hours. Results
showed that macrophages from starved mice had only a minimal
inhibitory effect on <u>T</u>. <u>gondii</u> multiplication compared to macrophages from fed mice. Addition of LPS did not enhance this
inhibitory capacity.

DISCUSSION

The antimicrobial and antitumor activity of macrophages from
starved mice, which was demonstrated in these experiments, indicated that these cells were at least partially activated. The
<u>in vivo</u> inhibition of bacterial multiplication in livers and
spleens of starved mice during the first 48 hours of infection
suggested that resident macrophage activity was enhanced (4).
In addition, the <u>in vitro</u> listeriacidal and tumoristatic assays
demonstrated directly that macrophages from starved mice had
enhanced function compared to those from fed mice.

Full macrophage effector function, however, was not induced
by starvation. The listeriacidal and tumoristatic activity of
macrophages from starved mice was significantly less than that of

macrophages from BCG-infected mice. Furthermore, macrophages from
starved mice did not inhibit intracellular multiplication of
T. gondii, a characteristic usually associated with activated
macrophages (3).

Interestingly, the addition of small quantitites of LPS
significantly enhanced the activity of macrophages from starved
mice in both assays. Other workers have suggested previously that
several different signals are required to induce macrophage
activation. This process is thought to occur in a step-wise
fashion (5,6,7). One of the substances that has potential as an
activating signal is LPS, which, in synergy with other molecules
such as macrophage activating factor, may activate macrophages.
In our model the addition of LPS did stimulate additional macrophage activity in two of the three assays. Whether this combination
is active in vivo remains to be determined.

The experiments in this study are one of the first demonstrations of enhancement of immunological effector function by nutritional deprivation. Although enhanced antibody production and
increased resistance to certain infections has been associated
with protein calorie malnutrition (8), the effect of acute starvation has not been studied previously. Since antigenic and chemical
stimulation were not factors in our model, the results suggest the
physiological mechanisms exist which, under sufficient stress,
enhance immunological effector mechanisms including macrophage
function.

ACKNOWLEDGMENT

This work was supported by a grant from the Health Research
and Services Foundation, Pittsburgh, PA.

REFERENCES

1. Wing, E.J. and Young, J.B. (1980). Acute stravation protects
        mice against Listeria monocytogenes. Infect. Immun.
        28:771-776.
2. Cole, P. (1975). Activation of mouse peritoneal cells to kill
        Listeria monocytogenes by T-lymphocyte products.
        Infect. Immun. 12:36-41.
3. Wing. E.J., Gardner, I.D., Ryning, F.W. and Remington, J.S.
        (1977). Dissociation of effector functions in popula-
        tions of activated macrophages. Nature 268:642-644.
4. Mitsuyama, M., Takeya, K., Nomoto, K. and Shimotori, S. (1978).
        Three phases of phagocyte contribution to resistance
        against Listeria monocytogenes. J. Gen. Microbiol.
        106:165-171.

5. Hibbs, J.B., Jr., Taintor, R.R., Chapman, H.A., Jr., and
        Weinberg, J.B. (1977). Macrophage tumor killing:
        influence of the local environment. Science 197:279-282.
6. Weinberg, J.B., Chapman, H.A., Jr., and Hibbs, J.B., Jr. (1978).
        Characterization of the effects of endotoxin on macro-
        phage tumor cell killing. J. Immunol. 121:72-80.
7. Meltzer, M.S., Ruco, L.P., Boraschi, D. and Nacy, C.A. (1979).
        Macrophage activation for tumor cytotoxicity: analysis
        of intermediary reactions. J. Reticuloendothel Soc.
        26:403-415.
8. Good, R.A., Jose, D., Cooper, W.C., Fernandes, G., Kramer, T.
        and Yunis, E. (1977). Influence of nutrition on anti-
        body production and cellular immune responses in man,
        rats, mice and guinea pigs. pp. 169-183. In: Malnutri-
        tion and the Immune Response. Suskind R.M. (Ed.).
        Raven Press, NY.

PILUS-MEDIATED CLEARANCE OF <u>SALMONELLA</u> <u>TYPHIMURIUM</u> BY THE
PERFUSED MOUSE LIVER

Robert J. Moon and Robert D. Leunk

Michigan State University
East Lansing, MI   48824

INTRODUCTION

Most <u>Salmonella</u> isolated from animal hosts are piliated. While
data are available to suggest that pili do not play a major role in
the disease process <u>per se</u> (1), the role of these organelles in
adherence of organisms to intestinal mucosa or deep tissue endothelium has not been systematically explored. For several years
our laboratory has utilized a perfused liver model to explore the
basic mechanisms involved in the early trapping and killing of
selected microorganisms. We have established that the liver can
trap <u>Salmonella</u> <u>typhimurium</u> and <u>Candida</u> <u>albicans</u> equally well in
the presence and absence of plasma (2,3). Plasma is a requirement
for the bactericidal activity of this organ and, at least for
<u>S</u>. <u>typhimurium</u>, killing is dependent on a functional alternate
pathway of complement (4). The results presented here describe
the importance of type 1 pili in the trapping of <u>S</u>. <u>typhimurium</u> by
liver. Type 1 pili are characterized as those which can agglutinate guinea pig red blood cells in a reaction which is sensitive
to mannose. Both <u>in</u> <u>vitro</u> and <u>in</u> <u>vivo</u> data suggest that, while
probably not the only surface antigen involved in trapping, pili
are certainly one of the main adherence factors on the bacterial
surface.

MATERIALS AND METHODS

<u>Bacteria</u>. <u>S</u>. <u>typhimurium</u> strain SR-11 was used in most experi-
ments except where otherwise noted. It was subcultured daily in
Brain Heart Infusion (BHI) broth (Difco Laboratories, Detroit, MI)
and incubated at 37°C. A nonpiliated, nonflagellated mutant which

arose spontaneously was used in some experiments. It was subcultured
daily in BHI broth at 37°C.

Mice. Female CD-1 mice (24-28 g) were purchased from Charles
River Laboratories, Portage, MI. They were given food and water
ad libitum.

Liver Perfusions. Our liver perfusion technique has been
described in detail elsewhere (2). Briefly, a mouse is heparinized
and anesthetized. The abdomen is opened, the portal vein canulated,
and sterile Medium 199 (M199, GIBCO) allowed to flow through the
liver. The inferior vena cava is cut below the kidneys to allow
media to drain. The chest cavity is then opened and the vena cava
cannulated. The inferior vena cava is tied off just above the
kidneys and the liver washed free of blood by 40-50 ml of M199.

Bacteria were harvested by centrifugation, washed once in M199,
and resuspended in M199 to 1 x 10^9 cfu/ml. One milliliter of the
bacterial suspension was perfused followed by a wash of 50 ml of
sterile M199 over a 30 minute period. The liver was then excised,
homogenized, and the bacteria in the homogenate counted by diluting
in distilled water and plating on Tryptose agar pour plates. The
perfusate and the initial bacterial suspension were blended and
plate counted in a similar way.

Data are expressed as percent of bacteria in the liver homogenate, percent in the perfusate, and percent total recovery,
using the initial bacterial count as 100%. Percent killing equals
100% minus the percent total recovery. Percent trapping equals
the percentage in the liver homogenate plus the percentage killed.

Hemagglutination assay. Hemagglutination assays were performed as previously described (5,6) with slight modification.
Guinea pig erythrocytes (RBC) were washed three times in sterile
saline and resuspended to 3% (V/V) in either saline alone or
saline plus 0.5% D-mannose. Bacteria were harvested by centrifugation (8,000 rpm, 10 min) and washed once with sterile distilled
water. They were plate counted as above to determine cell concentration. Bacteria were serially diluted 1:2 in 0.1 ml amounts of saline
in wells of a porcelain-enamel plate. One tenth ml of 3% guinea
pig RBC in saline was added to each well. After 10 min incubation
at room temperature with mixing, the weels were read for agglutination. If necessary, 0.1 ml of the suspension was spread on a glass
slide to facilitate reading. Guinea pig RBC plus saline and
guinea pig RBC in 0.5% mannose plus bacteria were always included
as controls. The minimum hemagglutinating dose (MHD) was the
smallest number of bacteria/ml which caused hemagglutination.

Blood clearance of bacteria. Mice were injected intravenously
in a tail vein with approximately 1 x 10^9 bacteria in 0.2 ml saline

or with 1 x 10^9 bacteria plus 0.125 gm D-mannose in 0.35 ml saline.
At timed intervals after injection, a 0.1 ml blood sample was
obtained by bleeding from the retro-orbital plexus and was plate
counted in Tryptose agar as above. The number of bacteria remaining
in the bloodstream was calculated assuming blood volume to be 8%
of body weight.

<u>Chemicals</u>. All sugars used in perfusion media were purchased
from Sigma Chemical Co., St. Louis, MO.

<u>Statistics</u>. Statistical evaluation was carried out using the
Student's "t" test.

RESULTS

<u>Trapping of piliated and nonpiliated Salmonella typhimurium
SR-11 by perfused mouse livers</u>. Transmission electron microscopy
of both piliated and nonpiliated <u>S</u>. <u>typhimurium</u> was performed
(figures not shown). Until recently it was thought that our
mutant was a true chromosomal mutant and not simply a nonpiliated
phase variant of SR-11 since a number of independent TEM observations showed the absence of pili. There is now some doubt of this,
since after several months of laboratory passage, we are able to
detect both low levels of hemagglutination as well as some pili
on certain organisms in culture.

Liver perfusion of 1 x 10^9 piliated wild type SR-11 yields
ca. 67% trapping on a single pass (Table 1). By contrast, only
1.2% of the nonpiliated variant is trapped. Total recovery
approximates 100% indicating no killing of the organisms as would
be expected in the absence of plasma.

Table 1. Perfusion of piliated <u>S</u>. <u>typhimurium</u> SR-11 and a nonpiliated mutant strain.

	% Recovery in[a]		
	Liver	Perfusate	Total Recovery
S. typhimurium (piliated phase)	66.7 ± 10.6	25.8 ± 7.7	92.5 ± 8.8
Nonpiliated mutant	1.2 ± 0.6	98.7 ± 9.2	99.8 ± 9.2

[a]Mean ± standard deviation of at least 7 experiments.

Table 2.   Saccharide inhibition of trapping of *S. typhimurium* strain SR-11 by perfused livers.

Saccharide[b]	% Recovery in[a]		
	Liver	Perfusate	Total Recovery
None	66.7 ± 10.6	25.8 ± 7.7	92.5 ± 8.8
Mannose	16.2 ± 9.2[c]	82.3 ± 16.6[c]	98.5 ± 9.5
α-methyl-D-mannoside	23.2 ± 6.1[c]	74.8 ± 10.9[c]	98.0 ± 9.1
Fructose	40.4 ± 15.3[c]	50.1 ± 13.4[c]	90.7 ± 11.6
Galactose	73.4 ± 8.1	25.3 ± 2.3	98.9 ± 7.2
Glucose	69.7 ± 7.7	36.4 ± 6.5	106.6 ± 5.4
Lactose	82.8 ± 2.2	22.6 ± 4.7	105.4 ± 5.9
Mannitol	62.5 ± 6.3	33.8 ± 5.0	96.2 ± 6.5
Ribose	63.1 ± 12.2	27.5 ± 4.3	90.5 ± 14.1
Sucrose	69.9 ± 6.9	25.3 ± 7.9	95.2 ± 8.8

[a] Mean ± standard deviation of at least 6 experiments.

[b] All sugars were present at 1% (w/v) in M199 for the entire perfusion.

[c] P < .001 vs control by Student's t test.

Inhibition of trapping of *S. typhimurium* SR-11 by mannose and other selected sugars. It is well established that mannose inhibits the ability of Type 1 piliated organsims to agglutinate guinea pig red blood cells (7). Table 2 evaluates the ability of mannose and a variety of other sugars to inhibit bacterial trapping by the liver. Both mannose and α-methyl-D-mannoside are very effective, while a variety of other sugars including galactose, glucose, lactose, mannitol, ribose, and sucrose have no effect. Both mannose and α-methyl-D-mannoside reduce trapping by more than 40%, while not altering total recovery. Fructose shows some limited ability to inhibit trapping while the remaining sugars are without significant effect.

The inhibitory effect of mannose is dose-dependent (Table 3). Maximum inhibition is seen when 1% (w/v) of D-mannose is added to perfusion medium. Ten-fold dilutions result in progressively less trapping to the point where 0.001% is without statistical inhibitory effects on clearance.

Table 3. Dose-dependence of inhibition of trapping by D-mannose.

Mannose Concentration (% w/v)	% Recovery[a]		
	Liver	Perfusate	Total Recovery
None	66.7 ± 10.6	25.8 ± 7.7	92.5 ± 8.8
1.0	16.2 ± 9.2	82.3 ± 16.6	98.5 ± 9.5
0.1	37.1 ± 9.6	62.5 ± 11.2	99.6 ± 6.1
0.01	50.6 ± 8.0	42.7 ± 8.0	93.3 ± 6.4
0.001	58.1 ± 4.9	37.8 ± 8.8	96.0 ± 5.4

[a]Mean ± standard deviation of at least 7 experiments.

<u>Mechanism of D-mannose inhibitory action</u>. It was of interest
to determine whether mannose binds to receptors on the bacteria or
on the hepatic tissue. Table 4 suggests the former mechanism.
When <u>S</u>. <u>typhimurium</u> were incubated for 30 min at 37°C with either

Table 4. Liver clearance of <u>S</u>. <u>typhimurium</u> following preincubation
with mannose, α-methyl-D-mannoside, and glucose.

Experimental[b]	% Recovery[a]		
	Liver	Perfusate	Total Recovery
Control	66.7 ± 10.6	25.8 ± 7.7	92.5 ± 8.8
1% Mannose	22.8 ± 8.7	71.6 ± 12.1	93.9 ± 6.0
1% α-methyl-D-mannoside	22.5 ± 4.5	85.2 ± 13.6	107.8 ± 13.0
1% Glucose	80.5 ± 6.8	17.0 ± 3.4	97.4 ± 7.9

[a]Mean ± standard deviation of at least 6 experiments.

[b]Bacteria were preincubated with the appropriate sugar for 30 min
at 37°C followed by centrifuge washing in M199 prior to perfusion.

1% mannose or α-methyl-D-mannoside followed by washing to remove
unbound sugar, the ability of the livers to trap bacteria was
significantly inhibited. A similar treatment using glucose had
no effect. By contrast, perfusion with 30 ml of 1% mannose
followed by 20 ml of normal M199 did not significantly alter trapping by normal livers (data not shown). When bacteria are initially
trapped and 2% mannose or α-methyl-D-mannoside is added only to the
perfusion media following bacterial trapping, substantial numbers
of trapped organisms are washed from the liver (Table 5). Continual
washing with M199 without additives or M199 with glucose or mannitol
had no effect, as the bacterial remained trapped in the liver.

<u>In vivo</u> correlation of trapping with piliation. Figure 1
compares the rate of clearance of the <u>S. typhimurium</u> SR-11 with or
without mannose and the nonpiliated mutant. It is readily apparent
that the piliated organism is removed more rapidly than the nonpiliated variant. Likewise injection of 125 mg of mannose simultaneously with the piliated strain decreased the rate of removal of
the bacteria from the blood.

DISCUSSION

The role of specific pili in adherence of enterotoxigenic <u>E.
coli</u> to intestinal epithelium is well established (8,9,10). Many
other models of pili mediated adherence are also available (11,12,
13) but most relate to adherence to mucosal linings of the GI or

Table 5. Ability of sugars to elute trapped bacteria from the liver

	% Recovery[a]		
Experimental[b]	Liver	Perfusate	Total Recovery
None	69.1 ± 8.3	17.7 ± 4.2	86.8 ± 6.7
Mannose – 2%	37.6 ± 11.6	54.5 ± 14.2	92.1 ± 11.9
– 1%	54.4 ± 9.8	37.9 ± 10.3	92.2 ± 9.1
α-Methyl-D-mannoside – 2%	33.7 ± 13.8	63.7 ± 15.3	97.3 ± 8.4
Glucose – 2%	60.6 ± 11.8	40.2 ± 11.3	99.4 ± 4.1
Mannitol – 2%	77.5 ± 4.9	15.3 ± 6.0	92.8 ± 6.8

[a]Mean ± standard deviation of at least 6 experiments.

[b]The sugar was in the perfusate only (after bacteria).

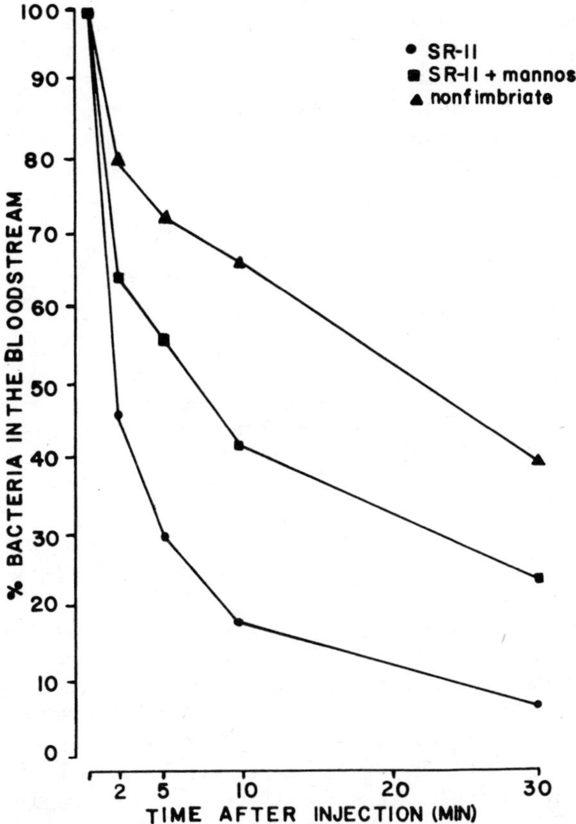

Figure 1.   *In vivo* clearance of *S. typhimurium* SR-11 in the
presence (■) and absence (●) of mannose, and clearance of the nonpiliated mutant (▲).

GU tracts. Very little is known with respect to the role of pili
in adherence to vascular epithelium. Using a nonpiliated spontaneous mutant of *S. typhimurium* our data in Table 1 and Figure 1
strongly suggest a role for these cellular appendages in vascular
clearance. The inability of perfused livers to trap these
organisms *in vitro* combined with data showing a significant
reduction in *in vivo* clearance strongly supports a role for pili
in this process. Since *in vivo* clerance does eventually occur it
is clear that pili are not the only surface component important
in trapping. The absence of plasma in the perfusion system might
lead one to suspect that *in vivo* the interaction with blood
modifies the bacterial surface in some way to enhance its affinity
for vascular epithelium.

The ability of mannose and α-methyl-D-mannoside to inhibit trapping in the perfused liver suggests that Type 1 pili are the main pilus type important in this study. Saccharide inhibition of trapping by the perfused liver was dose-dependent for mannose, and also occurred with α-methyl-D-mannoside and to a lesser extent with fructose (Tables 2 and 3). This same pattern of saccharide inhibition has been observed for other activities believed to be mediated by Type 1 pili, such as hemagglutination (7,14) and in vitro cell adherence (15). Old (7) has emphasized the importance of unmodified hydroxyl groups at C-2, C-3, and C-4 of the six carbon chain and a specific requirement for the α configuration at C-1 for a saccharide to inhibit Type 1 pilus-mediated hemagglutination. This might explain why the closely related sugars, glucose and mannitol, do not inhibit trapping by the perfused liver. Fructose is thought to possess inhibitory activity because the stereochemistry of the β-D-fructopyranosyl form somewhat resembles that of D-mannose.

Previous studies by us (3,16) have clearly shown that both Kupffer and endothelial cells possess the ability to trap Salmonella. Whether this implies that receptors for such an interaction are available on all endothelial cell types awaits further investigation. A recent review by Steer (17) clearly shows that carbohydrate receptors show differential distribution among the various hepatic cell types. The relationship between this body of evidence and ours is unclear at present, particularly since mannose receptors appear to be localized on hepatic endothelial cells, and yet pre-incubation of bacterial cells with mannose (cf Table 4) inhibits binding to these cells in the perfusion model.

There is growing evidence that pilus-mediated attachment phenomena observed in vitro actually reflect in vivo events and that pili do mediate attachment in vivo. Silverblatt has observed with electron microscopy that Proteus mirabilis attaches to epithelial cells of the renal pelvis via pili in early experimental pyelonephritis (18). Aronson et al. reported that α-methyl-D-mannoside given with bacteria inhibits E. coli adherence to rat bladder epithelial cells in experimental cystitis (19). Fader and Davis also observed that piliated-phase Klebsiella pneumoniae are less able to adhere to the rat bladder surface in vivo in the presence of α-methyl-D-mannoside (20).

The data in this study provide preliminary evidence suggesting that Type 1 pili mediate trapping of bacteria by the liver. Subsequent experiments will further investigate the nature of the receptor and ligand binding and the in vivo relevance of these observations to the overall host parasite relationship.

REFERENCES

1. Duguid, J.P., Darekar, M.R. and Wheater, D.W.F. (1976). J. Med. Microbiol. 9:459-473.
2. Moon, R.J., Vrable, R.A. and Broka, J.A. (1975). Infect. Immun. 12:411-418.
3. Sawyer, R.T., Moon, R.J. and Beneke, E.S. (1976). Infect. Immun. 14:1348-1355.
4. Friedman, R.L. and Moon, R.J. (1980). Infect. Immun. 29: 152-157.
5. Atkinson, H.M. and Trust, T.J. (1980). Infect. Immun. 27: 938-946.
6. Duguid, J.P. (1959). J. Gen. Microbiol. 21:271-286.
7. Old, D.C. (1972). J. Gen. Microbiol. 71:149-157.
8. Källenius, G. and Möllby, R. (1979). FEMS Microbiol. Lett. 5:295-299.
9. Evans, D.G., Silver, R.P., Evans, Jr., D.J., Chase, D.G. and Gorbach, S.L. (1975). Infect. Immun. 12:656-667.
10. Burrows, M.R., Sellwood, R. and Gibbons, R.A. (1976). J. Gen. Microbiol. 96:269-275.
11. Cheney, C.P., Schad, P.A., Formal, S.B. and Boedeker, E.C. (1980). Infect. Immun. 28:1019-1027.
12. Woods, D.E., Straus, D.C., Johanson, Jr., W.G., Berry, V.K. and Bass, J.A. (1980). Infect. Immun. 29:1146-1151.
13. Fader, R.C., Avots-Avotins, A.E. and Davis, C.P. (1979). Infect. Immun. 25:729-737.
14. Salit, I.E. and Gotschlich, E.C. (1977). J. Exp. Med. 146: 1169-1181.
15. Salit, I.E. and Gotschlich, E.C. (1977). J. Exp. Med. 146: 1182-1194.
16. Friedman, R.L. and Moon, R.J. (1977). Infect. Immun. 16: 1005-1012.
17. Steer, C.J. (1980). Kupffer Cell Bull. 3:15.
18. Silverblatt, F.J. (1974). J. Exp. Med. 140:1696-1711.
19. Aronson, M., Medalia, O., Schori, L., Mirelman, D., Sharon, N. and Ofek, I. (1979). J. Infect. Dis. 139:329-332.
20. Fader, R.C. and Davis, C.P. (1980). Infect. Immun. 30: 554-561.

IMMUNITY TO <u>SALMONELLA</u> INFECTION

Toby K. Eisenstein and Barnet M. Sultzer

Department of Microbiology and Immunology, Temple
University School of Medicine, Philadelphia, PA 19140
and Department of Microbiology and Immunology, State
University of New York, Downstate Medical Center
Brooklyn, New York 11203

INTRODUCTION

In spite of over half a decade of research on the problems of
immunity to <u>Salmonella</u> infection, our understanding of the subject
is still incomplete, and in many cases, controversial. Interest in
the organism and host defenses to it stems from the considerable
morbidity and mortality associated with human infection by <u>S. typhi</u>,
the causative agent of typhoid fever (1). This organism, as well as
<u>S. paratyphi</u> <u>A</u> and <u>B</u>, produce a clinical syndrome in man which is
termed "enteric fever" (2). These bacteria gain access to the body
through the oral cavity, penetrate the intestinal epithelium, enter
the circulation, and are trapped by the elements of the reticuloendothelial system in the liver and spleen where they multiply,
presumably intracellularly, in macrophages. Thus, enteric fevers are
systemic illnesses, with dissemination of the organisms to the
internal organs. As the only natural host of <u>S. typhi</u> is man, a
suitable experimental model is not readily available. Chimpanzees
can be employed, but their cost and scarcity limits their utility.
Although <u>S. typhi</u> is not naturally virulent for mice, hog gastric
mucin has been used to coat the organism, and artificially increase
its mouse virulence. An alternative, widely employed model for
typhoid fever involves parenteral [intraperitoneal (ip); intravenous (iv), or subcutaneous (s.c.)] infection of mice with mouse
virulent strains of <u>Salmonella</u>, such as <u>S. typhimurium</u> or <u>S.
enteritidis</u>. Strains of the latter organisms are naturally
occurring pathogens of mice which cause systemic syndromes that in
their pathology resemble typhoid fever of humans (3,4,5). For this
reason, <u>Salmonella</u> infection of mice, using appropriate strains of
<u>S. enteritidis</u> and <u>S. typhimurium</u>, has become the most widely

accepted experimental model for studying enteric fever. (A somewhat confusing aspect of this model is that S. typhimurium and S.
enteritidis are also human pathogens, but in man they cause gastroenteritis, not enteric fever (2). In gastroenteritis, the
Salmonella penetrate the mucosal surface, but systemic disease is
not a regular feature of the human infection, and in general, the
organisms are confined to the gut.)

The mouse model has been used extensively to investigate the
basic mechanisms of Salmonella immunity, as well as to evaluate,
in the laboratory, the potential efficacy of various types of
vaccines for protection against typhoid fever. Vaccines are made
from the mouse virulent strains, and these strains are also used
for challenge. Results are interpreted, by analogy, to apply to
human infection and immunity against S. typhi. Using this mouse
system, an extensive literature has developed from which has
emerged two essentially dicotomous views of immunity to Salmonella
infection. One school of thought places Salmonella in the group
of facultative intracellular pathogens for which the main host
defense is the cellular arm of the immune system. As live
Salmonella trigger cellular immunity, but killed organisms or
bacterial extracts are usually poor at evoking this type of immune
response, the proponents of this point of view advocate living,
attenuated vaccines as the only highly effective protection
against enteric fevers in mice or humans.

Other groups of investigators have found that humoral immunity
and nonliving vaccines can provide excellent protection for mice
against Salmonella infection, and have expended considerable
effort in evaluating the optimal type of nonliving vaccine, and
in characterizing the nature of the protective antigens of
Salmonellae. The degree of controversy and the seemingly contradictory results from different laboratories on the subject of
immunity to Salmonella infection are indeed puzzling. Roantree
(6) recognized and documented much of this conflicting data in an
excellent review article, but the reasons for the discrepancies
in data from different laboratories have never been satisfactorily
explained.

In this paper, the main lines of evidence supporting each
point of view will be presented, and will be interpreted, with
some new data from our laboratory, to try to achieve an understanding of the parameters influencing immunity to Salmonella
infection in the mouse, and the import of these results for immunity to typhoid fever in the human.

LITERATURE REVIEW

Evidence for Cellular Immunity in Murine Salmonella Infections

The hypothesis that immunity to <u>Salmonella</u> infection is primarily vested in the cellular, as opposed to the humoral, arm of the immune system is supported by several different types of experiments from major laboratories in Japan, England, and the United States.

Japanese investigators have had a long-standing interest in mouse typhoid (7). Ushiba and his colleagues compared protection engendered by five subcutaneous injections of a heat-killed cell vaccine (1 hr, 56°C) with that induced by a living, attenuated rough strain of <u>Salmonella</u> which lacked O antigens (8). They found that the live strain protected (77% survival), whereas the killed vaccines, while prolonging survival, did not give significant protection (33%). Further, in animals vaccinated with killed vaccines, no correlation was found between protection and anti-O antibody titers, as mice with significant titers still succumbed to the infection (9). Passive serum, while prolonging survival somewhat, was not protective (8). It could be shown that in the blood, peritoneal fluid, and spleens of infected mice previously immunized with heat-killed vaccines there was a rapid reduction in the numbers of organisms in the blood, peritoneum, and internal organs, but after two days the bacteria began to multiply unchecked in the spleens, ultimately resulting in death. In contrast, mice receiving a living vaccine did not show the initial decrease of organisms in the fluids or the spleen, but were able to prevent the subsequent bacterial multiplication in the spleen (8). These experiments suggested that live vaccines, but not killed ones, stimulated an antibacterial capacity of the organs.

Similar kinds of results were obtained by Hobson in England (5). He tested the ability of a single ip injection of a live, virulent strain of <u>Salmonella</u> to protect mice, in comparison with heat-killed cells prepared from this strain. When given an intraperitoneal challenge, mice vaccinated with the live organisms showed between 85% and 95% survival, depending on the interval between the primary infection and the challenge, but mice receiving the heat-killed vaccine showed only 35% survival. Thus, killed cells were again shown to be markedly inferior to living ones in providing protection. Serum obtained from mice 28 days after infection with the low virulence strain when administered subcutaneously, was unable to passively protect recipient mice against an intraperitoneal or an intravenous challenge. Only 16% of the survivors of the primary infection had anti-O titers to <u>Salmonella</u> at 28 days post exposure to the organism, a time when they were highly protected against challenge. In contrast, 43% of the mice which had received the killed-cell vaccine had anti-O titers, even

though they were poorly protected. Thus, in Hobson's experiments,
there was also a lack of correlation between protection and humoral
factors (5).

      At the Trudeau Institute, Mackaness, Blanden and Collins (10)
followed the course of Salmonella infection in mice given various
vaccines, by determining the numbers of organisms in the homogenized livers and spleens of groups of animals at intervals after
intravenous infection. They found that mice vaccinated with heat-
killed cells were not protected against an intravenous infection
with Salmonella, whereas mice surviving infection with a virulent
strain could be protected. Neither serum from survivors of a
previous infection, nor serum from mice vaccinated with heat-
killed cells, protected recipient mice against an intravenous
challenge with virulent Salmonella in spite of anti-O antibody.
Some protection against intraperitoneal challenge was observed
using serum raised to heat-killed cells (11). The use of a strepto-
mycin-sensitive Salmonella for primary infection and a streptomycin-resistant derivative for the challenge dose, a technique first
described by Hobson (5), enabled these investigators to enumerate
separately the numbers of organisms of the infecting and the
challenge strain in liver and spleen homogenates. Using this
technique, it was found that mice harboring a primary infection
could eliminate a secondary challenge if both inoculations were
given by the intravenous route. As mice vaccinated with heat-
killed cells were unable to demonstrate this "immune elimination"
of organisms from the liver and spleen, the results were interpreted
to mean that live, but not killed Salmonella, induce an acquired
antibacterial immunity (10). Collins published an extensive
series of carefully carried out studies to evaluate further the
influence of variables, such as the routes of vaccination and
challenge, which are involved in immunization by living as compared
to killed vaccines, and the effect of serum on progressive infection (12,13). The results confirmed the earlier reports from this
laboratory showing that killed vaccines, even in conjunction with
immune serum, gave only marginal protection against challenge.
Furthermore, the enumeration of organisms in the liver and spleen
showed that even in groups of animals which showed significant
protection by mortality statistics, the organisms in the challenge
inoculum grew to appreciable levels in the organs. In contrast,
mice vaccinated with living, attenuated Salmonella, were able to
completely eliminate the virulent organisms in the challenge dose.
The superiority of the living vaccines with regard to "immune
elimination" was particularly evident when the intravenous route
of infection was used (12). These results laid the foundation for
the concept that true protection against Salmonella infection is
only engendered by vaccines which stimulate the antibacterial
capacity of the tissues and can be shown to prevent growth of
the challenge strain in the liver and spleen. Protection afforded
by killed vaccines was attributed to an initial bactericidal

action which slowed the course of the infection, and thus allowed
time for the host to develop a true antibacterial immunity in
response to the challenge dose (13). Additionally, the Trudeau
group questioned the validity of intraperitoneal challenge in the
mouse model, as this route seemed to potentiate protection by
killed vaccines due to enhanced phagocytosis of organisms in the
peritoneal cavity, and thus obscured the measurement of immunity
due to the antibacterial mechanisms of the reticulendothelial
system (10,12,13). The intravenous route of challenge was
postulated to be more physiologic than the intraperitoneal
route, and to be more analogous to oral infection (10).

The observations described above were seen to have parallels
with experiments carried out by investigators using other intracellular pathogens, such as Brucella, Listeria, and Mycobacterium
tuberculosis (11,14,15,16). In particular, the ability of a mouse
which had received live cells to eliminate organisms in a secondary
infection was reminiscent of the Koch phenomenon. In these other
infections, development of antibacterial immunity had been related
by Mackaness to the onset of delayed-type hypersensitivity (DTH),
as measured in the mouse by foot-pad swelling (17). The demonstration by Collins and Mackaness (18) that live, but not killed,
Salmonella vaccines induce a state of DTH which develops in close
temporal association with the onset of antibacterial immunity in
the liver and spleen, added additional evidence that cellular
immunity was involved in the protection of mice against Salmonella
infection by living organisms. Cross-protection studies also
showed that live vaccines could engender protection against
Salmonella with heterologous O serotypes (19,20,21) or against
Listeria (11,22). Japanese workers showed that even living rough
Salmonella lacking O antigens could induce immunity (8,23). The
demonstration of cross-protection in Salmonella infection was
similar to the nonspecific resistance induced by other intracellular bacteria against each other, as shown by Mackaness (17).

Further evidence supporting the importance of nonhumoral
mechanisms of host defense to Salmonella infection came from
studies on the interaction of peritoneal macrophages from live-
cell as compared with killed-cell immunized mice. The results
indicated that the macrophages which had been exposed to living
Salmonella had greater initial bactericidal activity, and also
were able to inhibit subsequent intracellular growth of Salmonella,
whereas macrophages from killed-cell vaccinated mice permitted
intracellular replication of the organisms (11,24,25,26). Macrophages from live-cell immunized mice were found to passively transfer bacteriostatic activity to the macrophages of recipient mice
(27). Ribosomal or RNA extracts from these macrophages were also
found to be active in passive transfer of a macrophage activating
property (28).

Taken together, these observations were consonant with the view that Salmonella are intracellular pathogens, and the theory was proposed that the primary host defense mechanisms in control of Salmonella infection is "cellular immunity" (5,10,29). The concept of a "cellular immunity" to intracellular pathogens, which resides in macrophages, had been put forth previously to explain alterations in macrophage function in animals infected with other intracellular pathogens, primarily Mycobacteria (14,15). However, as a result of a series of brilliantly designed in vivo experiments using Brucella, Listeria, and Bacillus Calmette-Guerin (BCG), Mackaness extended this model, and deduced that the evocation of the immune state was immunologically specific, but expression of the immunity was nonspecific (17). In later refinements of the model, he proposed that T-lymphocytes recognize antigen, and through lympho-kine release bring about macrophage activation, thus endowing the macrophage with enhanced bactericidal acapacity which can be expressed against many different intracellular bacteria (30,31). In the case of Salmonella, immune spleen cells were shown to trans-fer immunity under conditions where immune serum was inactive, and T-cell depleted mice were shown to be incapable of developing anti-Salmonella immunity or delayed hypersensitivity (32). Experiments showing transfer of immunity with T cells were carried out using Listeria and Mycobacteria, but not Salmonella (30,33). Neverthe-less, since for the aspects tested, Salmonella semmed to fit a pattern of infection and immunity established for the other intra-cellular organisms, it has been included as a microbe for which cellular immunity was thought to be paramount in host defense.

As a consequence of this literature, a point of view developed, which is widely held, that in immunity to Salmonella infection, killed vaccines are "poorly" protective, in contrast to live vaccines, which are deemed to be "excellent". This view emphasizes the role of cellular immunity, while disregarding the beneficial effects of humoral immunity. A corollary to this thesis is that the optimal vaccine for typhoid fever should be living, attenuated organisms (32).

Nonliving Vaccines and Humoral Immunity in Salmonella Infection

In contrast to the literature just cited, there are numerous reports from several major laboratories showing that killed vaccines can induce effective anti-Salmonella immunity, and that anti-O antibody is protective. Jenkin and Rowley immunized mice with acetone-killed cells, alcohol-killed cells, heat-killed cells (boiling for 2 hrs) or attenuated living cells, and showed that a single inoculation of acetone-killed or alcohol-killed cells provided substantial levels of protection, although not quite as good as that provided survivors or the live cell infections (34). The method of killed vaccine preparation was crucial for a successful outcome, as the acetone and alcohol-killed cells

protected, but the heat-killed cells did not. Dr. Herzberg and
his colleagues carried out extensive investigations on protection
against Salmonella infection by heat-killed cells and by a deoxycholate extract of live cells. Both vaccines, if injected twice,
resulted in a 100% protection of vaccinated mice against 100 LD_{50}
challenge doses (35). The group at Standford University has for a
number of years been actively involved in studies of Salmonella
pathogenesis and immunity. In the study of Ornellas, Roantree
and Steward, multiple inoculations of heat-killed cells were given,
spaced seven days apart, with the final challenge 21 days after the
last inoculation. Using this immunization regimen, protection
against as many 10,000 LD_{50} challenge doses was obtained (36).
Marecki et al. also demonstrated considerable protection by heat-
killed vaccines (70°C, 30 min), and some protection by passively
transferred immune serum (37). Cameron and Fuls were able to
protect mice with formalinized vaccines (38).

For some time Dr. Eisenstein has been involved in investigation
of the efficacy of a subcellular, ribosome-rich extract of Salmonella
as a vaccine, first, in the laboratory of Dr. L. Joe Berry, and
then in her own laboratory. The early studies showed that this
nonliving preparation gave high levels of immunity (39). A
rigorous comparison of this ribosomal vaccine with several other
nonliving vaccines, as well as with a sublethal dose of live,
virulent organisms, was made by Angerman and Eisenstein (40). All
vaccines were prepared from S. typhimurium, strain W118-2, and
administered ip to Swiss, CD-1 mice (Charles River Breeding
Laboratories). Challenge was three weeks later with the same
organism given ip. Table 1 shows that the ribosomal vaccine was
highly protective in this system, but surprisingly, so was the
acetone-killed cell preparation, which approached the living cells
in its capacity to protect against challenge doses as great as
1000 LD_{50}s. Furthermore, as little as 1 µg of phenol-water purified
lipopolysaccharaide (LPS) protected 70% of the mice against a
challenge of 500 LD_{50} doses, and 25 µg of LPS induced 100% protection. The results of the checkerboard titration of immunizing
dose versus challenge dose shows that partial protection could be
induced by killed vaccines over a wide dosage range. Further,
it was observed that use of suboptimal doses of potentially
excellent killed vaccines made them appear inferior to live cells
(i.e., 10 µg of acetone-killed cells with 1000 LD_{50} challenge).
Also interesting, is that intermediate challenge doses give 100%
protection by all vaccines, thus obscuring differences between
preparations. These studies, in conjunction with those cited
above, suggest that the method of vaccine preparation, the number
of injections, and the vaccine dosage are important variables in
whether a killed vaccine will be protective, and how it will
compare quantitatively in protective capacity with a living vaccine.
Another important factor is route of challenge, as high levels of
protection were seen in mice receiving killed vaccines and

Table 1. Titration of protective capacity of Salmonella vaccines[f]

| Vaccine type[a] | % Survival at various challenge doses | | | | |
	1×10^{6}[b] (100 LD$_{50}$)	5×10^{6} (500 LD$_{50}$)	1×10^{7} (1,000 LD$_{50}$)	5×10^{7} (5,000 LD$_{50}$)	1×10^{8} (10,000 LD$_{50}$)
Live cells					
6.5×10^{3}	100[c]		90	90	30
1.3×10^{3}	100		90	50[d]	40[e]
AKC (μg)					
1.0	90		0		
10	100		60		
50	100		90		
100	100		100	10	0
250	100		100	60	0
500	100		100		
RIB (μg)					
0.1	60		0		
1.0	90		0		
25	100		30		
100	100		80	10	0
250	100		80		
1000	100		90	30	0
LPS (μg)					
0.1		30	0		
0.5		40	0		
1.0		70	0		
25		100	0		
100		100	0	0	0
150		100	10		
200		100	38		

[a]Vaccines given intraperitoneally in 0.5 ml of saline. AKC, Acetone-killed cells; RIB, ribosomes.

[b]Number of viable W118-2 given intraperitoneally 3 weeks after immunization.

[c]10 mice per group

[d]1.3×10^{3} live cells versus 250 μg of acetone-killed cells or 1,000 μg of ribosomes, not significant; 1.3×10^{3} lives cells or 250 μg of acetone=killed cells versus 100 μg of LPS, P < 0.01.

[e]1.3×10^{3} live cells versus other vaccines, P < 0.05.

[f]Reprinted by permission of Infection and Immunity. Angerman, C.R. and T.K. Eisenstein (1978). 19:575.

challenged intraperitoneally. However, it has been reported, that in contrast to live vaccines, nonviable vaccines are not highly protective against intravenous challenge (13). In support of this finding, Angerman observed that mice which received 250 µg of acetone-killed cells were protected against 1000 LD_{50} doses when the route of challenge was ip, but they were only protected against 30 LD_{50} doses when the challenge was intravenous (41). Considerable debate in the literature has centered about which of the systemic routes, ip or iv, is more like the natural oral route of infection (10,13,32,35,42). Those who favored the iv route argued it was more like the oral route because there is no peritonitis in a natural infection. Those favoring the ip route noted that the organisms had to pass through the lymphatic channels to enter the circulation from the peritoneum, which is also true for organisms entering through the gut, but not for those injected iv. No resolution of this issues seems possible, as clearly no parenteral route is physiologic. It is most interesting, however, that the model of immunity is altered simply by changing the route of challenge, indicating the probable complexity of the parameters involved in protection against systemic <u>Salmonella</u> infection. The oral route has not been routinely used in most of the studies cited so far, as many mouse strains are highly resistant to oral infection. Angerman titrated the oral LD_{50} in CD-1 mice at 1.6×10^{10} cells (41).

The demonstration that a nonviable preparation can provide acceptable levels of protection against <u>Salmonella</u> infection raises the question of the mechanism of the immunity and the nature of the protective antigen(s). As outlined in the preceding section on cellular immunity, the persistent reports of failure to find protection against <u>Salmonella</u> in the face of substantial titers of anti-O antibody, and the failure of passive serum to protect, cast doubt on the importance of humoral immunity and of O antigens as immunogens. Yet, review of the literature shows that many laboratories have published data showing the importance of these factors in host defense. As far back as 1929, Topley showed that the protective power of heat-killed, formalin preserved cells was attributable to the O antigens, but not the flagellar or the rough antigens (4). Jenkin and colleagues found that multiple injections of immune serum raised to whole cells could passively protect (43), as did Blanden et al. (11). Ornellas et al. (36) reported that if <u>Salmonella</u> were preopsonized with hyperimmune mouse serum raised to heat-killed organisms of the same O antigenic composition as the challenge strain (0:9, 12), the ip LD_{50} was increased 1000- to 10,000-fold. Only 10- to 100-fold protection was seen if the serum was raised against a <u>Salmonella</u> strain with a different major immunodeterminant (0:4,12). Thus, serum evidenced specificity of protection based on the O antigens of the immunizing and challenge strains. Similar O-dependent specificity of protection could be demonstrated using a double challenge procedure in which immunized

animals were infected with equal numbers of two strains of
Salmonella that differed in their 0 antigenic composition, but
which had similar in vivo growth rates. The strains displayed
nutritional markers that allowed each to be enumerated separately
on differential agars, so that if mice were sacrificed at various
times after infection, the livers and spleens could be homogenized,
plated, and the numbers of viable bacteria of each of the infecting
strains counted. The group at Stanford constructed pairs of strains
in which one partner was nearly identical to the other except for
its 0 atingens, which had been altered by transduction, to yield an
organism with the body of the original strain, but expressing a new
0 antigen. If one of these strains were used for immunization, and
the mice received a double challenge with equal numbers of both
organisms, any protection specific for the 0 antigens would be
revealed by suppression of growth of one strain, while the sister
strain continued to proliferate. Using this type of double-challenge
procedure, it was shown that a whole, heat-killed cell vaccine
induced specific suppression of growth of the challenge organism
which shared the same 0 antigens as the immunizing strain (44).
Similar types of experiments carried out in my laboratory using
ribosomal vaccines as immunogens, also showed that a major
portion of these preparations was 0 antigen dependent, although
some cross-serotype protection was also noted (45).

The importance of 0 antigens in protection has been further
demonstrated by some imaginative genetic manipulations of these
microbes. Two groups of investigators have prepared hybrid strains
of Salmonella. Diena and colleagues, by conjugation, produced a
bacillus with an S. typhimurium body, but the 0:9 determinants of
S. typhi. If mice were vaccinated with whole killed-cell vaccines
of S. typhi they were protected against challenge with the hybrid
strain expressing the same 0 antigens as the vaccinating organism,
but not against the original S. typhimurium strain which had the
heterologous 0:4 antigen (46). Kiefer et al., constructed hybrid
strains with E. coli bodies, but expressing Salmonella 0 antigens.
Live and killed hybrid vaccines protected against challenge with
Salmonella of the same 0 antigenic type, but not against challenge
with the E. coli strain (47).

All of these studies using whole organisms or subcellular
fractions point to the conclusion that 0 antigens are major protective immunogens in the mouse model. The fact that purified LPS,
itself, will protect (see Table 1) is additional support for the
role of 0 antigens in immunity to Salmonella infection. Definitive
proof of this issue has involved some elegant immunochemical studies
carried out by Lindberg and his colleagues (48). This group prepared the synthetic disaccharide 0 antigenic determinant, tyvelose-
mannose, which has the 0:9 specificity, coupled it to bovine serum
albumin (BSA), and used it in complete Freund's adjuvant as an

immunogen in mice. High titered bactericidal antibodies specific
for S. enteriditis, but not cross-reacting with S. typhimurium were
obtained. Animals double-challenged with equal numbers of an S.
enteriditis strain (0:9), and an S. typhimurium strain (0:4), showed
selective suppression of growth in livers and spleens of S. enteriditis, which has the 0:9 antigen in common with the synthetic
vaccine. Immune serum raised to the synthetic conjugate passively
protected against 100 LD_{50} doses of S. typhimurium (48). Using
another immunochemical approach, this group extracted lipopolysaccharide from a strain of S. typhimurium and then treated it with
bacteriophage P22 which contains an endorhamnosidase capable of
splitting the lipopolysaccharide to yield an octasaccharide, which
contains the immunogenic determinants for 0 antigens 4 and 12.
This octasaccharide was obtained from dialyzed culture filtrates
by column chromatography, and then coupled to BSA. Mice immunized
with this conjugate showed protection against challenge with live
S. typhimurium. Rabbit immune serum raised to the octasaccharide-
BSA vaccine was capable of passively protecting normal animals
against a challenge of 100 LD_{50} doses of S. typhimurium, but gave
little protection against S. enteriditis challenge (49). Thus,
the protective effect of 0 antigens in a mouse model, using an ip
route of inoculation, has been unequivocally demonstrated. Some
degree of 0 antigenic specificity has even been demonstrated in
the case of protection by living organisms, particularly during
recall of immunity upon reinfection (43,50,51,52).

The 0 antigens are not the only protective immunogens in nonliving vaccines of Salmonella. Outer membrane proteins have been
reported to confer protection (53), and the octasaccharide, when
complexed to an outer membrane protein gave protection superior
to that observed when complexed to albumin (49). The nature of
the protective antigens in ribosomal vaccines from Salmonella is
not certain, but some investigators have found protection in
protein-rich subfractions (54,55,56). Barber and Eylan showed
that the protein fractions extracted from whole Salmonella are
protective (57). Recent work from our laboratory has shown that
Endotoxin Protein (EP), a hydrophobic group of proteins tightly
associated with lipopolysaccharide in the outer membrane of Gram
negative bacteria, is highly protective in CD-1 mice (58,59). A
protein factor obtained from culture supernatants in complex with
0 antigens has also been shown to be protective (60).

Two other lines of evidence can be brought to bear in support
of the importance of the role of humoral immunity in host defense
to Salmonella infection. A study of passive cell transfer by
Hochadel and Keller (61) showed that mice given B cells from
animals vaccinated with live cells, had better survival rates than
those receiving T cells. Unexpected results were obtained by
Morris et al. (62) who showed that if mice were vaccinated with an
avirulent rough strain of Salmonella, and then treated with the

B-cell immunosuppressive drug, cyclophosphamide, the vaccinating
strain multiplied and eventually was lethal. Treatment of infected
mice with a T-cell suppressive regimen of anti-thymocyte serum did
not allow growth of the vaccine strain. The results of these two
studies suggest that B cells are more important than T cells in
ultimately controlling in vivo replication of Salmonella.

An explanation for the importance of B cells in the face of
only weak effects of passive immune serum may possibly be found in
an antibody that has been described by several laboratories to be
cytophilic for macrophages. Rowley et al. first presented evidence
for the existence of a macrophage cytophilic antibody as an explanation for the enhanced bactericidal activity of macrophages
taken from animals immunized with live Salmonella (42,63). The
antibody was in serum but could also be extracted from the macrophages (63). A similar 19S cytophilic antibody has also been described by Japanese investigators (64,65,66) and by Hsu's laboratory (37). Margolis and Bigley reported that a Salmonella ribosomal extract induced a cytophilic macroglobulin (67).

All of the studies cited above leave no doubt that nonliving
vaccines, whether whole-cell or bacterial extracts, protect
against Salmonella infection, that antibody has some protective
effect, and that O antigens are important immunogens.

<u>Commentary on the Literature</u>

Possible Explanations for Divergent Results

The literature review presented above leaves unexplained the
intriguing question of why different laboratories, all reputable,
have gotten such dicotomous results. Some laboratories have
achieved substantial levels of protection with whole-killed cells
or bacterial extracts and can protect with passive serum, whereas
others can demonstrate little or no protection with any nonliving
vaccine, and find serum ineffective. Thus, the relative contribution of cellular versus humoral immunity to Salmonella immunity,
and the efficacy of various vaccines is uncertain.

The degree of disagreement among investigators in the field is
reflected in a recent paper from the laboratory of Dr. Hsu,
questioning, on pathological grounds, whether Salmonella can even
be classified as intracellular pathogens (68). It is important to
recognize that not all laboratories have found only live or only
killed vaccines to be protective. In some cases only one type of
vaccine was tested, precluding a comparison (36,48,49). In our
own laboratory (40, and see Table 1), both types of vaccines were
found to be effective. Other laboratories have also found protection by both living and killed vaccines (34,38,42,52,69,70),

although their relative potency has differed depending on the
particular experimental conditions employed. A number of different
explanations have been advanced to try to account for why results
from different laboratories do not agree. A major variable was
hypothesized to be the method of preparation of killed cells,
particularly the effect of heating. Auzins and Rowley proposed the
existence of a heat-labile antigen as a reason for the variability
in the protection afforded by killed vaccines (71), and carried
out further studies to purify and characterize the antigen (34).
However, Collins and Milne (72,73) were unable to protect even
with whole-cell vaccines that preserved heat-labile antigens. It
is noteworthy that Ornellas et al. (36) demonstrated excellent
protection with heat-killed cells. Thus, while the method of
vaccine preparation is a variable, it cannot be the only major one.
As already discussed, the route of infection and the dose of
vaccine can also be important factors in the outcome of a protection experiment.

Recent work from our laboratory suggests that the genetic
constitution of the mouse is a major variable in the outcome of
protection experiments in murine Salmonellosis which could account
for the differences among laboratories in protection experiments
with this organism. In the next section of this paper, data will
be presented to substantiate this hypothesis, and the results
will be intrepreted to develop a comprehensive model for host
defense to <u>Salmonella</u> in the mouse.

*Critical Evaluation of the Value of Cellular versus Humoral
Immunity and Implications for Vaccine Strategy*

Based on mouse models using <u>S</u>. <u>typhimurium</u>, in which some
investigators showed that killed vaccines were much less protective
than live cells, the use of a live vaccine for protection against
<u>S</u>. <u>typhi</u> has been strongly urged (10,32,74). The rationale for
this proposal is that live cells can induce cellular immunity (75)
but killed vaccines cannot. Since only previous infection with a
living organism has been found capable in mice of bringing about
"immune elimination" of organisms in a secondary challenge dose,
it has been argued that only live vaccines can prevent "infection"
as measured by bacterial proliferation in the host (13). Thus,
merely demonstrating survival of vaccinated animals has been considered an insufficient measure of protection, as true protection
has been defined as a condition of immunity which gives the host
the capacity to prevent multiplication of infecting bacteria in
the organs (5,10,13,25,29). This redefinition of the criteria for
when a vaccine is protective has generated considerable confusion
in the literature, because even though killed vaccines can markedly
increase the number of survivors, they do not qualify as protective
by the criterion of immune elimination.

In evaluating the validity of this restricted definition of
protection, the following points should be considered. First, in
order to establish the conditions under which immune elimination
of a secondary challenge will occur, it is necessary to give a
primary infecting dose. The organisms of the primary dose may
persist for some time (5,19), and, in fact, if they do not persist,
the live vaccine will not protect (20,51). The persistence of the
vaccine strain in the host's tissues is not without potential
hazards, as revertants might occur, or if the person were immunosuppressed, as in the case of the cyclophosphamide-treated mice (62),
the vaccine strain might multiply. Furthermore, even when live
cells are used for vaccination, immune elimination of the secondary
challenge organism is not always rapid or complete. Thus, in
Hobson's experiments, the challenge organism was kept from multiplying extensively, but it was not eliminated (5). Probably, there is
a dosage effect, so that at low challenge doses, complete and
rapid elimination of organisms in the secondary challenge may occur,
whereas at high doses multiplication of the secondary organisms is
inhibited, but they are not eliminated. Immune elimination is also
most effective while the mouse still harbors the vaccine strain
(51). Thus, by sixty days after the primary infection, the immune
elimination mechanism has a lag period and has lost potency (51).
Therefore, on theoretical grounds, it is possible that a live,
parenteral vaccine against S. typhi might not be able to totally
prevent bacterial proliferation in the organs. Factors such as
the infecting dose of the virulent strain and the time elapsed
since vaccination might be crucial for the efficacy of a live
vaccine. Doubts about a priori assumption of superiority of live
vaccines are strengthened by reports that even recovery from
typhoid fever does not ensure permanent immunity. Marmion reported
well-documented second attacks of typhoid fever and cites other
investigators who reported similar findings (76).

When bacterial organ counts are determined in mice receiving
killed vaccines (35,39,45) one finds that bacteria multiply to
about 10^5 organisms in the whole carcass (about 10^4 in the liver
and spleen). They maintain this level for several weeks and then
the number of organisms begins to decline, although at 30 days
post-challenge many mice still harbor live bacteria. The kinetics
of this infection is not different from what happens to the live
vaccine strain which is used to create a primary infection. Mice
which would be protected by a killed vaccine do not show unrestricted multiplication of bacteria in their livers and spleens, for if
they did they would die. We have no evidence that immunized Swiss
mice would eventually succumb to infection, although in other
mouse strains such as the C3HeB/FeJ, some killed vaccines merely
prolong life, but are not fully protective (see Results). Thus,
while the demonstration of immune elimination is clearly a marker
for cellular immunity in hypersusceptible mice, and in these mice
may sometime separate effective from ineffective vaccines, solid

immunity can be demonstrated in CD-1 mice (see Table 1) without
the necessity to show immune elimination. The probable mechanism
of the protection is that initial humoral immunity, combined with
a high level of innate resistance, allows the host time to develop
cellular immunity to the organisms in the challenge dose (12,35).
Since live-cell vaccinated mice harbor the organisms of the
primary infection, and in the hands of some investigators even
organisms of the secondary challenge, the use of immune elimination as a requisite criterion to demonstrate vaccine potency
seems artificial.

Another point to be considered with regard to the mechanism
of protection afforded by living, rough organisms, is that part
of their efficacy may actually be attributable to antibody.
Kenny and Herzberg (69) found that protection with at least one
such vaccine strain was specific for smooth organisms with 0
antigens similar to those from which the rough strain had been
derived. Furthermore, immunization with the live, rough organism
primed the mouse to make an antibody response 100-fold greater than
that shown by a control mouse. One interpretation of the data
which they suggested was that the rough mutant synthesized small
amounts of 0 antigen. Their observations are in agreement with
those of Germanier (52) who showed that a live, rough mutant of
the gal E type was only protective if it synthesized 0 antigens
in vivo. Thus, protection by live cells may in some cases be
mediated by priming of the antibody-forming system, and may still
be related to an immune response to the 0 antigens.

Although the role of the macrophage may be crucial in allowing
the host to inhibit multiplication of intracellular bacteria, it is
well to remember that other factors of the defense system besides
the macrophage are certainly important in influencing the outcome
of an infection. For example, it has been shown that A/J mice are
Listeria susceptible, but Salmonella resistant, whereas C57BL/6J
mice are naturally Listeria resistant but Salmonella susceptible
(77). In this case it is clear that factors other than the macrophage must be responsible for these differences in natural
resistance, leaving open the distinct possibility that factors
other than the macrophage may be involved in expression of immunity
as well.

RESULTS

Most of the experiments in our laboratory, which involved
testing vaccines for protection against Salmonella infection had
been carried out using CD-1, outbred, Swiss-Webster mice, a line
purchased from Charles River Breeding Laboratories. Our interest
in the role of lipopolysaccharide (LPS) in immunity to Salmonella
infection prompted us to try C3H/HeJ mice, a strain which has a

genetic defect that makes it poorly responsive to the toxic and
mitogenic effects of LPS (78,79). In the course of routine
experiments to standardize the challenge dose of $\underline{S}$. $\underline{typhimurium}$,
strain W118-2 in this mouse strain, we discovered that the ip
LD_{50} for C3H/HeJ mice was less than 6 cells, whereas the very
same organism had an ip LD_{50} of 5×10^4 cells in CD-1 mice (80).
It seemed possible that the extraordinary hypersusceptibility of
the C3H/HeJ strain might be linked to its genetic defect to LPS.
To test this hypothesis, C3HeB/FeJ mice were obtained, and their
LD_{50} tested, as they had been reported in the literature to have
normal immunological responses to LPS (81). Again, to our surprise,
we found these mice to be as hypersusceptible to $\underline{Salmonella}$
infection as the LPS-nonresponder C3H/HeJ strain, with an ip
LD_{50} of less than 2 cells (82). A closely related strain,
C3H/HeNCr1BR, was found to be resistant to $\underline{Salmonella}$ infection
with an ip LD_{50} of 1.2×10^3 cells (83).

As these three mouse strains differed so markedly in their
susceptibility to $\underline{Salmonella}$, an investigation was undertaken to
see if there was a correlation between susceptibility to $\underline{Salmonella}$
infection and sensitivity to the toxicity of LPS (endotoxin). As
shown in Table 2, C3HeB/FeJ mice were found to be highly sensitive
to endotoxin toxicity (TD_{50} < 250 µg) but were susceptible to
$\underline{Salmonella}$ infection, whereas C3H/HeJ mice which were also hypersusceptible were resistant to endotoxin TD_{50} > 2000 µg. The
C3H/HeNCr1BR mice were $\underline{Salmonella}$ reistant but endotoxin sensitive (82).

As there was no correlation between sensitivity to endotoxin
and susceptibility to $\underline{Salmonella}$ infection, the results argue
against the hypothesis that endotoxin is an important factor in the
pathogenesis of $\underline{Salmonella}$ infection in mice (82,83).

Table 2. $\underline{Salmonella}$ susceptibility and LPS sensitivity in three
mouse strains.

Mouse strain	Intraperitoneal LD_{50}	TD_{50} LPS (µg)
C3H/HeJ	< 7 cells	<2000
C3HeB/FeJ	< 2 cells	250 < 150
C3H/HeNCr1BR	1.2×10^3 cells	500 < 250

In our initial study investigating immunity to lipopolysaccharide and ribosomal vaccines in C3H/HeJ mice, we used
acetone-killed cells as a control preparation. We found that this
vaccine, which had given such solid protection in CD-1 mice, was
only partially protective in C3H/HeJ mice, in spite of the fact
that it was able to induce anti-Salmonella agglutinins and anti-0
antibody as assessed by passive hemagglutination (80). In fact,
the protection results obtained using the acetone-killed cells
in the C3H/HeJ mice were very similar to those described by
laboratories which had reported that nonliving vaccines did not
protect against Salmonella, in spite of the presence of antibodies.
Since C3H/HeJ mice have a known genetic defect in LPS responsiveness, we thought that their poor protection by killed vaccines
might have been attributable to some aspect of their defect. To
test this possibility we decided to carry out a series of protection experiments in the three mouse strains in the C3H lineage
shown in Table 2. The C3HeB/FeJ mouse was used as a control to
assess the contribution of the LPS defect to any lack of immunizability in the C3H/HeJ mice, because it has normal responses to
LPS, but like the C3H/HeJ strain is hypersusceptible to Salmonella
infection.

A panel of vaccines was chosen that included highly purified
bacterial extracts, such as phenol-water purified lipopolysaccharide (PW-LPS), trichloroacetic acid extracted LPS (TCA-LPS),
and Endotoxin Protein (EP), as well as less characterized preparations, such as a subcellular ribosomal vaccine (RIB), and acetone-
killed cells (AKC). A single large batch of each vaccine was prepared to provide enough material to carry out immunizations in all
of the strains, thus eliminating any differences between lots of
immunogens as a variable. Animals were given a primary injection
and a booster injection two weeks later. Large doses of vaccine
were used in order to be in the optimal dosage range based on
previous experiments. Mice were challenged three weeks after the
booster injection. Immunization and challenge were both by the
intraperitoneal route. As shown in Tables 3 to 5, the most
important variable governing the ability of a nonliving vaccine
to protect was the mouse strain used for immunization. Thus, PW-
LPS and EP were protective in the inherently resistant C3H/HeNCr1BR
mice, but gave no protection in C3H/HeJ (58) or C3HeB/FeJ mice,
even to very low challenge doses, although mean time to death was
significantly prolonged in both strains. The TCA-LPS, which is
the naturally occurring combination of LPS and EP failed to protect C3H/HeJ mice, but protected all of the C3H/HeNCr1BR mice, and
100% effective in the C3H/HeNCr1BR strain. The most complex
vaccine, the acetone-killed cells, gave partial protection in
C3H/HeJ mice, and 100% protection in the other two strains. These
results show that the same vaccines, in comparable doses, behave
very differently with respect to their ability to immunize
different mouse strains in the C3H lineage. Differences in ability

Table 3. Protection in C3H/HeJ mice immunized with various
          Salmonella vaccines.

Mice Immunized With:	Dose (μg)[a]		%	60 Day Survival[b] Alive/Total	MTD[c]
	1	2			
PW-LPS	100	100	0	0/10	21
TCA-LPS	100	100	21	3/14	31
Ribosomes	250	250	50	7/14	–
Acetone-killed cells	60	60	36	5/14	29
Saline	–	–	0	0/14	10

[a] Mice immunized ip with dose 1, and 14 days later with dose 2.
[b] Mice infected ip 21 days after dose 2 with 6 cells.
[c] MTD = Mean Time to Death (Days)

Table 4. Protection in C3HeB/FeJ mice immunized with various
          Salmonella vaccines.

Mice Immunized With:	Dose (μg)[a]		%	60 Day Survival[b] Alive/Total	MTD[c]
	1	2			
PW-LPS	100	100	0	0/7[d]	19
Endotoxin Protein	100	100	20	2/10	17
TCA-LPS	100	100	0	0/5[e]	31
Ribosomes	100	100	60	6/10	–
Acetone-killed cells	60	60	100	10/10	–
PBS	–	–	0	0/10	12

[a] Mice immunized ip with dose 1, and 14 days later with dose 2.
[b] Mice infected ip 21 days after dose 2 with 24 cells of W118-2.
[c] MTD = Mean Time to Death (Days)
[d] 3 Mice died from toxicity of immunization.
[e] 5 Mice died from toxicity of immunization.

Table 5. Protection in C3H/HeNCrlBR mice immunized with various
          Salmonella vaccines.

Mice Immunized With:	Dose (µg)[a]		%	60 Day Survival[b] Alive/Total	MTD[d]
	1	2			
PW–LPS	100	100	63	5/8[d]	–
Endotoxin Protein	100	100	100	10/10	–
TCA–LPS	100	50	100	4/4[e]	–
Ribosomes	100	100	100	10/10	–
Acetone-killed cells	60	60	100	10/10	–
PBS	–	–	11	1/9	7

[a] Mice immunized ip with dose 1, and 14 days later with dose 2.

[b] Mice infected ip 21 days after dose 2 with 2.3×10^4 cells of
W118-2 ($\cong$ 19 LD_{50} doses).

[c] MTD = Mean Time to Death (Days).

[d] 2 mice died from toxicity of immunization.

[e] 6 mice died from toxicity of immunization.

to be immunized cannot be attributed to the Lps gene, for the
C3HeB/FeJ mouse, which is a normal LPS responder, behaves more
like the C3H/HeJ mouse, which is an LPS nonresponder, than it does
like the responder C3H/HeNCrlBR strain. Immunizability seems to
correlate best with the inherent susceptibility or resistance of
the mouse strain to Salmonella infection. Thus, the inherently
resistant, C3H/HeNCrlBR mice yielded results similar to those
obtained in the inherently resistant CD-1 mice (see Table 1),
whereas the more susceptible strains (C3H/HeJ and C3HeB/FeJ) were
hard to immunize with the simpler vaccines. Yet, there is a
gradation in the ability to be protected by the various vaccines
that is more complex than the LD_{50} data on inherent susceptibility
would suggest, for the C3H/HeJ and C3HeB/FeJ mice appear equally
susceptible based on their LD_{50} s, but the C3HeB/FeJ mouse is
somewhat more responsive to vaccination with acetone-killed cells.
In vivo studies of the growth kinetics of Salmonella in the livers
and spleens of these two strains showed that the organisms
multiplied somewhat slower in the C3HeB/FeJ mice (83).

In order to explore further the variables involved in
Salmonella immunity in C3H/HeJ mice, a live, avirulent strain was

used as a vaccine. The organisms, SL3235, is a smooth strain which
expresses O antigens, but is deficient in aromatic amino acid
synthesis. The mutant was prepared and characterized by Hoiseth
and Stocker (84) and kindly provided for these studies. An ip
vaccination of $5x10^4$ cells of this avirulent organism in C3H/HeJ
mice gave 91% protection against a challenge dose of 900 cells
and 73% protection against a challenge dose of 9000 cells. These
results are in striking contrast to the lack of protection, or
marginal protection, provided by all of the other vaccines tested
in this mouse strain (Table 3).

When these observations are viewed in terms of the literature
on host defense to murine salmonellosis, they provide a simple
explanation for the mass of conflicting data that has emanated from
different laboratories. It is proposed that it is the mouse
strain which is the major factor in determining whether living or
nonliving vaccines protect, and in influencing the relative
importance of humoral versus cellular immunity.

To test this hypothesis further, a study of passive serum
transfer was undertaken. Hyperimmune serum was raised in CD-1
mice by 3 injections of 2.5×10^8 acetone-killed cells. Serum
was obtained one week after the last injection. The immune
serum (0.2 ml), at a dilution of 1:10 was injected iv into normal
recipients. Other groups of mice received normal serum (1:10
dilution) or saline. Two hours later, all mice received virulent
Salmonella, strain W118-2, intravenously, after the organisms had
been incubated for 30 min on ice, with a 1:10 dilution of either
normal or immune serum, or with saline. Recipient mice were
either the inherently resistant CD-1 strain, or the hypersusceptible
C3HeB/FeJ strain. As shown in Table 6, the immune serum protected
the CD-1 mice, but did not protect the C3HeB/FeJ mice, in spite of
the fact that the C3HeB/FeJ mice received only 17 cells iv. CD-1
mice which received a comparable number of LD_{50} doses, but a much
larger number of cells ($1.5x10^5$), showed excellent protection,
thus supporting the hypothesis that inherently resistant mice can
be protected by antibody, whereas inherently susceptible mice
cannot.

To see if the protection induced by the SL3235 strain was
related to development of cellular immunity, studies were carried
out to assess delayed hypersensitivity as measured by a footpad
response to a culture-filtrate antigen of Salmonella. The data
are presented in another paper in this volume (see Killar, L.
and T. K. Eisenstein). It was found that C3H/HeJ and C3HeB/FeJ
mice infected with the smooth, avirulent mutant, SL3235, exhibited
immunity, but no delayed hypersensitivity.

Antibody studies carried out using an Enzyme-Linked Immunosorbent Assay (ELISA) showed that C3H/HeJ and C3HeB/FeJ mice which

Table 6. Passive serum protection in <u>Salmonella</u> resistant and
          susceptible mice.

Mouse Strain	Pretreatment	Challenge Dose (LD_{50})	60 Day Survival	
			Survivors/Total	%
CD-1	Immune Serum		12/14	86
	Normal Serum	15	0/14	0
	Saline		0/14	0
C3HeB/FeJ	Immune Serum		3/14	21
	Normal Serum	17	0/14	0
	Saline		0/14	0

had been infected with SL3235 had little or no antibody to LPS at
21 days post-infection, a time when they were known to be well
protected.

Thus, in the case of these hypersusceptible strains in the
C3H lineage, we have been able to protect them using a living,
attenuated vaccine, but this protection occurred in the absence
of detectable delayed hypersensitivity, a marker for cellular
immunity, and in the absence of antibody to the O antigens.
Experiments are currently in progress to further assess the
mechanism(s) for the resistance displayed by SL3235 immunized
C3H/HeJ mice, particularly the possibility that there is a population of sensitized T-cells which mediates immunity, but fails to
give perceptible delayed hypersensitivity responses.

DISCUSSION

The most apparent feature of the data presented from our
laboratory is the immense contribution of genetic factors to the
outcome of <u>Salmonella</u> infection in normal or vaccinated mice.
Although it was recognized as far back as 1937 that there were
genetic factors controlling innate susceptibility to <u>Salmonella</u>
infection (85,86,87) this older literature was not often cited
by researchers in the fifties and sixties, and is not proposed by
them as a potential explanation for disparate results from different
laboratories. Recently, however, interest has awakened in the
nature of genetic control of innate resistance to <u>Salmonella</u>
infection (88-93). Three genes have been identified which play
a role in natural susceptibility to this organism (94) and
others may be forthcoming.

With regard to genetic contributions to the ability
to be protected by vaccination, there is a limited literature.

Gowens carried out extensive studies in this area in the forties, but because of initial publication in obscure journals (95,96) his work seems to have gone largely unnoticed by the major laboratories working on mechanisms of immunity to <u>Salmonella</u>, until a review article appeared in 1967 (97). More recently, Robson and Vas (98) have shown differences in the ability of a panel of mouse strains to be protected by phenol-preserved whole cells. In a subsequent study, an RNA-rich subcellular vaccine was compared with the phenolyzed cells in two mouse strains, one inherently resistant to <u>Salmonella</u> infection and one inherently susceptible. Only the resistant mice were protected by the vaccines (99). Misfeldt and Johnson (100) also described differences in immunizability among several inbred strains when comparing a ribosomal vaccine with whole killed cells or living attenuated <u>Salmonella</u>.

In light of these reports and our own data, we propose a comprehensive set of hypotheses within which to view immunity to <u>Salmonella</u> infection.

<u>Axioms</u>

1. Nonliving vaccines induce humoral immunity.
2. Living vaccines induce humoral and cellular immunity.
3. There is a wide spectrum (10,000-fold range) of genetically determined innate susceptibilities of different mouse strains to <u>Salmonella</u> infection.

<u>Hypotheses</u>

1. Unlike host defense to the pneumococcus, which is primarily humorally mediated, or host defense to <u>Mycobacterium tuberculosis</u>, which is primarily cell mediated, immunity to <u>Salmonella</u> can be mediated by both arms of the immune system in conjunction with macrophages (Fig. 1).

2. The ability of the mouse to develop protective immunity to <u>Salmonella</u> infection in response to various vaccines is primarily genetically determined.

   <u>Corollary a)</u> There is good correlation, although imperfect between ability to be vaccinated by nonliving vaccines and inherent susceptibility or resistance of a mouse strain to <u>Salmonella</u> infection.

   <u>Corollary b)</u> The degree of protection afforded by vaccines in the hypersusceptible mice will be dependent on unknown factors which correlate to some extent with the rate of growth of <u>Salmonella</u> in unvaccinated mice of the particular strain.

3. In mouse strains which are inherently resistant to
    <u>Salmonella</u> infection, antibody alone can be protective,
    but in strains which are inherently hypersusceptible,
    antibody gives poor protection and cellular immunity is
    required.

    <u>Corollary a)</u> Nonliving vaccines which induce only anti-
    body will protect inherently resistant mice, but will
    give variable protection, usually poor, in inherently
    susceptible mice.

4. The route of infection is a major variable in determining
    the relative contribution of humoral versus cellular
    immunity in mice. The intravenous route accentuates
    protection by cellular immunity, whereas the intraperi-
    toneal route potentiates the action of antibody.

Some of the elements of the proposed model have been recognized
by other investigators and their contribution to its formulation is
readily acknowledged (6,10,11,29,69,70,75,92,97,98,100). It is
hoped, however, that there has indeed been a synthesis of all of
the results and ideas accumulated in this large literature to
provide a new and unifying way of interpreting data in murine
salmonellosis.

The model proposes that there are three major variables, the
mouse strain, the route of <u>Salmonella</u> infection, and the vaccine
(live or nonliving), which determine the nature of the protective
immune response. Of these, the most important is the mouse strain.
Since it would be an occult variable when comparing results from
different laboratories, the nature of the mouse seems to have been

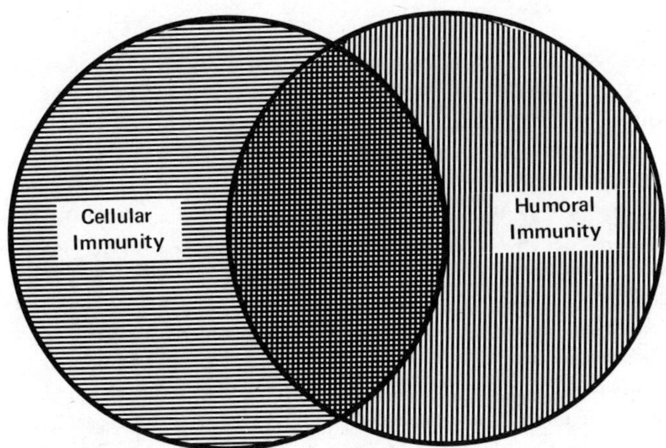

Figure 1. Model of mechanisms of immunity to <u>Salmonella</u> infection.

the least appreciated variable, even though it probably has the greatest influence on the outcome of an experiment. If one reviews the literature with this model in mind, examples can be found which fit each of the three situations illustrated in Fig. 1. Thus, there exist strains like the C3H/HeJ which cannot be well protected by any nonliving vaccine or by antibody, but require living cells which induce cellular immunity. Use of the intravenous route of challenge further obscures any potential protection afforded by the nonliving vaccine. In inherently resistant mice, such as the CD-1, passive antibody alone is sufficient for protection. Killed vaccines in resistant mice protect via antibody, and live vaccines induce both cellular and humoral immunity. Our results with the C3HeB/FeJ mice suggest that between the two extremes of very susceptible and very resistant mice there may be a spectrum of strains which are intermediate in ability to be immunized. The gross LD_{50} may be insufficient to distinguish subtle differences in immunizability, as C3H/HeJ and C3HeB/FeJ appear equally susceptible, but the C3HeB/FeJ can be 100% protected by acetone-killed cells and the C3H/HeJ cannot. We have also found mice with LD_{50} s of between 50 and 100 cells (83), and their ability to be immunized has yet to be investigated.

The proposed model also explains why so frequently there has been a lack of correlation between antibody titers to the 0 antigens and protection. Such a correlation can only exist in a resistant strain given a nonviable vaccine. If a hypersusceptible strain is given a nonviable vaccine, even if antibody is induced, it will not protect. If a resistant or a hypersusceptible strain is given live cells, the contribution of humoral immunity to 0 antigens cannot be assessed due to the simultaneous induction of cellular immunity. When whole organisms or crude cell extracts are used, imperfect correlations with anti-0 titers and protection may be due to protective humoral responses to other antigens, such as protein moieties (58,59,101).

The nature of the protection induced by ribosomal extracts prepared from _Salmonella_ perhaps requires additional commentary (102). Although such ribosomal extracts are nonliving preparations, some investigators have presented evidence that they are unique among nonliving vaccines in inducing cellular immunity (54,67,103). In our laboratory, when these extracts have been used in either CD-1 (40) or C3H/HeJ mice (80), the ribosomal extracts seem to behave similarly to acetone-killed cells in the degree of immunity they engender, suggesting that they are not unique among nonliving preparations in their mechanism of action (104). However, this field, like the whole subject of _Salmonella_ immunity has been fraught with contradictory results from different laboratories (102), and the literature is too extensive to review here. Suffice it to say that these subcellular preparations are thought by some investigators to provide a counter-example to

axiom #1. If this is true, then the picture of <u>Salmonella</u> immunity
is even more complicated that is presently appreciated.

In summary, the proposed set of axioms and hypotheses suggest
that instead of there being one model of murine <u>Salmonella</u>
infection, there are actually multiple models. In fact, it should
be possible to "engineer" a model to display the desired degree of
contribution of cellular or humoral immunity. If one uses a hypersusceptible mouse and an intravenous route of challenge, only
cellular immunity, induced by live cells, will be protective. On
the other hand, use of a highly resistant mouse such as the CD-1,
with an intraperitoneal route of challenge, will show levels of
protection by nonliving vaccines which approach that engendered
by live cells. In this model passive antibody alone will protect.
By the use of other mouse strains, with varying degrees of innate
resistance to <u>Salmonella</u>, and employing different routes of
challenge, it should be possible to weight the system one way or
the other, either towards cellular or humoral immunity.

IMMUNITY TO TYPHOID FEVER IN MAN

The crucial question which arises from all of these studies
using the mouse model of <u>Salmonella</u> infection is what we can learn
from them about human immunity to <u>S. typhi</u>. Which murine model of
salmonellosis is most analogous to the human situation? Is the
human immune response to vaccines more analogous to the hypersusceptible mice receiving an intravenous infection, or more like
resistant mice receiving an ip infection, or somewhere in between?
Little direct data is available on this subject, so it is clear
at the outset that no definitive answer is possible at the
present time. However, there are some pieces of information
which bear evaluation with the foregoing concepts in mind.

In 1897 Wright and Semple developed the first heat-killed
whole cell typhoid vaccine for use in the British military (105).
As it seemed to be efficacious in troops vaccinated during the
Boer War, the preparation was adopted for routine vaccination of
British and American troops during World War I (106,107). The
vaccine consisted of approximately 1 x 10^9 killed cells (heated
at 56°) of <u>S typhi</u>, plus 7.5 x 10^8 similarly killed cells of <u>S.
paratyphi</u> <u>A</u> and <u>B</u>. Organisms were preserved in phenol and the
preparation was designated "T.A.B." (1). The vaccine, although
widely used, had deficiencies. First, it had considerable
toxicity due to the high endotoxin content of the large number of
killed organisms. Redness, swelling, and tenderness of the arm
were common, as well as generalized symptoms such as fever and
malaise (108). Secondly, there were reports of epidemics of
typhoid fever in vaccinated populations, which brought into
question its efficacy (109,110). In 1934 Felix and Pitt discovered

the Vi antigen (111) and found that alcohol-killed organisms
retained more Vi antigen than heat-killed ones (112). Although
alcohol-preserved organisms were tried as a vaccine, the preparation proved inferior to the heat-killed cells in a field trial
(113). In 1953 Landy and his colleagues introduced acetone-
killed and dried organisms, the method used in the currently
marketed typhoid vaccines (114).

Over the years a number of field trials of the various vaccine
preparations have been carried out, and their efficacy has been
correlated with the induced antibody titers to the Vi, O and H
antigens, and to mouse protection tests. Although the results
are frequently confusing, and often seem contradictory, one fact
clearly emerges - in double-blind, controlled field trials,
whole killed cells have been shown to protect humans against
typhoid fever. Table 7 shows the results of the trials in
British Guiana carried out by Ashcroft and his colleagues over a
seven-year period post-vaccination (115). Groups of school
children, 5 to 15 years of age, were vaccinated with 1) tetanus
toxoid as the control, 2) acetone-killed cells of _Salmonella_
typhi, or 3) heat-killed, phenol-preserved whole cells of
Salmonella _typhi_.   There were approximately 24,000 vaccinees in
each group. As shown in Table 7, over the seven-year period post-
vaccination, the acetone-killed cell vaccine provided 88% protection, and the heat-killed, phenol-preserved vaccine, although
somewhat inferior, provided 65% protection. Thus, acceptable and
significant protection of seven years duration was provided by
these killed vaccines. Similar results have been obtained in
the field trials carried out in Yugoslavia, in Poland and the
U.S.S.R. (108,113).

One difficulty in interpreting these results is that, of
necessity, they had to be carried out in a typhoid endemic area to

Table 7. Efficacy of typhoid vaccines: field trials in Guiana,
          7 year results.

Vaccine[b]	Number Vaccinated	Number Typhoid Cases	% Protection
Tetanus Toxoid	24,241	146	–
Acetone-killed Typhoid	24,046	16	88.4
Heat-killed Phenol Preserved Typhoid	24,431	49	65.0

[a]Ashcroft et al. Lancet (1967) 2:1056.
[b]Dose given in 0.5 ml s.c., 2x.

provide enough cases in the control group to measure the reduction
in cases in the vaccine group. It has been argued that the killed
vaccine merely boosted an already existing state of cellular
immunity engendered by exposure to the living organism (115,116).
Vaccination and oral challenge of human volunteers who had no
history of typhoid infection has been carried out in a laboratory
situation by Drs. Hornick and colleagues (116). They have reported
that killed vaccines are 67% effective against a challenge dose of
S. typhi which causes disease in 25% of the controls. Thus nonviable vaccines can protect typhoid virgin volunteers, but we do not
know how effective they are compared with recovery from infection
with the live organism. The volunteer studies did show that protection could be overcome by increasing the dose of live organisms.
However, as already noted, there are reports in the literature of
second attacks of typhoid fever, so even the living organism does
not confer absolute protection (76).

    The nature of the protective antigens in the killed typhoid
vaccines is still an unsettled question. Two approaches have
been taken to this question. One is to try to correlate O, H, and
Vi antibody titers with protection in vaccinees participating in
field trials. The second has been the use of a mouse model employing active immunization with the test vaccine and challenge intraperitoneally with mucin-coated or saline-suspended S. typhi.
Correlations between antibody levels and protection have been
made in the immunized mice as has been done for humans. Passively
administered rabbit serum raised to the various vaccines has
also been employed.

    In field trials in humans in Yugoslavia, alcohol-preserved
vaccines were compared with the heat-killed, phenol-preserved
vaccines (113). The alcohol was used to preserve the Vi antigen,
which is poorly preserved in the heat-killed vaccine. The results
showed that the heat-killed vaccine gave better protection, in
spite of the fact that the Vi titer was higher in vaccinees receiving the alcohol-preserved organisms (113).   Active mouse protection tests showed the phenol-preserved vaccine to be better also,
but passive tests using rabbit sera raised to the two vaccines
showed that mice were better protected by sera against the
alcoholized vaccine (113). It was concluded that the Vi antigen
may not be as important to man as it is to the mouse.

    A similar lack of correlation between anti-Vi titer and pro-
tection was reached by Beneson in a study on sera of vaccinees in
the British Guiana and Yugoslavie field trials (119). He found
that vaccinees receiving acetone-killed cells, while showing better
protection in the field trial to typhoid, had no higher Vi titers
than those receiving heat-killed vaccines who were less well
protected (119). In conjunction with the field trial carried out
in British Guiana (see Table 7), and the ones in Yugoslavia,

extensive laboratory tests were carried out in 20 laboratories in 12 countries to try to correlate field trial results with mouse active or passive protection assays. It was found that there was a poor correlation in most of the tests. Route of vaccination of the mouse was an important variable, with subcutaneous vaccination giving poorer correlation than the intraperitoneal route (118,119). In general, no good correlations emerged with any of the laboratory tests except the anti-H titer in rabbit serum (120). The interpretation of the latter observation was not that H antibody was protective, but that it measured the gentleness of the procedure used to produce the vaccine, as H antigens are easily destroyed by harsh treatment (109). In an extensive series of experiments, Landy and colleagues sought to investigate the relative importance of the O and Vi antigens in protection of mice. They concluded that S. typhi O antigens, either purified or on whole bacilli, were poorly immunogenic for the mouse strain used (121). However, rabbit anti-O antibody was potent at passively protecting mice, showing a linear correlation between O agglutin in titer and mouse protective potency. Thus, it could be concluded that O antibody is protective in mice, but the typhoid vaccine is a poor inducer of this antibody in the mouse strain tested (121). Purified Vi antigen gave excellent active immunity to the mouse. Further, this immune mouse serum showed high potency in passively protecting unvaccinated mice to S. typhi infection (122). Since the O antigens were poor in inducing active immunity in his mice, but the Vi antigens were excellent, he concluded that his mouse potency assay was essentially an assay for Vi antigen (122). Later studies by Landy et al. in chimpanzees orally inoculated with S. typhi led to the conclusion that purified O antigen was not protective, but purified Vi antigen afforded some resistance to infection (123). It is clear from these results that the Vi antigen in their animal models proved an important protective immunogen. However, the anti-O antibody was also protective as shown by the passive serum transfer studies. Pistole and Marcus (124) have more recently shown that purified Vi antigen will protect mice against mucin-coated S. typhi. Evidence that the Vi antigen is not the only mouse protective factor was obtained by Standfast (125) who isolated a Vi-negative strain of S. typhi and showed that live cells of this strain protected mice against challenge with a Vi-containing organism (125). Also, in contrast to Landy's results, extensive studies carried out at the Walter Reed Army Institute for Research have shown no protective ability of the Vi antigen in a mouse model which uses as a challenge strain a genetically engineered hybrid strain of S. typhimurium which expresses the Vi antigen (126,127).

The Vi antigen used by Landy were purified from either E. coli or S. ballerup (121), whereas the Vi antigen used by the Walter Reed group was extracted from Citobacter freundii (127). The purified Vi vaccine of Landy did not protect human volunteers

against typhoid fever (116). Although these Vi antigens were
thought to be identical to the Vi of S. typhi, it has been
suggested that there are minor structural differences between
them, which could account for divergent results in different
systems (John Robbins, presentation at a conference on typhoid
vaccines, Walter Reed Army Institute for Research, July, 1979).
Recently, Vi antigen has been prepared from S. typhi, but has
not been tested for ability to protect humans (128).

A different approach to typhoid vaccination has been the use
of living, attenuated Salmonella typhi given by the oral route.
A streptomycin-dependent mutant of this type proved protective
in studies in human volunteers when freshly harvested cells were
used, by lyophilization resulted in loss of potency (129), thus
making it unsuitable for use in the field.

A mutant of S. typhi of the gal E type has been isolated by
Germanier and Furer (130) and shown to give excellent protection
in human volunteers (131). Interestingly, this organism is Vi
negative, but to be effective it must express O antigens. Preliminary field trial results indicate it has excellent efficacy
(132).

Critical Commentary on the Typhoid Literature

Results of mouse studies using S. typhi appear to be as complex and confusing as those using S. typhimurium. The S. typhi
mouse model measures only humoral immunity, but as the foregoing
literature review attests, even without the added dimension of
cellular immunity, the relative importance of the O as compared
with the Vi antigen has not been settled. One might suggest that
differences in the ability of O and Vi antigens to protect mice
seen in various laboratories may be related to genetic differences
in the ability of different mouse strains to respond to these
antigens. Certainly, Landy's observation of poor immunogenicity
of O antigens in mice is not true for the Swiss CD-1 mice used in
our laboratory (40). As already noted variation in the chemistry
of different Vi preparations may contribute to inadequate comparisons between animal models and human immunity to S. typhi.

It is also well to consider the possibility that the antigens
which induce humoral immunity may be different from those that
induce cellular immunity. Thus, one must clearly keep in mind the
nature of the mouse model used, whether it measures cellular or
humoral immunity, and if it weighs the contribution to protection
of various antigens in a manner similar to the human response.

The mechanisms of protection using oral vaccines are obviously
complex and must involve both inhibition of attachment to the
mucosal surface, which may be mediated by secretory antibody as

well as systemic humoral and cellular immunity for those organisms
which are able to penetrate the intestine, gain access to the
circulation, and proliferate in the liver and spleen (133,134).
Thus, the relative contributions of cellular and humoral immunity
may be altered under these conditions from those seen with
parenteral infection.

In spite of all of the conflicting data when comparing
different mouse models, and in spite of all the inconsistencies
between mouse and human protection studies, the following conclusions do emerge with regard to human immunity to typhoid fever.

There is no doubt from the field trial data and experiments
in human volunteers, that properly prepared acetone-killed cells
give excellent protection in humans. Thus, man does not appear
to be like the hypersusceptible mouse model of S. typhimurium
infection, in which only cellular immunity protects. Further, as
evidenced by second attacks following natural infection, cellular
immunity, if it is induced in man in a manner analogous to the
mouse, is not completely effective, and does not always induce
immune elimination. Thus, the value of killed, as compared with
live vaccines, for human vaccination against typhoid fever must
rest on practical considerations rather than on theoretical ones.
It is probable that only field trials can determine whether a
live vaccine will be superior to the present killed one. Judgment
will rest on which preparation gives the best protection, of
longest duration, with least toxicity.

SUMMARY

The foregoing literature review and data presentation have been
set forth in the hope of clarifying some complex and confusing
issues in regard to Salmonella infection. From a practical point
of view, the information presented has implications for the
direction to take with regard to improving the current typhoid
vaccine, as the presently used acetone-killed cell preparation has
considerable toxicity. The issues are important from a theoretical
standpoint, because they have bearing on the nature of the concepts
researchers and clinicians carry as working hypothesis with regard
to the mechanisms of immunity to Salmonella infection. An incomplete appreciation of the literature seems to have led many
scientists to believe that only cellular immunity can protect a
mouse, and by analogy a human, against Salmonella. The logical
deduction from such a premise is that only live vaccines will be
effective in humans against S. typhi. Such a conclusion would
appear unfounded, as documented in this review, for killed vaccines
have been shown to be highly effective in vaccinating many mouse
strains, as well as humans, against enteric fever.

A detailed consideration of the literature on mouse models of
Salmonella infection, in light of data from our laboratory, has
led to the formulation of the hypothesis that genetic variations
among mouse strains can account for much of the discrepancies in
results of protection studies from different laboratories. Further,
the various mouse models of Salmonella infection have been explored
to show that, depending on the mouse, the vaccine, and the route
of infection, the relative contribution of cellular versus humoral
immunity is altered. An attempt has been made to evaluate results
of typhoid vaccine field trials and protection studies in human
volunteers to discern which mouse model most resembles the human
situation. It is concluded that since both humoral and cellular
immunity contribute to human immunity to typhoid fever, the intravenous challenge of hypersusceptible mice is not representative of
the human situation.

With regard to the nature of the protective antigens, it is
clear that O antigens are protective in many mouse strains and
appear to be important in both killed and live attenuated typhoid
vaccines in man. The Vi antigen does not seem to be a prerequisite
for a protective vaccine in man, although whether or not it makes
any contribution to protection is at present difficult to evaluate.
The possibility of developing purified protein antigens to use as
vaccines is feasible, as shown by animal studies using Endotoxin
Protein and also outer membrane proteins. Such preparations seem
to be less toxic than lipopolysaccharide, and could provide a
useful alternative to live vaccines.

The mechanisms of protection of orally administered live
vaccines are yet to be elucidated, but results from experiments
using mouse models must be interpreted with caution, for all of
the variables affecting immunity to parenteral challenge will
apply to oral challenge, with the added complexities of secretory
immunity and bacterial adherence.

The extensive investigations of immunity to Salmonella in
the mouse and man, which have been reviewed in this paper, illumi-
nate the magnitude of the complexity of host defense to this
pathogen.

ACKNOWLEDGMENTS

This work was supported by Grants #AI-15613 and AI-16782
from the Institute of Allergy and Infectious Diseases of the
National Institutes of Health.

REFERENCES

1. Huckstep, R.L. (1962). Typhoid Fever and Other Salmonella
        Infections, E.S. Livingstone, Ltd., London.
2. Bornstein, S. (1943). J. Immunol. 46:439.
3. Ørskov, J., A.K. Jensen, and K. Kobayashi (1928). Z. Immuni-
        tätsforsch. 55:34.
4. Topley, W.W.C. (1929). Lancet. 1:1337.
5. Hobson, D. (1957). J. Hyg. 55:334.
6. Roantree, R.J. (1967). Annu. Rev. Micro. 21:443.
7. Kobayashi, R. and D. Ushiba (1952). Keio J. Med. 1:35.
8. Ushiba, D., K. Saito, T. Akiyama, M. Nakano, T. Sugiyama, and
        E. Shirono (1959). Japan J. Microbiol. 3:231.
9. Ushiba, D., M. Yoshioka, R. Iwahata, and S. Yamagata (1953).
        Keio J. Med. 2:75.
10. Mackaness, G.B., R.V. Blanden, and F.M. Collins (1966). J.
        Exp. Med. 124:573.
11. Blanden, R.V., G.B. Mackaness, and F.M. Collins (1966). J.
        Exp. Med. 124:585.
12. Collins, F.M. (1969). J. Bacteriol. 97:667.
13. Collins, F.M. (1969). J. Bacteriol. 97:676.
14. Lurie, M.B. (1964). Resistance to Tuberculosis, Boston,
        Harvard University Press.
15. Elberg, S.S. (1960). Bact. Rev. 24:67.
16. Pomales-Lebrón, A. and W.R. Stinebring (1957). Proc. Soc.
        Exp. Biol. 94:78.
17. Mackaness, G.B. (1964).   J. Exp. Med. 120:105.
18. Collins, F.M. and G.B. Mackaness (1968). J. Immunol. 101:830.
19. Collins, F.M., G.B. Mackaness, and R.V. Blanden (1966). J.
        Exp. Med. 124:601.
20. Collins, F.M. (1968). J. Bacteriol. 95:1343.
21. Howard, J.G. (1961). Nature 191:87.
22. Zinkernagel, R.M. (1976). Infect. Immun. 13:1069.
23. Nakano, M. and K. Saito (1969). Nature. 222:1085.
24. Saito, K., T. Akiyama, M. Nakano, and D. Ushiba (1960).
        Japan. J. Microbiol. 4:395.
25. Mitushashi, S., I. Sato, and T. Tanaka (1961). J. Bacteriol.
        81:863.
26. Furness, G. and I. Ferriera (1959). J. Inf. Dis. 104:203.
27. Saito, K., M. Nakano, T. Akiyama, and D. Ushiba (1962). J.
        Bacteriol. 84:500.
28. Saito, I. and S. Mitsuhashi (1965). J. Bacteriol. 90:1194.
29. Ushiba, D. (1965). Keio J. Med. 14:45.
30. Mackaness, G.B. (1969). J. Exp. Med. 129:973.
31. Mackaness, G.B. (1971). J. Infect. Dis. 123:439.
32. Collins, F.M. (1974). Bact. Rev. 38:371.
33. North, R.J. (1975). Infect. Immun. 12:754.
34. Jenkin, C.R. and D. Rowley (1965). Aust. J. Exp. Biol. Med.
        Sci. 43:65.

35. Herzberg, M., P. Nash, and S. Hino (1972). Infect. Immun. 5:83.
36. Ornellas, E.P., R.J. Roantree, and J.P. Steward (1970). J.
          Inf. Dis. 112:113.
37. Marecki, N.M., H.S. Hsu, and D.R. Mayo (1975). Brit. J.
          Exp. Path. 56:231.
38. Cameron, C.M. and W.J.P. Fuls (1974). Onderstepoort J. Vet.
          Res. 41:81.
39. Eisenstein, T.K. (1969). Ph.D. Thesis, Bryn Mawr College, Pa.
40. Angerman, C.R. and T.K. Eisenstein (1978). Infect. Immun.
          19:575.
41. Angerman, C.R. (1979). Ph.D. Thesis, Temple University,
          Philadelphia, Pa.
42. Rowley, D., I. Auzins, and C.R. Jenkin (1968). Aust. J. Exp.
          Biol. Med. Sci. 46:447.
43. Jenkin, C.R., D. Rowley, and I. Auzins (1964). Aust. J. Exp.
          Biol. Med. Sci. 42:215.
44. Lyman, M.B., B.A.D. Stocker, and R.J. Roantree (1979).
          Infect. Immun. 26:956.
45. Eisenstein, T.K. (1975). Infect. Immun. 12:364.
46. Diena, B.B., E.M. Johnson, L.S. Baron, R. Wallace, and L.
          Greenberg (1973). Infect. Immun. 7:5.
47. Kiefer, W., P. Gransow, G. Schmidt, and O. Westphal (1976).
          Infect. Immun. 13:1517.
48. Lindberg, A.A., L.T. Rosenberg, A. Ljunggren, P.J. Garegg,
          S. Svenson, and N.-H. Wallin (1974). Infect. Immun.
          10:541.
49. Svenson, S.B. and A.A. Lindberg (1978). J. Immunol. 120:1750.
50. Collins, F.M. (1968). J. Bacteriol. 95:1343.
51. Collins, F.M. (1968). J. Bacteriol. 95:2014.
52. Germanier, R. (1972). Infect. Immun. 5:792.
53. Kuusi, N., M. Nurminen, H. Saxen, M. Valtonen and H. Mäkelä
          (1979). Infect. Immun. 25:857.
54. Smith, R.A. and N.J. Bigley (1972). Infect. Immun. 6:377.
55. Johnson, W. (1973). Infect. Immun. 8:395.
56. Misfeldt, M. and W. Johnson (1979). Infect. Immun. 24:808.
57. Barber, C. and E. Eylan (1972). Revue d'Immunologie 36:77.
58. Eisenstein, T.K. and B.M. Sultzer (1982) in Bacterial Vaccines,
          ed. J.B. Robbins, J.C. Hill and J. Sadoff, Thieme-
          Stratton, N.Y. p. 423.
59. Sultzer, B.M., G.W. Goodman, and T.K. Eisenstein (1980). in
          Microbiology - 1980, Am. Soc. Microbiol., p. 61.
60. Plant, J.E., A.A. Glynn, and M.V. Valtonen (1980). Parasite
          Immunol. 2:293.
61. Hochadel, J.F. and K.F. Keller (1977). J. Infect. Dis.
          135:813.
62. Morris, J.A., C. Wray, and W.U. Sojka (1976). Brit. J. Exp.
          Path. 57:354.
63. Rowley, D., K. Turner, and C.R. Jenkin (1964). Aust. J. Exp.
          Biol. Med. Sci. 42:237.

64. Mitsuhashi, S., M. Kawakami, and S. Kurashige (1965). Proc.
        Japan. Acad. 41:635.
65. Kurashige, S., N. Osawa, M. Kawakami, and S. Mitsuhashi
        (1967). J. Bacteriol. 94:902.
66. Mitsuhashi, S., K. Saito, N. Osawa, and S. Kurashige (1967).
        J. Bact. 94:907.
67. Margolis, J.M. and N.J. Bigley (1972). Infect. Immun. 6:390.
68. Nakoneczna, I. and H.S. Hsu (1980). Brit. J. Exp. Path. 61:76.
69. Kenny, J. and M. Herzberg (1968). J. Bacteriol. 95:406.
70. Mitsuhashi, S., M. Kawakami, Y. Yamaguchi, and N. Nagai (1958).
        Japan. J. Exp. Med. 28:249.
71. Auzins, I. and D. Rowley (1963). Aust. J. Exp. Biol. Med.
        Sci. 41:539.
72. Milne, M. and F.M. Collins (1966). J. Bacteriol. 92:543.
73. Collins, F.M. and M. Milne (1966). J. Bacteriol. 92:549.
74. Collins, F.M. (1970). Infect. Immun. 1:243.
75. Collins, F. (1979). in CRC Critical Reviews in Microbiology,
        7:27.
76. Marmion, D.E., G.R.E. Naylor, and I.O. Stewart (1953). J.
        Hyg. 51:260.
77.   Vas, S.I. (1978). in Infection, Immunity and Genetics,
        University Park Press, Baltimore, p. 39.
78. Sultzer, B.M. (1969). J. Immunol. 103:32.
79. Sultzer, B.M. (1976). Infect. Immun. 13:1579.
80. Angerman, C. and T.K. Eisenstein (1978). J. Immunol. 121:1010.
81. Watson, J. and R. Riblet (1975). J. Immunol. 114:1462.
82. Eisenstein, T.K., L.W. Deakins, and B.M. Sultzer (1980). in
        Genetic Control of Natural Resistance to Infection and
        Malignancy, ed. E. Skamene, P.A.L. Kongshaven and M.
        Landy, Academic Press, N.Y. p. 115.
83. Eisenstein, T.K., L.W. Deakins, L. Killar, P.H. Saluk, and
        B.M. Sultzer (1982). Infect. Immun. 36:696.
84. Hoiseth, S.K. and B.A.D. Stocker (1981). Nature 291:238.
85. Webster, L.T. (1937). J. Exp. Med. 65:261.
86. Hill, A.B., J.M. Hatswell, and W.W.C. Topley (1940). J. Hyg.
        40:538.
87. Gowen, J. (1948). Annu. Rev. Microbiol. 11:215.
88. Plant, J. and A.A. Glynn (1976). J. Infect. Dis. 133:72.
89. von Jeney, N., E. Gunther, and K. Jann (1977). Infect.
        Immun. 15:26.
90. O'Brien, A.D., D.L. Rosenstreich, I. Scher, G.H. Campbell,
        R.P. MacDermott, and S.B. Formal (1980). J. Immunol. 124:20.
91. O'Brien, A.D., I. Scher, G.H. Campbell, R.P. MacDermott, and
        S.B. Formal (1979). J. Immunol. 123:720.
92. Hormaeche, C.E. (1979). Immunology. 37:319.
93. Hormaeche, C.E. (1979). Immunology. 37:371.
94. O. Brien, A.D., D.L. Rosenstreich, and I. Scher (1981). in
        Immunomodulation by Bacteria and Their Products, ed.
        H. Friedman, T.W. Klein and A. Szentivanyi, Plenum
        Press, N.Y., p. 37.

95. Gowens, J.W. (1943). Rep. Iowa Agric. Exp. Sta., p. 178.
96. Gowens, J.W. (1945). Ann. Missouri Bot. Garden. 32:187.
97. Gowens, J.W. and J. Stadler (1967). J. Infect. Dis. 117:129.
98. Robson, H.G. and S.I. Vas (1972). J. Infect. Dis. 126:378.
99. Medina, S., S. I. Vas, and H. G. Robson (1975). J. Immunol. 114:1720.
100. Misfeldt, M.L. and W. Johnson (1976). Infect. Immun. 14:652.
101. Angerman, C.R. and T.K. Eisenstein (1980). Infect. Immun. 27:435.
102. Eisenstein, T.K. (1978). in New Trends and Developments in Vaccines, MTP Press, Ltd., Lancaster, Eng., p. 211.
103. Smith, R.A. and N.J. Bigley (1972). Infect. Immun. 6:384.
104. Eisenstein, T.K., C.R. Angerman, and L.W. Deakins (1981). in Immunomodulation by Bacteria and Their Products, ed. H. Friedman, T.W. Klein, and A. Szentivanyi, Plenum Press, N.Y., p. 199.
105. Wright, A. and D. Semple (1897) Brit. Med. J. 1:256.
106. Siler, J.F., et al. (1941). Immunization to Typhoid Fever, Johns Hopkins Press, Baltimore, MD.
107. Cockburn, W.C. (1955). J. Roy. Army Med. Corps. 101:171.
108. Joó, I. (1971). Pan Am. Health Organ. Sci. Publ. 226:329.
109. Jordan, J. and H.E. Jones (1945). Lancet 2:333.
110. Syverton, J.T., R.E. Ching, F.S. Cheever, and A.B. Smith (1946). J. Am. Med. Assoc. 131:507.
111. Felix, A. and M. Pitt (1934). Lancet 1:186.
112. Felix, A. (1941). Brit. Med. J. 1:391.
113. Yugoslav Typhoid Commission (1964). Bull. W.H.O. 30:623.
114. Landy, M. (1953). Am. J. Hyg. 58:148.
115. Ashcroft, M.T. (1967). Lancet 2:1056.
116. Hornick, R.B., S.E. Greisman, T.E. Woodward, H.L. DuPont, A.T. Dawkins, and M.J. Snyder (1970). N. Engl. J. Med. 283:739.
117. Beneson, A.S. (1964). Bull. W.H.O. 30:653.
118. Standfast, A.F.B. (1960). Bull. W.H.O. 23:37.
119. Pittman, M. and H.J. Bohner (1966). J. Bacteriol. 91:1713.
120. Spaun, J. and K. Uemura (1964). Bull. W.H.O. 31:761.
121. Landy, M., A.G. Johnson, M.E. Webster, and J.F. Sagin (1955). J. Immunol. 74:466.
122.   Landy, M. (1957). Am. J. Hyg. 65:81.
123. Gaines, S., M. Landy, G. Edsall, A.D. Mandel, R.-J. Trapani, and A.S. Beneson (1961). J. Exp. Med. 114:327.
124. Pistole, T.G. and S. Marcus (1974). Immunol. Commun. 3:227.
125. Standfast, A.F.B. (1960). Bull. W.H.O. 23:4752.
126. Diena, B.B., L.S. Baron, E.M. Johnson, R. Wallace, and F.E. Ashton (1974). Infect. Immun. 9:1102.
127. Diena, B.B., A. Ryan, R. Wallace, F.E. Ashton, E.M. Johnson and L.S. Baron (1975). Infect. Immun. 12:1470.
128. Wong, K.H., J.C. Feeley, R.S. Northup, and M.E. Forlines (1974). Infect. Immun. 9:348.

129. Levine, M.M., H.L. DuPont, R.B. Hornick, M.J. Snyder, W.
         Woodward, R.H. Gilman, and J. Libonati (1976). J.
         Infect. Dis. 133:424.
130. Germanier, R. and E. Furer (1975). J. Infect. Dis. 131:553.
131. Gilman, R.H., R.B. Hornick, W.E. Woodward, H.L. DuPont,
         M.J. Snyder, M.M. Levine, and J.P. Libonati (1977).
         J. Infect. Dis. 136:717.
132. Germanier, R. (1982). in Seminars in Infectious Disease,
         Vol. 4, ed. J.B. Robbins, J.C. Hill and J.C. Sadoff,
         Thieme-Stratton, N.Y. p. 419.
133. Collins, F.M. and P.B. Carter (1972). Infect. Immun. 6:451.
134. Hohmann, A.W., G. Schmidt and D. Rowley (1978). Infect.
         Immun. 22:763.

STRAIN DEPENDENT VARIATION OF DELAYED-TYPE HYPERSENSITIVITY IN

SALMONELLA TYPHIMURIUM INFECTED MICE

Loran Killar and Toby K. Eisenstein

Department of Microbiology and Immunology, Temple
University School of Medicine, Philadelphia, PA

INTRODUCTION

Development of delayed-type hypersensitivity (DTH) in
Salmonella infection was reported as early as 1923, when Zinsser
showed that guinea pigs injected with live typhoid bacilli became
skin test positive to typhoid antigen (1). A close relationship
between DTH and immunity to Salmonella infection was demonstrated
in mice by Collins and Mackaness (2). They showed that mice
immunized with live Salmonella developed DTH as measured by an
increase in footpad thickness 24 hours after injection of a
culture supernatant antigen. The appearance of footpad responsiveness correlated with the onset of immunity to challenge with
virulent Salmonella (2).

The purpose of the studies reported here was to measure the
DTH response of several inbred mouse strains in the C3H lineage,
whose innate susceptibilities to Salmonella infection are widely
different, following immunization with a live, smooth avirulent
Salmonella mutant.

MATERIALS AND METHODS

Animals. Female CD-1 and C3H/HeNCr1BR mice, 19 to 21g, were
purchased from Charles River Breeding Laboratories, Wilmington,
Mass. Female C3H/HeJ and C3HeB/FeJ mice, 19 to 21g, were purchased
from Jackson Laboratories, Bar Harbor, Maine. They were housed in
plastic cages with Absorb-Dri for bedding. Purina Mouse Chow and
fresh water were available ad libitum. The CD-1 and C3H/HeNCr1BR
mice are innately resistant to Salmonella infection, with ip LD_{50}'s

of 1×10^4 and 1×10^3 cells, respectively. The C3H/HeJ and C3HeB/FeJ
mice are <u>Salmonella</u> hypersusceptible, with ip LD_{50}'s of 7, and
less than 2 cells, respectively. Their LD_{50} is assumed to be
theoretically one cell.

<u>Organisms</u>. <u>Salmonella typhimurium</u>, strain W118-2, which is
virulent in mice, was originally obtained from Dr. Samuel Formal,
Walter Reed Army Institute of Medical Research, and has been used
previously in our laboratory (3). <u>S</u>. <u>typhimurium</u>, strain SL3235,
s smooth <u>aroA</u>⁻ mutant, was obtained from Dr. Bruce A.D. Stocker,
Department of Medical Microbiology, Stanford University, Stanford,
Ca. It was derived by inserting the <u>aroA554</u>::Tn <u>10</u> transposon into
the <u>aroA</u> gene of <u>S</u>. <u>typhimurium</u> LT2 and later selecting for trans-
formants that had excised part of the <u>aroA</u> gene along with the
transposon (4). Strain SL3235 is relatively avirulent for mice
(4). For immunization or challenge, lyophilized organisms were
rehydrated and grown to log phase as previously described (3).

<u>Preparation of a Salmonella DTH Antigen</u>. An antigen for eliciting
footpad responses was prepared according to the method of Collins
and MacKaness by growing strain SL3235 for 72 hours in Syncase
medium, and ammonium sulfate precipitating the cell-free culture
supernatant (2). Doses were based on the amount of protein as
determined by the method of Lowry (5).

<u>Immunization and Challenge</u>. Mice were immunized with a single,
iv injection of live SL3235 in 0.1 ml of saline. Controls received
0.1 ml of saline iv. In protection studies, mice were challenged
three weeks after immunization with an ip injection of live W118-2
suspended in 0.5 ml of saline. Protection was assessed by 60 day
survival and evaluated statistically using the Fisher 2x2 test (6).
In the footpad reactivity studies, mice were tested 3 weeks after
iv immunization with SL3235. The eliciting antigen was suspended
in saline and 0.04 ml (2.8 ug of protein) was injected into the
right, hind footpad. An equal volume of slaine was injected into
the left, hind footpad. Control mice which had been injected iv
with saline were treated in the same manner. Prior to, and at 3, 6,
24, 30 and 48 hours after injection of the antigen, swelling of the
footpads was measured using dial gauge calipers (Schnelltaster,
Kroplin)(2). Triplicate measurements of each foot were made at each
time point and averaged. For each time point, the average measurement of the saline injected foot was subtracted from the average
measurement of the antigen injected foot, yielding the difference
in footpad thickness for each mouse. The mean and standard
deviation for the difference in footpad thickness for each group
of mice was then calculated.

TABLE 1. Protection of mice by the avirulent strain S. typhimurium SL3235

Mouse Strain	Immunized with:[a]	Challenge Dose (Number of cells)	Survivors[b] / Total
C3H/HeNCr1BR	SL3235	1×10^7	0/6
		1×10^6	6/6
		1×10^5	6/6
	Saline	1×10^5	0/6
C3HeB/FeJ	SL3235	1.1×10^4	2/6
		1.1×10^3	3/6
		1.1×10^2	4/6[c]
	Saline	1.1×10^2	0/6
C3H/HeJ	SL3235	1.1×10^4	0/6
		1.1×10^3	2/6
		1.1×10^2	4/6[c]
	Saline	1.1×10^2	0/6
CD-1	SL3235	1×10^8	0/6
		1×10^7	3/6
		1×10^6	6/6
	Saline	1×10^6	0/6

[a]All mice were immunized with 0.1 ml of live SL3235 iv. C3H/HeNCr1BR and CD-1 mice received 7×10^4 organisms and C3H/HeJ and C3HeB/FeJ mice received 1.3×10^5 organisms.

[b]All mice were challenged ip three weeks after immunization with live S. typhimurium, strain W118-2. Mortality was tabulated 60 days after challenge.

[c]$p \leq 0.03$ when compared with saline controls.

RESULTS

The ability of intravenously administered, live organisms of the avirulent strain SL3235 to protect mice against ip challenge with the virulent strain W118-2 was assessed. As shown in Table 1, all four mouse strains, including the innately hypersusceptible C3H/HeJ and C3HeB/FeJ mice, were significantly protected on day 21 after infection by live SL3235 when challenged with 100 to 1000 LD_{50} doses of W118-2.

TABLE 2. Footpad responses in mice three weeks after immunization with _S. typhimurium_, Strain SL3235.

mouse strain	Immunized with:[a]	Difference in Footpad Thickness (mm)					
		Time After Footpad Injection (hours)					
		0	3	6	24	30	48
C3H/HeJ	SL3235	$-.02\pm.05$	$.33\pm.06$	$.12\pm.10$	$.17\pm.08$	$.11\pm.11$	$.11\pm.06$
	Saline	$.03\pm.06$	$.18\pm.05$	$.13\pm.04$	$.02\pm.03$	$-.03\pm.08$	$.12\pm.09$
C3HeB/FeJ	SL3235	$.06\pm.07$	$.16\pm.08$	$.30\pm.11$	$.21\pm.11$	$.18\pm.19$	$.17\pm.03$
	Saline	$.01\pm.10$	$.01\pm.03$	$.02\pm.09$	$.09\pm.08$	$-.01\pm.10$	$.13\pm.03$
C3H/HeNCr1BR	SL3235	$-.03\pm.07$	$.19\pm.16$	$.29\pm.08$	$.67\pm.14$	$.54\pm.21$	$.32\pm.17$
	Saline	$.05\pm.04$	$.03\pm.02$	$.05\pm.02$	$.05\pm.05$	$.13\pm.07$	$.04\pm.10$
CD-1	SL3235	$-.03\pm.03$	$.25\pm.11$	$.24\pm.10$	$.96\pm.12$	$1.00\pm.09$	$.35\pm.04$
	Saline	$-.04\pm.01$	$-.01\pm.07$	$-.01\pm.01$	$.07\pm.05$	$.10\pm.07$	$.04\pm.01$

[a]Live SL3235 were injected in 0.1 ml saline iv. C3H/HeJ mice received 5.5×10^5 cells; C3HeB/FeJ mice received 7×10^5 cells; C3H/HeNCr1BR mice received 3.6×10^4 cells, and CD-1 mice received 4.0×10^5 cells.

[b]Each mouse received 0.04 ml of SL3235 culture supernatant antigen (2.8 ug of protein) in the right,hind footpad and 0.04 ml saline in the left, hind footpad. At each time point, three measurements of each footpad were made and the difference between the means of the right and left footpads for each mouse were calculated. Values given are the average increase in footpad thickness for three mice at each time point.

Table 2 shows the footpad responsiveness of separate groups
of three mice of each strain 21 days after immunization with SL3235.
Only the innately resistant strains, C3H/HeNCr1BR and CD-1, show
significant footpad swelling at 24 and 30 hours, a characteristic
of a delayed-type hypersensitivity response. The C3H/HeJ and
C3HeB/FeJ mice failed to show significant footpad swelling at any
time point, even though they were significantly protected by
SL3235 (see Table 1). Therefore, the innately hypersusceptible
C3H/HeJ and C3HeB/FeJ mice displayed immunity in the absence of a
footpad (DTH) response, while the inherently resistant C3H/HeNCr1BR
and CD-1 mice displaed both immunity and footpad responsiveness.

DISCUSSION

These studies reveal two interesting phenomena: 1) a
difference in the ability of mice in the same lineage with varying
innate susceptibilities to Salmonella to develop DTH, and (2) a
lack of correlation between the development of resistance and
delayed-type hypersensitivity in the hypersusceptible C3H/HeJ and
C3HeB/FeJ mice.

There have been previous reports showing differences in the
ability of mice of disparate lineages, when injected with live
Salmonella, to develop DTH, and the differences seemed to correlate
with the innate susceptibility of the mouse strain (7,8). It has
been suggested that innate Salmonella susceptibility and the ability
to mount a DTH response might be Ir gene dependent (7). However,
the three strains in the C3H lineage used in the present study are
known to have identical Ir genes, yet have different Salmonella
susceptibilities and display different footpad responsiveness.
Hormaeche et al. have observed that innately hypersusceptible
mouse strains can mount a footpad response, but that this response
occurs two to three weeks after the onset of immunity which occurs
at one week (8). We have tested the three different C3H strains for
their footpad reactivity at intervals of 1 to 4 weeks after immunization (data not shown), but the hypersusceptible C3H/HeJ and
C3HeB/FeJ mice showed no footpad reactivity at any of the times
tested.

C3H/HeJ mice are refractory to the effects of LPS, and it
may be tempting to hypothesize that these mice fail to give a footpad response because the reaction is dependent on LPS in the
eliciting antigen. However, C3HeB/FeJ mice, which are normal
responders to LPS(3) also do not show footpad responsiveness.
Thus, it is unlikely that the lack of response is due to
refractoriness to LPS. Perhaps a different eliciting antigen is
needed to show responses in the C3H/HeJ and C3HeB/FeJ mice.
However, the ability to demonstrate a significant footpad response

in immunized, inherently resistant C3H/HeNCrlBR and CD-1 mice shows
that the antigen used in this study for elicitation was active.
Alternatively, the lack of a footpad response in these strains
may be due to a sequestration of immune cells in lymphoid tissues,
so that they are not recruited to the injection site of the
eliciting antigen.

SUMMARY

    Inherently hypersusceptible C3H/HeJ and C3HeB/FeJ mice show
increased resistance to challenge with virulent S. typhimurium
after immunization with a live, avirulent S. typhimurium mutant
in the absence of a delayed-type hypersensitivity response, while
innately resistant C3H/HeNCrlBR and CD-1 mice show both immunity
and positive footpad reactions after immunization. Understanding
the mechanism for this specific anergy in the hypersusceptible
strains may provide clues for understanding the immune defect that
causes these mice to be so exquisitely susceptible to Salmonella
infection.

ACKNOWLEDGMENT

    This work was supported by grant #AI 15613 from the National
Institute of Allergy and Infectious Diseases.

REFERENCES

1. Zinsser, H. and J.T. Parker. (1923). J. Exp. Med., 37:275.
2. Collins, F.M. and G.B. Mackaness. (1968). J. Immunology, 101:
        830.
3. Eisenstein, T.K., L. Deakins, L. Killar, P. Saluk, and B.
        Sultzer. (1982). Infect. and Immunity, 36:696.
4. Hoiseth, S.K. and B.A.D. Stocker (1981). Nature, 291:238.
5. Lowry, O.H., N.J. Rosebrough, A.L. Farr and R.J. Randall
        (1951). J. Biol. Chem. 193:265.
6. Bliss, C.J. (1967). Statistics in Biology, vol. 3, McGraw-Hill,
        New York, p. 63-65.
7. Plant, J. and A.A. Glynn (1974). Nature, 248:345.
8. Hormaeche, C.E., MC. Fahrenkrog, R.A. Pettifor and J. Brock
        (1981). Immunology, 43:547.

MONOCLONAL ANTIBODIES TO <u>SALMONELLA</u> <u>TYPHIMURIUM</u> AND <u>ESCHERICHIA</u>

<u>COLI</u> LIPOPOLYSACCHARIDES

Jack L. Komisar and John J. Cebra

Department of Biology, University of Pennsylvania
Philadelphia, Pennsylvania

INTRODUCTION

Monoclonal antibodies reacting to lipopolysaccharides (LPS)
of <u>Salmonella</u> <u>typhimurium</u> and a slow-lactose-fermenting strain of
<u>Escherichia</u> <u>coli</u> (0 serotype 81) were made by immunization of
BALB/c mice with trichloroacetic acid-extracted LPS and fusion of
lymphoid cells with Sp2/0-Ag14 cells. Fusion supernatants were
screened by radioimmunoassay against phenol/water-extracted LPS.
Some of the antibodies reacted with LPS from one bacterial species
but not the other, suggesting that the antibodies are directed
against the 0-specific side chains of the LPS. The fine specificity of one of the antibodies was studied by inhibition of
radio-immunoassay by monosaccharides. Rhamnose and mannose
inhibited best, followed by galactose and glucose, while the
weakest inhibition was given by <u>N</u>-acetylglucosamine, which is not
part of the <u>S</u>. <u>typhimurium</u> 0 antigen. Agglutination and immunofluorescence results suggest that the determinant is on the
bacterial surface. Two other hybridomas were tested by Alf Lindberg
and Stefan Svenson in an enzyme-linked immunosorbent assay against
LPS and oligosaccharides. One antibody is directed against an
α-galactose (1→2) mannose disaccharide, while the other appears to
recognize factor 5 (2-acetyl-abequose).

The <u>E</u>. <u>coli</u> strain was originally isolated from the gastrointestinal tract of mice by Russell Schaedler, René Dubos, and co-
workers. Twelve anti-<u>E</u>. <u>coli</u> hybridomas, all secreting antibody of
the IgM class, were derived from fusion of spleen cells, while five
monoclonal antibodies, all of the IgA class, were obtained from
fusion of a Peyer's patch-mesenteric lymph node pool.

Lipopolysaccharide found in the outer membrane of Gram-negative bacteria can cause the mitosis of murine B-lymphocytes and their maturation into antibody-secreting plasma cells (reviewed in 10). LPS is classified as a type I thymus-independent antigen (11), but in the intact bacterial cell it is associated with other membrane components, some of which might influence the antibody response to the LPS molecule by interacting with T-cells (1). This antibody response might also differ depending on whether the bacteria carrying the LPS are present throughout life (chronic stimulation) or are encountered for the first time. We are investigating the immune response of mice to the LPS of Salmonella typhimurium and of a slow-lactose-fermenting strain of Escherichia coli. The latter was isolated from the gastrointestinal tract of normal mice by R. W. Schaedler, et al. (12). The former species can provide an acute bacterial challenge while the latter provides a chronic challenge. The E. coli strain has the serotype 081:H21 (F. Ørskov and I. Ørskov, personal communication, 1981).

We are adapting the Klinman clonal precursor assay (5) to study the response to 0-antigenic determinants of LPS. This assay is designed to measure the frequency and isotype potential of antigen-sensitive B cells. Limiting dilutions of lymphoid cells are transferred into irradiated recipients and spleen chunks from these mice are stimulated in vitro by 0-antigenic determinants linked to Limulus polyphemus hemocyanin or by LPS. Hybridomas secreting monoclonal antibodies reactive with S. typhimurium and E. coli LPS have been prepared for use as standards in a radio-immunoassay (RIA). We report here some of the characteristics of the monoclonal antibodies and early results from the clonal precursor assay.

The mice used to provide the normal antibody-secreting cells for preparation of anti-S. typhimurium hybridomas were injected twice intravenously with 10^7 heat-killed S. typhimurium cells (strain PUR A155 obtained from P. Hartman) and twice intraperitoneally with 50µg of S. typhimurium trichloroacetic acid-extracted LPS (TCA-LPS). The mouse used to make the anti E. coli hybridomas was immunized once with 10^7 heat-killed E. coli cells intravenously and six times with 50µg of E. coli LPS intraperitoneally. Fusions were performed according to the procedure of Kennett et al. (4) using Sp2/0-Ag14 cells (13) as the myeloma parent time.

HYBRIDOMAS

Nine hybridomas were obtained from a fusion using two BALB/c mice immunized with S. typhimurium (Table 1). The antibodies were detected by RIA using phenol-water extracted LPS (PW-LPS) adsorbed to a polyvinylchloride plate and radioiodinated rabbit anti-mouse immunoglobulins. Six of the antibodies reacted better with LPS

Table 1. Radioimmunoassay of hybridomas obtained by immunization
          with _Salmonella_ _typhimurium_ LPS.

Hybri-doma Number	_Proteus_ _morganii_	Schaed-ler's _E. coli_	_Salmon-ella_ _typhi-murium_	_E. coli_ 055:B5	Iso-type	CPM After Clon-ing
			Counts per minute vs.			
12D4	171	111	254	117	IgG3	3339 (Salmonella)[a]
12F4	156	155	2227	136	IgM	438 (Salmonella)
12F6	3178	100	159	126	IgG2b	448 (Proteus)
15G11	250	144	94	133	IgM	12000 (Proteus)
16F11	164	119	788	133	IgG2b	1713 (Salmonella)
16G11	193	106	1595	115	IgG1	4797 (Salmonella)
17F3	65	102	658	109	IgG3	6471 (Salmonella)
20E10	170	118	11183	115	IgG3	12500 (Salmonella)
16H1	3779	NT	696	136	IgG2b	11144 (Proteus)
						2212 (Salmonella)

[a] Species in parentheses indicate the source of LPS used to test
    the antibodies after cloning.

NT = Not Tested.

from _S. typhimurium_ than with LPS from Schaedler's _E. coli_, _Proteus_
morganii or _E. coli_ 055:B5. Since the 0 antigen varies greatly
among different species of bacteria while the core oligosaccharide
and lipid A portions of LPS are relatively conserved among _Entero-_
bacteriaceae (9), these antibodies are probably directed against
0-antigenic determinants. The specificity of five of the monoclonal antibodies was studied further, as discussed below.

     Seventeen hybridomas were made using lymphoid cells from one
BALB/c mouse immunized with Schaedler's _E. coli_ (Table 2). Twelve
were obtained by fusion of spleen cells with myeloma cells. All
of the antibodies produced were of the IgM class. Five hybridomas
producing monoclonal antibodies were obtained by fusion of a
Peyer's patch and mesenteric lymph node cell pool with myeloma
cells. All of these antibodies were of the IgA isotype. IgA is
an unusual isotype for monoclonal antibodies derived from the
spleen, but the fusion of cells from Peyer's patches and mesenteric
lymph nodes might be expected to yield a higher proportion of IgA
hybrids than the spleen does, since IgA is commonly expressed by

Table 2.   Radioimmunoassay of hybridomas obtained by immunization
of mice with Schaedler's E. coli LPS.

Hybridoma Number	CPM vs. Schaedler's E. coli	Salmonella typhimurium	Isotype	CPM After Cloning	CPM with Anti-IgM
27E8	1004	1434	IgM	1458	
27G6	1032	84	IgM	1048	
27G7	681	84	IgM	535	28821
28E9	NT	859	IgM	795	
28F9	274	5369	IgM	1814	
29C10	234	3088	IgM	523	
30B11	927	51	IgM	977	26217
30F4	450	1763	IgM	108	
30F11	208	1276	IgM	507	
32B4	730	574	IgA	2687	
32B6	196	3045	IgA	890	
32G2	306	3074	IgA	1079	
32E10	707	724	IgA	1718	
32E11	978	282	IgA	386	

NT = Not Tested.

those specificities of B cells that are chronically stimulated by
environmental antigens in the gut (3).

Hybridomas 12D4 and 17F3 were tested by A. Lindberg and S.
Svenson using an enzyme linked immunosorbent (ELISA) assay (2).
The 12D4 antibody reacted most strongly with an α-gal (1 $\to$ 2) - mannose disaccharide (Fig. 1) which is not present at the nonreducing
terminus of the native O antigen. We have found that the monoclonal antibody reacts with α-gal bovine serum albumin (α-gal-BSA)
by RIA, and with an even higher avidity with β-gal-BSA, suggesting
that its specificity is for galactosyl residues. The 17F3 monoclonal antibody, tested by Lindberg and Svenson, reacted better
with the native LPS than with oligosaccharides obtained from it
(Fig. 2) suggesting that it is directed against factor 5 (O-acetyl-
abequose), since the O-acetyl group is lost in the alkaline
hydrolysis step used to prepare the oligosaccharides (14).

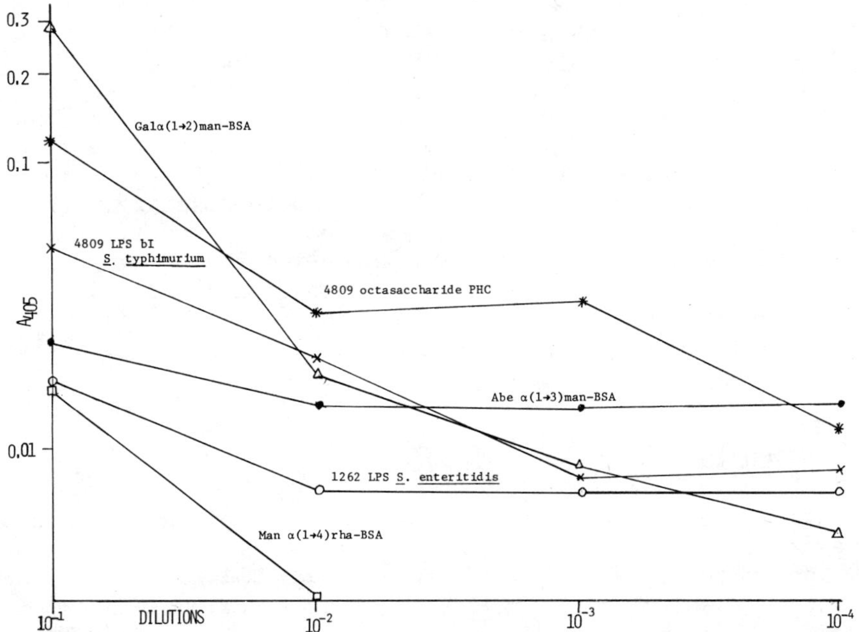

Figure 1.   12D4 Hybridoma Elisa Tests by Alf Lindberg and Stefan Svenson.

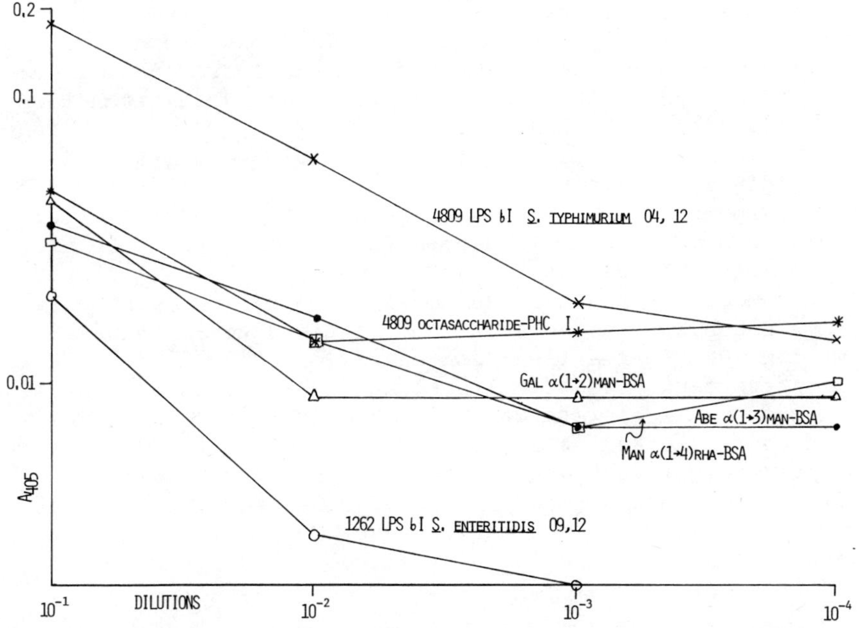

Figure 2.   17F3 Hybridoma Elisa Tests by Alf Lindberg and Stefan Svenson.

Tentative specificities for three of the monoclonal antibodies could be inferred from their reactions to LPS derived from S. typhimurium, S. minnesota, S. abortus equi and S. enteritidis. Thus, the 12F4 pattern suggests an anti-1 specificity, while 16G11 and 20E10 appear to recognize factor 4. The 20E10 antibody appears to recognize a determinant that is exposed on the surface of S. typhimurium, since it agglutinated heat-killed cells and stained alcohol-fixed cells as determined by immunofluorescence. The 20E10 antibody was purified by chromatography on protein A-Sepharose and iodinated by the chloramine-T method to measure its purity in an assay that determined the fraction of iodinated protein that would bind to antigen. This antibody was subsequently used as the standard for measurement of the antibody concentration of culture supernatants.

CLONAL PRECURSOR ASSAY

Work in our laboratory has shown that haptenated LPS can be used as a stimulating antigen in vitro in the Klinman clonal precursor assay (D. Zimmerman and J. Cebra, unpublished). By using concentrations of antigen in vitro that had previously been determined to be optimal for obtaining a response to haptenated LPS, a response was obtained to the O antigen of S. typhimurium LPS in the clonal precursor assay. The cell donors and recipients were BALB/c mice. The donors were 5-8 weeks old at the time of immunization and the spleens were removed seven weeks later, while the recipients were unprimed and were 8-11 weeks old at the time of adoptive transfer. Donors were given one intravenous injection of 5 µg of S. typhimurium TCA-LPS. Recipients were given 1400 rads of γ-irradiation 4-6 hours before transfer of 3.2×10^7 spleen cells. Three more irradiated mice were given no cells to determine whether radiation-resistant recipient cells contribute to the observed response. Twenty-four hours after adoptive transfer, the recipient spleens were each chopped (day 0) into about 50 pieces and these were cultured in individual wells of 96-well plates in Dulbecco's modified Eagle's medium with 4.5 g/liter glucose and 20% gamma globulin-free horse serum. The medium contained S. typhimurium PW-LPS at a final concentration of 15, 1.5, or 0.15 µg/ml. After four days of culture, antigen-containing medium was removed and the wells were washed with culture medium. Medium was replaced and collected on days 7, 10, and 13. Culture supernatants were tested for production of anti-S. typhimurium antibody by RIA. The 20E10 monoclonal antibody was used as a standard. Wells scoring higher than 1 ng of antibody per 20 µl (about 600 counts per minute) and for which an isotype could be determined were counted as positive.

The results are shown in Table 3. Since there were more than 10 clonal precursors per culture plate, the Poisson distribution

Table 3. Clonal precursor assay using <u>Salmonella</u> <u>typhimurium</u> PW-LPS <u>in</u> <u>vitro</u>.

Mouse	In vitro LPS Concentration	Total Clones	Some IgM	Some IgG3	Some IgG1	Some IgG2	Some IgA
1	15 μg/ml	13	0	6	9	12	4
2	15 μg/ml	12	3	5	4	10	3
3 (no cells transferred)	15 μg/ml	1	0	0	0	1	1
4	1.5 μg/ml	18	11	9	10	16	19
5	1.5 μg/ml	15	4	8	1	12	8
6 (no cells transferred)	1.5 μg/ml	0	0	0	0	0	0
7	0.15 μg/ml	4	1	1	2	4	2
8	0.15 μg/ml	4	1	2	2	4	0
9 (no cells transferred)	0.15 μg/ml	0	0	0	0	0	0
		67	20	31	27	69	37

predicts that several wells should have received more than one clonal precursor. Therefore, no conclusions can be drawn about the number of clones exclusively committed to any one isotype. However, it is possible to compare the relative amounts of each isotype produced and to calculate the frequency of clonal precursors from the fraction of negative wells by using the Poisson formula (8). Assuming that 4% of the injected cells lodge in the spleen (7), and that 40% of splenic lymphocytes are B cells, the precursor frequency in the once-primed mice was 3.9 per 10^5 B cells. Only one of the 67 clones observed came for a mouse that received no injected cells. This contribution of radioresistant recipient cells introduces a small error in the calculation of precursor frequencies. Another potential problem is the polyclonal B cell activating property of LPS. At the LPS concentrations used, the polyclonal activating activity was small since only three fragments scored positive for antibodies cross-reactive with DNP-BSA. Given the previously determined frequency of 20 DNP-sensitive cells per 10^5 B cells (6), if all DNP-sensitive cells were stimulated to antibody production, one would predict that most of the wells would have scored positively. Since all but three of the clones to <u>S</u>. <u>typhimurium</u> did not react to <u>Schaedler's</u> <u>E</u>. <u>coli</u> LPS in an RIA that used the 30B11 hybridoma antibody as a positive

control, it is concluded that almost all of the clones made antibodies to the 0 antigens.

The Klinman assay was originally designed to study the immune
response to thymus-dependent antigens. In this original form of
the assay, recipient mice were carrier-primed to ensure that every
antigen-sensitive B cell capable of a thymus-dependent response
would divide and give rise to a clone secreting antibody in vitro.
It is not known whether the conditions that we have used to
obtain a response to the 0 antigen of LPS allow every antigen-
sensitive cell to respond and express its complete isotype potential. Experiments are in progress to determine the cloning
efficiency of the assay and the extent to which the responding
clones reflect thymus-independent or thymus-dependent antigen-
sensitive cells.

ACKNOWLEDGMENT

This work was supported by Grant AI-17997 from NIAID.

REFERENCES

1. Ahlstedt, S. and Holmgren, J., (1975). Immunology 29:487-496.
2. Carlsson, H.E., Lindberg, A.A. and Hammarström, S., (1972).
         Infect. Immun. 6:703-708.
3. Cebra, J.J., Crandall, C.A., Gearhart, P.J., Robertson, S.M.,
         Tseng, J. and Watson, P.M. (1979). In: Immunology of
         Breast Milk (P. L. Ogra and D. H. Dayton, eds.) pp. 1-
         16, Raven Press, New York.
4. Kennett, R.H., Denis, K.A., Tung, A.S. and Klinman, N.R. (1978).
         Curr. Top. Microbiol. Immunol. 81:77-91.
5. Klinman, N.R. (1972). J. Exp. Med. 136:241-260.
6. Klinman, N.R., Pickard, A.R., Sigal, N.H., Gearhart, P.J.,
         Metcalf, E.S. and Pierce, S.K. (1976). Ann. Immunol.
         (Inst. Pasteur) 127C:489-502.
7. Klinman, N.R. and Press, J.L. (1975). Transplant. Rev. 24:
         41-83.
8. Lefkovits, I. and Pernis, B. (1979). Immunological Methods,
         Academic Press, New York, p. 364.
9. McCabe, W.R. (1980). Seminars Infec. Dis. III 38-88.
10. Morrison, D.C. and Ryan, J.L. (1979). Adv. Immunol. 28:293-
         450.
11. Mosier, D.E., Mond, J.J., Zitron, I., Scher, I. and Paul, W.E.
         (1977). pp. 699-706 In: Immune System: Genetics and
         Regulation. (E.E. Sercarz, L.A. Herzenberg and C.F.
         Fox, eds.) Academic Press, New York.
12. Schaedler, R.W., Dubos, R. and Costello, R. (1965). J. Exp.
         Med. 122:59-66.

13.   Shulman, M., Wilde, C.D. and Köhler, G. (1978).   Nature 276: 269-270.

14.   Svenson, S.B. and Lindberg, A.A. (1978).   J. Immunol. 120: 1750-1757.

CHARACTERIZATION OF MONOCLONAL ANTIBODIES WHICH RECOGNIZE SPECIFIC

CELL SURFACE DETERMINANTS ON SALMONELLA TYPHIMURIUM

Eleanor S. Metcalf, Alison D. O'Brien, Moira A. Laveck
and William E. Biddison

Department of Microbiology, Uniformed Services
University School of Medicine, and Immunology Branch
NCI, NIH, Bethesda, MD

INTRODUCTION

Murine typhoid, a naturally occurring disease in mice caused
by Salmonella typhimurium, is a good experimental model with which
to study the interactions between host and parasite and is especially important since the pathogenesis of this disease is similar
to typhoid fever in man (1). Although the role of the immune
system in the regulation of this disease has not been studied
extensively, previous studies have suggested that the murine immune
response to S. typhimurium was primarily regulated by T cells and
macrophages (2). However, recent observations suggest that B cells,
hence protective antibodies, also play an important role in
resistance (3-5). Nevertheless, the identification of the salmonella cell surface antigens towards which these protective antibodies are directed remains controversial. One approach to the
identification of these antigenic determinants would be to generate
large quantities of antibody directed against a specific bacterial
cell surface determinant and to use these antibodies in passive
transfer experiments to determine their role in the protection of
the host from murine typhoid.

The somatic cell hybridization technology which was originally
developed by Köhler and Milstein (6) represents one approach to the
production of large quantities of antibodies derived from a single
clone of antibody-producing cells. This technique can provide a
definable reagent which is monospecific and consists of a single
molecular species. Therefore, we have used this technology to
establish hybridoma cell lines which secrete antibodies specific
for S. typhimurium cell surface antigenic determinants. In this

313

report, the characteristics of the first series of these hybridoma
antibodies are described.

MATERIALS AND METHODS

     Bacterial strains and LPS preparations were obtained as pre-
viously described (7). The procedures of Ozato et al. (8) for
cell fusion and establishment of cell lines were followed. Immunoglobulin heavy chain class analysis was determined by Ouchterlony
double diffusion in agar. The class-specific antisera were the
kind gift of Dr. Richard Asofsky, NIAID, NIH, Bethesda, MD.

RESULTS

Immunization and screening procedures. In these studies only one
immunization protocol was employed. BALB/c mice were immunized
i.p. with 5×10^6 acetone killed and dried (AKD) S. typhimurium,
strain TML (TML). The mice were boosted with 5×10^6 AKD TML i.p.
2 weeks before and i.v. 3 days before cell fusion. In the fusion,
10^8 spleen cells from 2 of the hyperimmune mice were fused with
10^7 SP2/0 cells and distributed into microtiter wells at 2.5×10^5
cells/well. Two weeks after fusion, macroscopic examination of
the microtiter wells indicated that 43 out of 168 wells contained
growing hybrids. The culture fluids from these wells were subsequently screened by RIA on TML coated and S. pneumoniae coated
plates (7). Culture fluids from three wells were shown to contain
specific anti-TML antibody activity. Hybridoma cells from these
three wells were cloned by limiting dilution at 1 cell/well. The
three hybridoma antibodies to be characterized were termed 5D5,
3D9, and 7A11.

Specificity of the Anti-TML hybridoma antibodies. To determine
whether the hybridoma antibodies were specific for the LPS molecule
or for other TML outer membrane components, all three hybridomas
were assayed using either TML LPS or E. coli LPS as immunoadsorbents.
The results presented in Table 1 demonstrate that all three hybridomas were specific for TML LPS but not E. coli LPS. The results
in Table 1 also show the isotype of the three anti-LPS hybridoma
antibodies: IgG_1 (5D5), IgG_{2b} (3D9), and IgG_1 (7A11).

     To examine whether the three hybridoma antibodies were specific
for the O antigen region of the LPS molecule, the antibodes were
assessed for their capacity to bind to a Rough B mutant of S.
typhimurium, strain 1764, kindly provided by Dr. Bruce A. Stocker,
Standford University (9). This bacterial strain is O negative but
the core polysaccharides and Lipid A region remain intact. The
results presented in Table 2 demonstrate that all three anti-TML
hybridoma antibodies were specific for the O antigen of the TML LPS
molecule.

Table 1. Characterization of Three Anti-S. *typhimurium* Hybridoma Antibodies

Hybridomas or Immune Mouse Sera Tested[a]	Heavy Chain Class	Antigen Used to Coat Plates			
		S. typhimurium	S. typhimurium LPS	E. coli LPS	Streptococcus pneumoniae
5D5	IgG_1	16,020[b]	12,791	480	576
3D9	IgG_{2b}	15,797	14,056	538	676
7A11	IgG_1	13,524	13,419	348	592
SP2/0	–	340	398	290	
PA 2.6	IgG_1	484			
7G7.B6	IgG_{2a}	451			
Dilutent	–	389			
Normal Mouse Serum	–	2,694	786	801	5,017
anti-S. tyhpimurium	–	6,842			5,388
anti-S. pneumoniae	–	2,627			10,601
anti-E. coli LPS	–	–	794	2,385	
anti-S. typhimurium LPS	–	–	3,970	785	

[a] SP2/0 is myeloma variant, SP2/OAg14, used as a fusion partner. PA2.6 is a hybridoma antibody specific for HLA framework determinants (10) and 7G7.B6 is a hybridoma antibody specific for human activated T cells (W.E.B., manuscript in preparation).

[b] Counts bound per minute in the RIA.

Table 2.   Specificity of Three Anti-S. typhimurium Hybridoma
           Antibodies

Bacteria Used to Coat Plates	O-Specific Antigens Expressed	Counts Bound per Minute by Anti-Salmonella Hybridomas		
		5D5	3D9	7A11
S. typhimurium strain TML	1,4,12	16,020	15,797	13,524
S. typhimurium LPS	1,4,12	12,791	14,056	13,419
S. typhimurium strain 1764	O negative	446	555	498
S. derby	4,12	467	572	313
S. enteritidis	9,12	265	238	210
S. dublin	1,9,12	3,259	2,654	323
S. newport	6,8	344	493	302

To determine the O-specific antigen towards which the hybridoma antibodies were directed, each of the three hybridoma anti-TML antibodies was tested for its capacity to bind to a panel of Salmonella species which express a variety of other O-specific antigens. The results presented in Table 2 demonstrate that these antibodies did not bind to S. derby which shares O antigens 4 and 12 with S. typhimurium, strain TML, nor did they bind to S. enteritidis, which shares O:12. Hybridoma 7A11 also did not bind to S. dublin, which is O:1,9,12. Thus, these results suggest that at least two of the hybridoma antibodies isolated in this study appear to be specific for the O:1 antigen.

DISCUSSION

This study was undertaken to generate antibodies in large quantity which were specific for antigenic determinants on the outer membrane of S. typhimurium. Three anti-S. typhimurium hybridoma antibodies were produced which were specific for the O antigen of the S. typhimurium LPS molecule. Although the hybridoma antibodies were directed against the O antigen, two of the antibodies were IgG_1 and one was an IgG_{2b}.

The ultimate aim of these studies is to determine the role of antibodies directed against specific Salmonella cell surface determinants in protective immunity. The murine model which will be used in these studies is the CBA/N mouse. Recent studies from this laboratory demonstrated that mice which express the xid gene,

CBA/N and their F_1 male progeny, are salmonella-susceptible and that these mice have an anti-salmonella antibody defect (5). These studies showed that F_1 males could be protected from a lethal challenge with live S. typhimurium by passive transfer of F_1 female immune serum. The hybridomas which have been characterized in this study will be used to analyze the protective capabilities of anti-O antibodies and to determine whether the heavy chain class of anti-TML antibodies is important in protective immunity.

ACKNOWLEDGMENT

USUHS protocol numbers C07305 and R07313.

REFERENCES

1. Wilson, G.S. and Miles, A. (1975). In:Topley and Wilson's
        Principles of Bacteriology and Immunity. Vol. I.
        Academic Press, NY. p. 939.
2. Collins, F.M. (1974). Bacteriol. Rev. 38:371.
3. Kuusi, N., Nurminen, M., Saxen, H., Valtonen, M. and Makela,
        P.H. (1979). Infect. Immun. 25:857.
4. Svenson, S.B. and Lindberg, A.A. (1979). J. Immunol. Methods
        25:323.
5. O'Brien, A.D., Scher, I. and Metcalf, E.S. (1981). J. Immunol.
        126:1368.
6. Köhler, G. and Milstein, C. (1975). Nature 256:495.
7. Metcalf, E S. and O'Brien, A.D. (1981). Infect. Immun. 31:33.
8. Ozato, K., Mayer, N. and Sachs, D.H. (1980). J. Immunol.
        124:533.
9. Nikaido, H., Leventhal, M., Nikaido, K. and Nakane, K. (1967).
        PNAS 57:1825.
10. Brodsky, F.M., Parham, P., Barnstasle, C.J., Crumpton, M.J.
        and Bodmer, W.F. (1979). Immunol. Rev. 47:3.

MONOCLONAL ANTIBODIES AS PROBES FOR ANTIGENS OF <u>MYCOPLASMA</u>

<u>PULMONIS</u>

Faina Varshavsky Rose, Michael F. Barile and John J.
Cebra

Department of Biology, University of Pennsylvania
Philadelphia, Pennsylvania 19104; and Division of
Bacterial Products, Bureau of Biologics, Food and
Drug Administration, Bethesda, Maryland 20205

INTRODUCTION

Mycoplasma organisms are a major cause of respiratory, uro-
genital and joint infections in man and animals. The spectrum
of the disease is wide and includes a number of complications.

<u>Mycoplasma pulmonis</u>, which is primarily a rodent pathogen, is
representative of the class <u>Mollicutes</u>. Like <u>M</u>. <u>pneumoniae</u> and
<u>M</u>. <u>gallisepticum</u>, it has no rigid wall, grows in colonies and
exists in two forms, globular and filamentous (8). Both forms
are motile and pathogenic. An important event in developing
mycoplasma respiratory infection is the attachment of the organism
to the epithelium. This attachment results in ciliostasis and
slow destruction of host tissues (3,7). Although an immune
response is usually generated during the course of infection, it
does not eliminate or reduce the inflammatory process but, on the
contrary, may contribute to tissue damage (2). Limited information is available on the mechanisms of pathogenesis and of host
immune responses to mycoplasma. In order to analyze mycoplasma-
host interactions, the surface components of mycoplasma must be
identified. Here, we report on the use of monoclonal antibodies
to <u>Mycoplasma pulmonis</u> to identify antigenic components including
surface determinants of that organism.

METHODS AND RESULTS

   Monoclonal antibodies to M. pulmonis antigens produced by hybridomas generated using cells for BALB/c mice.   The procedures for producing monoclonal antibodies have been well described in the current literature (5).   Briefly, BALB/c mice were immunized intraperitoneally with M. pulmonis in complete Freund's adjuvant. Three weeks later the mice were restimulated intravenously with a suspension of M. pulmonis in PBS.   Three days later the animals were sacrificed, their spleens removed and a single cell suspension of splenocytes was prepared.   The spleen cells were fused with mouse myeloma cells (SP2/0-Ag14 cell line) in the presence of 30% polyethylene glycol (m.wt 1000) and aliquoted into 96 well microtiter plates in media containing hypoxanthine-aminopterin-thymidine (HAT).   Two weeks later supernatants from positive cultures were tested for presence of anti -M. pulmonis antibody by solid phase radioimmunoassay (9).   Only cultures that scored at least ten fold above the background signal were considered positive.

   Use of monoclonal antibodies to identify antigenic components including surface determinants on M. pulmonis.   From two fusions, fourteen hybridoma clones were studied.   Characteristics of these are shown in Table 1.   There are five IgM, three IgG1, one IgG2a, four IgG2b and one IgG3 secreting clones.   There were no IgA secreting clones, a common observation when spleen cells are used to provide the normal immunoblasts for fusion.   Supernatants from these clones were collected, dialyzed against PBS and analyzed for specificity by using them for immunoprecipitation followed by resolution of the reacting antigens by SDS-PAGE (6).   The immunoprecipitation procedure is depicted in Figure 1, and has been used in all of the following experiments.   Since little is known about the solubility of M. pulmonis antigens, we used two different detergents (0.5% deoxycholate (DOC) and 0.5% NP40) to attempt solubilization of M. pulmonis antigens.   As shown in Figures 2 and 3, the non-ionic detergent NP40 seems to solubilize more antigens of M. pulmonis than the ionic detergent DOC, since of eight monoclonal antibodies used, five precipitated molecules from ^{3}H-leucine labelled M. pulmonis solubilized in 0.5% NP40 (Fig. 2) and only one precipitated molecules from ^{3}H-leucine labelled M. pulmonis solubilized in 0.5% DOC (Fig. 3).   Normal serum and immune serum from the mice used as cell donors were used to determine the range of antigenic specificities that could encompass the monoclonal antibodies.   Monoclonal IgG1 antibody is the only one that precipitated the same component after the solubilization with both types of detergents.   To increase specific activity of products precipitated by anti-M. pulmonis monoclonal antibodies, organisms were grown in media containing ^{35}S-methionine.   SDS-PAGE analysis of ^{35}S-met labelled M. pulmonis antigens is shown in Figure 4.

Table 1.   The Distribution of Heavy Chain Isotypes of Anti-$\underline{M}$.
           $\underline{pulmonis}$ Hybridomas

Hybridoma Clone	Anti-$\underline{M}$. $\underline{pulmonis}$ Heavy Chain Isotype	C.P.M. obtained with Rabbit-anti-mouse Immunoglobulin to:	
		Fab	Specific Isotype
			IgM
14G6.6	μ	11,200	8,994
13E4.6	μ	31,738	29,540
12G9.7	μ	1,026	4,888
10D9.3	μ	25,828	30,742
1E2.6	μ	1,180	5,842
			IgG1
9F8.3	γ_1	5,190	2,562
5B5.1	γ_1	2,816	1,456
2B8.2	γ_1	4,108	2,970
			IgG2b
13D9.0	$\gamma_2 b$	1,958	8,490
11F2.2	$\gamma_2 b$	1,886	7,282
8E5.3	$\gamma_2 b$	1,856	3,414
7C7.1	$\gamma_2 b$	1,772	5,640
			IgG2a
9C8.3	$\gamma_2 a$	20,004	46,730
			IgG3
12B3.8	γ_3	25,842	34,096

Counts per minute obtained with ^{125}I-labelled rabbit anti-mouse
Fab or specific anti-isotype after incubation of 20μl of hybridoma
supernatant with $\underline{M}$. $\underline{pulmonis}$ organisms adsorbed to microtiter plate.

Of the eighteen hybridoma antibodies used in that immunoprecipitation, fourteen have precipitated detectable macromolecules and
among those, some had similar patterns of reactivity. Approximate
molecular weights of the precipitated molecules were determined by
comparing their relative migration distances with those of known
standard proteins. As illustrated in Table 2, the precipitated
macromolecules fall into a number of groups: 30k, 40k, 60k, 76k
and 90k daltons. Next, to determine whether any of the precipitated

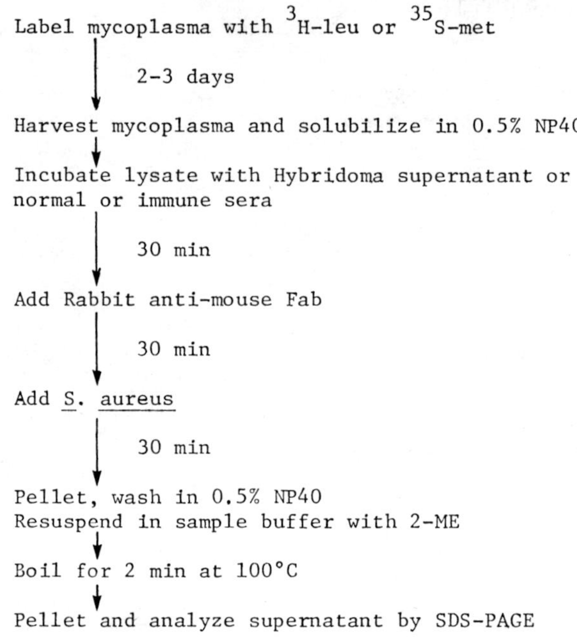

Figure 1.   Immunoprecipitation technique to analyze monoclonal antibodies specificities.

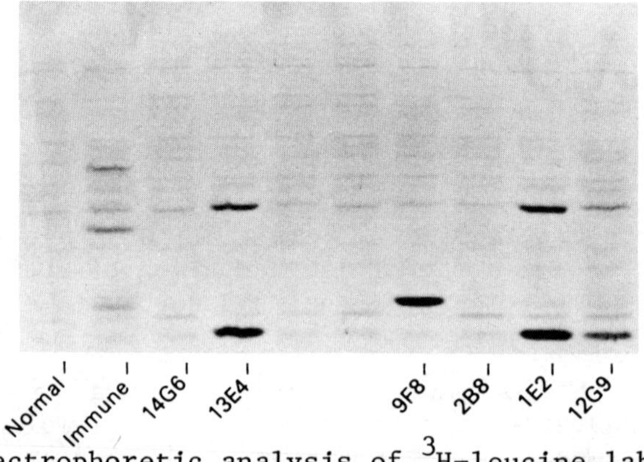

Figure 2.   Electrophoretic analysis of ^{3}H-leucine labelled, 0.5% NP40 solubilized M. pulmonis with monoclonal antibodies. Lanes: 1-normal serum; 2-immune serum; 3-10 monoclonal anti-M. pulmonis antibodies.

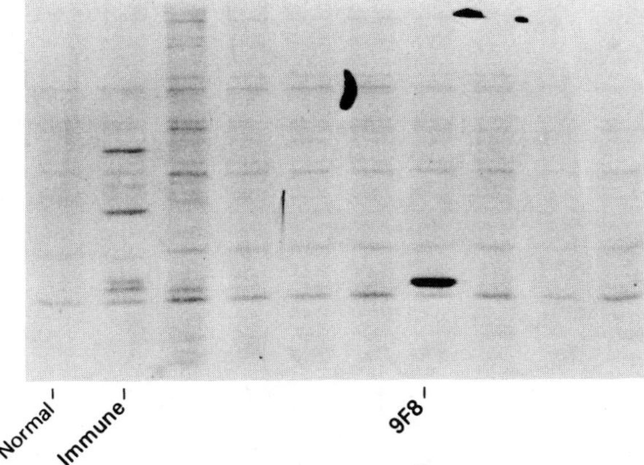

Figure 3.   Electrophoretic analysis of ^{3}H-leucine, labelled, 0.5%
DOC solubilized M. pulmonis with monoclonal antibodies.
1-normal serum; 2-immune serum; 3-10 monoclonal anti-
M. pulmonis antibodies.

components were surface antigens, intact M. pulmonis was labelled
with ^{125}I using the lactoperoxidase method (4).   Figure 5 illustrates the SDS-PAGE patterns of ^{125}I-labelled, NP40 solubilized
M. pulmonis antigens using eight different monoclonal antibodies
for precipitation.   Two of these monoclonal antibodies (2B4 and
11F2) precipitated ^{125}I-labelled molecules.   Other monoclonal
antibodies are currently being tested for their ability to precipitate other surface determinants from ^{125}I-labelled M. pulmonis.

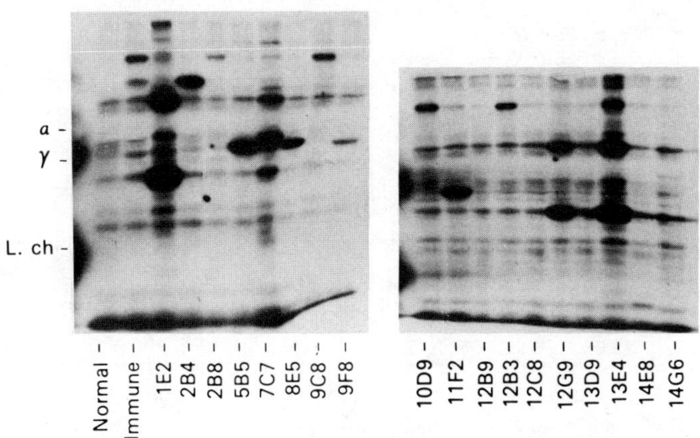

Figure 4. Immunoprecipitation and electrophoretic analysis of
^{35}S-methionine labelled, 0.5% NP40 solubilized M.
pulmonis with monoclonal antibodies.   1-normal serum;
2-immune serum; 3-20 monoclonal anti-M. pulmonis antibodies.

Table 2. Molecular weights of macromolecules precipitated by
anti-<u>M</u>. <u>pulmonis</u> hybridoma antibodies

Anti-<u>M</u>. <u>pulmonis</u> Hybridoma		Molecular Wts. of Molecules Precipitated by the Hybridoma Antibodies		
Clone	Isotype			
1E2	μ	64,500	36,500	
2B4	–	76,000		
2B8	γ_1	91,000		
5B4	γ_1	45,500		
7C7	$\gamma_2 b$	66,000	49,000	37,000
8E5	$\gamma_2 b$	45,500		
9C8	$\gamma_2 a$	91,000		
9F8	γ_1	45,500		
10D9	μ	93,500		
11F2	$\gamma_2 b$	40,000		
12G9	μ	63,000	33,000	
13E4	μ	93,500	60,000	33,000

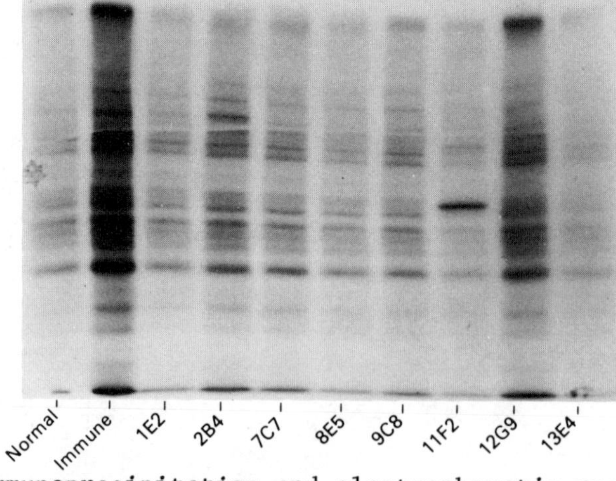

Figure 5. Immunoprecipitation and electrophoretic analysis of
^{125}I-labelled <u>M</u>. <u>pulmonis</u> with monoclonal antibodies.
1-normal serum; 2-immune serum; 3-10 monoclonal anti-
<u>M</u>. <u>pulmonis</u> antibodies.

SUMMARY

We have established a library of anti-M. pulmonis monoclonal
antibodies with different specificities and heavy chain isotypes.
These monoclonal antibodies have been used in immunoprecipitation
studies to isolate and partially characterize antigens including
surface determinants from radiolabelled M. pulmonis. The precipitated macromolecules, as visualized by SDS-PAGE, are proteins, and
at least two of those molecules are on the surface of M. pulmonis.
It should be noted that lack of radioiodination by the lactoperoxidase method does not necessarily indicate that an antigen is
not on the surface of the M. pulmonis organism.

These results are in agreement with the results obtained by
Asa et al. (1) using ^{125}I-labelled M. pulmonis and immune sera to
that organism. Antigens precipitated by antisera to M. pulmonis
had molecular weights ranging from 28,000 to 150,000 and were
proteins in nature.

Using our approach we can now identify individual determinants
on a given mycoplasma organism and try to assess the significance
of each antigen (in particular an attachment moiety) to the
infection process or to host resistance. In our case, monoclonal
antibodies to M. pulmonis will be used in in vitro attachment
inhibition and in vivo protection assays. Correlation of
specificities and isotypes of these and other monoclonal antibodies with protection will be sought.

This approach should help to elucidate the mechanism of
interaction between mycoplasma and host, and the nature of the
immune response to individual antigens.

ACKNOWLEDGMENT

This work was supported by Grant AI-17997 from NIAID.

REFERENCES

1. Asa, P., Acton, R., Cassell, G., Wise, K. (1980). J. Immunol.
        124:997.
2. Cassell, G., Lindsey, J., Overcash, R. and Baker, H. (1973).
        Ann. N.Y. Acad. Sci. 225:395.
3. Gabridge, M., Gunderson, H., Schaeffer, S. and Barden-Stahl, D.
        (1978). Infect. Immun. 21:533.
4. Hubbard, A., Cohn, Z. (1975). J. Cell. Biol. 69:438.
5. Kennett, R., McKearn, T. and Bechtol, K. (1980). Monoclonal
        Antibodies. Plenum Press, NY.

6.   Laemmli, U. (1970).   Nature 227:680.
7.   Muse, K., Poswell, D. and Collier, A. (1976).   Infect. Immun.
          13:229.
8.   Smith, P. (1973).   The Biology of Mycoplasmas.   Academic Press,
          NY.
9.   Taylor, G. (1979).   Infect. Immun. 24:701.

ELECTRON MICROSCOPIC EXAMINATION OF THE INFLAMMATORY RESPONSE OF
GUINEA PIG NEUTROPHILS AND MACROPHAGES TO LEGIONELLA PNEUMOPHILA

Sheila Moriber Katz and Shahab Hashemi

Department of Pathology, Hahnemann Medical College and
Hospital, Philadelphia, PA   19102

We studied the ultrastructural morphology of neutrophils and
macrophages from splenic and pulmonary lesions of guinea pigs
inoculated with Legionella pneumophila. Both organs exhibited a
similar mixture of these two inflammatory cells. We used 10 male
Hartley guinea pigs. Four were infected intranasally with 0.5 ml
of inoculum containing $5x10^6$ organisms. Four were infected intraperitoneally with 0.5 ml of inoculum containing $5x10^6$ organisms.
The remaining two were inoculated intranasally with 0.5 ml of
sterile diluent. All animals were autopsied three days after
inoculation, and the pulmonary and splenic tissues were prepared
for electron microscopic studies.

Neutrophils and macrophages differed in their morphologic
response to Legionella pneumophila. Neutrophils actively phagocytized Legionella pneumophila (Fig. 1). The microbes were seen
within the cytoplasm of these inflammatory cells, both within and
outside of membrane bound phagosomes. In both of these locations,
microbes showed signs of degeneration (Fig. 2), suggesting that
the intracellular milieu of neutrophils is somewhat hostile to
these bacteria. Neutrophils also actively phagocytized infected,
injured cells and cellular debris (Fig. 3).

Macrophages differed from neutrophils in their response to
Legionella pneumophila. First, the macrophages were somewhat
unresponsive to extracellular Legionella pneumophila and did not
display pseudopods indicative of phagocytosis. Secondly, although
most of the microbes were intracellular, degenerating microbial
forms were seldom noticed. Thirdly, intracellular Legionella
pneumophila was often located within membrane bound vacuoles morphologically indistinguishable from rough endoplasmic reticulum

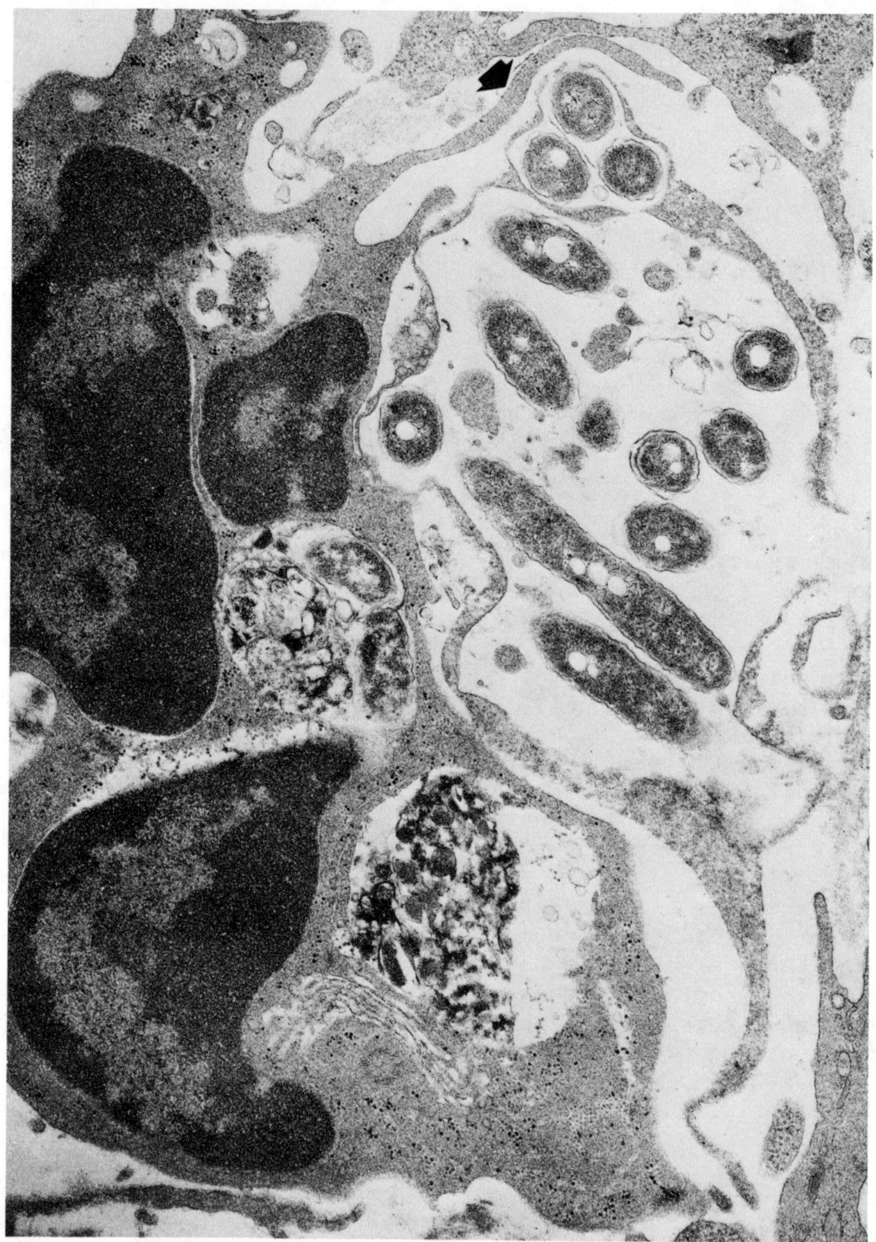

Figure 1.   Extracellular <u>Legionella</u> <u>pneumophila</u> are phagocytized
            by a neutrophil. Note pseudopods (arrow) and debris
            within the cytoplasm of the neturophil, (Uranyl
            acetate, lead citrate X 21,200).

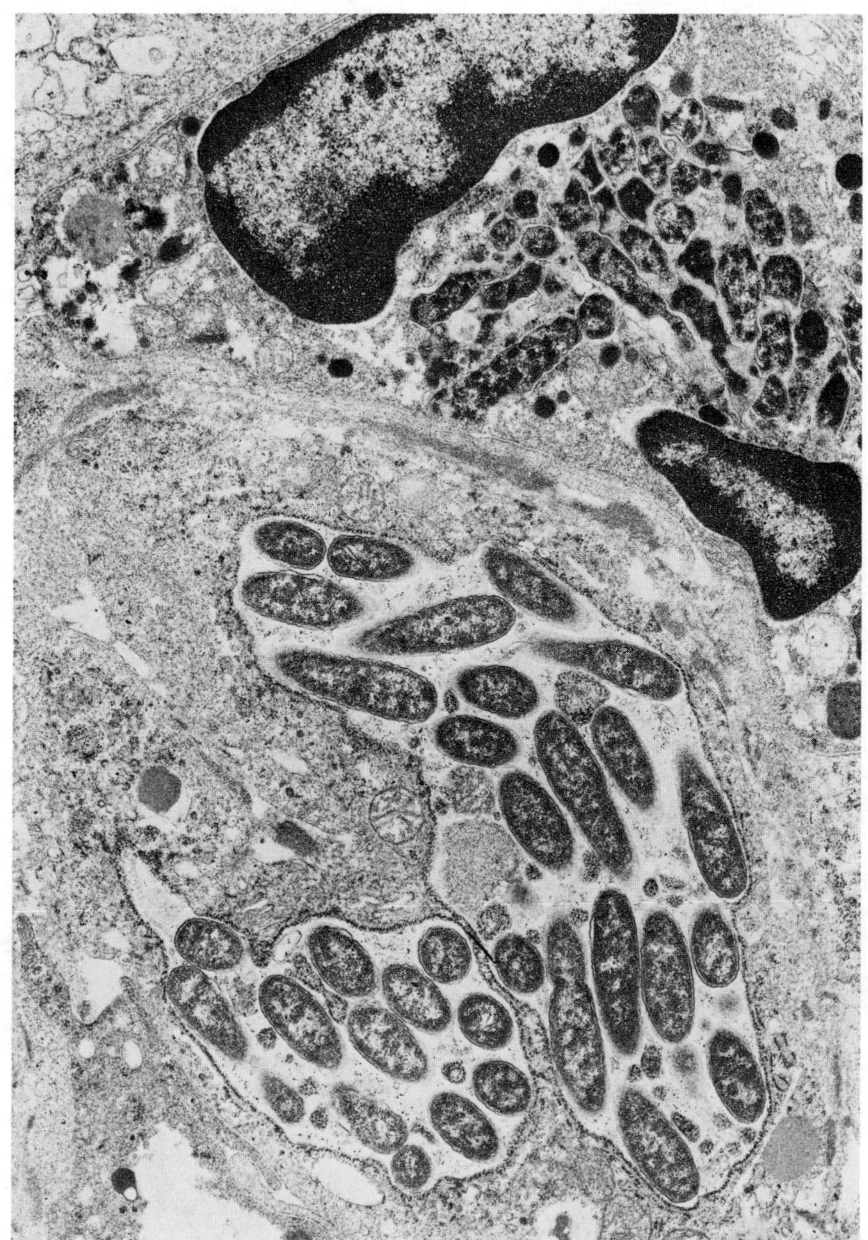

Figure 2. Contrast the intact <u>Legionella</u> <u>pneumophila</u> within a
macrophage (lower) with the degenerating <u>Legionella</u>
<u>pneumophila</u> within a neutrophil (upper) (Uranyl
acetate, lead citrate X 12,100).

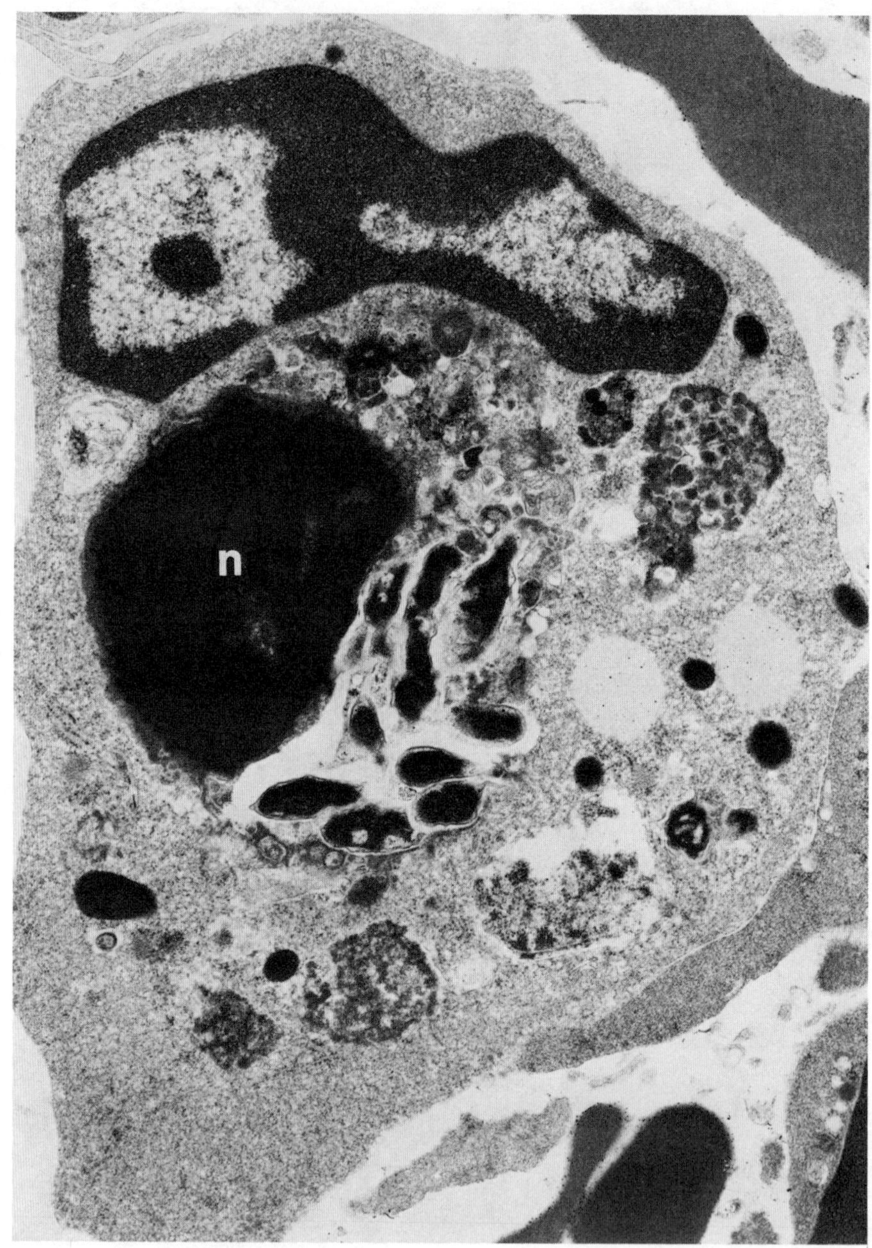

Figure 3. The cytoplasm of a neutrophil contains a pyknotic nucleus (n) and adjacent degenerating osmiophilic bacteria. (Uranyl acetate, lead citrate X 9,700).

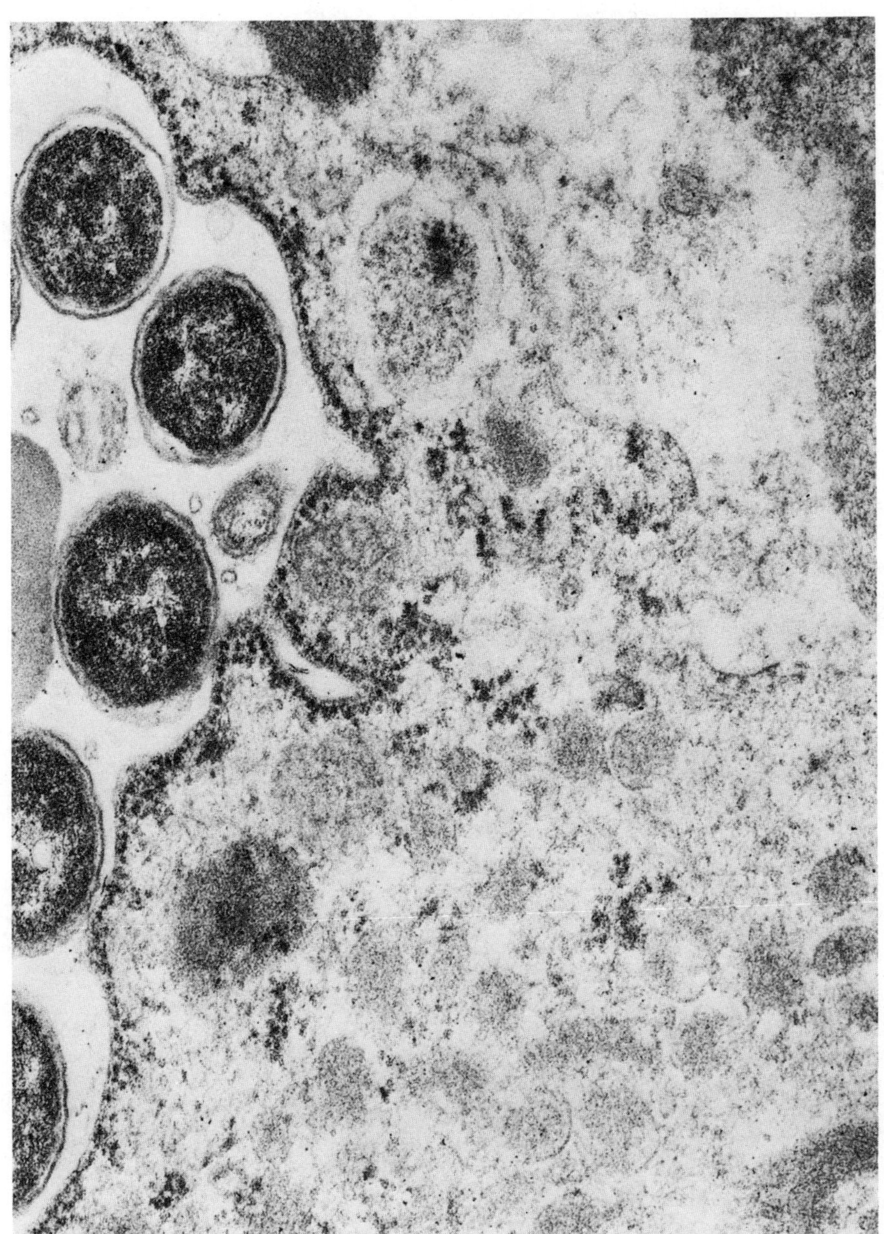

Figure 4. High magnification of the limiting membrane of a vacuole containing <u>Legionella</u> <u>pneumophila</u> particulate electron dense structures on the outer rim that are the same size and configuration as adjacent poly-ribosomes. (Uranyl acetate, lead citrate X 53,500).

(Figs. 2 and 4). The thickness of the limiting membrane of the
vacuole was in the range of that described for rough endoplasmic
reticulum, and its outer surface was studded with particles
morphologically identical to ribosomes (Fig. 4). Moreover, in
some sections, more segments of rough endoplasmic reticulum were
continuous with the dilated segment containing Legionella pneomo-
phila. In some zones, only one microbe was enveloped by rough
endoplasmic reticulum, whereas, in other segments, large vacuoles
contained greater than 20 microbes per section. Cells containing
the largest number of microbes frequently exhibited disruption of
cytoplasmic membranes, suggesting that they rupture and release
newly formed microbes into the extracellular compartment.

Localization of Legionella pneumophila to the rough endoplasmic reticulum makes this organism unique among bacteria and
parasites. The reason for the homing to this location, and the
mechanism of the entry into the cisternae remains a mystery.
Perhaps (1) Legionella pneumophila is phagocytized by macrophages
and the phagosome fuses with the rough endoplasmic reticulum, or
(2) Legionella pneumophila is phagocytized by macrophages and the
parasitized phagosome is transformed into rough endoplasmic
reticulum by attachment of free ribosomes to the outer surface of
the phagosome, or (3) Legionella pneumophila actively invades
rough endoplasmic reticulum of macrophages either through phagosomal membranes or directly through the cytoplasmic membrane.

Horwitz and Silverstein (1) have demonstrated that Legionella
pneumophila is a facultative intracellular parasite of human monocytes. The study presented herein confirms those findings in
guinea pig macrophages and shows, by morphological techniques,
that Legionella pneumophila remains intact in guinea pig macrophages and localized to what appears to be the rough endoplasmic
reticulum of these cells. In contrast to our findings of infected
neutrophils, we seldom found morphological evidence of death and
disintegration of Legionella pneumophila within the rough endoplasmic reticulum of macrophages. Rather, the Legionella pneumophila appeared to be thriving and replicating at this site.
Legionella pneumophila requires amino acids for growth (2) and
perhaps the agent lodges in the rough endoplasmic reticulum, which
is the site of cellular protein synthesis, for some of its raw
nutrients. Moreover, rough endoplasmic reticulum might shelter
Legionella pneumophila from the bactericidal activities of macrophages.

Our morphologic findings indicate that Legionella pneumophila
is an intracellular parasite of guinea pig macrophages, and that
guinea pig neutrophils have an adverse effect on most of the
microbes. The homing of Legionella pneumophila to macrophage
cytoplasmic organelles which are morphologically indistinguishable
from rough endoplasmic reticulum has no bacteriologic parallel.

It remains to be determined how <u>Legionella pneumophila</u> enters this organelle, and whether this structure is, in fact, functional rough endoplasmic reticulum.

REFERENCES

1. Horwitz, M.A. and Silverstein, S.C. (1981). Interaction of
        the Legionnaires' disease bacterium (<u>Legionella
        pneumophila</u>) with human phagocytes. J. Exp. Med.,
        153:386-397.
2. Feeley, J.C., Gorman, G.W. and Gibson, R.J. (1979). Primary
        isolation media and methods. In: "Legionnaires'"
        the disease, the bacterium, the methodology. Ed:
        Jones, G.L. and Hebert, G.A., U.S. Dept. of Health,
        Education and Welfare.

ACTIVATION OF MACROPHAGES FOR KILLING OF RICKETTSIAE: ANALYSIS OF
MACROPHAGE EFFECTOR FUNCTION AFTER RICKETTSIAL INOCULATION OF
INBRED MOUSE STRAINS

Carol A. Nacy, Monte S. Meltzer, Edward J. Leonard,
Mary M. Stevenson and Emil Skamene

Department of Immunology, Walter Reed Army Institute
of Research, Washington, DC 20009; National Cancer
Institute, National Institutes of Health, Bethesda,
Maryland 20205; and Montreal General Hospital Research
Institute, Montreal, Canada

INTRODUCTION

The rickettsiae are obligate intracellular bacteria that infect
a variety of cells in vitro, including peritoneal macrophages of
mice (1). The organisms replicate freely in the cytoplasm of
infected macrophages and, within 24-30 hr of onset of replication,
migrate from infected macrophages to initiate infection of other
cells (Fig. 1). Both numbers of infected macrophages and numbers
of intracellular rickettsiae increase with time in culture. The
rickettsiae, because of their obligate intracellular parasitism,
have evolved unique ways to avoid most mechanisms of intracellular
destruction. The very location of rickettsial replication, in the
cytoplasmic matrix of the cell, protects these organisms from
potent lytic enzymes of phagolysosomal systems (1,2). That this
destruction does occur in vivo, however, is documented by development of both natural resistance and acquired immunity following
rickettsial inoculation of mice (3,4,5).

Analysis of early immunologic responses to R. tsutsugamushi
infection in resistant and susceptible mice suggests that vigorous
development of nonspecific immunity, particularly events that
affect macrophage function, is essential for resolution of this
infection (6,7). In the absence of immune intervention, rickettsiae
infect and replicate in the cytoplasm of macrophages (Fig. 2).
Since this does not occur in macrophages of mice resistant to

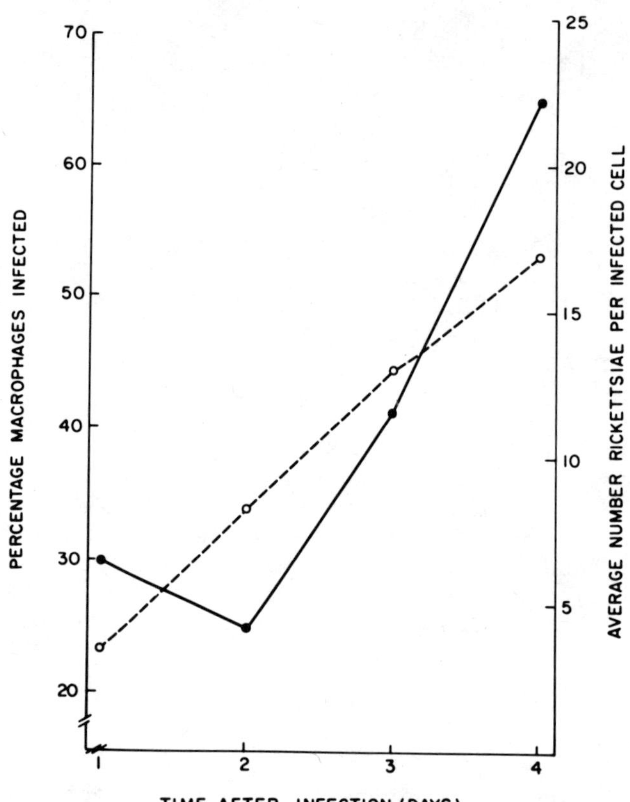

Figure 1.   Macrophages were exposed to 5 plaque-forming units (PFU)
Gilliam strain R. tsutsugamushi for 1 hr, washed, and
incubated at 37C for 24–96 hr. Samples were removed
every 24 hr, and cell smears were examined microscopically for percent infected macrophages and number
of intracellular rickettsiae/infected macrophage.

R. tsutsugamushi infection, interaction of macrophages and
rickettsiae during immune responses must be quite different (6).
Certain properties of activated macrophages, cells with enhanced
nonspecific microbicidal and tumoricidal activities, can be
induced in resident or inflammatory peritoneal macrophages of
mice in vitro with soluble products of antigen- or mitogen-
stimulated spleen cells (lymphokines) (8,9,10,11,12). The
rickettsiae are ideal organisms for quantitative analysis of macrophage microbicidal activities: (a) these bacteria do not survive
in an extracellular environment (thus eliminating problems associated with removal of inocula) (13);   (b) the organisms enter cells
within 15–30 min (13);   (c) generation time of rickettsiae in
macrophases ranges from 18–22 hr. and intracellular killing events

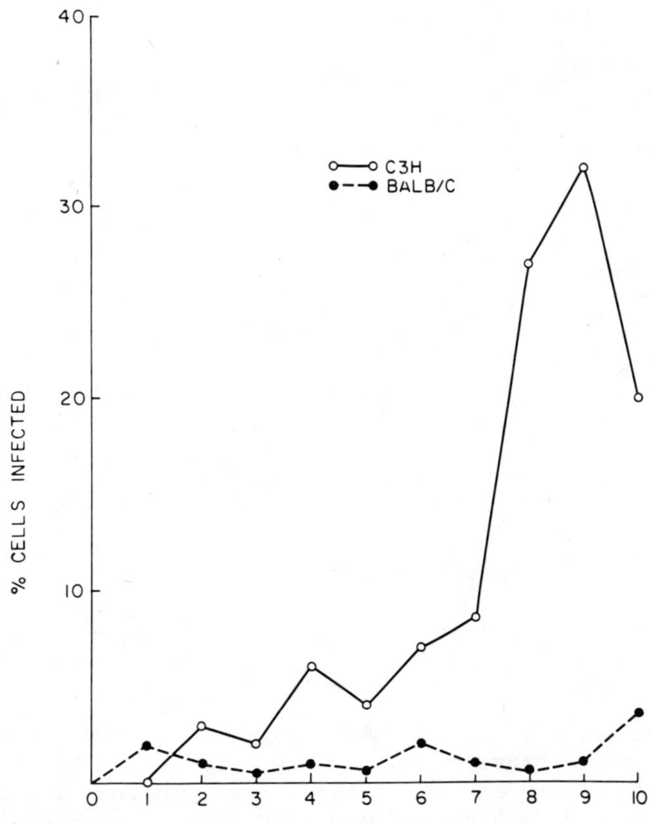

Figure 2. Percent macrophages containing intracellular rickettsiae
at the time of macrophage harvest. Susceptible C3H and
resistant BALB/c mice were inoculated with 1000 LD_{50}
Gilliam R. tsutsugamushi and macrophages were harvested
from infected mice 1-10 days after inoculation of
rickettsiae. Peritoneal cell smears were observed
microscopically for percent macrophages containing
rickettsiae.

are not obscured by rapid intracellular replication in the few
residual cells (1,14).

RESULTS AND DISCUSSION

To analyze the interaction of rickettsiae and immunologically
activated macrophages, we treated resident peritoneal macrophages
from C3H/HeN mice with lymphokines in vitro, and exposed treated
and untreated control macrophages to the Gilliam strain of

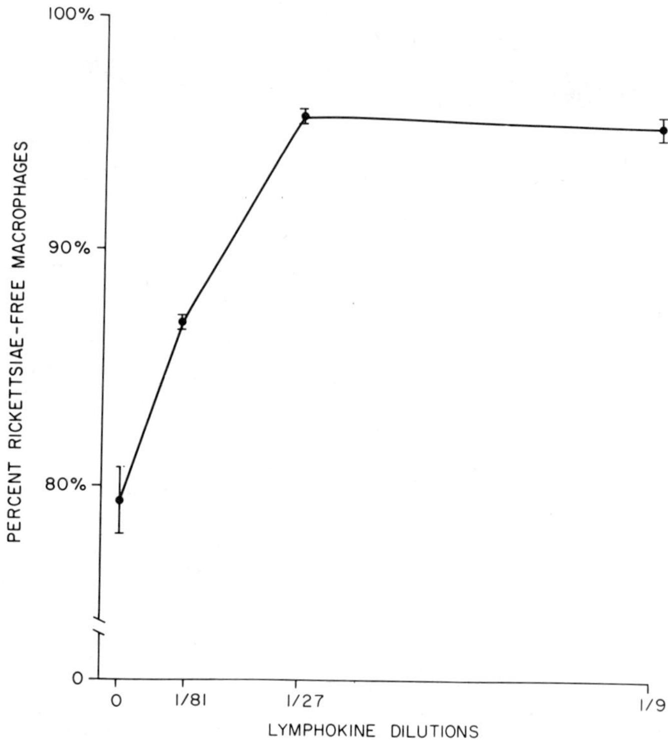

Figure 3.   Dose response of macrophages treated with lymphokines
            before exposure to rickettsiae. Macrophages were
            treated with medium or dilutions of lymphokines for
            4 hr, washed and exposed to 5 PFU R. tsutsugamushi for
            1 hr. Infected macrophages were incubated 24 hr, and
            cell smears were examined for percent infected macro-
            phage.

R. tsutsugamushi (12). Lymphokine effects on rickettsiae-macro-
phage interactions were both dose and time dependent. Treatment
of macrophages with lymphokines before exposure to rickettsiae
markedly affected the ability of R. tsutsugamushi to infect and
replicate in mouse macrophages (Fig. 3). The number of cells
free of intracellular rickettsiae 24 hr after infection increased
with increasing lymphokine concentration. The time for initiating
lymphokine effects was relatively short (Fig. 4). An increase
in rickettsiae-free macrophages was observed after as little as
1 hr of lymphokine treatment, and maximal effects occurred 6-12
hr after treatment. Cells treated longer than 12 hr returned to
untreated control levels. A similar time course and decay of
activity has been observed for lymphokine-induced macrophage
tumoricidal activity, another effector function of activated
macrophages (15). A slight, but constant, increase in

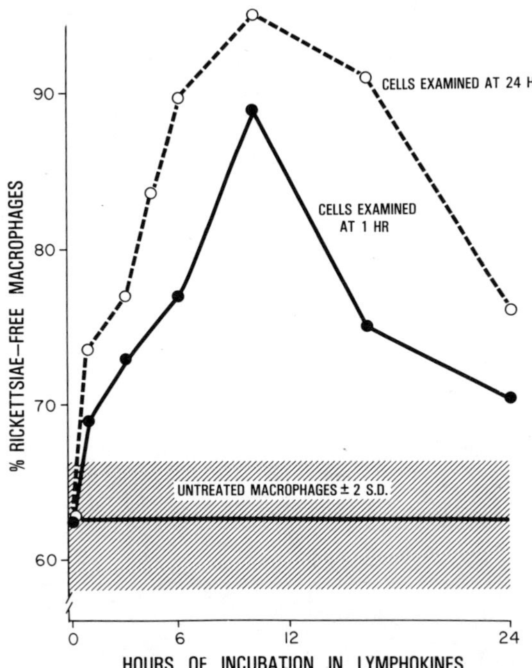

Figure 4.   Kinetics of the response of macrophages treated with lymphokines before exposure to rickettsiae. Macrophages were treated with 1/10 lymphokines for various times (0–24 hr) and then exposed to 5 PFU R. tsutsugamushi for 1 hr. Cells were washed and sampled immediately (1 hr) or incubated an additional 24 hr. Cell smears were examined microscopically for percent infected macrophages.

rickettsiae-free macrophages was observed at 24 hr. This observation suggested that, in addition to effects on internalization of rickettsiae, intracellular killing of the bacteria also occurred in cultures treated with lymphokines before infection. To analyze macrophage intracellular destruction of rickettsiae, we infected macrophages with R. tsutsugamushi and treated cultures with lymphokines after infection. Seventy-five percent of infected macrophages were free of intracellular rickettsiae by 24 hr in cultures treated with as little as 1/30 of the original concentration of lymphokines after infection (Fig. 5). Induction of macrophage activation for intracellular killing of rickettsiae was essentially complete after 4 hr of lymphokine treatment (Fig. 6).

Two distinct microbicidal activities could be documented in macrophage cultures treated with lymphokines in vitro: (1) macrophages treated with lymphokines before exposure to rickettsiae

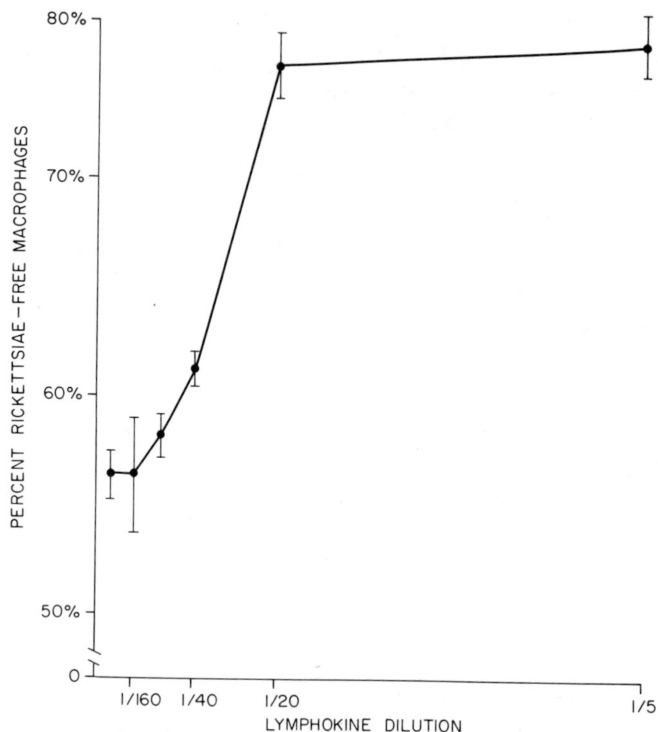

Figure 5.   Dose response of macrophages treated with lymphokines
             after infection. Macrophages were exposed to 5 PFU
             R. tsutsugamushi for 1 hr, washed, and treated with
             various concentration of lymphokines for 24 hr.   Cell
             smears were examined for percent infected macrophages.

developed increased resistance to infection with R. tsutsugamushi
(35% fewer macrophages contained intracellular rickettsiae at
1 hr compared to control cultures), and (2) macrophages treated
with lymphokines after infection developed the capacity to kill
the intracellular bacteria (80–90% fewer macrophages contained
intracellular rickettsia at 24 hr).   The number of viable intracellular rickettsiae in macrophage cultures could also be measured
by modification of a plaque assay for R. tsutsugamushi (16).
Infected macrophage cultures treated with medium or lymphokines
were lysed, diluted in medium, and dilutions were inoculated onto
irradiated L-929 cell monolayers (Table 1). Lymphokine treatment
of macrophages before infection induced a 39% decrease in viable
intracellular rickettsiae at 1 hr. Lymphokine treatment after
infection induced 75–90% reduction in viable organisms. Plaque
titration of viable intracellular R. tsutsugamushi confirmed the
microscopic analysis of macrophage microbicidal activities, as
the percent reduction in viable rickettsiae correlated closely with

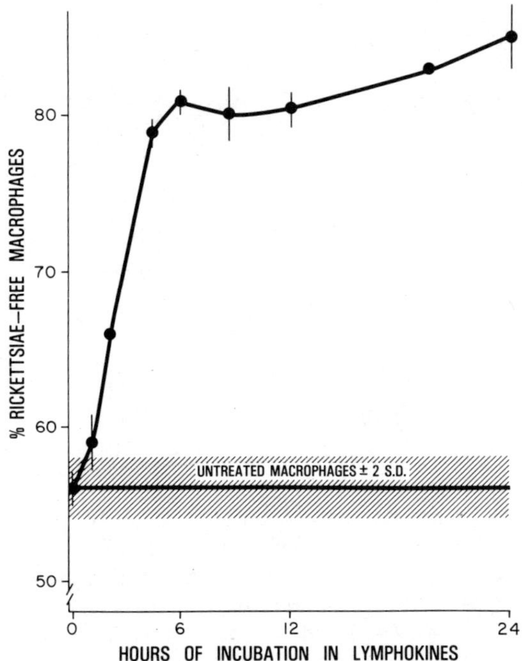

Figure 6.   Kinetics of the response of macrophages treated with
lymphokines after infection. Macrophages were exposed
to 5 PFU R. tsutsugamushi for 1 hr, washed and exposed
to 1/10 lymphokines for various times (1-24 hr) after
infection. Cell smears were examined microscopically
for percent infected macrophages.

figures obtained from microscopic observation of percent infected
macrophages (see Table 1 and Figs. 3 and 5).

The lymphokines that induce the two different microbicidal
activities were also distinct (Fig. 7). Fractionation of lymphokine supernatants on Sephadex G-200 revealed three different
activity peaks for intracellular killing of rickettsiae (130,000,
45,000 and ≤ 10,000 daltons) (17). One of these activity peaks,
the 45,000 MW molecule(s), induced increased resistance to
infection in macrophages treated with lymphokines before exposure
to rickettsiae, as well. Activity for induction of macrophage nonspecific tumor cytotoxicity also eluted in this region (Fig. 7)
(18). The lymphokines that induced macrophage microbicidal activities were sensitive to treatment at 56C for 30-60 min. Heat-treated
whole lymphokine supernatants (Fig. 8), as well as pooled active
fractions, lost activity for both microbicidal functions. Since
bacterial lipopolysaccharides (LPS) are not affected by heating

Table 1. Titration of viable <u>R</u>. <u>tsutsugamushi</u> on irradiated L-929
          cell monolayers.

Assay Time	Macrophages Treated With [1]	Plaque Titer (10^4)	Decrease in Viable Rickettsiae
1 hr	Medium	10.4	
	LK Before Infection	6.0	39%
24 hr	Medium	9.2	
	LK After Infection	2.2	76%
	LK Before and After Infection	0.9	91%

[1] Macrophages were exposed to medium or a 1/10 dilution of lymphokines for 4 hr, washed, and exposed to 5 PFU <u>R</u>. <u>tsutsugamushi</u>
for 1 hr. Infected cells were washed, and lysed immediately,
or incubated in medium or 1/10 lymphokines an additional 24 hr.
Lysed samples were serially diluted in medium (10-fold dilutions)
and plated on irradiated L-929 cell monolayers. Plaques were
read at 16 days.

at 56C, these data suggest that the activity of lymphokine supernatants was not due to contaminating LPS.

    One mechanism of host elimination of rickettsiae during
infections might be microbicidal activity by nonspecifically
activated macrophages. Certainly, <u>in</u> <u>vitro</u> analysis of macrophage
microbicidal activity suggested that lymphokine activated macrophages effectively destroyed intracellular rickettsiae. We analyzed
the development of nonspecifically activiated macrophages during
<u>R</u>. <u>tsutsugamushi</u> infection of BALB/c mice, a mouse strain resistant
to lethal intraperitoneal inoculation of this organism (1,6).
Macrophages obtained from mice inoculated 4-9 days previously with
1000 LD_{50} of Gilliam strain, <u>R</u>. <u>tsutsugamushi</u> were tested for their
capacity to kill rickettsiae <u>in</u> <u>vitro</u> (Table 2). Forty-six
percent fewer macrophages contained intracellular rickettsiae than
control macrophages from uninfected mice at 1 hr; 82% fewer cells
contained intracellular rickettsiae at 24 hr. Macrophages from
mice with regressing scrub typhus infections had microbicidal
activities against rickettsiae <u>in</u> <u>vitro</u> similar to that of resident
peritoneal macrophages treated with lymphokines (Table 2). This microbicidal activity was, however, entirely nonspecific, since macrophages from mice treated <u>in</u> <u>vivo</u> with macrophage activating agents
such as <u>Mycobacterium</u> <u>bovis</u>, strain BCG, or <u>Corynebacterium</u> <u>parvum</u>

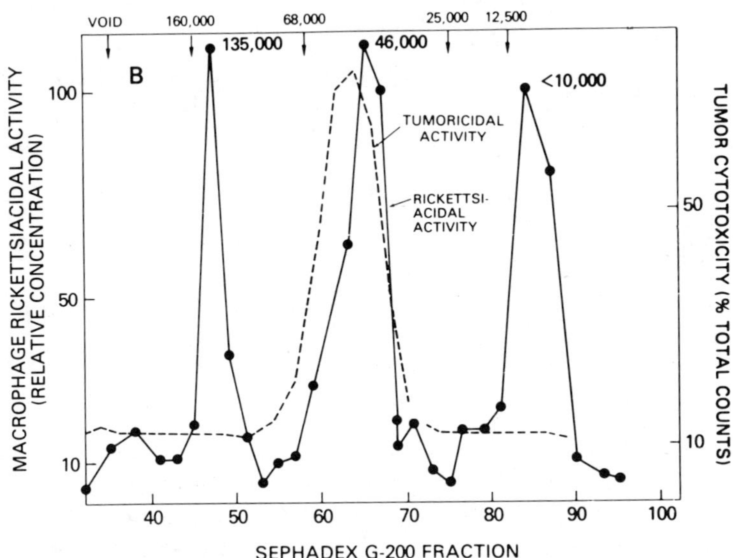

Figure 7. Induction of macrophage nonspecific tumoricidal and microbicidal activities by fractionated lymphokine supernatants. Whole lymphokine supernatants were concentrated 100-fold and fractionated by Sephadex G-200 column chromatography. Macrophages were treated with 1/20 lymphokine fractions for 4 hr before exposure to prelabeled mKSA-TU5 tumor target cells (9), or exposed to 5 PFU R. tsutsugamushi for 1 hr, washed, and treated with 1/20 lymphokine fractions after infection for 24 hr (17). Activity for these two effector functions was compared to dose responses for unfractionated lymphokine material, and graphed as relative concentration for each effector activity.

were as rickettsiacidal as macrophages treated in vitro with lymphokines. Induction of macrophage microbicidal activities in vivo required development of an immune response. Macrophages from mice injected with a variety of sterile inflammatory agents, such as thioglycollate, fetal calf serum, or latex beads were not activated to kill intracellular rickettsiae unless exposed further to lymphokines in vitro (Table 3). Thus, neither resident peritoneal cells nor inflammatory cells spontaneously restricted intracellular replication of rickettsiae. Both resident and inflammatory cells could be induced to eliminate the intracellular bacteria, however, in the presence of immune reactions in vivo or lymphokines in vitro.

Although development of microbicidal macrophages during the course of rickettsial infection certainly suggested a role for

Table 2.   Effector functions of macrophages activated *in vivo*.

Treatment *in vivo* [1]	Macrophage Microbicidal Activity at [2]		Macrophage Tumor Cytotoxicity [3]
	1 hr	24 hr	
None	0	0	8
+ LK *in vitro*	40	75	47
BCG	42	86	75
C. parvum	46	92	64
R. Tsutsugamushi	46	82	86

[1] Peritoneal cells were harvested from mice injected 8 days previously with BCG (10^6 viable organisms), C. parvum (75 mg/kg), or R. tsutsugamushi (1000 LD_{50}). Macrophages were exposed to 5 PFU R. tsutsugamushi for 1 hr, washed, and sampled immediately or incubated an additional 24 hr.

[2] Microbicidal activity was determined by:

$$100 \times \frac{(\% \text{ infected control macrophages} - \% \text{ infected treated macrophages})}{\% \text{ infected control macrophages}}$$

[3] Tumoricidal activity was determined by the release of ^{3}H-thymidine from prelabeled mKSA-TU5 tumor target cells (9).

Table 3.   Microbicidal activity of macrophages from mice inoculated with sterile inflammatory agents.

Treatment *in vivo* [1]	Spontaneous Macrophage Microbicidal Activity at		Lymphokine-Induced Macrophage Microbicidal Activity at	
	1 hr	24 hr	1 hr	24 hr
None	0	0	40	80
Phosphate-Buffered Saline	2	4	50	80
Thioglycollate	0	6	42	68
Latex Beads	2	0	46	60
Starch	0	2	46	74

[1] Macrophages were harvested from mice inoculated with thioglycollate broth 7 days previously; with starch, 5 days previously; with latex beads, 3 days previously; or with PBS, 1 day previously. Cells were treated *in vitro* with medium or 1/10 lymphokines for 4 hr before infection or 24 hr after infection with 5 PFU R. tsutsugamushi.

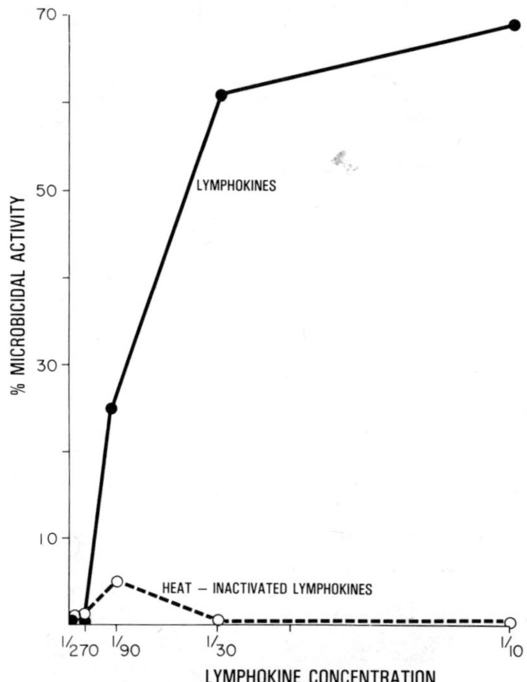

Figure 8.   Heat-treatment of lymphokine supernatants. Whole
lymphokine supernatants were subjected to heating at
56C for 30 min. Macrophages were exposed to 5 PFU
R. tsutsugamushi for 1 hr, washed, and treated with
dilutions of heat-treated and untreated lymphokines for
24 hr. Microbicidal activity was determined by:

$$100 \times \frac{(\%\ \text{infected control macrophages} - \%\ \text{infected treated macrophages})}{\%\ \text{infected control macrophages}}$$

rickettsiacidal macrophages in resolution of this disease, the
essential requirement for nonspecifically activated macrophages
was not recognized until we performed protection tests in animals
treated with nonspecific macrophage activation agents. Here, the
tool of mouse genetics enabled us to rigorously assess the contribution of activated macrophages in resistance to R. tsutsugamushi
infection. C3H/HeN and C3H/HeJ mice are equally susceptible to
the lethal effects of intraperitoneal inoculation of Gilliam strain
R. tsutsugamushi - one organism equals 1 LD_{50} in both mouse strains
(3). C3H/HeJ mice, however, are genetically different from C3H/HeN
mice at the Lps gene locus, which controls expression of at least
one effector function of activated macrophages, nonspecific tumor
cytotoxicity (19,20). C3H/HeJ mice do not respond to activation
signals in vivo or in vitro for this effector activity (20,21,22).

Table 4.   Survival of mice treated with BCG and challenged with
           R. tsutsugamushi.

Mouse Strain	R. tsutsugamushi Challenge Dose LD_{50}	Percent Survival of Mice Treated with[1]	
		Medium	BCG
C3H/HeN	10^4	0	75
	10^3	0	100
	10^2	0	100
	10^1	0	100
C3H/HeJ	10^4	0	0
	10^3	0	0
	10^2	0	0
	10^1	0	0

[1]Mice were inoculated with medium or 10^6 viable BCG, and challenged
with graded doses of R. tsutsugamushi. There were 20-30 mice per
group.

BCG-treated and control untreated mice were inoculated with graded
doses of R. tsutsugamushi, ranging from 10^1 to 10^4 viable plaque-
forming units (Table 4). The untreated mice all died within 8-11
days of rickettsial inoculation. BCG-treated C3H/HeN mice, however,
survived rickettsial challenges of up to 10^4 LD_{50}. In contrast,
C3H/HeJ mice, the strain that cannot respond to in vivo BCG treatment with activated tumoricidal macrophages, were not protected
against even low numbers (10^1 LD_{50}) of rickettsiae. Treatment
of mice with C. parvum gave similar results. C3H/HeN mice survived
rickettsial challenge, while C3H/HeJ mice succumbed to rickettsial
infection.

Several lines of evidence now suggest a role for activated
macrophages in both natural resistance and acquired immunity to
experimental scrub typhus infection: (a) macrophages from mice
naturally resistant to infection were minimally and transiently
infected with rickettsiae throughout the acute phase of disease,
(b) activated macrophages, cells with enhanced nonspecific microbicidal and tumoricidal activities, developed during regressing
scrub typhus infections, (c) protection from lethal rickettsial
infection could be induced by pretreatment of certain mouse strains
with nonspecific macrophage activating agents, such as BCG and C.
parvum, and (d) macrophages activated in vivo or in vitro had
enhanced resistance to infection with rickettsiae and enhanced

intracellular killing of the organisms in vitro (1,6,12,23).
Nevertheless, immunity to experimental scrub typhus infection is
very complex, and efforts to determine the gene product(s) or cell
type(s) that are influenced by the single gene that confers
resistance to this infection have been largely unsuccessful (6,24).

Several mouse strains have been identified with macrophage
defects in response to activation signals in vivo and in vitro for
nonspecific tumor cytotoxicity (Table 5): (a) strains derived from
the A mice, (b) strains with the defective Lps gene, Lps^d, on either
the C3H or C57BL backgrounds, and (c) the P strain mice (25).
Susceptibility patterns of inbred mice to intraperitoneal inoculation of Gilliam strain, R. tsutsugamushi suggested that genetic
control of natural resistance to this rickettsial infection was not
influenced by genes that control macrophage activation for tumor
cytotoxicity (3,25,26). Mouse strain susceptibility to R. akari
(etiologic agent of rickettsialpox), however, was very similar to
that observed for defective development of macrophage tumoricidal
activities (Table 6, see Table 5). Five of six strains susceptible

Table 5. Response of macrophages from inbred mouse strains to
          lymphokines in vitro or macrophage activating agents
          in vivo for development of nonspecific tumor cytotoxicity

Responsive	Variable	Nonresponsive
C3H/HeN	BALB/cAnN	A/J
AKR/N	A/WySnJ	A/HeJ
CBA/CaHN	AL/N	A/HeN
CBA/N	RIII/AnN	
C3HeB/FeJ		C3H/HeJ
C57L/N		C57BL/10ScCR
C57BL/6N		C57BL/10ScN
C57BL/10SnJ		
C57BL/10J		P/J
DBA/1JN		P/JN
DBA/2N		
NZB/N		
NZW/N		
NIH Swiss (outbred)		
B10.A/SgSnJ		

Table 6.   Infection of inbred mouse strains with R. akari.

Resistant	Susceptible
AL/N	* A/HeN
BALB/cN	* A/J
CBA/N	
C3H/HeN	* C3H/HeJ
C57BL/6J	* C57BL/10ScN
C57BL/10J	* C57BL/10ScCR
C57L/J	
DBA/2J	NZW/N
* P/J	

Mice were inoculated intraperitoneally with graded doses of R. akari, ranging from 10^1 to 10^5 LD_{50}. Mouse strains that survived a challenge greater than 10^3 LD_{50} dose were designated resistant.

* Mouse strains with characterized macrophage defects for development of nonspecific tumor cytotoxicity.

to R. akari infection had defective macrophage functions (A/J, A/HeJ, C3H/HeJ, C57BL/10ScN, C57BL/10ScCR). These strains were 1000-10,000 fold more susceptible to the lethal effects of R. akari than resistant strains (27). These results suggested that the activated macrophage may play a determinative role in yet another host-rickettsial interaction.

To analyze the apparent correlation between macrophage function and resistance to R. akari infection, we inoculated mice with 1000 LD_{50} R. akari and harvested macrophages from resistant (BALB/ cN, C3H/HeN, C57BL/10J) and susceptible (A/J, C3H/HeJ, C57BL/10ScN) mice at various days after rickettsial inoculation (27). Peritoneal macrophages from susceptible mice contained intracellular rickettsia as early as 2 days after infection. Levels of infected macrophages in peritoneal cells harvested from these animals were too high to analyze development of microbicidal macrophages by further in vitro exposure to rickettsiae. Macrophages from susceptible mice were not, however, activated for nonspecific tumor cytotoxicity at any time (Fig. 9). Susceptible mice died of rickettsial infection on days 7-9. In contrast, macrophages from resistant mice developed significant levels of microbicidal activity against rickettsiae and nonspecific tumoricidal activity 2-4 days after inoculation of rickettsiae (Fig. 9). Activated microbicidal and tumoricidal

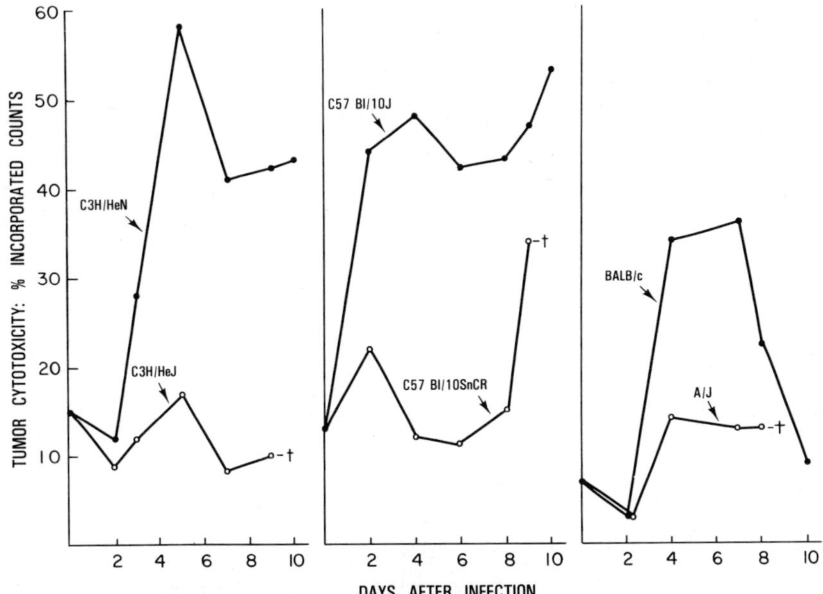

Figure 9.   Development of nonspecifically activated macrophages
after rickettsial inoculation of resistant and susceptible mouse strains.   Mice were inoculated with 1000 LD_{50}
R. akari and macrophages were harvested from infected
mice 1-10 days after rickettsial inoculation.   Cells
were incubated with ^{3}H-thymidine prelabeled mKSA-TU5
tumor target cells for 48 hr (9).   Results are expressed
as percent release of total nuclear label.

macrophages were present throughout the acture phase of the infection
(3-9 days).   These mice survived R. akari infection.

Although we could not analyze microbicidal activities of macrophages from R. akari-infected susceptible mouse strains, we could
treat resident peritoneal macrophages from these strains with
lymphokines and determine the capacity of cells to become activated
for microbicidal activities in vitro (Table 7).   Lymphokine-
treated macrophages from resistant mice developed increased
resistance to R. akari infection (quantified by treatment of
macrophages before infection and determination of percent infected
cells at 1 hr) and increased intracellular killing of rickettsiae
(quantified by treatment of infected macrophages and determination
of number of infected cells at 24 hr).   In contrast, lymphokine-
treated macrophages from mice susceptible to infection showed no
increased resistance to infection, and levels of intracellular
killing were 1/3 to 1/10 that by cells from resistant strains.
Macrophages from susceptible mice treated in vivo with the macrophage activating agent, BCG, were also not rickettsiacidal in

Table 7. Microbicidal activity of macrophages treated <u>in vitro</u>
with lymphokines.

Mouse Strain	Microbicidal Activity of Macrophages Treated with Lymphokines	
	1 hr	24 hr
C3H/HeN	25	39
C3H/HeJ	0	11
C57BL/10J	35	55
C57BL/10ScCr	0	20
BALB/cN	29	70
A/J	0	8

Macrophages were treated with medium or 1/10 lymphokines for 4 hr,
exposed to 5 PFU <u>R</u>. <u>akari</u>, and sampled at 1 hr or treated with
medium or lymphokines for 24 hr.

<u>vitro</u> (Table 9). Again, increased resistance to infection was
absent in these cells, and levels of intracellular killing were
a fraction of levels attained in activated macrophages from
resistant strains of mice. BCG treatment did not protect
susceptible C3H/HeJ mice from the lethal effects of <u>R</u>. <u>akari</u>
infection.

     That activated macrophages play an important role in resolution
of rickettsial disease is suggested by studies with both <u>R</u>. <u>tsutsu-</u>
<u>gamushi</u> and <u>R</u>. <u>akari</u>. Moreover, these studies support the concept
that macrophage activation for microbicidal and tumoricidal
activities occurs simultaneously, and may be under regulation
of the same, or closely linked genes. We tested this hypothesis
by analyzing the genetic regulation of these two macrophage
effector functions. F_1, F_2 and backcross hybrids of a susceptible
(A/J) and a resistant (B10.A) strain of mice were either infected
with 1000 LD_{50} <u>R</u>. <u>akari</u> (Table 9) or sacrificed, and their peritoneal macrophages were analyzed for development of tumoricidal
activity with lymphokine treatment <u>in vitro</u>. Patterns of resistance
to infection in these hybrid mice were consistant with a single
dominant autosomal gene. It was interesting to note that
regulation of macrophage activation for nonspecific tumoricidal
activity in AxB10.A mice also suggested single gene control. To
determine if the genes that regulate resistance to infection and
development of tumoricidal macrophages were in fact the same, we
infected AxB10.A, F_1, F_2, and backcross hybrids with 1000 LD_{50}

Table 8. Microbicidal and tumoricidal activities of macrophages
          from mice treated with BCG.

| Mouse Strain | Effector Functions of BCG-Treated Macrophages | | |
| | Microbicidal Activity at | | Tumoricidal Activity |
	1 hr	24 hr	
BALB/cN	30	72	55
C3H/HeN	32	89	40
C57BL/10J	35	76	46
A/J	0	32	11
C3H/HeJ	0	28	14
C57BL/10ScCR	2	14	8

Mice were inoculated with 10^6 viable BCG for 8 days; macrophages
were harvested from BCG-treated and control untreated mice,
infected with 5 PFU R. akari for 1 hr, and sampled immediately or
24 hr later. Tumor cytotoxicity was determined by release of
^{3}H-thymidine from prelabeled tumor target cells, mKSA-TU5 (9).

Table 9. Summary of R. akari genetics

| Mouse Strain | Survivors/Total | Percent Survival | |
		Observed	Expected
A/J	4/35	11	0
B10.A	23/23	100	100
(B10.A x A)F_1	44/47	93	100
F_2	50/74	67	75
F_1 x A backcross	83/175	47	50
F_1 x B10.A backcross	50/50	100	100

Mice were inoculated with 10^5 PFU R. akari intraperitoneally;
deaths were recorded for 30 days. All susceptible mice died of
rickettsial disease days 8-15 after inoculation of rickettsiae.

<u>R</u>. <u>akari</u>, and tested macrophages from individual mice that survived
rickettsial infection for development of nonspecific tumoricidal
activity following lymphokine treatment <u>in</u> <u>vitro</u>. Twenty-five
percent of macrophages from surviving mice did not develop tumoricidal activity <u>in</u> <u>vitro</u>, suggesting that two different genes
regulate resistance to <u>R</u>. <u>akari</u> infection and development of
activated tumoricidal macrophages.

There may be a molecular basis for the <u>in</u> <u>vivo</u> dissociation of
macrophage effector functions since activity in lymphokine supernatants for tumor cytotoxicity eluted from Sephadex G-200 in a
single peak of 45,000 MW, but activity for intracellular killing
of rickettsiae eluted in three separate peaks of 130,000, 45,000,
and ≤ 10,000 daltons. Microbicidal and tumoricidal activities
that develop simultaneously may reflect the ability of a single
species of lymphokine (the 45,000 MW molecule) to induce both
effector activities. Analysis of the contribution of the other
two intracellular killing lymphokines may provide important
insights into the regulation of resistance to rickettsial disease.

ACKNOWLEDGMENT

In conducting research described in this report, the investi-
gators adhered to the Guide for Laboratory Animal Facilities and
Care as promulgated by the Committee of the Guide for Laboratory
Animals and Care of the Institute of Laboratory Animal Resources,
National Academy of Sciences, National Research Council.

REFERENCES

1. Nacy, C.A. and Osterman, J.V. (1979) Infect. Immun. 26:744.
2. Rikihisa, Y. and Ito, S. (1980) Infect. Immun. 30:231.
3. Groves, M.G. and Osterman, J.V. (1978) Infect. Immun. 19:583.
4. Anderson, G.W., Jr., and Osterman, J.V. (1980) Infect.
        Immun. 28:132.
5. Catanzaro, P.J., Shirai, A., Agniel, L.D. and Osterman, J.V.
        (1977) Infect. Immun. 18:118.
6. Nacy, C.A. and Groves, M.G. (1981) Infect. Immun. 31:1239.
7. Catanzaro, P.J., Shirai, A., Angiel, L.D., Jr. and Osterman,
        J.V. (1976) Infect. Immun. 13:861.
8. Anderson, S.E., Jr., and Remington, J.S. (1974) J. Exp. Med.
        139:1154.
9. Mauel, J., Buchmuller, Y. and Behin, R. (1978) J. Exp. Med.
        148:393.
10. Noguera, N. and Cohn, Z.A. (1978) J. Exp. Med. 148:288.
11. Ruco, L.P. and Meltzer, M.S. (1977) J. Immunol. 119:889.
12. Nacy, C.A. and Meltzer, M.S. (1979) J. Immunol. 123:2544.
13. Cohn, Z.A., Bozeman, F.M., Campbell, J.M., Humphries, J.W.
        and Sawyer, T.K. (1959) J.Exp. Med. 109:271.

14. Nacy, C.A. (1980) in Manual of Macrophage Methodology: Collec-
         tion, Characterization, and Function. Marcel Dekker, Inc.,
         New York, p. 268.
15. Ruco, L.P. and Meltzer, M.S. (1977) J. Immunol. 119:889.
16. Oaks, S.C., Jr., Osterman, J.V. and Hetrick, F.M. (1977)
         J. Clin. Microbiol. 6:76.
17. Nacy, C.A., Leonard, E.J. and Meltzer, M.S. (1981) J. Immunol.
         126:204.
18. Leonard, E.J., Ruco, L.P. and Meltzer, M.S. (1978) Cell.
         Immunol. 41:347.
19. Watson, J., Riblet, R. and Taylor, B.A. (1977) J. Immunol.
         118:2088.
20. Ruco, L.P., Meltzer, M.S. and Rosenstreich, D.L. (1978) J.
         Immunol. 121:543.
21. Ruco, L.P. and Meltzer, M.S. (1978) J. Immunol. 120:329.
22. Ruco, L.P. and Meltzer, M.S. (1978) Cell.Immunol. 41:347.
23. Nacy, C.A., Leonard, E.J. and Meltzer, M.S. (1981) in Proc.
         Int. Congress on Phagocytosis: Metchnikoff Centenial
         (in press).
24. Jerrells, T.R. and Osterman, J.V. (1981) Infect. Immun. 31:
         956.
25. Boraschi, D. and Meltzer, M.S. (1979) Cell. Immunol. 45:188.
26. Groves, M.G., Rosenstreich, D.L., Taylor, B.T. and Osterman,
         J.V. (1980) J. Immunol. 125:1395.
27. Meltzer, M.S. and Nacy, C.A. (1980) Cell. Immunol. 54:378.
28. Nacy, C.A. and Meltzer, M.S. (Infect. Immun., accepted, June
         Issue).

PARAMETERS OF CELLULAR IMMUNITY IN ACUTE AND CHRONIC RICKETTSIA

TSUTSUGAMUSHI INFECTIONS OF INBRED MICE

T. R. Jerrells and J. V. Osterman

Department of Rickettsial Diseases, Walter Reed Army
Institute of Research, Washington, D.C.   20012

INTRODUCTION

The role of cell-mediated immunity (CMI) has been shown to be
a major factor in acquired resistance to the obligate intracellular
parasite R. tsutsugamushi. Specifically, it has been shown that
immune animals possess spleen cells with the characteristics of
thymus-derived lymphocytes (T-Cells) which confer protection against
subsequent challenge when transferred to naive mice (9). The
interaction of antigens with specifically sensitized T-cells can
be demonstrated in vivo by a delayed-typed hypersensitivity (DTH)
response (6), or in vitro by a specific antigen induced lymphocyte
proliferative (LP) response (7).

The present study was designed to assess the development of
DTH and LP responses in inbred mice as a result of R. tsutsugamushi
infection. The severity of infection was modulated by selection
of inbred strains of mice known to be genetically susceptible
or resistant to intraperitoneal infection (3) and utilization of
alternative routes of inoculation known to elicit either a progressive, lethal infection (intraperitoneal) or a chronic,
immunizing infection (subcutaneous) in susceptible mice (2).

MATERIALS AND METHODS

Animals and rickettsiae. C3H/HeDub, BALB/c Dub, and C3H/RV mice
were obtained from Flow Laboratories (Dublin, Va.). Propagation
and storage of rickettsiae have been described previously (3).

<u>Delayed-Type Hypersensitivity Testing</u>. Mice were tested for DTH responses using the method of Lefford (5). Animals were pre-treated with 1 µci of tritiated thymidine 24 hrs. prior to antigen inoculation intradermally in the ear. The mice were sacrificed by CO_2 asphyxiation 24 hrs later and a circular tissue specimen was removed from each ear. After digestion, counts per minute (CPM) and disintegrations per minute (DPM) were obtained by scintillation counting. The data were expressed as ear ratios, and calculated by dividing the DPM of the left ear by the DPM of the right ear.

<u>Lymphocyte Proliferation Assay</u>. The ability of splenic lymphocytes to proliferate in response to rickettsial antigens was assayed using a standard microculture lymphocyte proliferation assay as previously described (3).

<u>Cell Transfer Studies</u>. Transfer of DTH by spleen cells was assessed by the method described by Youdim (11). Recipient animals were given 1.0 ml (50 x 10^6) of the cell suspension intravenously (i.v.). In further experiments, immune and normal spleen cells were passed over nylon wool columns according to the procedure of Julius (4) and 50 x 10^6 nonadherent cells, presumably T-cells, were given i.v. To deplete lymphocytes with Thy 1.2 antigen on their surface, immune spleen cells were treated with anti-Thy 1.2 serum and rabbit complement. Recipient animals were given 1.0 ml (50 x 10^6) of the resulting cells. All recipients were tested within 2 h of cell transfer.

<u>Detection of Suppressor Cells</u>. Spleen cells from mice immunized with Keyhole Limpet Hemocyanin (KLH) were mixed with spleen cells from mice infected 9 days earlier with Gilliam rickettsiae or with spleen cells from noninfected mice. These mixtures, as well as KLH immune cells and normal cells, were injected i.v. into normal mice which were ear tested with KLH antigen.

RESULTS

    C3H/HeDub mice inoculated s.c. with 1,000 MLD_{50} of Gilliam were evaluated at various times after infection for their DTH response, as well as acquired resistance to secondary challenge (Table 1). In these experiments animals became resistant to rechallenge between 14 and 21 days after rickettsial inoculation and DTH reactivity also reached peak levels at about the same time. Interestingly, the DTH response declined 21 days after initial infection, although the mice remained completely immune to re-challenge. It should be noted that although the DTH reactivity had declined by day 28 these mice had a significantly greater response than unimmunized animals. When these mice were tested for DTH reactivity 3 and 7 days after a secondary i.p. challenge

Table 1.   Development of delayed-type hypersensitivity, acquired
           resistance, and lymphocyte proliferative responses in
           mice immunized with viable R. tsutsugamushi.

Day Post s.c. Immunization	DTH Response	Percent Survival	nCP[a]	SI[b]
3	1.08 ± .06	0	9,0C9 ± 554	1.6
7	1.27 ± .08	40	16,229 ± 2,289	2.9
14	1.41 ± .03	90	10,311 ± 1,823	4.1
21	1.56 ± .05	100	11,878 ± 802	3.5
28	1.37 ± .08	100	49,406 ± 3,428	8.2
45	Not Done	100	68,869 ± 3,533	21.4

[a] Experimental Cpm – Control Cpm

[b] Stimulation Index = Experimental Cpm/Control Cpm

a rapid and elevated response was noted (data not presented)
suggesting a specific memory response. Noninfected mice tested
in parallel did not demonstrate a response to the rickettsial
antigens used in this study. The development of a LP response
was also examined following s.c. infection. The LP response of
BALB/c mice following s.c. inoculation with Karp rickettsiae
differed from the DTH response in that there was a continuous
increase in response throughout the 45 day period of animal
observation.

In experiments designed to determine the nature of the cell
population responding to rickettsial antigens, spleen cells from
immune mice 21 days post primary infection, T-cell enriched spleen
cells (nylon wool passed), and T-cell depleted spleen cells (Anti-
Thy 1.2 treated) were transferred to naive animals which were
subsequently tested for their DTH response. It was found (Table 2)
that the transfer of DTH activity was accomplished by immune
spleen cells and T-cell enriched spleen cells but the DTH response
was abrogated by complement-mediated lysis of spleen cells with
anti-Thy 1.2 serum.

The development of DTH reactivity during acute and chronic
infection was examined using congenic strains of C3H mice which
are sensitive (C3H/HeDub) or resistant (C3H/RV) to acute lethal
infection with Gilliam rickettsiae. Both strains of mice were
found to evidence a DTH response within 3-5 days postinfection
but the reactivity remained relatively high in the resistant
C3H/RV mice compared with a drastic decline in the C3H/HeDub
mice (Table 3).

Table 2. Nature of cell populations mediating passive transfer
          of DTH to rickettsial antigens.

Cell Transferred	Ear Ratio ± SEM
NonImmune spleen cells	0.93 ± 0.05
Immune spleen cells	1.41 ± 0.09
Nylon wool passed	1.59 ± 0.15
Immune spleen cells + C'	1.30 ± 0.06
Immune spleen cells + anti-Thy 1.2 + C'	0.96 ± 0.08

Table 3. DTH responses of C3H mice infected intraperitoneally
          with R. tsutsugamushi Gilliam.

Day Post-infection	Ear Ratio	
	C3H/RV	C3H/HeDub
1	1.01	1.06
3	1.03	1.18
7	1.19	1.29
7	1.32	1.05
9	1.20	1.01
11	1.22	–
15	1.18	–

Table 4. Suppression of transfer of DTH reactivity to KLH by
          spleen cells from R. tsutsugamushi infected mice.

Cell transferred	Ear Ratio ± SEM[a]
10×10^7 KLH Immune	1.28 ± .06
5×10^7 Immune + 5×10^7 Normal	1.29 ± .08
5×10^7 KLH Immune + 5×10^7 Infected	1.06 ± .04
10×10^7 Normal	1.05 ± .03

[a] Response to KLH antigen.

In experiments designed to further investigate the mechanism(s)
of the anergy noted in acute infections, spleen cells from infected
mice were mixed with KLH immune spleen cells and transferred to
recipient animals. It was found (Table 4) that spleen cells from
infected mice completely abrogated the transfer of KLH reactivity.

DISCUSSION

In this study we have demonstrated that mice infected with R. tsutsugamushi develop a protective immunity, and that the development of this immunity was paralleled by the development of DTH activity and LP responses. Although these two parameters initially developed at about the same rate, a striking lack of correlation was noted after 28 days postinfection. Although the DTH activity waned, the LP responses continued to increase through 45 days postinfection and remained at this level through 90 days postinfection (data not shown). In this system, both assays have been shown to be mediated by T-cells, but the DTH response is apparently under more complex control mechanisms than in vitro LP responses. It has been shown that s.c. infection of mice with viable R. tsutsugamushi results in a chronic infection with a long lasting rickettsemia (Groves and Osterman, unpublished observation). It is possible that the rickettsiae are in the form of antigen-antibody complexes which are either not stimulatory for DTH effector cells or act to suppress the expression of DTH effector cells. It has also been shown that chronic microbial infections lead to the development of suppressor cells which affect a number of aspects of CMI (1,8,10). In this system, if suppressor cells develop they would have to be specific for DTH effector cells as the cells responding in culture do not appear to be suppressed.

In acute infections of susceptible mice (C3H/HeDub) the DTH reactivity waned to unreactive levels prior to death of the animals. It is possible that the infection of lymphocytes and macrophages (3), or the appearance of suppressor cells may be responsible for this decline. To partly address this question, we have performed cell transfer experiments which demonstrated that a population of cells was present which was capable of suppressing DTH activity to an unrelated antigen, suggesting that the decline of DTH activity in acute infections was due at least in part to the appearance of suppressor cells.

We have shown that both in vivo and in vitro parameters of CMI initially correlate with protective immunity and should provide rapid and reliable methods to assess the immune status of a host prior to and following immunization. Further, these assays also provide systems which will allow study of immunoregulatory mechanisms involved in the development of chronic infections as well as the pathogenesis of acute, lethal infections.

<u>REFERENCES</u>

1. Collins, F.M. and Watson, S.R. (1979). Suppressor T-cells in BCG infected mice. Infect. Immun. 25:491-496.

2. Groves, M.G. and Osterman, J.V. (1978). Host defenses in
       experimental scrub typhus: Genetics of natural
       resistance to infection. Infect. Immun. 19:583-588.
3. Jerrells, T.R. and Osterman, J.V. (1981). Host defenses in
       experimental scrub typhus: Inflammatory response of
       congenic C3H mice differing at the ric gene. Infect.
       Immun. 31:1014-1022.
4. Julius, M.H., Simpson, E. and Herzenberg, L.A. (1973). A
       rapid method for the isolation of functional thymus-
       derived murine lymphocytes. Eur. J. Immunol. 3:645-649.
5. Lefford, M.J. (1974). The measurement of tubercular hyper-
       sensitivity in rats. Int. Arch. 47:570-585.
6. Mackaness, G.B. (1967). The relationship of delayed hyper-
       sensitivity to acquired cellular resistance. Br. Med.
       Bull. 23:52-54.
7. Patel, P.J. and Lefford, M.J. (1978). Antigen specific:
       lymphocyte transformation, delayed hypersensitivity and
       protective immunity. I. Kinetics of the response. Cell.
       Immunol. 37:315-326.
8. Riglar, C. and Cheers, C. (1980). Macrophage-activation during
       experimental murine Brucellosis, II. Inhibition of
       in vitro lymphocyte proliferation by Brucella-activated
       macrophages. Cell. Immunol. 49:154-167.
9. Shirai, A., Cantanzaro, P.J., Phillips, S.M. and Osterman,
       J.V. (1976). Host defenses in experimental scrub
       typhus: Role of cellular immunity in heterlogous
       protection. Infect. Immun. 14:39-46.
10. Tamura, S., Kojima, A. and Egashira, Y. (1980). Regulatory
       mechanism of delayed-type hypersensitivity in mice.
       II. Effect of suppressor cells on the development of
       memory cells for delayed-type hypersensitivity. Cell
       Immunol. 51:250-261.
11. Youdim, S., Stutman, O. and Good, R.A. (1973). Studies of
       delayed hypersensitivity to L. monocytogenes in mice:
       Nature of cells involved in passive transfers. Cell.
       Immunol. 6:98-109.

LYMPHOKINE STIMULATED MACROPHAGES INHIBIT INTRACELLULAR <u>CHLAMYDIA</u>

<u>PSITTACI</u> REPLICATION BY MECHANISMS DISTINCT FROM INTRACELLULAR

INHIBITION OF <u>TOXOPLASMA</u> <u>GONDII</u> REPLICATION

Gerald I. Byrne and Henry W. Murray

Cornell University Medical College
New York, NY 10021

<u>Chlamydia</u> and <u>Toxoplasma</u> are obligate intracellular parasites
that replicate within macrophages and other eucaryotic host cells.
Although chlamydia are procaryotes and <u>Toxoplasma</u> are eucaryotes,
these organisms have several features in common during their
intracellular growth and development. Each of these parasites is
taken into the host cell by an endocytic mechanism that resembles
phagocytosis (1,7), and each subverts normal phagocytic function by
inhibiting fusion of host cell lysosomes with parasite-containing
vacuoles (3,5). The mechanisms involved in this inhibition are
not known, although Eissenberg and Wyrick (2) have reported that
inhibition of fusion in peritoneal macrophages challenged with
<u>C. psittaci</u> and either <u>Saccharomyces</u> <u>ceriviciae</u> or <u>Escherichia</u>
<u>coli</u> was restricted to chlamydiae-containing vesicles. Chlamydiae
and <u>Toxoplasma</u> remain within membrane bound vesicles during their
entire intracellular development. <u>Toxoplasma</u> replicate by a
process called endodyogeny. Chlamydiae first differentiate from
a metabolically inactive infective form called an elementary body
to an intracellular reticulate body, then grow and divide by binary
fission forming a microscopically visible inclusion. At some point
the intracellular microcolony receives a signal that results in a
second round of differentiation and a new population of elementary
bodies are released from lysed host cells.

Mechanisms whereby the host limits intracellular growth are
not understood. Natural chlamydial infections generally do not
result in overt, fulminate parasite replication, but rather, are
characterized by a low grade persistence, latency or chronicity.
This is also often true for <u>Toxoplasma</u> infections in immunologically
mature hosts.

Activation of macrophages _in vitro_ by lymphocyte products is a well characterized aspect of the normal immune response. Lymphokine-activated macrophages have been shown to kill or inhibit replication of a variety of obligate intracellular parasites (4,6, 10,11). Similar mechanisms have been postulated to occur _in vivo_ in response to these infections. Extracellular _Toxoplasma_ are sensitive to toxic intermediates of oxygen metabolism (8) and inhibition of _Toxoplasma_ replication by lymphokine-activated macrophages is thought to involve superoxide radical (O_2^-) and hydrogen peroxide (H_2O_2) (9). We report here that peritoneal macrophages activated _in vitro_ by lymphokines inhibit chlamydial replication. However, in contrast to the role of O_2^- and H_2O_2 in the lymphokine-induced inhibition of intracellular replication of _T. gondii_, impairing the oxidative activity of activated macrophages had no inhibitory effect on chlamydial growth.

Lymphokines (LK) were prepared by incubating spleen cells (5×10^6/ml) from _C. psittaci_-immune A/J mice with 5 µg of concanavalin A (ConA) per ml of culture medium for 24 h at 37°C. For lymphocyte proliferation experiments 5×10^5 spleen cells in 0.2ml were incubated in 96-well Linbro plates for 1 to 5 days in the presence or absence of ConA. One µCi of ^{3}H-thymidine was added during the last 6h of incubation. Precipitable counts were collected with a cell harvester. Peritoneal macrophages (MAC's), elicited with thiogly-collate or stimulated _in vitro_ with heart infusion broth (HIB) (8mg/ml) were plated onto coverslips (4×10^5 cells per slip) and washed free of the non-adherent populations after 1 h at 37°C. Cells were then incubated in 10% LK-containing or control medium for 24 h prior to further treatments or infection. Impairment of oxidative activity in LK-treated MAC's was done in one of 3 ways. Cells were incubated for 4 h in glucose-free medium, for 4 h in medium containing 2 mg of catalase per ml, or for 1 h in medium containing 200 mg of phorbol myristate acetate (PMA) per ml prior and during the initial hour of infection. _Chlamydia_ (one ID_{50}) or _Toxoplasma_ (10^6 parasites) were added to the macrophage cultures for 1 h at 37°C, then washed off. Infected cells were incubated for 17 h at 37°C, fixed, stained and examined microscopically for evidence of parasite growth. Chlamydial replication is expressed as the product of the percent inclusion-containing MAC's and the average number of MAC's per 630X microscope field (Replication Index). _Toxoplasma_ replication is expressed by the number of parasites per infected MAC vacuole. Chlamydial uptake studies were done by incubating ^{3}H-labelled parasites (10 ID_{50}) with 3×10^6 MAC's in 60 mm diameter culture dishes. Cells were collected by scraping at specific intervals after infection, subjected to several rounds of differential centrifugation and MAC-associated counts were measured in a liquid scintillation spectrophotometer.

Replication of _C. psittaci_ was inhibited when thioglycollate elicited or resident macrophages were incubated in 5 to 20 percent

LK-containing supernatant fluids. Eight mg per ml HIB was required
for activation of resident macrophages by LK. Inhibition of C.
psittaci was always more pronounced in thioglycollate elicited,
LK-activated macrophages. LK parasite-inhibitory activity was
detected in spleen cell ConA supernatant fluids before lymphocyte
proliferation reached a maximum and therefore was not temporally
related to T cell blastogenesis LK. Activation of thioglycollate
elicited macrophages did not affect endocytosis of the parasite
as evidenced by similar uptake kinetics of radioisotopically
labelled preparations of C. psittaci. Inhibition of Chlamydial
growth in LK-activated macrophages was reversible after removal
of LK. Evidence of parasite replication was detected 24 to 34 h
after washing LK-treated cells and infecting them with one ID_{50}
of C. psittaci. Evidence for parasite replication is normally
observed 14 to 18 h after infection. Therefore, the replicative
cycle was interrupted or delayed for 6 to 10 h when macrophages
were pretreated with LK prior to infection. Inhibition of O_2^- and
H_2O_2 had no effect on LK-induced inhibition of chlamydial replication. This result is in direct contrast to results obtained when
glucose deprivation, addition of exogenous catalase or pretreatment
with PMA are employed prior to an during T. gondii infection
(Table 1). These data are interpreted to indicate that toxic
intermediates of oxygen metabolism are important for inhibition of
T. gondii but not C. psittaci replication in macrophages pretreated
with LK prior to infection. The results suggest that diverse
molecular mechanisms may underlie the macrophage antimicrobial
activity induced by LK.

Table I. Effect of Lymphokines and Inhibitors of O_2^- and H_2O_2
          Production on the Replication of Toxoplasma and Chlamydia
          in Macrophages

Host Cell Treatment Prior to Infection	Effect on Intracellular Parasite Replication	
	Chlamydia	Toxoplasma
None	No inhibition	No inhibition
Lymphokine	Inhibition	Inhibition
Lymphokine, then glucose deprivation	No reversal of inhibition	Reversal
Lymphokine, then exo-genous catalase	No reversal of inhibition	Reversal
Lymphokine, then PMA pretreatment	No reversal of inhibition	Reversal

ACKNOWLEDGMENT

    This work was supported by Public Health Service Grant AI16459
from the National Institute of Allergy and Infectious Disease.

REFERENCES

1. Byrne, G.I. and Moulder, J.W. (1978). Parasite-specified
       phagocytosis of Chlamydia psittaci and Chlamydia
       trachomatis by L and Hela cells. Infect. Immun. 19:598-
       606.
2. Eissenberg, L.G. and Wyrick, P.B. (1981). Inhibition of
       phagolysosome fusion is localized to Chlamydia psittaci-
       laden vacuoles. Infect. Immun. 32:889-896.
3. Fries, R.P. (1972). Interaction of L cells and Chlamydia
       psittaci: entry of the parasite and host response to its
       development. J. Bacteriol. 110:706-721.
4. Hinricks, D.J. and Jerrells, T.R. (1976). In vitro evaluation
       of immunity to Coxiella burnetii. J. Immunol. 117:996-1003.
5. Jones, T.C. and Hirsch, J.G. (1972). The interaction between
       Toxoplasma gondii and mammalian cells. II. The absence of
       lysosomal fusion with phagocytic vacuoles containing
       living parasites. J. Exp. Med. 136:1173-1194.
6. Jones. T.C., Len, L. and Hirsch, J.G. (1975). Assessment of
       in vitro immunity against Toxoplasma gondii. J. Exp. Med.
       141:466-482.
7. Jones, T.C., Yeh, S. and Hirsch, J.G. (1972). The interaction
       between Toxoplasma gondii and mammalian cells. I. Mechanism
       of entry and intracellular fate of the parasite. J. Exp.
       Med. 136:1157-1173.
8. Murray, H.W. and Cohn, Z.A. (1979). Macrophage oxygen-
       dependent antimicrobial activity. I. Susceptibility of
       Toxoplasma gondii to oxygen intermediates. J. Exp. Med.
       150:938-949.
9. Murray, H.W., Juangbhanich, C.W., Nathan, C.F. and Cohn, Z.A.
       (1979). Macrophage oxygen dependent antimicrobial
       activity. II. The role of oxygen intermediates. J. Exp.
       Med. 150:950-964.
10. Nacy, C.A. and Meltzer, M.S. (1979). Macrophages in resistance
       to rickettsial infection: Macrophage activation in vitro
       for killing Rickettsia tsutsugamushi. J. Immunol. 123:
       2544-2549.
11. Nogueira, N., Gordon, S. and Cohn, Z. (1977). Trypanosoma
       cruzi: The immunological induction of macrophage
       plasminogen activator requires thymus-derived lymphocytes.
       J. Exp. Med. 148:172-183.

NATURAL AND ACQUIRED RESISTANCE TO <u>TRYPANOSOMA</u> <u>CRUZI</u>

Thomas M. Trischmann

Tropical Medicine Center, The Johns Hopkins University
School of Hygiene and Public Health, Baltimore, MD

INTRODUCTION

Since the identification in 1909 of <u>Trypanosoma</u> <u>cruzi</u> as the
causative agent of Chagas' disease, a large body of literature has
arisen relating to the nature of the disease. Two observations in
particular initially stimulated my interest in the immunology of
Chagas' disease. The first was that the severity of the disease
that occurs following the initial infection with <u>T</u>. <u>cruzi</u> (called
the acute stage of the disease) can range from asymptomatic to
lethal (1). This variation in what can be termed natural resistance
may be the result of multiple genetic and environmental factors,
but it is easy to imagine how differences in the immune responses
made by individuals could be critical for influencing the ensuing
course of the disease.

The second observation was that an individual experiences an
acute stage of the disease only one time. This acquired resistance
has been attributed to the development of a protective immunity
(2) which is obviously quite effective once induced. As in many
parasitic diseases, premunition is found to occur, i.e., after an
initial high parasitemia, parasites in the blood and tissues decline
to almost undetectable levels but still persist despite the onset
of immunity. The immune mechanisms responsible for the clearance
and destruction of parasites are unknown as are the tricks used
by the parasite to evade the immune response of the host.

Fortunately for the immunologist, the mouse appears to provide
a model for the acute stage of Chagas' disease. Inbred strains of
mice have been reported to differ in their susceptibility to <u>T</u>.
<u>cruzi</u> (3) and even susceptible strains of mice were found to be

easily immunized and protected against challenge with a variety of
virulent strains of the parasite (4,5). The murine model thus
offers the possibility of studying both natural resistance and
acquired resistance to T. cruzi in a well-defined system.

NATURAL RESISTANCE

   Characterization of mouse strains. Our initial characterization of the resistance of inbred mouse strains (6) was done by
challenging 9 inbred strains of mice with different doses of the
Brazil strain of T. cruzi, which has been well characterized by
Hanson (7,8). A continuum of resistance was found among the mouse
strains ranging from highly susceptible strains, e.g., C3H, which
developed high parasitemias and died, to resistant strains, e.g.,
C57BL/10, which developed low parasitemias and survived. The
absence of clearly separable groups of resistant and susceptible
mice suggested that multiple factors may be contributing to the
genetic basis of resistance to T. cruzi. It has long been recognized that both the sex and age of mice influence their resistance,
female mice being more resistant than males (9) and younger mice
being more susceptible than older mice (10). The biological bases
of these differences have not been determined. We have, therefore,
throughout our studies, primarily used 7-8 week old female mice.

   While our interest was in the contribution of the immune
response to the genetic basis of natural resistance, it was
possible that the basis of resistance was unrelated to any immunological phenomena, but was rooted instead in the inability or
decreased ability of parasites either to invade cells or to multiply
once inside. This possibility was dealt with by challenging
resistant C57BL/10 mice after impairment of their immune responsiveness by X-irradiation, splenectomy, or injection of silica.
Each of these treatments led to high parasitemias, demonstrating
the ability of the parasite to multiply in a resistant strain and
implicating the immune response in the low parasitemias and survival of resistant mice (6). This was even more evident when
athymic Nu/Nu mice with a BALB/c background were challenged.
BALB/c mice display an intermediate level of resistance to T.
cruzi with mice developing a moderate parasitemia that is subsequently cleared. In Nu/Nu mice, the parasitemia is evident much
sooner, it continually increases, and it reaches exceptionally high
levels before death (6). Parasitemias in Nu/Nu mice usually approach
or exceed 10^8/ml while in C3H mice a high parasitemia is in the
range of 10-20 x 10^6/ml of blood.

   The immune response is clearly necessary for survival and plays
a critical role in the elimination of parasites and control of
parasitemia. However, it cannot be inferred that the immune response

Table 1. Survival of female mice derived from crosses of C3H/AN
          and C57BL/10 parental mice challenged with 10^4 trypo-
          mastigotes

C3H/An	C57BL/10
0/31[a] (0%)	67/74 (91%)

F1
67/67 (100%)

F2
174/243 (72%)

Backcross to C3H
49/197 (25%)

[a] Number surviving at 8 weeks/number infected.

is necessarily different between reistant and susceptible strains
or that it forms the genetic basis of natural resistance.

<u>Genetics of natural resistance</u>. Despite the suggestion that
multiple factors influenced resistance, it was possible that a
single genetic locus was of primary importance in governing
resistance. We therefore chose to study the genetics of resistance
to <u>T. cruzi</u> using C3H/An mice as the susceptible parental strain
and C57BL/10 mice as the resistant parental strain. F1, F2, and
backcross mice to the C3H parent were bred and the female offspring
challenged with 10^4 trypomastigotes, the infective form of the
parasite found in the blood of the vertebrate host. The results,
shown in Table 1, gave close to the 75% survival among F2 mice that
would be expected for a single gene, but the survival of the backcross mice was far below the 50% that would be indicative of a
single gene. One reason for the discrepancy appeared when the
influence of the H-2 region was examined.

We had previously found using H-2 congenic strains that
resistance was associated with the strain background rather than
with the H-2 haplotype (6), but the congenic strains used had the
genetic background of either highly susceptible or highly resistant
strains. Thus, a secondary influence of the H-2 region on resistance could have been masked. This indeed turned out to be the
case when H-2 congenic mice were used which had a BALB/c background,
BALB/c mice being of intermediate resistance. A striking influence
of H-2 haplotype was evident (Table 2), with mortality ranging from
14% to 96% among female mice depending on the H-2 locus present.

Table 2. Mortality of H-2 congenic strains with BALB/c backgrounds
          after challenge with 10^4 trypomastigotes

Strain	Male	Female
BALB/c	22/31[a] (71%)	3/21 (14%)
BALB.B	26/26 (100%)	18/37 (49%)
BALB.K	18/18 (100%)	25/26 (96%)

[a] Number dead at 8 weeks/number infected.

Table 3. H-2 distribution among F2 female mice and survival
          after challenge with 10^4 trypomastigotes[a]

	$H-2^k$	$H-2^{k/b}$	$H-2^b$	Total
Number of mice	30 (25%)	65 (53%)	27 (22%)	122
Survived[b]	12 (40%)	52 (80%)	23 (85%)	87 (71%)
Died	18 (60%)	13 (20%	4 (15%)	35 (29%)

[a] F2 mice were derived from crosses of C3H and C57BL/10 parental
mice.

[b] Survival and mortality were determined at 8 weeks after
challenge.

This influence of the H-2 locus was confirmed by H-2 typing of
F2 mice (Table 1) before challenge. Survival among mice of differing H-2 haplotypes is seen in Table 3, and shows again the increased
susceptibility of mice with the $H-2^k$ haplotype compared to the $H-2^b$
and $H-2^{k/b}$ haplotypes. The 72% survival seen for the F2 mice
(Table 1) is thus an average of the widely varying survival rates
of mice with different H-2 haplotypes. The net survival of backcross mice would be even more affected since half of the mice
would carry the $H-2^k$ haplotype as opposed to only one quarter of
the F2 mice.

In studies of the genetics of resistance to a variety of bacterial, rickettsial, and parasitic organisms, the H-2 region has not been found to act as the major determinant of resistance. However, more careful analyses of the role of the H-2 locus have shown an influence of the H-2 complex on resistance to <u>Leishmania donovani</u> (11) and to <u>Toxoplasma gondii</u> (12). One reason for the failure to implicate the H-2 region more often might be that the influence of the H-2 locus is often at a secondary level and, as in our initial work, can be overlooked if the congenic strains examined have backgrounds of high resistance or susceptibility.

The genetics of resistance to <u>T</u>. <u>cruzi</u> was further complicated when recombinant inbred (RI) strains of mice were used to see if a major locus for resistance could be mapped. The BXD RI strains were used which were derived from the parental C57BL/6J and DBA/2J strains (13). DBA/2 mice had initially been found to be susceptible (6) and a large number of BXD RI strains were available. When challenged, the RI strains showed a spectrum of resistance as shown in Table 4, but the distribution of strains did not correlate with any known, mapped trait. While challenging the RI strains, the parental strains were also challenged as controls. Quite surprisingly, the DBA/2J mice proved to be resistant, and it is unknown why the small group initially tested were susceptible, although they had been bred in our animal facility rather than purchased. We therefore had the unusual result of susceptible strains derived from two resistant parental strains.

One possible explanation is that different genes are responsible for the resistance of each strain and that failure to inherit either of these genes leads to susceptibility. While in such a situation both genes could be effecting resistance via the same mechanism, it is clear that an understanding of the genetics of resistance to <u>T</u>. <u>cruzi</u> will become increasingly difficult as more genetic loci capable of influencing resistance are identified. Even if a major locus governing resistance is identified, it will be necessary to show whether the same locus governs resistance in all strains of mice, and whether the strain of <u>T</u>. <u>cruzi</u> is related to the results obtained.

<u>Reciprocal bone marrow transfers</u>. While direct comparisons of the immune responses of resistant and susceptible mice have not yet identified any significant differences, this could be due to the <u>in</u> <u>vitro</u> nature of the assays done with macrophages, to qualitative rather than quantitative differences in the antibody produced, or to the lack of a comparison of cell mediated immunity. The situation is further complicated by a lack of data comparing the developing immune response early in the course of infection.

Table 4. Resistance of BXD recombinant inbred strains after
          challenge with 10^4 trypomastigotes

BXD strain	Mortality[a]	Resistance to *T. cruzi*	BXD strain	Mortality	Resistance to *T. cruzi*
1	0/8	R[b]	18	5/9	I
2	4/10	I	19	8/8	S
5	4/8	I	21	0/6	R
6	1/8	R	22	0/8	R
8	6/7	S	23	7/9	S
9	9/9	S	24	5/9	I
11	7/7	S	25	0/8	R
12	4/6	I	27	1/6	R
13	1/7	R	28	2/6	I
14	5/7	I	29	7/8	S
15	6/8	I	30	2/9	R
16	5/5	S			

[a] Number dead at 8 weeks/number infected.

[b] S, susceptible; I, intermediate; R, resistant.

     Another means of assessing the role of the immune response in
natural resistance is by reciprocal bone marrow transfers between
resistant and susceptible mice. If resistance is related to
differences in the immune responses of mice, then lethal irradiation
of susceptible mice followed by reconstitution with bone marrow from
resistant animals should confer resistance on the susceptible mice
as was found in the case of *Leishmania tropica* infections in mice
(14). If the immune responses of both strains are similar and
natural resistance is unrelated to the immune response, then
resistance and susceptibility of mice would be unaffected by the
source of their bone marrow. Naturally, neither of these clearly
defined situations occurred!

     C3H/HeJ mice were used as the susceptible strain and F1
(C3H/HeJ x C57BL/6J) mice as the resistant strain. Mice were
lethally irradiated (C3H, 900 rads; F1, 960 rads) and reconstituted
several hours later with 2 x 10^7 bone marrow cells which had been

Table 5. Survival of bone marrow reconstituted mice following
          challenge with trypomastigotes[a]

Lethally irradiated recipient mice[b]	Source of bone marrow cells	Challenge dose	Maximum parasitemias $(\times 10^6/\text{ml})$	Mortality[c]
C3H	C3H	2.5×10^3	5.1, 4.2, 3.25 2.9, 2.35, 1.0	6/6
		10^3	14.1, 10.2, 10.1 8.25, 3.85, 3.75 3.1	7/7
F1	F1	2.5×10^3	0.325, 0.325, 0.25 0.25, 0.25, 0.15 0.10	0/7
		10^3	0.225, 0.125, 0.10 0.10, 0.10, 0.075 0.05	0/7
C3H	F1	2.5×10^3	4.8, 2.125, 2.025 2.00, 1.375	4/5
		10^3	1.80, 1.20, 1.15 1.10, 0.825, 0.75 0.65, 0.55	8/8
F1	C3H	2.5×10^3	11.9, 9.4, 8.5 4.05, 2.90	5/5
		10^3	14.9, 10.9, 7.70 6.40, 2.25, 1.55	6/6

[a] Mice were reconstituted with 2×10^7 bone marrow cells depleted
of mature T-cells and challenged 3 months later.

[b] C3H mice received 900 rads; F1 mice received 960 rads.

[c] Number dead at 8 weeks/number infected.

treated with anti-Thy 1.2 plus complement to remove mature T-cells.
Mice were challenged 3 months after reconstitution.

The results in Table 5 show that C3H mice reconstituted with
syngeneic bone marrow were completely resistant, even at higher
challenge doses. C3H and F1 mice reconstituted respectively with
F1 and C3H bone marrow were both susceptible. The most noticeable
difference between these latter two groups was that mice receiving

F1 bone marrow usually had maximum parasitemias lower than those
seen in mice receiving C3H bone marrow.

Uninfected mice from each of the groups were challenged with
sheep red blood cells and the hemagglutination titers measured to
determine if any group of mice was deficient in its humoral
responsiveness. While mice reconstituted with syngeneic bone
marrow had slightly higher titers than mice receiving allogeneic
cells, there was little difference between the titers of the latter
two groups.

Our results may be more understandable in light of work on
the genetics of resistance to Salmonella typhimurium (15). Overall
resistance to this organism is polygenic, but a single gene has
been found to determine the net growth rate in the liver and spleen
in the first few days of the infection. A mouse strain with a
fast net growth rate is always susceptible, but a slow net growth
rate alone does not insure survival. Other factors, possibly
involving the cellular immune response, also are involved in the
control of bacterial growth and play a role later in the course
of the infection.

A similar situation in the case of T. cruzi is possible. Non-
immunologically related factors may limit the ability of the intracellular parasite to multiply in certain strains of mice, but
survival may also require a qualitatively or quantitatively better
specific immune response, which may occur in the F1 mouse. A
suggestion of differing immune responses came from the observation
of lower maximum parasitemias in C3H mice receiving F1 bone marrow
cells than in C3H mice given C3H bone marrow cells (Table 5).

It is obviously difficult to quantitate parasite proliferation
within the first few days of infection since T. cruzi invades a
wide variety of cells and tissues and blood parasitemia is very
low. In athymic Nu/Nu mice, the cellular and humoral immune
responses are greatly impaired by the absence of T-cells. Following
challenge, parasitemia in Nu/Nu mice is observable earlier, and is
even much higher than in the most susceptible mouse strains. The
parasitemia seen in Nu/Nu mice may thus be a reflection of parasite
proliferation in the absence of an immune response. To compare the
course of parasitemia in F1 and C3H mice lacking immune responsiveness, we used mice that had been lethally irradiated and bone marrow
reconstituted with syngeneic bone marrow and then challenged 4 days
later. As seen in Table 6, C3H mice had parasitemias by day 12
which were higher than those usually found in Nu/Nu mice at that
time. Parasitemias in F1 mice took at least a week longer to reach
comparable levels. If the immune responses of these mice have been
sufficiently impaired, the differences in the levels of parasites
in the blood may be a reflection of differing rates of growth in the
hosts. More complete studies will need to be carried out using
neonatally thymectomized mice.

Table 6. Parasitemias in mice lethally irradiated, reconstituted
          with syngeneic bone marrow cells and challenged 4 days
          later with 10^4 trypomastigotes

		Parasitemia $(\times 10^6/ml)$			
		Day			
Mice[a]		12	13	18	22
F1 reconstituted with	1		0.050		4.30
F1 bone marrow	2		0.350		2.65
	3		0.475		1.90
	4		0.275		5.05
	5		0.150		
C3H reconstituted with	1	2.90		28.0	
C3H bone marrow	2	2.20		7.2	
	3	2.15		14.2	
	4	3.60		17.4	
	5	4.25		10.6	
	6	4.05			

[a] Mice were reconstituted with 2×10^7 bone marrow cells depleted
of mature T-cells. C3H mice received 900 rads; F1 mice received
960 rads.

We are particularly interested in continuing to study the
host-parasite interaction early in the infection, and also in a
qualitative and quantitative comparison of the immune responses
in resistant and susceptible strains of mice. By focusing on
more defined events during the course of the disease, it may be
possible to discern which events differ between resistant and
susceptible strains and which are under simple genetic control.
Whether the immune response is an integral part of the genetic
basis of natural resistance remains uncertain.

It might be argued whether natural resistance, particularly
if nonimmunological in origin, should be considered a host defense
mechanism in the sense that, for example, the inability of an
organism to penetrate or multiply within a cell may appear to be
more a "passive" than an "active" or "induced" defense. However,
with two other intracellular organisms, _Listeria monocytogenes_
and _Mycobacterium lepraemurium,_ mouse strain susceptibility is
not determined at the level of the macrophage alone. Instead,
extrinsic factors have been implicated in regulating macrophage
activity. If such factors are produced as a response to the

invading organism, their mechanism of action, when protective,
would seem to be a true host defense. An understanding of the
precise mechanisms by which natural resistance to organisms is
effected may be useful as a means of identifying strategies for
increasing host resistance to parasites like T. cruzi for which
there are no vaccines or suitable drugs.

ACQUIRED RESISTANCE

     All strains of mice appear to be capable of developing a pro-
tective immune response to T. cruzi. Mice develop immunity after
injection with live epimastigotes (5) which are found only in the
insect vector and are easily grown in cell-free culture, avirulent
trypomastigotes (4), or sublethal doses of virulent trypomastigotes
(5). The immune mechanisms responsible for protection have not
been clearly identified and any impairment of the immune response
usually leads to increased parasitemias and higher mortality.
Using in vitro systems, parasite killing has been achieved by
antibody plus complement (16), by antibody-dependent cell-mediated
cytotoxicity (17), by macrophages (18), and by cellular immunity
(19). The relative importance of these mechanisms in vivo is
uncertain.

     We were able to transfer protection to C3H mice with immune
spleen cells, a T-cell population, a T-cell depleted population,
and an Fc+ population, all obtained from spleens of C3H mice which
had been immunized with epimastigotes and later challenged with
blood forms (20). While informative, these studies had two
limitations. Firstly, it is unknown if T-cells are acting in
protection through their role as helper cells in antibody production or through a direct cytotoxic mechanism. And secondly,
the immune cells are being transferred to immunocompetent animals
and thus the transferred cells are acting in consort with the
immune response of the host. What we wished to know was if
either the cellular or the humoral immune response in the absence
of the other was capable of protecting.

     Humoral immunity. The transfer of protection with immune
serum has a long history and a variety of antibody-mediated
mechanisms have been found in vitro to kill T. cruzi (21). To
evaluate the ability of antibody to protect in the absence of
T-cells, we utilized the Nu/Nu mouse in passive antibody studies.
Immune serum was obtained from C3H mice after two injections of
50×10^6 epimastigotes three weeks apart followed 3 weeks later by
challenge with 10^4 trypomastigotes. Mice were given 0.5 ml 1 day
before challenge and every 2 or 3 days after challenge as described
in the legend to Figure 1. As long as antibody was being given, no
parasitemia or signs of illness were observed in the recipients

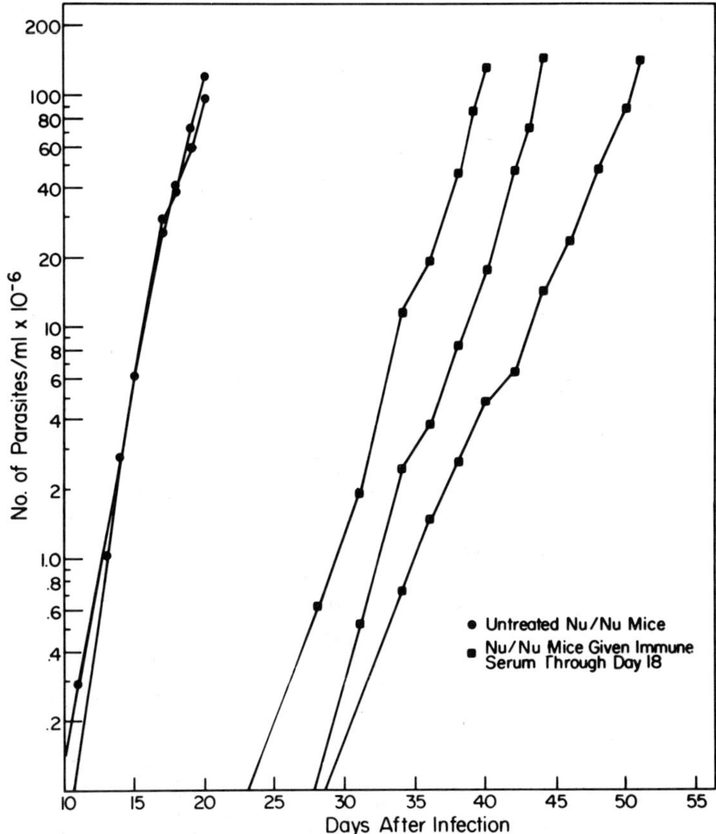

Figure 1.   Immune serum was raised in C3H mice as described in the text.   Three 8 wk old female Nu/Nu mice were each given 1/2 ml of immune serum on days -1, 1, 3, 5, 7, 9, 12, 15, and 18.   Intraperitoneal challenge with 10^4 Brazil strain trypomastigotes obtained from the blood of an infected C3H mouse was done on day 0.

while control mice developed the usual early, rapid rise in parasitemia that we have described for the Nu/Nu mouse.

After transfer of immune serum was stopped, mice developed observable parasitemias within several weeks and the parasites were able to increase to the very high levels seen in Nu/Nu mice. Parasites were obviously being harbored by the host which were able to multiply once the level of specific antibody reached low enough levels.   Similar findings were reported by Kierszenbaum using the Tulahuen strain of <u>T. cruzi</u> (22).   Antibody thus appears to be capable of protecting in the absence of T-cells.   In our procedure, antibody was given before challenge and it would be

of great interest to know if in the normal course of infection in
a resistant mouse the humoral response alone was capable of protecting. Of course, it is not possible to eliminate T-cells and
get a normal humoral response.

To begin to look at the mechanisms by which antibody is destroying parasites, we made use of an observation by Hanson that immune
mice, but not normal mice, could rapidly clear from the blood intravenously injected parasites (8). Trypomastigotes of T. cruzi can
be grown in large numbers in Vero cell cultures starting with
infected blood. The parasites replicate in the Vero cells, lyse the
cells, and can then be harvested from the media. When 20×10^6
parasites are injected intravenously into a normal mouse, a parasitemia of around 10×10^6/ml is found after 15 min. In immune
mice, less than a few hundred thousand parasites/ml are found after
15 min. A few experiments utilizing this procedure are summarized
in Table 7.

The rapid clearance of parasites seen in immune mice can be
duplicated in naive C3H mice by the prior transfer of immune serum,
but not of immune spleen cells. The ability of passive antibody
to clear parasites in Nu/Nu mice indicates that T-cells are not
involved in the process. If there is a cellular component to the
clearance by antibody, the cells involved clearly are not required
to be specifically immune. In fact, immune spleen cells did not
enhance the ability of antibody to clear parasites when the amount
of antibody transferred was not sufficient by itself to clear all
parasites. In immune mice, depletion of C3 with cobra venom factor
did not impair clearance of parasites but intravenous injection of
silica did, suggesting that phagocytic cells may be involved.

What the data show quite clearly is the efficiency of immune
serum in clearing parasites from the blood. The very high parasitemias reached in Nu/Nu mice may be a reflection of their lack
of a humoral response. Occasionally, susceptible C3H mice will
live longer than four weeks after challenge and parasitemias in
these animals drop to subpatent levels. A maturing immune response
may well be able to eliminate parasites, although too late to
prevent death.

Cellular immunity. To evaluate the protective ability of the
cellular immune response, mice were prevented from making antibodies
by continued injections of rabbit anti-mouse IgM from within 12 hrs
of birth until the death of the animals (23). Fl (C3H x C57BL/6)
mice were used since the high resistance of these mice might be
due in part to their making a better immune response than susceptible
mice. Fl mice were treated with the anti-IgM antiserum and
challenged at 7 weeks of age with 10^4 trypomastigotes. The course
of parasitemia in these mice and controls is shown in Figure 2.
At three weeks post-infection, the anti-IgM treated mice had

Table 7. Parasitemias in mice measured 15 minutes after intra-
          venous injection of parasites

Mice	Parasitemias relative to controls			
	<5%	5-35%	35-65%	65-100%
Challenged with Brazil strain trypomastigotes on day 0[a]				
C3H				X
Immunized with epimastigotes	X			
10^8 immune spleen cells given day -6[b]				X
1 ml hyperimmune serum given day -1[b]	X			
1/10 ml hyperimmune serum given day -1			X	
10^8 immune spleen cells given day -6 plus 1/10 ml hyperimmune serum given day -1			X	
Lethal irradiation given day -6[c]				X
Lethal irradiation given day -6, 1 1/2 ml hyperimmune serum given day -1	X			
Nu/Nu mice given 1 ml hyperimmune serum day -1	X			
5 U cobra venom factor given to immune mice days -3, -2, -1	X			
3 mg silica given to immune mice days -3, -2, -1			X	

[a] 20×10^6 trypomastigotes harvested from Vero cell cultures were
  injected per mouse.

[b] Immune spleen cells and hyperimmune serum were from mice
  immunized with epimastigotes and challenged with blood forms.

[c] 900 rads.

parasitemias similar to those of the control mice. After three
weeks, however, the parasitemias in the anti-IgM treated mice did
not decline as in the controls but continued in a steady rise
until the death of the mice.

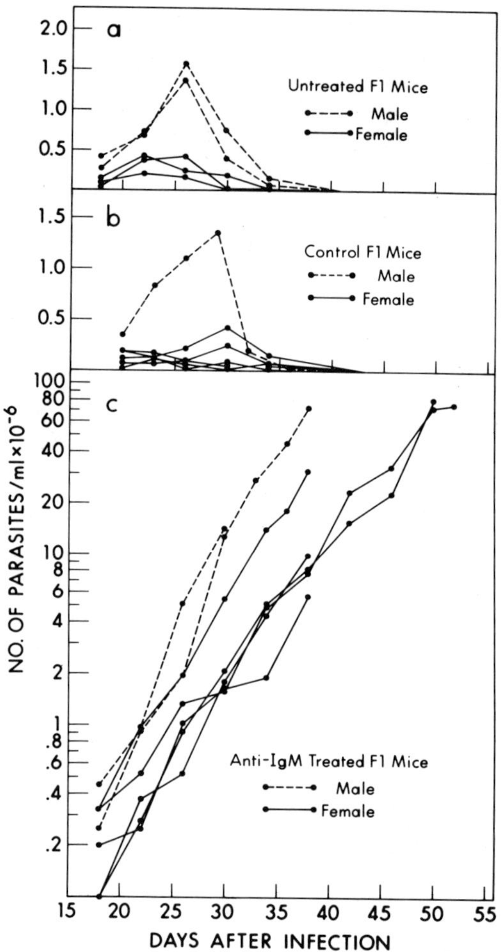

Figure 2.   Anti-IgM treated F1 mice (C3H x C57BL/6) were given
purified rabbit IgG containing rabbit anti-mouse IgM
from within 12 hrs of birth and later ever other day
until death.   Untreated F1 mice received nothing.
Control F1 mice were given normal purified rabbit IgG
following the same regimen as for mice receiving the
anti-IgM antibody.   All mice were challenged intraperitoneally at 7 wks of age with 10^4 Brazil strain
trypomastigotes.

Clearly the cellular immune response alone, even in resistant
mice, is not sufficient to confer full protection, although the
mice survive from 5 to 7 weeks after challenge which is longer than
a susceptible C3H mouse lives and may reflect some effect of cell-
mediated immunity (CMI).   The longer survival, however, could

also be related to the high natural resistance of these mice. Also
interesting were the high parasitemias attained in these mice
suggesting again the important role of antibody in the clearance
of parasites.

While a naive animal may not be able to survive infection
without its humoral response, it was possible that an immune animal
could. C3H female mice, when given 10^7 epimastigotes, survive
challenge 1 week later with 10^4 trypomastigotes. In the experiment
shown in Figure 3, the anti-IgM treated C3H mice turned out to be

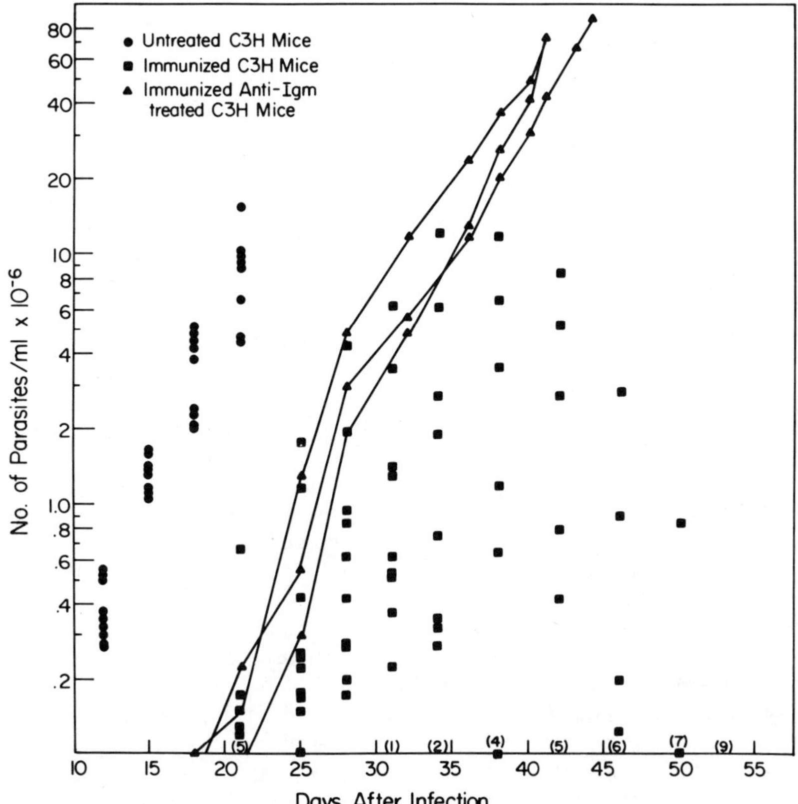

Figure 3. Anti-IgM treated C3H mice were given rabbit anti-mouse
IgM prepared by isolation of the specific anti-IgM antibodies from an immunoadsorbent column. Mice received
their first injection within 12 hrs of birth and later
every other day until death. The 3 anti-IgM treated
male mice and 10 normal male C3H mice were given 10^7
Brazil strain epimastigotes when 7 wks old. One week
later, these mice and 10 additional untreated male mice
were challenged with 10^4 Brazil strain trypomastigotes.

males so males were also used in the control groups. Untreated
mice developed parasitemias and died as expected. The immunized
mice developed observable parasitemias later than the immunized
mice. The parasitemias in the immunized mice varied and some were
much higher than would be seen in female mice since among vaccinated
mice, males have been found to resist infection less readily than
females (24). One mouse did in fact die while the other nine mice
all cleared their parasitemias and lived.

In the immunized, anti-IgM treated mice, the early parasitemias
were comparable to those of the immunized controls and much lower
than those of the unimmunized mice. But whereas the immunized
controls were able to clear their parasites and live, parasitemias
in the anti-IgM treated mice continued to increase until the mice
died. Once again, very high parasitemias occurred in the absence
of a humoral response. For the first three weeks post-infection,
the immune response of the mice appears to have been effective,
but then to have failed to control the ensuing parasitemia. Whether
CMI is acting on the parasites directly or on parasite infected
cells is unknown. Infected cells do not appear to express parasite antigens on their surface (25,26), and only low levels of
killing of infected cells in vitro by immune cells have been
reported (19).

Unlike passive antibody which was capable of protecting Nu/Nu
mice (Figure 1), specifically activated cellular immunity failed to
confer total protection. Additional work will need to be done to
determine if a higher immunizing dose of epimastigotes or a longer
period between immunization and challenge will increase the
capacity of CMI to protect. Cellular immunity does appear able to
effect protection during the first three weeks of the infection,
but after this period the lack of a humoral response leads to a
high, fatal parsitemia.

Hoff has reported that macrophages taken from infected mice
showed an enhanced ability to destroy parasites beginning 3 weeks
after infection (18). The level of specific serum antibodies
naturally increases during this period and immune serum has been
demonstrated in vitro to lead to increased uptake of parasites by
macrophages (27). A combination of these events may be important
in the clearance of parasites after peak parasitemia, which occurs
in mice infected with the Brazil strain between the third and fifth
weeks after infection (6,8). Either impairment of macrophage
function with silica (Table 7) or a failure to produce antibody
(Figs. 2 and 3) would lead to a lessened ability to clear parasites
or limit their proliferation.

There is no reason to believe that only one specific mechanism
is involved in the killing of T. cruzi, and each of the demonstrated
in vitro systems of killing may contribute to acquired immunity to
some extent. Verification of the occurrence and extent of these

mechanisms in vivo remains to be done. T. cruzi has been included
in this symposium since it is, of course, an intracellular parasite.
Yet there is no convincing evidence that the immune response can
recognize or destroy an infected cell, which might explain the
persistence of parasites in the immune host. And while the protective aspects of the immune response to T. cruzi are evident,
the possibility has been raised of the chronic form of the disease
being a consequence of an immune response to a parasite antigen
cross-reactive with an antigen of heart tissue (28). Thus, even
the immunologists goal of a vaccine is hindered by the many unanswered questions concerning the immunology of Chagas' disease.

CONCLUSIONS:

When challenged with the Brazil strain of T. cruzi, inbred
strains of mice were found to vary in their natural resistance to
the parasite. Variation of the H-2 haplotype of susceptible C3H
or resistant C57BL/10 mice did not affect the resistance of these
strains, indicating that the H-2 locus is not the principal genetic
determinant of resistance. Challenge of H-2 congenic mice on a
strain background of intermediate resistance, BALB/c, showed
clearly an influence of H-2 haplotype on resistance. This was
confirmed by following the survival of H-2 typed F2 mice derived
from parental C3H and C57BL/10 strains. Using reciprocal bone
marrow transfers, neither C3H mice reconstituted with bone marrow
from highly resistant F1 (C3H x C57BL/6) mice, nor F1 mice reconstituted with C3H bone marrow were able to survive challenge
although parasitemias were consistently lower in mice receiving
F1 bone marrow. Both immunological and nonimmunological factors
may contribute to natural resistance.

Immune spleen cells from C3H mice, a purified T-cell population, a T-cell depleted population, and Fc+ cells were all capable
of protecting C3H recipients when transferred 1 day prior to
challenge. Passive transfer of immune serum to athymic Nu/Nu
mice suppressed parasitemias as long as the immune serum was being
administered, demonstrating the protective potential of the humoral
response in an absence of T-cells. Antibody production in F1
(C3H x C57BL/6) mice was suppressed by continued injection of
rabbit anti-mouse IgM antibodies. Slowly rising parasitemias
occurred after challenge which eventually led to very high levels
of parasitemia and death, indicating that the cellular immune
response was not capable of protecting in the absence of the
humoral response.

ACKNOWLEDGMENTS

This work was supported by Public Health Service Grant AI-
16113 and by a grant from the UNDP/World Bank/WHO Special Program
for Research and Training in Tropical Diseases.

REFERENCES

1. Segovia, S.A., et al. (1974). Bull. WHO, 50:459.
2. Teixeira, A.R.L. (1977). In "Immunity to Blood Parasites of·
      Animals and Man" (L. Miller, J. Pino, and J. McKelvey,
      eds.), p. 243. Plenum, New York.
3. Pizzi, T., Agosin, M., Christen, R., Hoecker, G., and Neghme,
      A. (1949). Boletin Informaciones Parasitarias Chilenas,
      4:48.
4. Seah, S., and Marsden, P.D. (1969). Ann Trop. Med. Parasitol.,
      63:211.
5. Hauschka, T.S., Goodwin, M.B., Palmquist, J. and Brown, E.
      (1950). Am. J. Trop. Med.,30:1.
6. Trischmann, T., Tanowitz, H., Wittner, M. and Bloom, B. (1978).
      Exp. Parasitol.,45:160.
7. Hanson, W.L., and Roberson, E.L. (1974). J. Protozool.,21:512.
8. Hanson, W.L. (1977). In "Chagas' Disease". Pan American
      Health Organization.Scientific publication No. 347,p. 22.
9. Hauschka, T.S. (1947). J. Parasitol. 33:399.
10. Culbertson, J.T., and Kessler,W.R. (1942).J. Parasitol., 28:155.
11. Blackwell, J., Freeman, J., and Bradley, D. (1980). Nature,
      283:72.
12. Williams, D.M., Grumet, F.C., and Remington, J.S. (1978).
      Infect. Immun., 19:416.
13. Taylor, B.A., Bailey, D.W., Cherry, M., Riblet, R., and
      Weigert, M. (1975). Nature, 256:644.
14. Howard, J.G., Hale, C., and Liew, F.Y. (1980). Nature,
      288:161.
15. Hormaeche, C.E. (1979). Immunology, 37:319.
16. Budzko, D.B., Pizzimenti, M.C., and Kierszenbaum, F. (1975).
      Infect. Immun., 11:86.
17. Lopez, A.F., Bunn Moreno, M.M., and Sanderson, C.J. (1978).
      Int. J. Parasitol., 8:485.
18. Hoff, R. (1975). J. Exp. Med., 142:299.
19. Kuhn, R.E., and Murnane, J.E. (1977). Exp. Parasitol., 41:66.
20. Trischmann, T.M., and Bloom, B.R. (1980). Exp. Parasitol.,
      49:225.
21. Brener, Z. (1980). Adv. Parasitol., 18:247.
22. Kierszenbaum, F. (1980). J. Parasitol. 66:673.
23. Gordon, J. (1979). J. Immunol. Meth. 25:277.
24. McHardy, N. (1978). Trans. R. Soc. Trop. Med. Hyg., 72:201.
25. Ribeiro Dos Santos, R., and Hudson, L. (1980). Parasite
      Immunol.,2:1.
26. Abrahamsohn, I.A., and Kloetzel, J.K. (1980).Parasitology,
      80:147.
27. Nogueira, N., Chaplan, S. and Cohn, Z. (1980). J. Exp. Med.,
      152:447.
28. Santos-Buch, C.A., and Teixeira, A.R.L. (1974). J. Exp. Med.,
      140:38.

IMMUNITY TO FUNGAL INFECTIONS

Judith E. Domer and Emily W. Carrow

Tulane University School of Medicine
New Orleans, LA

INTRODUCTION

Medically-important fungi potentially capable of initiating
life-threatening disease can be categorized roughly into two
groups, viz., primary pathogens and opportunists. The primary
pathogens are those fungi which regularly cause disease in individuals with no known underlying clinical conditions, while the
opportunists seldom create problems for the healthy person, but
instead attack the patient whose normal defenses are compromised
by iatrogenic factors or disease.

The primary pathogens that will be dealt with in this presentation include the dimorphic fungi, _Histoplasma capsulatum_,
Blastomyces dermatitidis, _Paracoccidioides brasiliensis_, and
Coccidioides immitis, and the monomorphic fungus, _Crypto-
coccus neoformans_. The latter is considered by many to be an
opportunist, but at least 50% of cases of crytococcal meningitis
occur in persons with no known underlying disease (1). The true
opportunists, on the other hand, such as selected members of the
genera _Candida_, _Aspergillus_, _Mucor_, _Rhizopus_, or _Absidia_ are
virtually always seen as secondary invaders.

The study of mechanisms of resistance to fungi as a whole
remains relatively primitive, many times limited to vaccine considerations. Three of the organisms mentioned above have been rather
extensively studied, however, and have provided some insight into
general, as well as specific, mechanisms of resistance. Those
organisms, _H. capsulatum_, _C. neoformans_, and _C. albicans_, will be
covered in detail in the following sections. Mechanisms of
resistance to fungi causing primarily superficial or subcutaneous

disease will not be covered. For information relative to the superficial fungi, at least through 1974, there is a review by Grappel
et al. (2).

PRIMARY PATHOGENS

      Common features of the primary pathogens include the fact that
they are all soil saprophytes so that infection is initiated by
inhalation, and that, with one exception, they occur in vivo as
unicellular or budding yeast forms. The exception, Coccidioides
immitis occurs in vivo either as unicellular endospores or multi-
cellular spherules, the latter of which rupture to release the
endospores. Spherules are quite large, e.g. 50μ or more, and are
a formidable opponent for the immune system, but the newly released
endospores are small and could potentially be phagocytized and
dealt with as the other yeast forms. Four of the five primary
pathogens mentioned above, viz., H. capsulatum, B. dermatitidis,
P. brasiliensis, and C. immitis are dimorphic, i.e., they occur as
molds on artificial media at ±25°C or in soil, but as yeasts or
endospores (spherules) in tissue or under special conditions in
vitro, including a temperature of 35-37°C. C. neoformans occurs
in soil and tissue as a yeast. In its most common form it is encapsulated and therefore presents different problems to the immune
system of the host compared to the unicellular forms of the dimorphic fungi. Because of its uniqueness, it will be treated separately below, while the dimorphic fungi will be treated as a unit.

The Dimorphic Fungi

      In vitro phagocytic studies. The dimorphic fungi, at some
point in their life cycle, occur in tissue in a unicellular form,
so that the potential for interaction with phagocytic cells occurs
with all of them. Accordingly, they have been studied in vitro
in various phagocytic assay systems. It should be remembered,
however, that the tissue forms of B. dermatitidis, P. brasiliensis
and C. immitis are usually observed extracellularly in infected
tissue. H. capsulatum, on the other hand, is a facultative intracellular parasite and is found predominantly within mononuclear
cells of the reticuloendothelial system.

      Since attraction to and contact with the phagocytic cell is
a crucial first step in the phagocytic process some effort has
gone into a search for fungal chemotactic factors as well as the
possibility of complement activation. Sixbey et al. (3), have
detected factors in the culture filtrates of B. dermatitidis and
H. capsulatum which were chemotactic for human polymorphonuclear
leukocytes (PMN). That of B. dermatitidis was more active on a
dilution basis. Although not specifically stated, the chemotactic
activity appeared to be serum-independent. On the other hand,

mycelial or spherule filtrates of C. immitis were only chemotactic
in the presence of fresh unheated serum, implicating complement-
mediated chemotaxis (4). Complement-activation involving both the
classical and alternative pathways, in fact, was reported with
similar extracts in a later publication (5). Whole yeasts or
isolated cell walls of P. brasiliensis are chemotactic for human
PMN and do activate the alternative pathway of complement, but
culture filtrates alone were not active (6). H. capsulatum was
shown to activate the alternative pathway of complement as well,
but no correlations were made with chemotactic responses (7).
H. capsulatum and B. dermatitidis, therefore, appear to produce
chemotactic factors capable of attracting cells in the absence
of serum, whereas chemotactic factors of P. brasiliensis and
C. immitis require a complement component as intermediary.

     Alveolar macrophages, peritoneal macrophages and blood leuko-
cytes have all been tested to varying degrees with the dimorphic
fungi for uptake and/or killing of fungal forms. In general, the
role of heat-labile factors at any stage of the process is controversial (8,9,10) and the presence of immune serum had little or no
effect when added to assay mixtures with Histoplasma (10,11),
Blastomyces (3) or Coccidioides (12). Alveolar macrophages from
normal animals phagocytized H. capsulatum yeasts (11) as well as
C. immitis endospores or arthrospores (12) to a limited extent,
e.g. 16–32%, but there was essentially no killing in either
system. Immunization of animals with yeasts or cell fractions
improved uptake of yeasts, but the yeasts were not killed in cells
from immunized animals either. Calderone and Peterson (13) in a
cell-free assay, however, were able to demonstrate the inhibition
of protein sythesis in H. capsulatum yeasts with lysosomal-rich
extracts of rabbit alveolar macrophages.

     Histoplasma yeasts can be phagocytized and killed by blood
monocytes (MN) or PMN from nonimmune humans (14). Guinea pigs
and mouse PMN are effective at uptake and killing of Histoplasma
yeasts as well, and Howard, in a series of papers (15,16,17,18),
has detailed the mechanisms by which PMN exert their effect.
Oxidative mechanisms appear to be more important than non-oxidative
mechanisms for killing both yeasts and conidia. In vitro, effective
killing can be mediated, for example, by incubation with 10^{-5} M
hydrogen peroxide (H_2O_2), 10^{-5} M potassium iodide (KI) and horse-
radish peroxidase (15) or human myeloperoxidase (MPO) (18). All
the ingredients were necessary for death to occur providing the
appropriate concentrations of each component were used. Initially
a granule lysate from guinea pig PMN was reported to be an
effective replacement for the peroxidase, but more recently
Howard (18) has noted that the granule lysate alone was inhibitory
to yeasts but not conidia, raising the possibility for involvement
of nonoxidative mechanisms in killing. Further, the yeast phases
of different strains of Histoplasma varied with respect to their

catalase content and a positive correlation was demonstrated between
the level of catalase production and sensitivity to oxidative
mechanisms of killing.

Studies of interactions of B. dermatitidis, P. brasiliensis,
and C. immitis with peripheral blood leukocytes have been quite
limited. Human PMN will phagocytize B. dermatitidis well, but kill
it poorly, e.g. 95% phagocytosis and 29% killing (3). Monkey
peripheral blood leukocytes were ineffective at killing endospores
or arthrospores of C. immitis (19). The degree of phagocytosis in
the in vitro system was not reported, but when C. immitis endospores,
C. albicans yeasts and Listeria monocytogenes were compared in the
same system, the percentage of inoculum killed was 13%, 53% and
97% respectively. Using killed endospores, Deresinski, et al.
(20) noted rapid uptake with no differences in phagocytic rate
between mononuclear cells from nonimmune and immune humans.

The mouse peritoneal macrophage, resident or stimulated with
various agents, has been used extensively in the study of Histoplasma yeasts, but there have been only a few reports with the other
dimorphic fungi. Since Histoplasma normally resides within mononuclear cells, it is not surprising that it is readily phagocytized
by nonimmune peritoneal macrophages, but is not killed (21,22).
To the contrary, macrophages from immune animals are inhibitory
to the yeasts (8,9,10,23) and Howard has demonstrated that the
effect is fungistatic (24) and probably mediated by lymphocytes
in the cell mixtures (25). Despite attempts to produce one,
however, a lymphokine could not be detected in supernatants from
lymphoid cells incubated with heat-killed yeasts (25).

Brummer et al. (26) have explored in vitro interactions
between mouse peritoneal macrophages activated by fetal calf
serum (FCS), concanavalin A (Con A) or by infection, and B.
dermatitidis yeasts. They could not distingush between fungistatic or fungicidal effects, but comparable levels of inhibition
were observed with macrophages from animals stimulated with FCS,
Con A or by infection. There was approximately a 50% reduction
in colony-forming units (CFU) between control and stimulated
cultures. Resident macrophages inhibited only to the extent of
about 25%. The greatest reductions occurred with freshly harvested
cells. In a second paper from the same laboratory (27) attempts
were made to correlate the inhibitory effects of macrophage
cultures with degree of virulence of the strain of Blastomyces
tested. Indeed, all strains were inhibited in cultures if CFU
were determined after 24 hours of coculture, but if the cocultures
were allowed to incubate for 72 hours and then CFU determined, the
virulent strain was no longer inhibited, whereas the attenuated
and avirulent strains were inhibited. Here again, activated macrophages and macrophages from infected animals were much more
effective than resident macrophages from uninfected animals.

Even less is known of the interaction of P. brasiliensis and
macrophages than is known of B. dermatitidis, but in a single study,
Calich et al. (6) examined the requirements for adherence and/or
uptake of the yeast form by mouse thioglycollate stimulated macrophages. Essentially no phagocytosis occurred when heat-inactivated
nonimmune serum was used to pretreat yeast cells prior to phagocytic assays. However, either fresh serum or immune heat-inactivated
sera restored competence to the system, suggesting that specific
antibodies and complement were equally opsonic. By using various
inhibitors the most likely pathway for complement involvement was
the alternative pathway.

In summary, it is clear that the PMN can deal effectively with
H. capsulatum but not with B. dermatitidis yeasts, and that the
nonimmune resident monocyte/macrophage has little inhibitory effect
on any of the dimorphic fungi. Moreover, only in the case of B.
dermatitidis yeasts has immune serum been shown effective and that
was simply a test of opsonic effect, not inhibition or killing.
To the contrary, macrophages from immune animals are fungistatic
(-cidal?) for H. capsulatum and B. dermatitidis yeasts, and in the
case of the former, the fungistatic effect was probably mediated
by lymphocytes.

Specific immunity in vivo. Clinical correlations over the
past 15 years or so, especially in histoplasmosis and coccidioidomycosis,have been suggestive of a major role for cell-mediated
immunity in successful defense against H. capsulatum and C. immitis.
Increased incidences of the two diseases in immunosuppressed
patients in recent years has supported the early evidence (28,29,
30). There have been only few reports of blastomycosis (31) or
paracoccidiodomycosis (32) in immunosuppressed patients, however.
In fact, in the few clinical studies reported, defects in PMN
function (33) or chemotaxis (34) have been highlighted. On the
basis of the histologic response to both B. dermatitidis and P.
brasiliensis, it is not surprising that defective PMN activity would
be deleterious to the host because PMN are much more prominent in
lesions throughout the course of disease with each of these fungi
than they are with H. capsulatum or C. immitis.

Experimentally, natural or induced immunosuppression,
especially that involving the T-lymphocyte, results in increased
susceptibility of the host to H. capsulatum. For example, nude
mice (35) or mice treated with antilymphocyte serum (36,37) were
much more susceptible to Histoplasma than normal littermates or
untreated mice respectively. Further, the implantation of a thymus
into nude mice restored resistance to normal levels (35). Nude
mice could be protected by whole spleens, purified T cells and
macrophages alone from immune animals as well (38). Using non-
immunosuppressed animals, successful transfer of protection to
either H. capsulatum (39,40) or C. immitis (41,42,43) could be

effected with T-enriched lymphoid cells and not with serum. Moreover, treatment of transfer suspensions to functionally or physic-
ally remove T lymphocytes resulted in the abrogation of transferred
resistance in both diseases (38,40,42,43).

Studies of mechanisms of specific immunity in experimental
blastomycosis have been few and have involved primarily attempts to
correlate the presence of delayed hypersensitivity with resistance
to disease (44) but such correlations are only suggestive, of course,
and indicate concomitant but not necessarily related immune phenomena. Similarly, although all untreated patients with paracoccidioidomycosis have depressed cellular immune function to both
homologous and heterologous antigens (45), specific mechanisms of
resistance have not been determined experimentally.

Cryptococcosis

Despite the occurrence of <u>Cryptococcus neoformans</u> in many
commonly encountered habitats, the majority of people who come in
contact with this agent do not develop clinically apparent disease
(46). Susceptibility is often correlated with conditions of lowered
host resistance, notably in patients suffering from neoplastic
disease (47,48,49), and/or undergoing immunosuppressive therapy
(50). The increasing numbers of such medically compromised persons
accounts in part for the markedly growing incidence of cryptococcosis seen in the last decade (51).

The common association of the disease with impaired reticuloendothelial or lymphatic function has prompted research in the
direction of defining the role of cellular mechanisms in host
defense. Evidence accumulated so far favors an important role for
phagocytic cells (both PMN and MN) in the initial stages of acute
infection, and the subsequent participation of T-lymphocytes (and
possibly macrophages) in the development of immunity. A significant role for antibody has not been found.

<u>In vitro phagocytic studies</u>. Early studies showed that
phagocytosis of cryptococci by human peripheral blood leukocytes
took place readily <u>in vitro</u> (52), although uptake was inhibited by
the polysaccharide capsule (53). Once engulfed, crytococci were
usually killed within four hours (54), but due to the antiphago-
cytic properties of the capsule, killing efficiency was affected by
the extent of encapsulation. It had previously been estimated (55)
that synthesis of capsular material by a nonencapsulated yeast
takes place within a few hours of entering host lung tissue, so it
is obvious that in order to prevent serious infection, the organism
must be taken care of swiftly, before it becomes refractory to
phagocytosis.

Diamond et al. (56) found that intracellular killing by human PMN was more efficient than by monocytes, although the latter cell type was not affected by capsule size, as the PMN was. Evidence for an MPO-peroxide-halide system of killing in the PMN was also presented. On the other hand, intracellular killing by mature macrophages cultured from human peripheral blood monocytes was not detected over a period of 2 days (57). Moreover, activated macrophages (SK-SD or cryptococcin) were no more effective than normal cells at killing the ingested yeasts.

Mitchell and Friedman (58) using glycogen-stimulated rat peritoneal macrophages in an in vitro system developed by them, confirmed the inverse relationship between capsular size and efficiency of phagocytosis. They too were unable to detect much killing. After 24 and 48 hours, most of the ingested yeasts remained viable, irrespective of capsule size. A requirement for a heat-labile serum factor for phagocytic activity was noted.

Bulmer and Tacker (59) obtained similar results with guinea pig alveolar macrophages. Nonencapsulated cryptococci were readily phagocytized in vitro, but the exogenous addition of partially purified capsular polysaccharide markedly reduced uptake, reconfirming the inhibitory nature of the capsular material. Serum was required for phagocytosis in this system. The amount of killing observed after 6 hours for engulfed yeasts was no different than that obtained when macrophages were left out of the system, however. The authors attributed any decrease in cell viability to an unidentified anticryptococcal serum factor, which had been described in a previous report (54).

More recent studies have shed some light on the nature of the factors influencing phagocytosis of C. neoformans. With regard to phagocytosis by PMN, Diamond et al. (60) determined that intact classical and alternative complement pathways were necessary for optimal phagocytosis by human cells. The classical pathway presumably functions to activate the alternative pathway, which in turn mediates opsonization. It also emerged from this study that adequate function of the classical pathway occurred only in the presence of specific anticryptococcal antibody. Cell walls of cryptococci were found to activate the alternative complement pathway and generate a chemotactic response by rabbit PMN (61). Capsule was not involved, and its inhibitory effect was thought not to be due to blocking the binding or release of complement components and cleavage products, but to masking the presence of cell-wall bound opsonins.

Working with mouse peritoneal macrophages, Kozel and Mastroianni (62) have shown that cryptococcal polysaccharide inhibits phagocytosis by preventing attachment of the yeast to leukocytes. It was subsequently demonstrated (63) that the principal opsonin for mouse macrophages is IgG. Cryptococcal polysaccharide does not prevent binding of IgG to the yeast cell surface but effectively masks its presence so that it is unable to participate in Fc-mediated phagocytosis.

These studies indicate that phagocytic clearing of cryptococci is probably most effective in the early stages of infection. The importance of a rapid, strong inflammatory response has been stressed by Gadebusch (64). The classical histopathological picture of a naturally acquired (respiratory) infection, as described by Gadebusch (64), is characterized by the initial presence of numerous PMN's in the alveoli, which are able to remove most of the organisms that deposit there (size range 1-5 µ). Within 24 hours the PMN's are joined by macrophages, which presumably are better able to handle those persistent yeast cells that have by then synthesized large capsules (diameter up to 50 µ). Over the next 7 days granulomas may develop, populated mainly by mononuclear phagocytes. This is a crucial stage in containment of infection, and the outcome may determine whether or not dissemination occurs (65,66).

If indeed a significant inoculum of cryptococci manages to escape the phagocytic activities of these cells, is the host then doomed to a serious, possibly lethal infection of extrapulmonary tissues (i.e. the CNS for which C. neoformans has a well-known predilection)?

Several lines of evidence suggest that other mechanisms exist to dispose of the large-sized parasite and prevent dissemination. Schneerson-Porat et al. (67) first recognized an unusual histo-pathological response in both in vivo and in vitro studies of infected mouse peritoneal exudate cells. Whereas small yeasts were readily phagocytized, the larger sized yeasts (>50 µ) became surrounded by histiocytes within 12 hours of intraperitoneal injection, if the yeast cells had been incubated previously in immune serum. In the absence of serum treatment, the ring response was delayed. The response was characterized by an early appearance of many PMN, soon replaced by concentric rings of macrophages which appeared to destroy the encircled yeast. Similar studies in the rabbit (68,69,70) confirmed this sequence of events. Yeasts were destroyed within 36-72 hours. Several hydrolytic enzymes were released from the phagocytic cells, which presumably acted upon the enclosed yeast, and pseudopods from both PMN's and macrophages could be seen penetrating into the yeast capsule. Disintegration of the capsule was observed, as well as phagocytosis of its material. The process was dependent upon the presence of a heat-labile opsonin.

Diamond (71) also described a nonphagocytic extracellular mechanism of killing, although the nature of the effector cell(s)

is not as clear. When incubations of the adherent leukocyte fraction
from human peripheral blood were carried out in the presence of
anticryptococcal antibody and cytochalasin B (to block phagocytosis),
up to 70% killing of cryptococci occurred within 24 hours. Similar
results were obtained using the IgG fraction of normal serum.
Attempts to identify more precisely the cells responsible for
killing (72) provided evidence that a mixed population of cells
(including monocytes, granulocytes and lymphocytes) were involved
in this antibody dependent cellular mechanism of cytotoxicity.

Specific Immunity In Vivo

Resistance to reinfection by C. neoformans has been studied
experimentally using a variety of inocula and routes of infection.
The idea that the major protective mechanisms probably reside at the
cellular level began to take shape about fifteen years ago.

Although the presence of serum antibody in cryptococcosis
patients is consistent with a favorable prognosis (73), attempts
to confer long term protection by passive immunization of experimental animals has been largely unsuccessful (74,75,76,77). The
first reports implicating a role for cell mediated mechanisms were
those by Abrahams (78), who demonstrated protection, in the absence
of serum antibody, in mice previously immunized with a mixture of
killed cryptococci and Bordetella pertussis adjuvant; Adamson and
Cozad (79) who showed that pretreatment with antilymphocyte serum
greatly reduced survival time in mice infected with C. neoformans;
and Gentry and Remington (80), who reported prolonged resistance
to cryptococcal challenge in mice chronically infected with intracellular protozoans or injected twice with Listeria monocytogenes.
It appeared that macrophages, activated by the prolonged infections,
were responsible for the heightened resistance. More recently,
varying degrees of protection against challenge have been demonstrated after intranasal infection with live yeast cells (81,82);
intraperitoneal infection with live avirulent pseudohyphae (83);
and subcutaneous inoculation with either a cell sap preparation
(84) or live yeast cells (85). In three of these studies, assays
for induction of cell mediated immunity correlated with protection,
i.e. delayed type hypersensitivity (DTH) (82,84) and lymphocyte
transformation (LT) (83).

Definitive evidence for the central role of T-lymphocytes in
protection against cryptococcal infection has been presented
recently in a number of reports. Parameters of protection include
mortality pattern, multiplication of cryptococci in tissues, and
DTH response. Treatment of immunized and unimmunized mice with
anti-thymocyte serum increased their susceptibility (86), and
nude mice have consistently been shown to be highly susceptible to
infection (87,88,89). Transplantation of thymus tissue from
normal mice significantly restored resistance, as well as ability

to mount a DTH response in nude mice (90). On the other hand,
treatment with cyclophosphamide, which among its several effects,
depresses the B-lymphocyte population, had little or no effect on
susceptibility (91), nor did administration of anti-mouse μ antiserum (92). Additional evidence for the importance of T-lymphocytes
has come from studies involving passive transfers of lymphocytes.
Protection was adoptively transferred to naive mice with T-enriched
splenic lymphocytes from mice previously infected with C. neoformans
(86,77,93) with a concomitant ability to produce DTH (86,77) or
LT (93). The ability of the sensitized T-enriched pools to transfer protection was ablated by treatment with anti-Thy 1.2 antiserum
plus complement (77).

Exactly how the T-lymphocyte functions and interacts with other
components of the immune system, such as B-lymphocytes, mononuclear
and polymorphonuclear cells, complement, antibody, and other humoral
factors, remains to be determined.

THE OPPORTUNISTIC FUNGI

The diseases and organisms in this section are grouped on the
basis of the fact that they attack the compromised host and occur
in tissue in the hyphal form, although lesions in candidiasis may
contain both yeasts and hyphae. It should be emphasized, however,
that clinical entities caused by the respective organisms are not
necessarily the same. The portal of entry to the host for Candida,
for example, is probably the gastrointestinal (GI) tract, while
that for Aspergillus is the respiratory tract, and that for the
agents of mucormycosis is usually a wound of some sort, resulting
from surgery, puncture by some foreign object or by abrasion of the
mucocutaneous surface. Candida is endogenous to the human host,
in that it can be isolated from the GI tract of a large percentage
of healthy humans, while Aspergillus and the agents of mucormycosis
are found as contaminants in the air.

Candida

Candidiasis may occur in a wide variety of clinical syndromes
ranging from only superficial involvement to severe systemic disease
resulting in death. Resistance to Candida appears to be more
complex than that of the other fungi, and the wide range of clinical
entities observed may be the result of different defects in the
overall defense mechanisms, including both innate and specific
components. PMN, for example, seem to be crucial as a prominent
defense mechanism, since patients with few or defective neutrophils
or defective neutrophil function (94,95,96) are more susceptible
to disease with Candida than individuals with normal neutrophil
function. On the other hand, individuals with defects in their
cellular immune system seem to be more susceptible to candidiasis

as well. The classic example of the latter is chronic mucocutaneous
candidiasis (CMC), a relatively rare but nonetheless important
disease seen in individuals with defective cellular immunity (97,
98). Interestingly, however, CMC patients do not develop systemic
disease so that spread from the cutaneous and mucocutaneous sites
is prevented by some as yet undefined mechanism. Although there
are several species of Candida capable of initiating human disease,
C. albicans is the species most often encountered and will be the
species highlighted in the studies presented below.

As with the primary pathogens, searches have been made for
chemotactic factors capable of attracting phagocytic cells,
particularly PMN, to the site of the candidal lesion. Soluble
chemotactic factors for the PMN have been described in three in
vitro systems (99,100,101), two of which were complement-dependent
(99,101), and one of which was complement-independent (100). Moreover, Ray and Wuepper (102) could generate chemotactic activity with
viable or killed C. albicans in the presence of complement but
could not demonstrate a soluble product in supernatants. The
studies of Ray et al. (101) are especially important in that
their chemotactic substance was mannan. Its importance lies in
the fact that mannan antigenemia does occur in candidiasis (103)
and free mannan can bind to MPO (104) and remove its potential for
interaction with the candidal surface. The presence of free mannan,
then, may have both helpful and detrimental effects, in that it may
attract PMN to the area but may then inhibit their potential for
exerting their influence by oxidative mechanisms. Indirect evidence for the activation of complement in vivo came from the work
of Sohnle et al. (105) who detected deposits of C3 and properdin
in the basement membranes of half of the patients with CMC that they
tested. As they could find no immunoglobulin or C4 they concluded
that the inflammatory lesions observed in CMC were the direct
result of the activation of the alternative pathway of complement
by soluble C. albicans components diffusing into the area.

The role of complement in uptake and killing of C. albicans
by human (106,107,108) or mouse leukocytes (109) is controversial.
Morrison and Cutler (110) however, recently published an in vitro
study involving the role of complement in phagocytosis and found
that mouse macrophages or PMN required intact complement for maximum ingestion. Furthermore, blocking the alternative pathway of
complement blocked virtually all uptake by PMN but only partially
blocked uptake by macrophages. C3 was needed by both macrophages
and PMN for ingestion. In vivo, Gelfand et al. (111) have
presented data indicating that both the alternative and classical
complement pathways are necessary to recovery from C. albicans
sepsis. Guinea pigs with a congenital deficiency of C4 or normal
guinea pigs treated with cobra venom factor have greatly increased
rates of mortality compared to normal animals.

Interactions with Phagocytic Cells In Vitro

Since _Candida_ can exist in two different forms, viz., yeast
or hyphal, in tissue as well as _in vitro_ under specific conditions
of incubation, phagocytic cells interact differently depending upon
the form studied. For yeasts, the degree of ingestion and the fate
of the cell once phagocytized varies widely with the cell type and
the investigator. There does seem to be agreement, however, that
the mouse peritoneal macrophage will phagocytize _C. albicans_
yeasts but that it subsequently has either no effect on the growth
of the fungus (112, 113) or else is simply fungistatic (114, 115).
Cutler and Poor (115) used a novel approach for their studies
which involved the implantation of chambers into the peritoneal
cavities of mice, the details of which they had reported earlier
(116), and found that whereas peritoneal macrophages from normal
mice were fungistatic, those from nude mice were fungicidal.
Interactions of PMN derived from nude or normal mice involved
killing in both cases. With _Candida_ yeasts, of course, one can
easily separate inhibitory effects from noninhibitory effects
because yeast cells that are not inhibited will form germ tubes
and eventually grow out of the phagocytic cell, resulting in
death to the latter cell. Germ tube formation is, in fact, a
common assay in these studies, the problem being, however, that
one cannot distinguish stasis from killing. Another problem in
comparing studies among laboratories is that different groups use
different ratios of phagocytic to fungal cell which may certainly
influence the end result. Cutler and Poor (115), for example, got
efficient killing in their chamber studies with a 40:1 phagocyte
to fungal cell ratio whereas others use 10:1 or 1:1 ratios.

Investigators have only recently begun looking at the interaction of alveolar macrophages, in this case of the rabbit, with
Candida yeasts. Peterson and Calderone (117,118) have shown that
normal alveolar macrophages inhibit ingested _Candida_, as evidenced
by an inhibition of macromolecular synthesis and less germ tube
formation compared to uningested yeasts, and that a lysosomal-rich
fraction from similar macrophages inhibits the uptake of amino
acids by _C. albicans_. Arai et al. (119) found that nonimmune
alveolar macrophages in the presence of nonimmune serum would
ingest 38% of the _Candida_ put into their system, whereas immune
macrophages in the presence of immune serum ingested 83%. Killing,
however, was constant regardless of the combinations, and remained
at about 20%.

The human PMN, and to a lesser extent the monocyte (MN) has
been used extensively in studies with _Candida_. In an early study,
Louria and Brayton (120) reported that PMN readily phagocytized
C. albicans but that the fungus survived the encounter and germi-
nated, destroying the PMN. In more recent studies, however,
investigators have shown that both the PMN and the MN can kill

yeast cells, but no more than 50% of the yeast cells ingested were
killed. For example, in one study (121) 96% of the Candida were
ingested by both PMN and MN, and PMN killed 58% while MN killed
50%. In another study (122) PMN phagocytized 92% and killed 29%.
Finally, in a third study (123) PMN phagocytized 70% and killed
30%, while MN phagocytized 90% and killed 60%. In the latter two
studies, the investigators also examined the uptake and killing of
pseudohyphal forms. Schuit (123) found that the PMN took up 80%
of the inoculum, killing only 10%, but Scherwitz and Martin (122)
only detected 10% uptake and 6% killing. The MN in Schuit's
system, on the other hand, killed 40% of the 90% pseudohyphae
ingested.

Diamond and his colleagues have carried out a series of
extensive investigations of the interaction of Candida hyphae and
human PMN and MN in vitro in the absence of serum (124,125,126,
127,128). No phagocytosis occurred but killing did occur. The
sequence of events was as follows. Contact occurred between PMN
and hyphae which resulted in degranulation of the PMN and the
deposition of lysosomes on the hyphal surface. Contact between
the hyphae could be slightly inhibited by mannan or strongly
inhibited by preincubation of the PMN with Candida or chymotrypsin.
Oxidative mechanisms involving the MPO-halide system seemed to play
a major role in killing. Additionally, dead or live hyphae could
liberate a substance, 2,500 to 3,000 daltons which also inhibited
contact between the PMN and the hyphae and impaired the function
of the PMN.

Both oxidative and nonoxidative mechanisms appear to be important as mechanisms of defense on the part of the PMN and MN.
Lehrer (129) demonstrated the activity of the MPO system from PMN
against C. albicans yeasts, and Diamond et al. (126) and Diamond
and Haudenschild (128) reported the importance of oxidative
mechanisms of PMN and MN against hyphae in the absence of serum.
The nonoxidative mechanisms involved seem to be lysosomal cationic
proteins of guinea pig (130) or human (131) PMN, as well as
alveolar macrophages (132,133).

It should be emphasized again that the PMN is an important
cell in innate defense against C. albicans. The clinical studies
supporting this were mentioned earlier, but there is also experimental evidence to support the role of the PMN. Cyclophosphamide
(CY), for example, a drug that will initially eliminate virtually
all circulating PMN, given in large doses just prior to challenge
with Candida, increases susceptibility of animals to the Candida
(134,135,136). Furthermore, granulocyte transfusions (135)
following CY treatment restored resistance.

Contrary to the studies with PMN, the role of the MN in in
vivo protection is much less clear. Nude mice are killed less

rapidly by an overwhelming dose of Candida than are their normal
littermates (137), and they inhibit Candida better in their tissues
when given a relatively small dose (138). This has been attributed
to more active macrophages in the nude mice. In the long term,
however, Miyake et al. (139) found nude mice more susceptible to
candidiasis than normal mice. "Conventionalized" nude mice are
more resistant (140), and germ free rats with understimulated
innate defenses are more susceptible (141). On the whole,
activation of macrophages either in vivo or in vitro with agents
other than C. albicans seems to elicit protective responses against
Candida. Rogers and Balish (142) could not induce protection in
vivo with BCG, but others using Listeria (143), lipids extracted
from Listeria (144), or glucan (145) could induce protection.
Further, tumor-bearing mice with hyperactive macrophages were able
to kill C. albicans more effectively than normal mice (146).
Macrophages activated with BCG or PMA (147), BCG or LPS (148) or
muramyl dipeptide (149) were all more effective at killing C.
albicans in vitro than unstimulated macrophages. Although Candida
can be killed in vitro and more readily handled in vivo by macrophages activated by heterologous agents, the significance of these
observations with respect to candidal disease in vivo is obscure.

Specific Immunity

The mechanism of specific immunity, that is, immune responses
induced by C. albicans which ultimately protect the animal from
another challenge with Candida, is not clearly defined. For some
time, even as late as the early 1970s, it was believed to be cell-
mediated. This conclusion was based primarily on the patients with
CMC as mentioned previously (97,98). Patients with CMC are
increasingly being shown to have other defects, e.g. in neutrophil
function (150,151) or antibody production (152,153) so that the
notion that cellular immunity is the effective arm for defense is
losing favor. Furthermore, experimental evidence regarding the
nature of immunity to Candida, although conflicting, is stronger
on the side of humoral than cellular defense mechanisms. The one
overriding difficulty with all the studies, however, is that the
protection obtained in experimental studies, when demonstrated,
has almost always been minimal or temporary.

Al-Doory (154), Mourad and Friedman (155) and Pearsall et
al. (156) all successfully transferred some degree of protection
with immune serum. When the administration of immune serum was
stopped, however, Mourad and Friedman's (155) mice immediately
resumed a normal pattern of dying. Furthermore, Pearsall et al.
(156) based their conclusion of protection on smaller thigh
lesions, and we have shown in our own work (157) that two lesions
of unequal size can have the same CFU within them. Indirect
evidence for antibody involvement was provided by Oblack and
Holder (159) also, who immunized mice to obtain antibody and noted

less damage to muscle upon challenge. Contrary to these studies,
Hurd and Drake (158) and Miyake et al. (139) were unable to
protect animals with immune serum.

Experimental evidence in support of a role for the T-lymphocyte
in the development of a protective response comes largely from the
studies of Miyake et al. (139) Sohnle et al. (160), Giger et al.
(161) and Grubek et al. (162). Miyake et al. (139) and Sohnle
et al. (160) both reported transfer of protection with lymphoid
cells, but the protection demonstrated by Miyake et al. was
minimal and sufficient experimental detail was lacking in the
studies of Sohnle et al., to critically evaluate their data.
Our own studies with T-depleted mice (161) and mice treated with
CY (157) implicate the T-lymphocyte in immunity, but that T-cell(s)
may well be involved in antibody production rather than cellular
immunity. CY-treated animals, for example, treated in such a way
as to abrogate the ability to produce antibody but retain the
ability for at least some cellular immune function, viz., delayed
hypersensitivity and in vitro lymphocyte stimulation with mitogens
and candidal antigens, did not develop resistance to reinfection
under the same conditions that normal (untreated) animals became
immune. Our T-depleted animals (161) became neither more susceptible nor more resistant to a first challenge with C. albicans, nor
did they develop resistance to reinfection.

In summary, resistance to Candida appears to be mediated by
innate, (primarily the PMN), and specific responses, the latter
of which are not clearly defined. Undoubtedly important also, but
as yet essentially unexplored, is the whole area of immunoregulatory
influences of antigen, antibody, and suppressor cells. Several
studies have been directed toward the demonstration of nonspecific
suppression in experimental candidiasis (163,164,165), but specific
regulation has not yet been reported.

Aspergillosis and Mucormycosis

The importance of cellular mechanisms in response to the
agents of aspergillosis and mucormycosis is indicated by the
clinical syndromes which correlate with both diseases. Humoral
factors appear to be less important, although nonspecific serum
resistance against agents of mucormycosis has been detected (166,
167,168). Immune responses to these fungi have not been well
studied, and their existence is mostly inferred from clinical
correlations.

Aspergillosis. Aspergillus spores are widespread in nature.
They are relatively small (<5 μ), so once inhaled they can easily
pass through the respiratory tract and deposit in the alveoli.
Spores must be cleared quickly before they germinate into invasive
hyphae, which are too large to be ingested by phagocytic cells (169).

PMN and macrophages appear to be important in the early phases of
acute infection, although their role seems to be more fungistatic
than fungicidal.

Liver and spleen were found to be highly efficient in clearing
virulent _Aspergillus flavus_ spores from blood of mice inoculated
intravenously (170). The spores remained viable but their growth
was checked. Brain and kidney, however, were unable to eliminate
or significantly inhibit growth of infecting spores. The lack of
a good inflammatory response in target organs of disseminated
disease suggests that the cellular response there is directed
against hyphae, not spores. The results of this study implied
that inflammatory cells function primarily to prevent germination,
whereas some type of immune mechanism may be necessary to impede
growth of hyphae.

Lehrer and Jan (171) drew similar conclusions from _in vitro_
studies with human leukocytes and _A. fumigatus_ spores. Phagocytosis
by PMN and macrophages took place readily in the presence of serum,
but 3 hours after ingestion most spores remained viable. The fate
of these spores within PMN appeared to be a function of their
relative resistance to oxidative mechanisms found previously to be
toxic for _C. albicans_ (129).

Because aspergillosis is so often associated with underlying
diseases such as cancer, leukemia, chronic granulomatous disease
and systemic lupus erythematosis, or immunosuppressive therapy (51,
172), several studies have been conducted to determine the effects
of immunosuppression on experimental infections. _In vivo_ phagocytosis of _A. flavus_ spores by alveolar macrophages was followed
ultrastructurally by Merkow et al. (173) in both normal and
steroid-treated mice. Results of this and previous studies (174,
175) demonstrated a marked inability to prevent germination in
macrophages from steroid-treated mice. This was attributed to
the failure of lysosomal membranes to interact significantly
with phagosomal membranes in these mice.

Turner et al. (176) also found that administration of
steroids (cortisone) allowed the development of more severe
granulomas in lymphoid tissue, and reduced the ability of rat
peritoneal macrophages to check hyphal growth.

Phagocytosis of spores is clearly a primary component of
resistance to Aspergillus infection. Both macrophages and PMN
are involved, and it appears that in most cases, serum opsonins
are required for maximal activity. Recently Diamond et al. (177)
have described a nonphagocytic mechanism of killing that could be
effective against hyphae _in vivo_. _In vitro_, human PMN damaged
and apparently killed hyphae of _A. fumigatus_, even in the absence
of serum. Oxidative mechanisms, i.e. MPO-peroxide-halide appeared

to be responsible for most of the damage, but a secondary role for
lysozyme was also indicated.

   Mucormycosis. Clinical conditions that predispose to mucormycosis include diabetes, leukemia, lymphoma, Hodgkin's disease,
immunosuppressive therapy, thermal burns, and surgery. Rhino-
cerebral and pulmonary forms are the most frequent, although several
cases of cutaneous and subcutaneous infections, attributed to
contaminated bandages, have been recently reported (168). The
three most common etiologic agents of mucormycosis, Mucor, Rhizopus
and Absidia, are fast-growing fungi, familiar to most laboratory
personnel as common contaminants. Deposition of spores in the nasal
mucosa of a susceptible host can lead to a rapid fulminating
infection that may subsequently disseminate to the lung or other
organs.

   Nonspecific host defense mechanisms, particularly those involving phagocytic cells, are probably of major importance in
protection, although experimental evidence to that effect is
lacking.

   Gale and Welch (166) first reported that normal human serum
inhibits growth of R. oryzae by a mechanism that involves neither
antibody, complement nor phagocytosis. The factor responsible
for the fungistatic activity was not identified further than being
characterized as a heat-stable, low molecular weight substance.
Notably, it was missing in the serum of some diabetics.

   Owens et al. (167) subsequently confirmed those findings,
noting the loss of inhibitory capacity in sera of ketoacidotic
diabetics. In addition, they found that the sera of some patients
with leukemia or nutritional cirrhosis also lacked the fungistatic
factor.

   Marchevsky (168) has recently noted the clinical correlation
between severity of disease and degree of serum fungistasis
in vitro, irrespective of antibody.

   Diamond et al. (177) have demonstrated a nonphagocytic
mechanism of killing similar to that described for Aspergillus.
The hyphae succumbed to substances which were presumably released
from the surrounding leukocyte. In addition to the fungicidal
action of the MPO-peroxide-halide system of the PMN, cationic
proteins may also have participated in damaging the hyphae.

   Experimental infections designed to elucidate cellular defense
mechanisms have dealt primarily with A. corymbifera. The effects
of immunosuppressive agents on the development of experimental
infections were studied by Corbel and Eades (178). They found
that inhibitors of complement and various components of the

reticuloendothelial system increased susceptibility in mice,
whereas T-cell or B-cell inhibitors had no effect. Furthermore,
New Zealand Black mice (T-cell deficient) were no more susceptible
than CBA mice (immunocompetent) to infection by A. corymbifera, M.
pusillus, or R. oryzae, although they were a great deal more
susceptible to infection by A. fumigatus, C. albicans and C. neoformans (179). The clinical and histopathological picture was
similar in both groups of mice except in the cases of C. albicans
and C. neoformans infections which were more severe in the New
Zealand Black mice.

Although these studies indicated little or no importance of
thymus-dependent processes in acute infection, resistance to reinfection by A. corymbifera was usually associated with a DTH
response and pathology strongly suggestive of cell mediated
immunity (180). Similarly, although athymic (nude) mice were no
more susceptible to A. corymbifera infection than their pheno-
typically normal counterparts, they did develop more extensive
lesions and eliminated viable spores less effectively (181).

SUMMARY

Extensive studies of either innate or specific defense against
the dimorphic fungi capable of causing life-threatening disease in
humans have been done primarily with H. capsulatum and C. immitis.
Both PMN and MN are most certainly important phagocytic cells in
defense against all of the dimorphic fungi, although in vitro the
PMN does not appear to be highly effective against Blastomyces.
In vivo, however, PMN are the hallmark cells in lesions of both
blastomycosis and paracoccidioidomycosis, and more severe forms of
disease have been correlated with defective neutrophil function.
Killing of dimorphic fungi by macrophages, on the other hand,
appears to require that they be specifically or nonspecifically
activated. In terms of specific immunity, both clinical and experimental evidence to date, especially that accumulated for histoplasmosis and coccidioidomycosis, favors a major role for T-
lymphocytes and cellular immunity. For example, experimentally or
naturally immunosuppressed hosts are highly susceptible to H.
capsulatum infections, and protection against H. capsulatum and
C. immitis can be adoptively transferred to experimental animals
with T-enriched immune cell suspensions, while passive transfer
with serum is ineffective.

For cryptococcosis, the common clinical association with
impaired RES or lymphatic function also suggests the importance
of cellular defense mechanisms in specific immunity. Again,
phagocytosis is important, but the cells involved are rebuffed
by the polysaccharide capsule, so mechanisms apparently evolved
to enable leukocytes to damage the yeast in the absence of

ingestion. A great deal of evidence has accumulated implicating
the involvement of lymphocytes, particularly T-lymphocytes, in the
development of specific immunity. As with H. capsulatum, passive
immunization with immune sera is ineffective, whereas T-enriched
cell transfers do confer protection. Immunosuppressed animals
are similarly highly susceptible to C. neoformans infection, but
as with histoplasmosis, resistance can be restored by transplantation of thymus tissue.

Candida, perhaps the most important of the opportunistic fungi,
complicates the clinical picture of patients with such a wide
variety of underlying diseases, that defense against it is probably
much more complex than that against most other fungi. Phagocytic
cells can ingest the yeast form of Candida, but killing is relatively inefficient. On the other hand, serum-independent nonphagocytic mechanisms of killing have been demonstrated and may be
especially important against invasive pseudohyphae which are too
large to be ingested. The mechanism of specific immunity remains
poorly understood. Although lymphocyte-depleted animals are less
resistant to infection, cell transfers are only minimally effective
in recipients. Animals derive more benefit, it seems from the
passive transfer of serum. Moreover, they cannot mount a protective
response when their ability to produce antibody is impaired. These
findings suggest that humoral responses may be crucial to the
development of specific immunity to Candida, and that thymus-
dependent processes, although one facet of the defense are
secondary.

Clinical syndromes associated with aspergillosis and mucormycosis also indicate the importance of cellular mechanisms.
Phagocytosis of these agents is important in initial stages of
infection, with the primary effect being to prevent spore germination, although the spores may not be killed. Opsonins are
generally required for maximal activity. As with Candida, nonphagocytic mechanisms may be important in dealing with the
invasive hyphal forms. Virtually no experimental data are available with regard to the role of cell mediated immunity in
acquired immunity.

In summary, although it was generally believed for many years
that adequate defense against most fungi was of a cellular rather
than humoral nature, it is becoming increasingly clear that innate
defenses and probably antibody may be of equal or greater importance in selected fungal diseases, or at the very least, that all
three components, innate, humoral and cellular, are necessary for
complete eradication of fungi from the host.

ACKNOWLEDGMENTS

This work was supported by National Institutes of Health Training Grant AI 07152.

REFERENCES

1.   Bennett, J.E. (1974). in P.D. Hoeprich, ed., Infectious diseases, Harper and Row, New York. 945.
2.   Grappel, S.F., Bishop, T., and Blank, F. (1974). Bacteriol. Rev. 38:222.
3.   Sixbey, J.W., Fields, B.T., Sun, C.N., Clark, R.A., and Nolan, C.M. (1979). Infect. Immun. 23:41.
4.   Galgiani, J.N., Isenberg, R.A., and Stevens, D.A. (1978). Infect. Immun. 21:862.
5.   Galgiani, J.N., Yam, P., Petz, L.D., Williams, P.L., and Stevens, D.A. (1980). Infect. Immun. 28:944.
6.   Calich, V.L., Kipnis, T.L., Mariano, M., Neto, C.F., and Dias da Silva, W. (1979). Clin. Immunol. Immunopathol. 12:20.
7.   Ratnoff, W.D., Pepple, J.M., and Winkelstein, J.A. (1980). Infect. Immun. 30:147.
8.   Hill, G.A. and Marcus, S. (1960). J. Immunol. 85:6.
9.   Wu, W.G. and Marcus, S. (1963). J. Immunol. 91:313.
10. Miya, F. and Marcus, S. (1961). J. Immunol. 86:652.
11. DeSanchez, S.B. and Carbonell, L.M. (1975). Infect. Immun. 11:387.
12. Beaman, L. and Holmberg, C.A. (1980). Infect. Immun. 28:594.
13. Calderone, R.A. and Peterson, E. (1979). J. Reticuloendothel. Soc. 26:11.
14. Holland, P. (1971). in Ajello, L., Chick, E.W., and Furcolow, M.L., ed., Histoplasmosis - Proceedings of the second national conference. Charles C. Thomas, Springfield, Illinois. 380.
15. Howard, D.H. (1973). Infect. Immun. 8:412.
16. Howard, D.H. (1975). Mycoses. PAHO Scientific Publication No. 304.
17. Howard, D.H. (1977). in Iwata, K., ed., Recent advances in medical and veterinary mycology. University of Tokyo Press, Tokyo, 197.
18. Howard, D.H. (1981). Infect. Immun. 32:381.
19. Beaman, L. and Holmberg, C.A. (1980). Infect. Immun. 29:1200.
20. Deresinski, S.C., Levine, H.B., and Stevens, D.A. (1978). Mycopathologia 64:179.
21. Howard, D.H. (1964). J. Bacteriol. 87:33.
22. Howard, D.H. and Otto, V. (1969). Sabouraudia 7:186.
23. Howard, D.H., Otto, V., and Gupta, R.K. (1971). Infect. Immun. 4:605.
24. Howard, D.H. (1973). Infect. Immun. 8:577.

25. Howard, D.H. and Otto, V. (1977). Infect. Immun. 16:226.
26. Brummer, E., Morozumi, P.A., and Stevens, D.A. (1980).
         J. Reticuloendothel. Soc. 28:507.
27. Brummer, E., Morozumi, P.A., Philpott, D.E., and Stevens,
         D.A. (1981). Infect. Immun. 32:864.
28. Kauffman, C.A., Israel, K.S., Smith, J.W., White, A.C.,
         Schwartz, J.C.,and Brooks, G.F. (1978). Amer. J. Med.
         64:923.
29. Davies, S.F., Khan, M., and Sarosi, G.A. (1978). Amer. J.
         Med. 64:94.
30. Fraser, D.W., Ward, J.I., Ajello, L., and Plikaytis, B.D.
         (1979). J.A.M.A. 242:1631.
31. Sarosi, G.A. and Davies, S.F. (1979). Amer. Rev. Resp. Dis.
         120:911.
32. Severo, L.C., Londero, A.T., Geyer, G.R., and Porto, N.S.
         (1979). Mucopathologia 68:171.
33. Goihman-Yahr, M., Essenfeld-Yahr, E., De Albornoz, M.C.,
         Yarzábal, L., De Gomez, M.H., San Martín, B., Ocanto, A.,
         Gil, F., and Convit, J. (1980). Infect. Immun. 28:557.
34. Repine, J.E., Clawson, C.C., Rasp, F.L. Jr., Sarosi, G.A.,
         and Hoidal, J.R. (1978). Amer. Rev. Resp. Dis. 118:325.
35. Williams, D.M., Graybill, J.R., and Drutz, D.J. (1978).
         Infect. Immun. 21:973.
36. Adamson, D.M., and Cozad, S.C. (1969). J. Bacteriol. 100:1271.
37. Berry, C.L. (1969). J. Pathol. 97:653.
38. Williams, D.M., Graybill, J.R., and Drutz, D.J. (1981).
         Sabouraudia 19:39.
39. Tewari, R.P., Sharma, D., Solotorovsky, M., Lafemina, R.,
         and Balint, J. (1977). Infect. Immun. 15:789.
40. Tewari, R.P., Sharma, D.K. and Mathur, A. (1978). J. Infect.
         Dis. 138:605.
41. Kong, Y.M., Savage, D.C., and Levine, H.B. (1966). J. Immunol.
         95:1048.
42. Beaman, L., Pappagianis, D., and Benjamini, E. (1977). Infect.
         Immun. 17:580.
43. Beaman, L., Pappagianis, D., and Benjamini, E. (1979). Infect.
         Immun. 23:681.
44. Cozad, G.C. and Chang, C.T. (1980). Infect. Immun. 28:398.
45. Restrepo, A., Restrepo, M., de Restrepo, F., Zristiz'abal,
         L.H., Moncada, L.H., and Velez, H. (1978). Sabouraudia
         16:151.
46. Newberry, W.M., Walter, J.E., Chandler, J.W., and Tosh, F.E.
         (1967). Ann. Int. Med. 67:724.
47. Zimmerman, L.E. and Rappaport, H. (1954). Amer. J. Clin.
         Pathol. 24:1050.
48. Miller, D.G. (1965). in Samter, M. and Alexander, H.L., ed.,
         Immunological diseases. Little, Brown and Company,
         Boston, 382.
49. Kaplan, M.H., Rosen, P.P., and Armstrong, D. (1977). Cancer
         39:2265.

50. Gordon, M.A. (1975). in Al-Doory, Y., ed., The epidemiology
       of human mycotic diseases. Charles C. Thomas, Springfield,
       Illinois. 141.
51. Fraser, D.W., Ward, J.L., Ajello, L., and Plikaytis, B.D.
       (1979). J.A.M.A. 242:1631.
52. Bulmer, G.S. and Sans, M.D. (1967). J. Bacteriol. 94:1480.
53. Bulmer, G.S. and Sans, M.D. (1968). J. Bacteriol. 95:5.
54. Tacker, J.R., Farhi, F., and Bulmer, G.S. (1972). Infect.
       Immun. 6:162.
55. Farhi, F., Bulmer, G.S., and Tacker, J.R. (1970). Infect.
       Immun. 1:526.
56. Diamond, R.D., Root, R.K., and Bennett, J.E. (1972). J.
       Infect. Dis. 125:367.
57. Diamond, R.D. and Bennett, J.E. (1973). Infect. Immun. 7:231.
58. Mitchell, T.G. and Friedman, L. (1972). Infect. Immun. 5:491.
59. Bulmer, G.S. and Tacker, J.R. (1975). Infect. Immun. 11:73.
60. Diamond, R.D., May, J.E., Kane, M.A., Frank, M.M., and
       Bennett, J.E. (1974). J. Immunol. 112:2260.
61. Laxalt, K.A. and Kozel, T.R. (1979). Infect. Immun. 26:435.
62. Kozel, T.R. and Mastroianni, R.P. (1976). Infect. Immun.
       14:62.
63. McGaw, T.G. and Kozel, R. (1979). Infect. Immun. 25:262.
64. Gadebusch, H.H. (1972). CRC Critical Reviews in Microbiology.
       311.
65. Gadebusch, H.H. and Gikas, P.W. (1965). Amer. Rev. Resp.
       Dis. 92:64.
66. Gadebusch, H.H. and Johnson, A.G. (1966). J. Infect. Dis.
       116:551.
67. Schneerson-Porat, S., Shahar, A., and Aronson, M. (1965).
       J. Reticuloendothel. Soc. 2:249.
68. Kalina, M., Kletter, Y., Shahar, A., and Aronson, M. (1970).
       Proc. Soc. Exp. Biol. Med. 136:407.
69. Aronson, M. and Kletter, J. (1973). in Weiss, D., and
       Zuckerman, A., ed., Dynamic aspects of host-parasite
       interrelationships. Vol. 1. Academic Press, New York,
       132.
70. Kalina, M., Kletter, Y., and Aronson, M. (1974). Cell. Tis.
       Res. 152:165.
71. Diamond, R.D. (1974). Nature 247:148.
72. Diamond, R.D. and Allison, A.C. (1976). Infect. Immun. 14:716.
73. Binschadler, D.D. and Bennett, J.E. (1968). Ann. Int. Med.
       69:45.
74. Gadebusch, H.H. (1958). Proc. Soc. Exp. Biol. Med. 98:611.
75. Gordon, M.A. and Lapa, E. (1964). J. Infect. Dis. 114:373.
76. Louria, D.B. and Kaminski, T. (1965). Sabouraudia 4:80.
77. Lim, T.S. and Murphy, J.W. (1980). Infect. Immun. 30:5.
78. Abrahams, I. (1966). J. Immunol. 96:525.
79. Adamson, D.M. and Cozad, G.C. (1969). J. Bacteriol. 100:1271.
80. Gentry, L.O. and Remington, J.S. (1971). J. Infect. Dis.
       123:22.

81. Karaoui, R.M., Hall, N.K., and Larsh, H.W. (1977). Mykosen
       20:409.
82. Lim, T.S., Murphy, J.W., and Cauley, L.K. (1980). Infect.
       Immun. 29:633.
83. Fromtling, R.A., Blackstock, R., Hall, N.K., and Bulmer, G.S.
       (1979). Mycopathologia 68:179.
84. Graybill, J.R. and Taylor, R.L. (1978). Int. Arch. Allergy
       Appl. Immunol. 57:101.
85. Dykstra, M.A. and Friedman, L. (1978). Infect. Immun. 20:446.
86. Graybill, J.R. and Mitchell, L. (1979). Mycopathologia
       69:171.
87. Graybill, J.R. and Drutz, D.J. (1978). Cell. Immunol. 40:263.
88. Cauley, L.K. and Murphy, J.W. (1979). Infect. Immun. 23:644.
89. Nishimura, K. and Miyaji, M. (1979). Mycopathologia 68:145.
90. Graybill, J.R., Mitchell, L., and Drutz, D.J. (1979). J.
       Infect. Dis. 140:546.
91. Graybill, J.R. and Mitchell, L. (1978). Infect. Immun. 21:
       674.
92. Monga, D.P., Kumar, R., Mohapatra, L.N. and Malaviya, A.N.
       (1979). Infect. Immun. 26:1.
93. Fromtling, R.A., Blackstock, R., and Bulmer, G.S. (1979).
       in Kuttin, E.S. and Baum, G.L., ed., Proceedings of the
       VII Congress of ISHAM. International Congress Series
       No. 480.
94. Lehrer, R.I. (1970). Infect. Immun. 2:42.
95. Laforce, F.N., Mills, D.M., Iverson, K., Cousins, R., and
       Everett, E.D. (1975). J. Lab. Clin. Med. 86:657.
96. Staples, P.J., Boujak, J., Douglas, R.G. Jr., and Leddy, J.P.
       (1977). Clin. Immunol. Immunopathol. 7:157.
97. Kirkpatrick, C.H., Rich, R.R., and Bennett, J.E. (1971).
       Ann. Int. Med. 74:955.
98. Valdimarsson, H., Higgs, J.M., Wells, R.S., Yamamura, M.,
       Hobbs, J.R., and Halt, P.J.L. (1973). Cell. Immunol.
       6:348.
99. Denning, T.J.V. and Davies, R.R. (1973). Sabouraudia 11:210.
100. Cutler, J.E. (1977). Infect. Immun. 18:568.
101. Ray, T.L., Hanson, S., Ray, L.F., and Wuepper, K.D. (1979).
       J. Invest. Dermatol. 73:269.
102. Ray, T.L. and Wuepper, K.D. (1976). J. Clin. Invest.
       67:700.
103. Weiner, M.H. and Yount, W.J. (1976). J. Clin. Invest.
       58:1045.
104. Wright, C.D., Herron, M.J., Gray, G.R., Holmes, B., and
       Nelson, R.D. (1981). Infect. Immun. 32:731.
105. Sohnle, P.G., Frank, M.M., and Kirkpatrick, C.H. (1976).
       Clin. Immunol. Immunopathol. 5:340.
106. Morelli, R. and Rosenberg, L.T. (1971). J. Immunol. 107:476.
107. Kernbaum, S. (1975). Ann. Microbiol. 126A:75.
108. Yamamura, M. and Valdimarsson, H. (1977). Scand. J. Immunol.
       6:591.

109. Ferrante, A. and Thong, Y.H. (1979). Sabouraudia 17:293.
110. Morrison, R.P. and Cutler, J.E. (1981). J. Reticuloendothel.
     Soc. 29:23.
111. Gelfand, J.A., Hurley, D.L., Fauci, A.S., and Frank, M.M.
     (1978). J. Infect Dis. 138:9.
112. Stanley, V.C. and Hurley, R. (1969). J. Pathol. 97:357.
113. Evron, R. (1980). Infect. Immun. 28:963.
114. Ozato, K. and Uesaka, T. (1974). Jap. J. Microbiol. 18:29.
115. Cutler, J.E. and Poor, A.H. (1981). Infect. Immun. 31:1110.
116. Poor, A.H. and Cutler, J.E. (1981). Infect. Immun. 31:1104.
117. Peterson, E.M. and Calderone, R.A. (1977). Infect. Immun.
     15:910.
118. Peterson, E.M. and Calderone, R.A. (1978). Infect. Immun.
     21:506.
119. Arai, T., Mikami, Y., and Yokoyama, K. (1977). Sabouraudia
     15:171.
120. Louria, D.B. and Brayton, R.G. (1964). Proc. Soc. Exp. Biol.
     Med. 115:93.
121. Leigh, P.C.J., van den Barselaar, M.T., and vanFurth, R.
     (1977). Infect. Immun. 17:313.
122. Scherwitz, C. and Martin, R. (1979). Dermatologica 159:12.
123. Schuit, K. (1979). Infect. Immun. 24:932.
124. Diamond, R.D., Krzesicki, R., and Jao, W. (1978). J. Clin.
     Invest. 61:349.
125. Diamond, R.D. and Krzesicki, R. (1978). J. Clin. Invest.
     61:360.
126. Diamond, R.D., Clark, R.A., and Haudenschild, C.C. (1980).
     J. Clin. Invest. 66:908.
127. Diamond, R.D., Oppenheim, F., Nakagawa, Y., Krzesicki, R.,
     and Haudenschild, C.C. (1980). J. Immunol. 125:2797.
128. Diamond, R.D. and Haudenschild, C.C. (1981). J. Clin.
     Invest. 67:173.
129. Lehrer, R.I. (1969). J. Bacteriol. 99:361.
130. Zeya, H.I. and Spitznagel, J.K. (1966). J. Bacteriol.
     91:750.
131. Lehrer, R.I., Ladra, K.M., and Hake, R.B. (1975). Infect.
     Immun. 11:1226.
132. Patterson-Delafield, J., Martinez, R.J., and Lehrer, R.I.
     (1980). Infect. Immun. 30:180.
133. Peterson, E.M. and Calderone, R.A. (1978). Infect. Immun.
     21:506.
134. Hurtrel, B., Lagrange, P.H., and Michel, J.C. (1980).
     Ann. D'Immunol. C131:105.
135. Ruthe, R.C., Anderson, B.L., Cunningham, B.L.. and Epstein,
     R.B. (1978). Blood 52:493.
136. Mukherji, A.K. and Basu Mallick, K.C. (1972). Indian J.
     Med. Res. 60:1584.
137. Cutler, J.E. (1976). J. Reticuloendothel. Soc. 19:121.
138. Rogers, T.J., Balish, E., and Manning, D.D. (1976). J.
     Reticuloendothel. Soc. 20:291.

139. Miyake, T., Takeya, K., Nomoto, K., and Muraoka, S. (1977).
         Microbiol. Immunol. 21:703.
140. Lee, K.W. and Balish, E. (1981). J. Reticuloendothel. Soc.
         29:71.
141. Rogers, T.J. and Balish, E. (1978). Infect. Immun. 19:737.
142. Rogers, T.J. and Balish, E. (1977). J. Reticuloendothel.
         Soc. 22:309.
143. Marra, S. and Balish, E. (1974). Infect. Immun. 10:72.
144. Jakoniuk, P., Borowski, J., Jablonska-Strynkowska, W., and
         Badmajew, W. (1978). Archivum immunologial et therapiae
         experimentalis 26:637.
145. Williams, D.L., Cook, J.A., Hoffmann, E.O., and DiLuzio, N.R.
         (1978). J. Reticulendothel. Soc. 23:479.
146. Mardon, D.N. and Robinette, E.H. (1978). Canad. J. Microbiol.
         24:1515.
147. Maiti, P.K., Kumar, R., and Mohapatra, L.N. (1980). Infect.
         Immun. 29:477.
148. Sasada, M. and Johnston, R.B. Jr. (1980). J. Exp. Med.
         152:85.
149. Cummings, N.P., Pabst, M.J., and Johnston, R.B. Jr. (1980).
         J. Exp. Med. 152:1659.
150. Djawari, D., Hornstein, O.P., Gross, J., and Meinhof, W.
         (1977). Arch. Derm. Res. 260:159.
151. Mobacken, H., Lindholm, L., and Olling, S. (1977). Acta
         Dermatovener 57:335.
152. Cahill, L.T., Ainhander, E., and Glade, P.R. (1974). Cell
         Immunol. 14:215.
153. Axelson, N.H., Kirkpatrick, C.H., and Buckley, R.H. (1974).
         Clin. Exp. Immunol. 17:385.
154. Al-Doory, Y. (1970). Sabouraudia 8:41.
155. Mourad, S. and Friedman, L. (1968). Sabouraudia 6:103.
156. Pearsall, N.N., Adams, B.L., and Bunni, R. (1978). J.
         Immunol. 120:1176.
157. Moser, S.A. and Domer, J.E. (1980). Infect. Immun. 27:376.
158. Hurd, R.C. and Drake, C.H. (1953). Mycopathol. et Mycol.
         Appl. 6:290.
159. Oblack, D.L. and Holder, I.A. (1979). J. Med. Microbiol.
         12:503.
160. Sohnle, P.G., Frank, M.M., and Kirkpatrick, C.H. (1976).
         J. Immunol. 117:523.
161. Giger, D.K., Domer, J.E., Moser, S.A., and McQuitty, J.T. Jr.
         (1978). Infect. Immun. 21:729.
162. Grubek, H., Szymanska, D., Trenkner, E., Weyman-Rzucidlo, D.,
         and Okon, M. (1970). Acta Microbiol. Pol. Ser. A 2:21.
163. Segal, E., Schwartz, J., Altboun, Z, Vardinon, N., and Eylan,
         E. (1977). Microbios. 19:79.
164. Rogers, T.J., and Balish, E. (1978). Clin. Immunol.
         Immunopathol. 10:298.
165. Vardinon, N. and Segal, E. (1979). Exp. Cell. Biol. 47:275.
166. Gale, G.R. and Welch, A.M. (1961). Amer. J. Med. Sci. 88:604.

167. Owens, A.W., Shacklette, M.H., and Baker, R.D. (1965).
      Sabouraudia 4:179.
168. Marchevsky, A.M., Bottone, E.J., Geller, S.A., and Giger,
      D.K. (1980). Human Pathol. 11:457.
169. Frenkel, J.K. (1962). Lab. Invest. 11:1192.
170. Ford, S., Baker, R.D., and Friedman, L. (1968). J. Infect.
      Dis. 118:370.
171. Lehrer, R.I. and Jan, R.G. (1970). Infect. Immun. 1:345.
172. Young, R.C., Bennett, J.E., Vogel, C.L., Carbone, P.P., and
      DeVita, V.T. (1970). Medicine 49:147.
173. Merkow, L.P., Epstein, S.M., Sidransky, H., Verney, E., and
      Pardo, M. (1971). Amer. J. Pathol. 62:57.
174. Sidransky, H., Verney, E., and Beede, H. (1965). Arch.
      Pathol. 79:299.
175. Sidransky, H. and Friedman, L. (1959). Amer. J. Pathol.
      35:169.
176. Turner, K.J., Hackshaw, R., Papadimitriou, J., Wetherall,
      J.D., and Perrott, J. (1975). Immunology 29:55.
177. Diamond, R.D., Krzesicki, R., Epstein, B., and Jao, W.
      (1978). Amer. J. Pathol. 91:313.
178. Corbel, M.J. and Eades, S.M. (1975). J. Med. Microbiol.
      8:551.
179. Corbel, M.J. and Eades, S.M. (1976). Sabouraudia 14:17.
180. Corbel, M.J. and Eades, S.M. (1976). J. Hygiene 77:221.
181. Corbel, M.J. and Eades, S.M. (1977). Mycopathologia 62:117.

ANTIBODY-INDEPENDENT MECHANISMS IN THE DEVELOPMENT OF ACQUIRED

IMMUNITY TO MALARIA

William P. Weidanz and James L. Grun

The Malaria Research Unit, Department of Microbiology
and Immunology, Hahnemann Medical College and Hospital
Philadelphia, PA   19102

The plasmodia are hemoprotozoan parasites which cause malaria
in a variety of animal species, including man. Infection with
these parasites stimulates acquired immune responses which are
directed primarily against blood stages and, when successful,
bring about a reduction in the number of circulating plasmodia
and/or protect the host against reinfection with homologous parasites (1,2). While the actual mechanisms of resistance are ill-
defined, it appears that immunity to malaria requires the participation of both T and B lymphocyte systems.

Antibodies reactive with plasmodia are readily demonstrated
in the sera of hosts which have recovered from acute disease.
Certain of these antibodies have been shown to have neutralizing
activity against merozoites, thus preventing the infection of new
target cells (3). The same or other antibodies may serve to
agglutinate merozoites and/or to opsonize them as well as parasitized erythrocytes for eventual clearance by phagocytic mechanisms
(4-7). Several reports have indicated that plasmodia or parasitized erythrocytes may be destroyed by antibody-dependent cell
cytotoxicity mechanisms (8,9). Further support for the protective
role of antibody in malaria comes from numerous passive transfer
studies in which infection could be delayed, prevented, or cured
following the injection of immune globulin or hyperimmune serum
(reviewed in reference 10). Also, experimental animals rendered
B-cell deficient prior to infection routinely died when injected
with avirulent strains of plasmodia (11-13). While similar results
were observed when T-cell deprived hosts were infected with the
same parasites, it was reasoned that the absence of T-helper cells
prevented these animals from making antibodies (14).

It should be noted that the mere presence of serum antibodies reactive with plasmodia does not insure that the host will resist infection with the homologous parasites. While monkeys immunized with P. knowlesi merozoites in complete Freund's adjuvant produced neutralizing antibodies and were immune to challenge infection, monkeys immunized with merozoites in other adjuvants produced comparable titers of neutralizing antibodies but remained suscept-ible to challenge infection (15). Also, Miller et al. (16) observed that there was no correlation between functional immunity in malaria and serum levels of merozoite neutralizing antibody.

Other findings suggest that mechanisms in addition to those mediated by antibody may have a protective role in malaria. For example, Cox (17) observed that a considerable degree of "hetero-logous immunity" existed among hemoprotozoan parasites; i.e., infection with one species of murine plasmodia subsequently protected the host against infection by heterologous species of plasmodia or babesia. The strength of the immunity induced far exceeded that which would be explained by levels of cross-reacting antibodies in the serum of the resistant host (18). More recently, Allison, Clark, and colleagues achieved a considerable degree of resistance against various hemoprotozoa by stimulating mice with assorted immunomodulators prior to challenge infection (19,20). The resistance displayed by the treated mice could not be attri-buted to antibodies. While it cannot be denied that the use of immunomodulators may induce nonspecific immunity, the fact that such treated mice appear to destroy hemoprotozoan parasites by nonantibody-mediated mechanisms is quite interesting. The same or similar mechanisms of resistance may be functional during natural infection in conjunction with or apart from antibody-dependent mechanisms.

Additional evidence suggesting that antibody-independent mechanisms contribute to acquired immunity in malaria has been obtained from experiments in which immunity to assorted plasmodia was induced in B-cell deficient hosts. Initially, chickens were rendered B-cell deficient by means of combined chemical bursectomy (21). When subsequently infected with P. gallinaceum, treated chickens uniformly died, whereas their immunologically intact hatchmates spontaneously resolved their acute infections and were immune to subsequent challenge infections (12). However, when acute P. gallinaceum infections in B-cell deficient chickens were controlled with subcurative chemotherapy, the birds developed chronic low-grade infections and resisted challenge infection with homologous parasites. These data suggest that drug-controlled acute malarial infections in B-cell deficient hosts activated a nonantibody-mediated mechanisms of immunity which was capable of limiting the numbers of parasites within the blood but which was not sufficient to sterilize the infection. Further, they suggest that whereas a B-cell product, presumably antibody, was essential

for the host to survive acute infection or to terminate a chronic
one, the maintenance of chronicity as well as "re-infection"
immunity was dependent upon nonantibody-mediated mechanisms of
immunity.

Because of difficulties inherent in experimenting with
chickens, a mouse model was developed to study further the phenomenon of antibody-independent immunity to malaria. To accomplish
this, goat antibody reactive with mouse IgM was raised against
the mouse myeloma protein, MOPC104E (μ_2, λ_2). Selected litters
of mice were subsequently rendered B-cell deficient by lifelong
treatment with goat anti-IgM serum (anti-μ). The characteristics
of the B-cell deficiency produced in anti-μ treated mice are
summarized in Table 1. As observed by others (22,23), treatment
with anti-μ produced a severe B-cell deficiency in mice. Prior
to experimentation, the sera of anti-μ treated mice were checked
for the absence of detectable mouse IgM and the presence of goat
anti-μ.

Table 1. The Effect of Anti-μ Treatment on Assorted Parameters
          of Immunity

Parameter Measured	Level or Response In Comparison With Immunologically Intact Mice
Immunoglobulin level	Decreased
Serum IgM	Undetectable
Membrane Bound	Undetectable
Antibody Response to:	
Goat Immunoglobulin	Undetectable
Sheep Erythrocytes	Undetectable
DNP	Undetectable
Mitogen Response to:	
LPS	Undetectable
Con A	Normal
Contact Dermatitis to:	
Oxazolone	Normal

     Experimental infections with P. yoelii, 17X strain,were
initiated with 10^5 to 10^6 parasites injected intravenously or
intraperitoneally. Infections were followed by microscopic examination of thin blood films prepared from tail blood and stained
with Giemsa. The kinetics of typical infections in immunologically
intact and B-cell deficient mice is diagrammed in Fig. 1. Whereas
this parasite produced a self-limiting infection in normal mice,
it uniformly killed B-cell deficient mice by the third week after
infection (13). Since this particular isolate of P. yoelii was
resistant to chloroquine, B-cell deficient mice infected with
P. yoelii were treated orally with clindamycin to terminate their
acute infections. Subsequently, B-cell deficient mice developed
recurrent malaria and died or, of greater significance, developed
low-grade chronic parasitemias, as shown in Fig. 1. The latter
mice controlled their chronic infections and resisted challenge
infection with fresh parasites (24). Thus, they resembled B-cell
deficient chickens rendered immune to P. gallinaceum. These
results were encouraging, since the mode of inducing the B-cell
deficiency was quite different from that employed in the chicken.
Why certain mice developed recurrent disease after drug therapy
while others became immune remains to be elucidated. Preliminary
results with different strains of mice suggest that this form of
immunity has a genetic basis. For example, it could not be
induced in C57BL/10 mice, whereas approximately 50% of B-cell
deficient BALB/c mice were rendered "immune" by this technique.

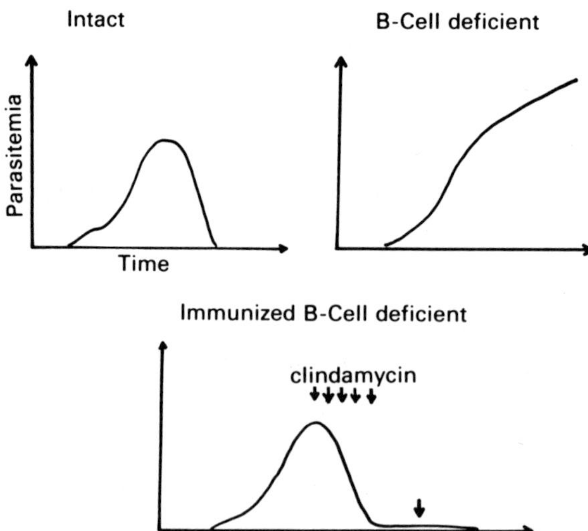

Figure 1. P. yoelii malaria in B-cell deficient mice.

The maintenance of antibody-independent immunity to P. yoelii
in B-cell deficient mice required the continued presence of parasites in the blood. When "immune" B-cell deficient mice were
divided into two groups on the basis of patent parasitemia two
months after the initial infection and then challenged with P.
yoelii, lethal infections occurred only in mice which lacked
detectable parasites prior to challenge infection.

In subsequent experiments, B-cell deficient mice were
infected with P. vinckei which is normally lethal in immunologically intact mice. When such B-cell deficient mice were rescued
from otherwise lethal disease by treatment with chloroquine, they
resisted challenge infection with homologous parasites. This
indicated that the phenomenon of antibody-dependent immunity to
malaria could be readily induced by a variety of plasmodial
species; however, in each instance, it was necessary to terminate
acute infections with drugs in order to observe the induced
immunity.

However, when B-cell deficient mice were infected with P.
chabaudi adami, an avirulent malarial parasite, the results were
most surprising. As shown in Fig. 2, acute P. chabaudi infections
are identical in B-cell deficient mice and in immunologically
intact controls. Thus, B-cell deficient mice infected with P.
chabaudi resolved their acute infections without drug intervention (25). Whereas B-cell deficient mice developed chronic low-
grade infections with parasitemias <0.1%, intact mice cleared
P. chabaudi from their blood. Both types of mice resisted
challenge infection with homologous parasites. These data
indicate that B lymphocytes are not required by the host to
resist acute P. chabaudi infections but would appear to be
necessary for the termination of chronicity.

In order to determine the specificity of antibody-independent
immunity, groups of "immune" B-cell deficient mice were challenged
with homologous as well as heterologous species of plasmodia. In
addition to the three species listed above, mice were tested for
their ability to resist P. berghei, NYU strain, which is lethal

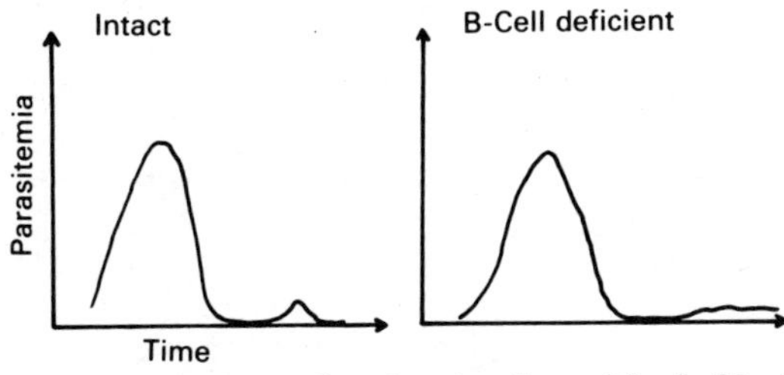

Figure 2.   P. chabaudi malaria in B-cell deficient mice.

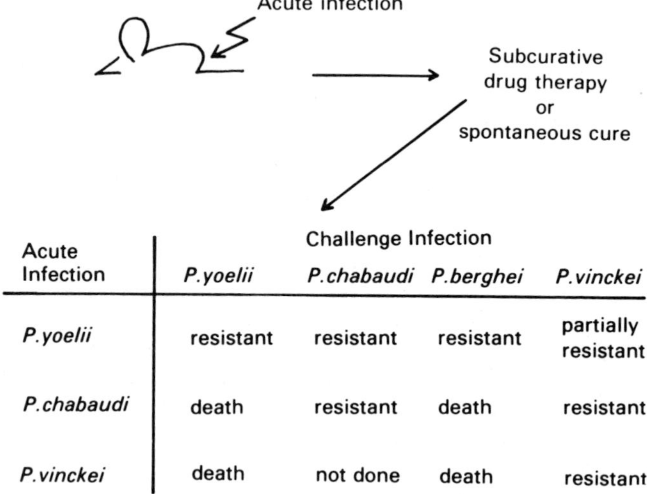

Acute Infection	P.yoelii	P.chabaudi	P.berghei	P.vinckei
P.yoelii	resistant	resistant	resistant	partially resistant
P.chabaudi	death	resistant	death	resistant
P.vinckei	death	not done	death	resistant

Figure 3.   Specificity of antibody-independent immunity to malaria.

for intact mice. The results summarized in Fig. 3 show that the
specificity of antibody-independent immunity to malaria is
dependent upon the immunizing strain. Whereas mice immunized with
P. yoelii showed significant resistance to the three heterologous
species of plasmodia, B-cell deficient mice immunized with either
P. chabaudi or P. vinckei died when infected with the avirulent
P. yoelii. Mice immune to P. chabaudi resisted P. vinckei. While
the reason for this difference in specificity remains to be deter
mined, it should be noted that P. chabaudi and P. vinckei infect
mature erythrocytes, whereas P. yoelii and P. berghei perfer
reticulocytes. It is possible that the difference in specificity
induced by the immunizing strain reflects differences in the
mechanisms of how parasitized erythrocytes are removed and/or
destroyed by the host.

It is apparant that the experimental murine infections
discussed thus far have three possible outcomes: (1) acute
infection is not terminated and death results; (2) acute infection
is terminated and a state of chronicity as evidenced by continued
low-grade parasitemia develops; or (3) acute infection is terminated
and the parasites are totally cleared from the blood. In addition,
animals which have survived acute infection may resist challenge
infection with homologous parasites as well as with certain heterologous parasites. These relationships and their association with
immune events mediated by T and B lymphocyte systems are shown in
Fig. 4. Acute malaria resulting from infection with either lethal
species, P. vinckei or P. berghei, is fatal in immunologically
intact mice, possibly because immune mechanisms are not activated
in time to prevent pathology resulting in death. While such

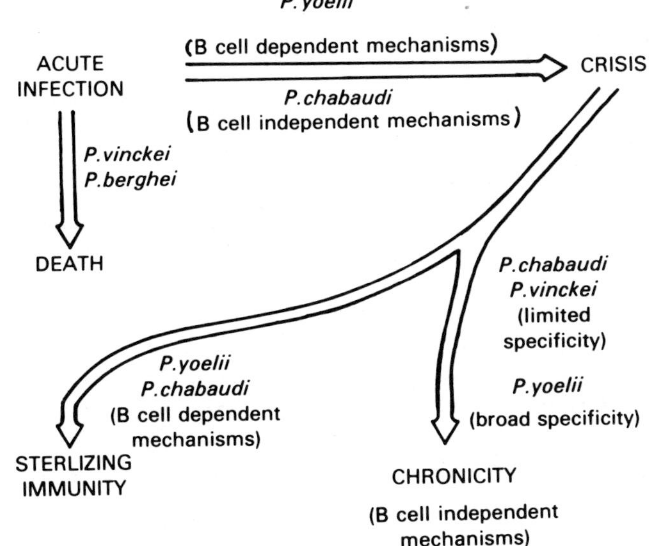

Figure 4. Patterns of acquired immunity to malaria.

mechanisms of resistance can be induced in B-cell deficient mice
by immunizing them with P. yoelii or P. chabaudi, it is apparent
from the specificity studies that different mechanisms of immunity
are required to prevent infections by each species of the lethal
parasites. For example, mice immunized by drug-cured P. yoelii
infections resisted both lethal parasites, whereas mice immunized
with P. chabaudi were immune to P. vinckei but not to P. berghei.
Thus, B lymphocytes and/or their products are necessary to control
acute P. yoelii infections. The termination of acute P. chabaudi
infections is B-cell independent; however, it appears that B cells
and/or their products (antibodies) are required for the complete
termination of parasitemia in both P. chabaudi and P. yoelii
infections. Whether these different parasite species are removed
by similar or different B-cell dependent mechanisms remains to be
determined. Likewise, the mechanisms of terminating acute P.
yoelii infections versus sterilization have not been elucidated.
While B cells may be required for both events, antibodies differing
in titer, isotype, specificity, or avidity might be responsible
for parasite clearance in different stages of the infection. Also,
the actual mechanisms of parasite injury could be different. It
seems that antibody-independent mechanisms induced by P. yoelii and
P. chabaudi are qualitatively rather than quantitatively different,
since the kinetics of P. yoelii infections in B-cell deficient mice
immune to P. chabaudi was identical to that in nonimmunized B-cell
deficient mice. This suggests that mechanisms of antibody-independent immunity in addition to those acting against P. chabaudi are
needed to control P. yoelii in blood.

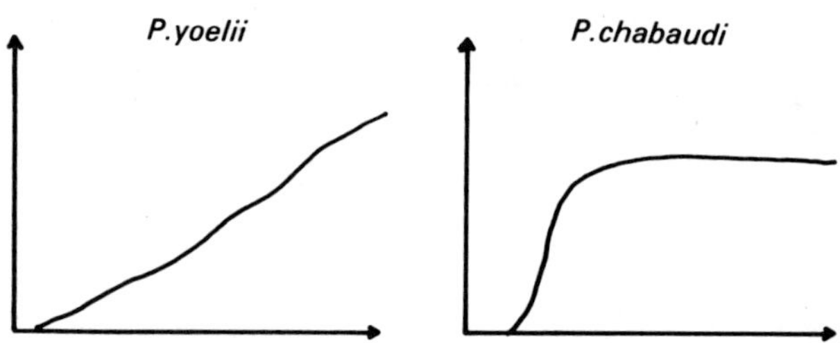

Figure 5. Malaria in athymic nude mice.

To determine whether antibody-independent immunity to malaria
was T-cell dependent, athymic nude mice were infected with either
P. yoelii or P. chabaudi. The kinetics of both infections is shown
in Fig. 5. None of the infected nude mice were capable of terminating their infections and had significant parasitemias at the
time of death (13,25). In addition, when nude mice were treated
for the P. yoelii infections, they uniformly developed recurrent
lethal malaria in the absence of continued drug therapy (13). As
shown in Table 2, lethal P. yoelii infections could be prevented
in nude mice by prior thymic grafting or by treatment with hyperimmune serum. While antiserum-treated nude mice did not develop
recurrent malaria, they were susceptible to challenge with P.
yoelii once the transferred antibodies had cleared from the blood.
These data suggest that antibody-independent immunity is T-cell
dependent.

The requirement for a functioning thymus-dependent lymphocyte
system in the development of antibody-independent immunity to
malaria is further supported by the observation that B-cell
deficient as well as immunologically intact mice, when treated
with rabbit and anti-thymocyte serum and infected with P. chabaudi,
were unable to terminate their acute infections as did mice
injected with normal rabbit serum (Fig. 6).

Table 2. P. yoelii Infection in Nude Mice: Modulating Effects
          Effects of Antibody or Thymic Grafting

Treatment	Outcome
None	Lethal malaria
Thymus Graft	Self-limiting infection
Hyperimmune Serum	Prevention of recurrent malaria in drug-treated mice

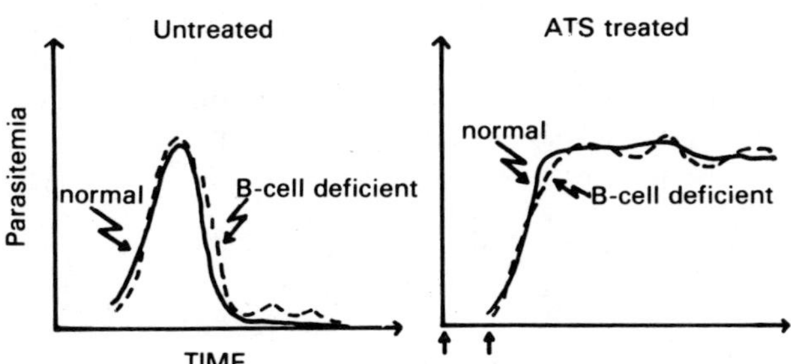

Figure 6.   P. chabaudi infections in ATX-treated normal and
B-cell deficient mice.

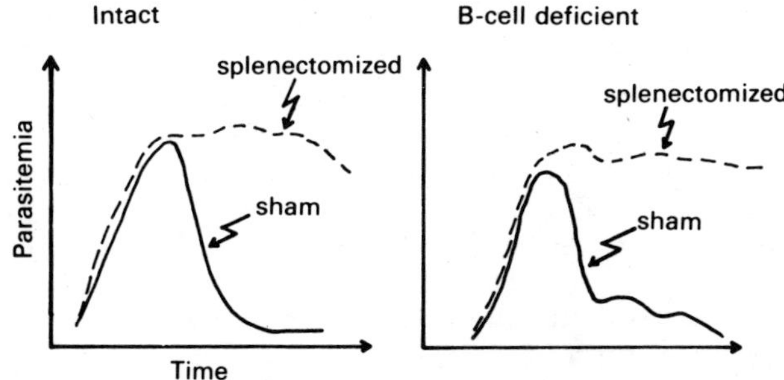

Figure 7.   Effects of splenectomy on P. chabaudi infections in
normal and B-cell deficient mice.

     The presence of an intact spleen also appeared to be necessary
for the development of antibody-independent immunity to malaria.
When B-cell deficient mice or intact mice were splenectomized
prior to infection with P. chabaudi, they were unable to terminate
their acute infections but instead developed chronic high-level
parasitemias (Fig. 7).   The injection of dispersed spleen cells
into such mice did not alter their parasitemias.   This suggests
that the structure of the spleen plays a significant role in the
induction of antibody-independent immunity to P. chabaudi malaria.

     What mechanisms might explain antibody-independent immunity
to malaria?   We can visualize that parasitic infection is a dynamic
process which activates a multitude of resistance mechanisms in
the repertoire of the host.   The array of these host responses may
change with time, since those which effect a reduction in parasite
numbers or which at least regulate parasitemia are probably prolonged in their duration and may assume a larger role as infection
progresses.   Unsuccessful responses are probably terminated.

Thus, a variety of nonprotective products of the activated
immune system may be detected in host tissues and, because assorted
cell types have been activated by the infection, the host may be
nonresponsive to heterologous antigens introduced at this time.
Failure to achieve protective immune responses results in the death
of the host. Similarly, different species of parasites may vary
in their susceptibility to host responses and may be controlled by
more than one resistance mechanism as our results suggest. The
sum total of the various host responses interacting with a particular parasite must then determine the course and the outcome of
infection.

Possible mechanisms of resistance which could contribute to
antibody independent immunity to malaria have been described by
various authors. Clark and his colleagues have proposed that intracellular hemoprotozoa may be destroyed by soluble factors released
from activated macrophages (26). Candidate substances include TNF,
and interferon (20,27), and recent studies by Playfair and
colleagues have shown that sera rich in TNF can bring about
reduction in parasite numbers _in vitro_ (28). Malaria is known to
activate macrophages (29,30) and macrophage activation during
malaria appears to be T-cell dependent (31). While it is possible
that such factors may destroy a significant proportion of circulating parasites at a particular time during infection, they were
not present in sufficient quantities to prevent chronic malaria
from occurring in B-cell deficient mice. If such factors are
responsible for crisis during acute malaria where the parasitemias
may exceed 20%, one wonders why they fail to terminate parasitemias
of less than 0.1% in chronically infected mice. The parasites
obtained from chronically infected mice do not show altered
infectivity and the acute infections which they initiate are
similarly terminated by crisis.

It has also been suggested by Allison and his colleagues
that NK cells may kill intracellular protozoa (32,33). This
concept was based in part on the correlation between high NK
activity and resistance to hemoprotozoa characteristic of certain
inbred strains of mice. Yet nude mice which have high NK activity (34) always died when infected with plasmodia (11,25,35). In
addition, recent genetic experiments conducted by Lyanga and
Skamene (36) showed that resistance to _P. chabaudi_ infection
could be disassociated from NK activity.

Enhanced clearance of parasites or parasitized erythrocytes
has been observed in malarious hosts (10,37). However, in most
instances the presence of antibodies was required for the uptake
of plasmodia by cells of the reticuloendothelial system. As
stated previously, activation of macrophages occurs in malaria,
and we have observed that B-cell deficient mice cleared colloidal
carbon from their blood in an accelerated manner as long as they

had patent parasitemia. Despite the fact that parasites and
malarial pigment can be demonstrated within the phagocytic cells
of such mice, it is not known whether intact parasitized erythrocytes are ingested or whether the parasites are released from
erythrocytes prior to their uptake. Recent studies by Quinn and
Wyler (38) suggest that alterations in the microcirculation of
the spleen, combined with the increased deformability of parasitized
erythrocytes, can be associated with the enhanced clearance of
parasitized erythrocytes which occurs at the time of crisis.
They point out that the close association of phagocytic cells
with slowly moving parasitized erythrocytes could allow for
intraerythrocytic parasite killing by soluble mediators released
from macrophages. Since crisis does not occur in nude mice, it
seems likely that this nonspecific mechanism requires the presence
of functioning T cells for its induction. It is interesting to
note that splenic macrophage recruitment as well as splenomegaly
resulting from malarial infection are T-cell dependent phenomena
(31,39).

    Several reports from John Playfair's laboratory suggest that
the liver may also play a significant role in resistance to
malaria (40,41). They observed that labeled bone marrow and
lymph node cells as well as parasitized erythrocytes accumulated
to a greater degree in the livers rather than in the spleens of
infected mice. Large numbers of T cells could be extracted from
the livers of these animals and increased numbers of activated
macrophages and monocytes were observed in the liver sinusoids of
infected mice. These changes were especially marked in the livers
of vaccinated mice infected with P. yoelii. Cells extracted from
the livers of vaccinated mice taken shortly after infection were
capable of killing P. yoelii in vitro. While the mechanism(s) of
killing was not determined, it is possible that nonspecific
soluble mediators produced by activated macrophages were involved.

    In addition to the foregoing T-cell dependent functions, i.e.,
macrophage recruitment and activation which might contribute to
antibody-independent immunity, it is possible that enhanced
erythropoiesis resulting from T-cell activation during malaria
may play a significant role in resolving nonlethal infections.
Activation of T cells does occur in these infections (42); and
recent evidence suggests that T cells are intimately involved in
the regulation of erythropoiesis (43,44). Erythropoiesis during
malaria appears to have a marked influence on the kinetics of
infection caused by plasmodial species which prefer reticulocytes.
The delayed progress of P. yoelii infections accompanied by increasing anemia which we have observed in nude mice could be correlated
with the availability of target cells in the blood of these
deficient mice. While nude mice are not anemic under normal
circumstances, they may fail to compensate for infection-related
erythrocyte loss due to the absence of a responding T cell which

ordinarily would modulate erythropoiesis. It is of interest to
note that Lu et al. (45), who studied Ia-bearing macrophages in
peritoneal exudates of euthymic and athymic mice after Listeria
infection, demonstrated two levels of control: a basal T-cell
independent level and a stimulated level dependent on the activation of T cells. Similarly, the activation of T cells during
malaria may be needed to stimulate erythropoiesis above a basal
T-cell independent level. Finally, crisis in _P. yoelii_ infection
occurs at a time when large numbers of reticulocytes appear in
peripheral blood. It is conceivable that the influx of new
reticulocytes into the blood could serve to dilute further the
number of parasitized erythrocytes at a time when parasitized
erythrocytes as well as altered but uninfected erythrocytes were
being selectively cleared from the blood by phagocytic mechanisms.

In conclusion, mechanisms of antibody-independent immunity to
malaria remain to be elucidated. One can visualize that plasmodia
are cleared from the blood by multiple components of resistance
which may be nonspecific or even nonimmunologic and which may
function individually or collectively, but all of which are
activated through the stimulation of T lymphocytes by plasmodia
and their products. The selective activation of particular
mechanisms as well as their suppression may be seen as a function
of the infecting species. Antibodies could serve to amplify
certain of these mechanisms and provide them with specificity.

Further insight into nonantibody-mediated mechanisms responsible for parasite elimination should prove useful in the design of
future malarial vaccines. Also, efforts now being undertaken to
identify and characterize parasite substances having biological
activity may provide a new generation of immunomodulators which
could be employed to alter the behavior of selected cell types
in a desired manner.

ACKNOWLEDGMENT

This research was supported in part by the U.N.O.P./World/
Bank/W.H.O. Special Program for Research and Training in Tropical
Diseases and U.S.P.H.S. research grant Al-12710. We wish to
thank our colleague Carole A. Long for her helpful comments.

REFERENCES

1. Cohen, S. (1979). Immunity to malaria. Proc. Roy. Soc. Lond.
         B. 203:323.
2. Brown, K.N. (1976). Resistance to malaria. In: Immunity to
         parasitic infections. (Eds. S. Cohen & E.H. Sadun),
         pp. 268-295. Blackwells: Oxford.

3. Cohen, S., Butcher, G.A. and Crandall, R.B. (1969). Action of
       malarial antibody in vitro. Nature, Lond. 223:368.
4. Zuckerman, A. (1945). In vitro opsonic tests with Plasmodium
       gallinaceum and Plasmodium lophurae. J. Infect. Dis.
       77:28.
5. Diggs, C.L. and Osler, A.G. (1975). Humoral immunity in rodent
       malaria. III: Studies on the site of action. J. Immunol.
       114:1243.
6. Miller, L.H., Aikawa, M. and Dvorak, J.A. (1975). Malaria
       (Plasmodium knowlesi) merozoites: Immunity and the
       surface coat. J. Immunol. 114:1237.
7. Hamburger, J. and Kreier, J.P. (1975). Antibody-mediated
       elimination of malaria parasites (Plasmodium berghei)
       in vivo. Infect. Immun. 12:339.
8. Coleman, R.M., Rencricca, N.J., Stout, J.P., Brissette, W.H.
       and Smith, D.M. (1975). Splenic mediated erythrocyte
       cytotoxicity in malaria. Immunology 29:49.
9. Brown, J. and Smalley, M.E. (1980). Specific antibody-dependent
       cellular cytotoxicity in human malaria. Clin. Exp.
       Immunol. 41:423.
10. Kreier, J.P. and Green, T.J. (1980). The vertebrate host's
       immune response to plasmodia. In: Malaria (J.P. Kreier,
       Ed.), Vol. 3, p. 111. Academic Press, New York.
11. Weinbaum, F.L., Evans, C.B. and Tigelaar, R.E. (1976).
       Immunity to Plasmodium berghei yoelii in mice. I. The
       course of infection in T-cell and B-cell deficient
       mice. J. Immunol. 117:1999.
12. Rank, R.G. and Weidanz, W.P. (1976). Nonsterilizing immunity
       in avian malaria; an antibody-independent phenomenon.
       Proc. Soc. Exp. Biol. & Med. 151:257.
13. Roberts, D.W., Rank, R.G., Weidanz, W.P. and Finerty, J.F.
       (1977). The prevention of recrudescent malaria in
       adult nude mice by thymic grafting or by treatment
       with hyperimmune serum. Infect. Immun. 16:821.
14. Cohen, S. and Butcher, G.A. (1972). The immunologic response
       to plasmodium. Am. J. Trop. Med. & Hyg. 21:713.
15. Butcher, G.A., Mitchell, G.H. and Cohen, S. (1978). Antibody-
       mediated mechanisms of immunity to malaria induced by
       vaccination with Plasmodium knowlesi merozoites.
       Immunology 34:77.
16. Miller, L.H., Powers, K.G. and Shiroishi, T. (1977).
       Plasmodium knowlesi: Functional immunity and anti-
       merozoite antibodies in rhesus monkeys after repeated
       infection. Exptl. Parasit. 41:105.
17. Cox, F.E.G. (1970). Protective heterologous immunity
       between malarial parasites and piroplasms in mice.
       Bull. Wld. Hlth. Org. 43:337.
18. _________ (1978). Heterologous immunity between piroplasms
       and malaria parasites: The simultaneous elimination of
       Plasmodium vinckei and Babesia microti from the blood
       of doubly-infected mice. Parasitol. 76:55.

19. Allison, A.C. and Clark, I.A. (1977). Specific and nonspecific
    immunity to hemoprotozoa. Am. J. Trop. Med. & Hyg.
    26:216.
20. Clark, I.A. (1979). Protection of mice against *Babesia microti*
    with cord factor, COAM, zymosan, glucan, *Salmonella*, and
    Listeria. Parasite Immunol. 1:179.
21. Lerman, S.P. and Weidanz, W.P. (1969). Antibody deficiency
    syndromes resulting from cyclophosphamide treatment of
    bursectomized chickens. Bact. Proc., p. 91.
22. Manning, D.D. (1975). Heavy chain isotype suppression: A
    review of the immunosuppressive effects of heterologous
    anti-Ig heavy chain antisera. J. Res. 18:63.
23. Gordon, J. (1979). The B lymphocyte-deprived mouse as a tool
    in immunobiology. J. Immunol. Meth. 25:227.
24. Roberts, D.W. and Weidanz, W.P. (1979). T-cell immunity to
    malaria in the B-cell deficient mouse. Am. J. Trop.
    Med. Hyg. 28:1.
25. Grun, J.L. and Weidanz, W.P. (1981). Immunity to *Plasmodium
    chabaudi adami* in the B-cell deficient mouse. *Nature*
    290:143.
26. Clark, I.A., Cox, F.E.G. and Allison, A.C. (1977). Protection
    of mice against *Babesia* spp. and *Plasmodium* spp. with
    killed *Corynebacterium* parvum. Parasitol. 74:90.
27. Clark, I.A., Vivelizier, J-L., Carswell, E.A. and Wood, P.R.
    (1981). Possible importance of macrophage-derived
    mediators in acute malaria. Infect. Immun. 32:1058.
28. Taverne, J., Dockrell, H.M. and Playfair, J.H.L. (1981).
    Endotoxin-induced serum factor kills malarial parasites
    in vitro. Infect. Immun. 33:82.
29. Taliaferro, W.H. and Mulligan, H.W. (1937). The histopathology
    of malaria with special reference to the function and
    origin of the macrophages in defense. Indian. Med. Res.
    Mem. 29:1.
30. Shear, H.L., Nussenzweig, R.S. and Bianco, C. (1979). Immune
    phagocytosis in murine malaria. J. Exp. Med. 149:1288.
31. Roberts, D.W. and Weidanz, W.P. (1978). Splenomegaly, enhanced
    phagocytosis and anemia are thymus-dependent responses
    to malaria. Infect. Immun. 20:728.
32. Allison, A.C., Christensen, J., Clark, I.A., Elford, B.C. and
    Eugui, E.M. (1979). The role of the spleen in protection
    against murine babesia infections. In: The Role of
    the Spleen in the Immunology of Parasitic Diseases.
    Schwabe & Co., A.G. Basal.
33. Eugui, E.M. and Allison, A.C. (1980). Differences in
    susceptibility of various mouse strains to haemoprotozoan
    infections: Possible correlation with natural killer
    activity. Parasite Immunol. 2:277.
34. Kindred, B. (1979). Nude mice in immunology. Prog. Allergy
    26:137.

35. Clark, I.A. and Allison, A.C. (1974). _Babesia microti_ and
    Plasmodium berghei yoelii infections in nude mice.
    Nature 252:328.
36. Lyanga, J.J. and Skamene, E. (1981). Genetic control of
    malaria infection in mice. Fed. Proc., p. 1115.
37. Quinn, T.C. and Wyler, D.J. (1979). Intravascular clearance
    of parasitized erythrocytes in rodent malaria. J. Clin.
    Invest. 63:1187.
38. Wyler, D.J., Quinn, T.C. and Chen, L-T. (1981). Relationship
    of alterations in splenic clearance function and micro-
    circulation to host defense in acute rodent malaria.
    J. Clin. Invest.
39. Wyler, D.J. and Gallin, J.I. (1977). Spleen-derived mono-
    nuclear cell chemotactic factor in malarial infections:
    A possible mechanism for splenic macrophage accumula-
    tion. J. Immunol. 118:478.
40. Playfair, J.H.L., DeSouza, J.B., Dockrell, H.M., Agomo, P.U.
    and Taverne, J. (1979). Cell-mediated immunity in the
    liver of mice vaccinated against malaria. Nature,
    282:731.
41. Dockrell, H.M., DeSouza, J.B. and Playfair, J.H.L. (1980).
    The role of the liver in immunity to blood-stage
    murine malaria. Immunol. 41:421.
42. Jayawardena, A.N., Targett, G.A.T., Leuchars, E., Carter, R.L.,
    Doenhoff, M.J. and Davies, A.J.S. (1975). T-cell
    activation in murine malaria. Nature 258:149.
43. Wiktor-Jedrzejczak, W., Sharkis, S., Ahmed, A., and Sell, K.
    (1977). Theta-sensitive cell and erythropoiesis:
    Identification of a defect in W/W^V anemic mice.
    Science 196:313.
44. Nathan, D.G., Chess, L., Hillman, D.G., Clarke, B., Breard, J.,
    Merler, E. and Housman, D.E. (1978). Human erythroid
    burst-forming unit requirement for proliferation _in
    vitro_. J. Exp. Med. 147:324.
45. Lu, C.Y., Peters, E. and Unanue, E.R. (1981). Ia-bearing
    macrophages in athymic mice: Antigen presentation and
    regulation. J. Immunol. 126:2496.

INTRACELLULAR DESTRUCTION OF <u>LEISHMANIA</u> <u>TROPICA</u> BY MACROPHAGES
ACTIVATED <u>IN</u> <u>VIVO</u> WITH <u>MYCOBACTERIUM</u> <u>BOVIS</u> STRAIN BCG[1]

Michael G. Pappas, Charles N. Oster and Carol A. Nacy

Department of Immunology, Walter Reed Army Institute
of Research, Washington, D.C. 20012

<u>Leishmania</u> <u>tropica</u>, an intracellular protozoan parasite,
infects and replicates in macrophages of C3H/HeN and C3H/HeJ mice
<u>in vitro</u> (1). Exposure of resident peritoneal macrophages to
amastigotes for 1 hr results in 20-25% infected macrophages.
Intracellular parasites increase from 1.7 to >3.5 per macrophage
over 72 hr.

C3H/HeJ mice differ from C3H/HeN mice at the <u>Lps</u> gene locus.
The <u>Lps</u>^{d}, or a closely linked gene, controls expression of at least
one effector function of activated macrophages, tumor cytotoxicity
(2-4). To determine whether this gene influences microbicidal
capacity of activated macrophages, we analyzed the effects of
lymphokine treatment <u>in vitro</u>, and BCG infection <u>in vivo</u>, on
infection and replication of <u>L. tropica</u> amastigotes in macrophages
of both C3H/HeN and C3H/HeJ mice.

Macrophage-mediated antimicrobial activities against <u>L. tropica</u>
were analyzed <u>in vitro</u> by treating C3H/HeN and C3H/HeJ macrophages
with lymphokines generated from concanavalin A-stimulated splenocytes for 4 hours before infection, or for both 4 hours before
infection and 72 hours after infection. Cells were observed for
percent infection and distribution of parasites (Fig. 1). From
the distribution of amastigotes in infected macrophages three
conclusions can be drawn. (1) Lymphokine pretreated C3H/HeN
macrophages demonstrated increased (28%) resistance to amastigote
infection without a change in amastigote distribution in lymphokine-
treated C3H/HeN or C3H/HeJ macrophages as compared to untreated
controls (p = NS, paired t test), (2) By 72 hr., intracellular
destruction of <u>L. tropica</u> was observed in lymphokine-treated
C3H/HeN macrophages (90% decrease in infected cells); 85% of the

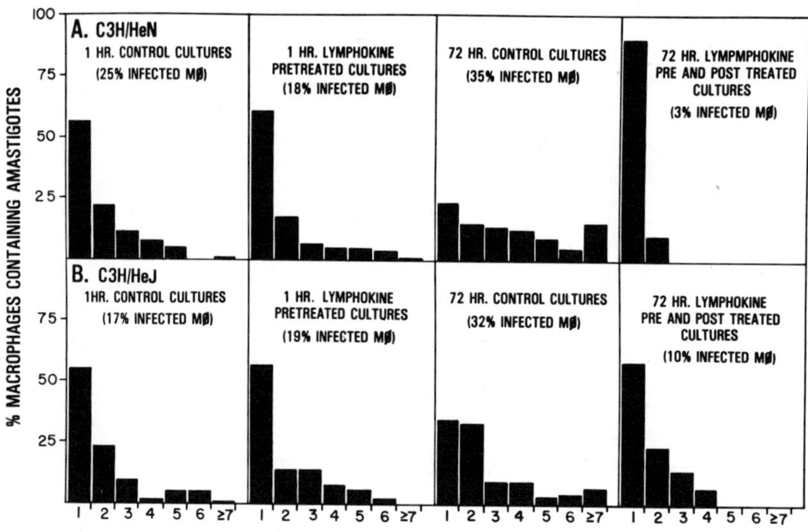

Figure 1.   Distribution of amastigotes in infected C3H/HeN (A) and
C3H/HeJ (B) resident peritoneal macrophages treated
with lymphokines or medium.   Number of intracellular
amastigotes in each of 100 infected macrophages was
observed at each sample time.   Results are mean percent
macrophages containing designated numbers of intracellular amastigotes, and are a composite of at least two
experiments.

infected cells contained only one parasite, (3) Intracellular
killing was also observed in lymphokine-treated C3H/HeJ macrophages,
although killing was less than in C3H/HeN macrophages (55% compared
to 90% microbicidal activity).   The distribution of amastigotes in
residual infected cells was similar to the distribution at 1 hr.

Activated macrophages from BCG-treated C3H/HeN mice also
demonstrated increased resistance to infection at one hour and
enhanced parasite killing at 72 hours as described with lymphokine
activation of macrophages in vitro (Table 1).   Mice were injected
intraperitoneally with 10^6 viable BCG.   Peritoneal macrophages
were harvested from control (untreated) and BCG-treated mice 8 days
after injection and cells were exposed to L. tropica amastigotes
in vitro.

Microbicidal activities of BCG-treated C3H/HeN macrophages
were similar to those described for macrophages activated by
lymphokines.   BCG-treated C3H/HeN cells were resistant to infection
at 1 hr (40% fewer infected cells compared to control macrophages)
and were capable of killing intracellular L. tropica (95% decrease

Table 1. <u>L</u>. <u>tropica</u> Infection of C3H/HeN and C3H/HeJ Macrophages

MOUSE STRAIN	BCG TREATMENT <u>in</u> <u>vivo</u>	PERCENT MACROPHAGE INFECTION[a] 1 Hour	72 Hours
C3H/HeN	−	20	33
	+	12	4
C3H/HeJ	−	20	33
	+	17	31

[a] Resident and BCG-treated macrophages were infected with 0.5-1
<u>L</u>. <u>tropica</u> amastigotes per macrophage for 1 hour. At least 400
macrophages were examined per group, per experiment. Results are
expressed as mean percent macrophages containing intracellular
parasites.

in infected cells) at 72 hr. In contrast, macrophages from BCG-
treated C3H/HeJ mice were not microbicidal. <u>L</u>. <u>tropica</u> infection
of C3H/HeJ BCG-treated macrophages was not different from control
macrophages of 1 hr or at 72 hr. No difference was noted in the
distribution of amastigotes in resident or BCG-stimulated macrophages of either C3H/HeN or C3H/HeJ mice at 1 hr. At 72 hr.,
however, markedly fewer parasites were observed in each infected
macrophage of BCG-treated C3H/HeN mice.

Calculation of the total amastigotes/100 macrophages clearly
identified microbicidal defects of C3H/HeJ macrophages (Table 2).

To confirm that BCG treatment of mice had induced activated
peritoneal macrophages, aliquots of resident and BCG-treated
macrophages were assayed for nonspecific tumor cytotoxicity at
the same time they were assayed for microbicidal activity (Fig. 2)
(2). Resident peritoneal macrophages from C3H/HeN mice did not
respond to lymphokines for nonspecific tumor cytotoxicity (5).
Macrophages from BCG-treated C3H/HeN mice, however, were activated
to kill tumor cells and their response was further augmented with
additional lymphokine treatment <u>in</u> <u>vitro</u>. In contrast, C3H/HeJ
macrophages were not tumoricidal following treatment <u>in</u> <u>vivo</u>
with BCG or <u>in</u> <u>vitro</u> with lymphokines (6). Macrophages from
BCG-treated C3H/HeJ mice could, however, be activated for nonspecific tumoricidal activity by further treatment with lymphokines
<u>in</u> <u>vitro</u> (7).

The defective response of C3H/HeJ macrophages to <u>in</u> <u>vivo</u>
BCG-activation for microbicidal activities against <u>L</u>. <u>tropica</u>

Table 2. Intracellular Replication of <u>L. tropica</u> in BCG-Treated
          and Resident Peritoneal Macrophages.

MOUSE STRAIN	BCG TREATMENT in vivo	AMASTIGOTES PER 100 MACROPHAGES[a]	
		1 Hour	72 Hours
C3H/HeN	−	39	87
	+	19	3
C3H/HeJ	−	30	87
	+	28	86

[a] Amastigotes/100 macrophages = Percent infection x average number
of amastigotes/infected macrophage. At least 400 macrophages
were viewed for intracellular parasites per group per experiment.
Numbers of intracellular amastigotes in macrophages from C3H/HeJ
mice treated with BCG were not different from numbers of
amastigotes in macrophages from untreated mice at 1 hr or 72 hr.

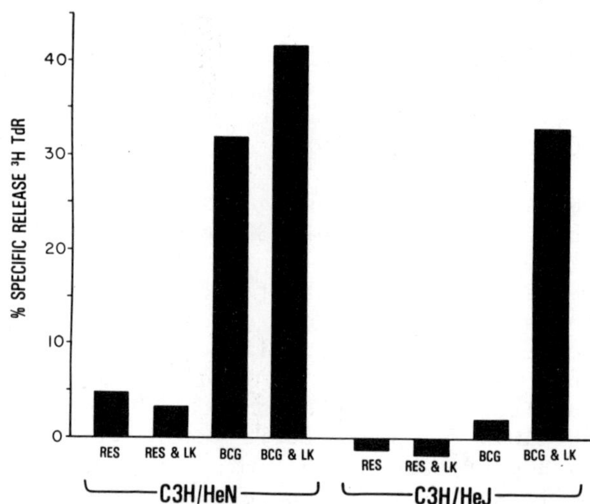

Figure 2. Nonspecific tumoricidal activity of C3H/HeN and C3H/HeJ
           macrophages from BCG-treated mice. Macrophages were
           incubated with ^{3}H-thymidine-labeled mKSA TU-5 tumor
           target cells at an effector to target ratio of 10:1 for
           48 hrs. Some macrophage groups were incubated with
           lymphokines 4 hrs before addition of tumor cells. Per-
           cent specific release was calculated by:

$$100 \ x \ \frac{\text{Experiment Release} - \text{Spontaneous Release}}{\text{Total Release} - \text{Spontaneous Release}}$$

Table 3. Intracellular Replication of <u>L. tropica</u> In Macrophages
          Treated with BCG and Lymphokines.

| | TREATMENT | | AMASTIGOTES/100[a] | % MICROBICIDAL |
MOUSE STRAIN	BCG	LYMPHOKINE	MACROPHAGES	ACTIVITY
C3H/HeN	−	−	103	0
	−	+	20	78
	+	−	6	86
	+	+	6	90
C3H/HeJ	−	−	97	0
	−	+	37	45
	+	−	41	34
	+	+	6	90

[a] After a 1 hr period of infection, macrophages were incubated with
or without lymphokines for 72 hrs, at which time intracellular
amastigotes were counted and microbicidal activity calculated.
At least 400 macrophages were viewed for intracellular parasites
per group, per experiment.

could be augmented by lymphokine treatment <u>in vitro</u> (Table 3).

Macrophages from BCG-treated C3H/HeJ mice could be induced to
express antimicrobial activity against intracellular <u>L</u>. <u>tropica</u> by
treatment with lymphokines <u>in vitro</u>. These data suggest that
several signals are required for induction of macrophage antimicrobial activities. Other macrophage effector functions (induction of macrophage tumor cytotoxicity, secretion of plasminogen
activator, release of H_2O_2) also require multiple activation
signals (7-9). In C3H/HeJ mice, BCG infection is not sufficient
to activate antimicrobial activity in macrophages, but does render
cells more receptive (primed) to activation signals present in
lymphokine supernatants.

Animals injected with BCG are nonspecifically resistant to a
number of intracellular bacteria and parasites (10-16). BCG
infection induces two antimicrobial activities in C3H/HeN macrophages <u>in vitro</u>: (1) increased resistance of activated macrophages
to infection with <u>L</u>. <u>tropica</u>, and (2) increased intracellular
killing of the parasite. These two effector functions are also
present in macrophages activated <u>in vitro</u> with lymphokines and
have been demonstrated in macrophages activated <u>in vivo</u> for

destruction of another obligate intracellular organism, rickettsia (13,17-19).

BCG or lymphokine-treated C3H/HeJ macrophages were defective in one or both antimicrobial activities. C3H/HeJ macrophages treated _in vitro_ with lymphokines did not show increased resistance to infection, but were capable of killing intracellular _L. tropica_. Macrophages from C3H/HeJ mice treated _in vivo_ with BCG were defective for both microbicidal activities. BCG-primed C3H/HeJ macrophages, however, were both tumoricidal and microbicidal after additional treatment with lymphokines _in vitro_. The functional activity of these sequentially activated macrophages was equivalent to that of activated C3H/HeN macrophages. Analysis of sequential activation steps for macrophage microbicidal activity against _L. tropica_ in C3H/HeJ macrophages may provide information on regulation of the effector functions important for host defense against this intracellular pathogen.

ACKNOWLEDGMENT

In conducting research described in this report, the investigators adhered to the Guide for Laboratory Animal Facilities and Care as promulgated by the Committee of the Guide for Laboratory Animals and Care of the Institute of Laboratory Animal Resources, National Academy of Sciences, National Research Council.

REFERENCES

1. Nacy, C.A. and Diggs, C.L. (1981). Infect. Imm. (in press).
2. Ruco, L.P. and Meltzer, M.S. (1977). Cell. Immunol. 32:203.
3. Ruco, L.P. and Meltzer, M.S. (1978). Cell. Immunol. 41:35.
4. Ruco, L.P., Meltzer, M.S. and Rosenstreich, D.L. (1978).
        J. Immunol. 120:329.
5. Ruco, L.P. and Meltzer, M.S. (1978). J. Immunol. 120:1054.
6. Boraschi, D. and Meltzer, M.S. (1979). Cell. Immunol. 45:188.
7. Ruco, L.P. and Meltzer, M.S. (1978). J. Immunol. 121:2035.
8. Gordon, S., Unkeless, J.C. and Cohn, Z.A. (1974). J. Exp.
        Med. 140:995.
9. Nathan, C.F. and Root, R.K. (1977). J. Exp. Med. 146:1648.
10. Youmans, G.P., Youmans, A.S. and Kanai, K. (1959). Am. Rev.
        Resp. Dis. 80:753.
11. Ratzan, K.R., Musher, D.M., Keusch, G.T. and Weinstein, L.
        (1972). Infect. Immun. 5:499.
12. Medina, S., Vas, S.I., and Robson, H.G. (1975). J. Immunol.
        114:1720.
13. Nacy, C.A., Radlick, G. and Meltzer, M.S. (1980). Genetic
        Control of Natural Resistance to Infection and Malignancy
        (Eds. Skamene, E., Kingshavn, P. and Landy, M.) Academic
        Press, New York.

14. Oritz-Oritz, L., Gonzalez-Mendoza, A., and Lamuyi, E. (1975).
         J. Immunol. 114:1424.
15. Maddison, S.E., Chandler, F.W., McDougal, J.S., Slemenda,
         S.B. and Kagan, I.G. (1978). Am. J. Trop. Med. Hyg.
         27:966.
16. Civil, R.H. and Mahmoud, J. (1978). J. Immunol. 120:1070.
17. Nacy, C.A. and Meltzer, M.S. (1979). J. Immunol. 123:2544.
18. Buchmuller, Y. and Mauel, J. (1979). J. Exp. Med. 150:359.
19. Nacy, C.A., Meltzer, M.S., Leonard, E. and Wyler, D. (submitted).

IN VITRO MACROPHAGE ANTIMICROBIAL ACTIVITIES AND IN VIVO

SUSCEPTIBILITY TO LEISHMANIA TROPICA INFECTION

Anne L. Haverly, Michael G. Pappas, Robin R. Henry
and Carol A. Nacy

Department of Immunology, Walter Reed Army Institute
of Research, Washington, D.C. 20012

L. tropica, an obligate intracellular parasite, replicates in
phagolysosomes of macrophages. Resistance to L. tropica infection
may depend upon alteration of this intracellular environment (1,2).
Soluble products of antigen- or mitogen-stimulated lymphocytes
(lymphokines) induce enhanced macrophage antimicrobial activity
against a number of intracellular organisms (3,4,5,6,7). These
activated macrophages may be effector cells during resolution of
L. tropica infections. In this study we analyzed the interaction
of L. tropica and macrophages treated with lymphokines in vitro,
and correlated these findings with susceptibility to L. tropica
infection in vivo.

We recently documented that treatment of resident peritoneal
macrophages of C3H/HeN mice with lymphokines in vitro induced two
antimicrobial activities against L. tropica (Fig. 1) (8). Control
medium-treated macrophages were 35% infected by 1 hr and supported
the replication of L. tropica over 72 hr. Numbers of intracellular
parasites increased 4 fold in these cells. Macrophages treated
with lymphokines 4 hr before infection showed a 30% decrease in
percent infected cells at 1 hr, and this decrease remained constant
over 72 hr. Macrophages treated with lympholines after infection
resulted in a 90% decrease in infected cells by 72 hr. This
decrease reflected intracellular killing of the parasite.

We analyzed lymphokine-induced macrophage antimicrobial
activities in a number of inbred mouse strains. These strains
included several with previously characterized macrophage defects
in response to activation signals for nonspecific tumor cytotoxicity (Table 1) (9).

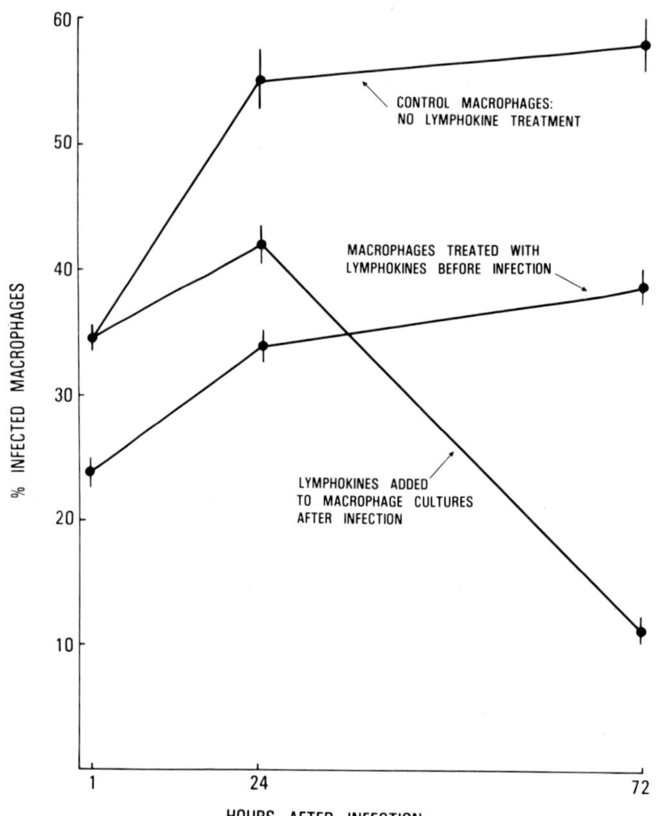

Figure 1.   L. tropica infection of resident peritoneal macrophages
            treated with lymphokines or medium. Macrophages were
            incubated in medium or 1/10 lymphokines for 4 hr,
            washed, and exposed to 1 amastigote L. tropica per
            macrophage for 1 hr. Infected cultures were washed
            and incubated in 1/10 lymphokines or medium an additional
            0-72 hr. Samples were observed microscopically for
            percent infected macrophages.

     Although macrophages from most mouse strains responded to
lymphokines in vitro for both microbicidal activities, C3H/HeJ,
A/J, C57BL/10ScN macrophages did not respond to lymphokines for
increased resistance to infection when macrophages were treated
4 hr before exposure to L. tropica. BALB/c, NZW/N, C57L/J, and
P/N macrophages did not respond to lymphokine activation signals
for intracellular killing.

     To determine if these defective macrophage responses influenced
the course of in vivo infection with L. tropica, mouse strains were

Table 1. Lymphokine-Induced Microbicidal Activity of Resident
          Peritoneal Macrophages Against <u>Leishmania</u> <u>tropica</u>

Peritoneal Macrophages From:	Microbicidal Activity of Macrophages Treated with Lymphokines	
	Before Infection	After Infection
C3H/HeN Mice	100	100**
C3HeB/FeJ Mice	135	96
*C3H/HeJ Mice	0	70
C57L/J Mice	119	25
C57BL/6J Mice	82	112
C57BL/10J Mice	100	135
*C57BL/10ScN Mice	31	132
*A/J Mice	43	50
BALB/cJ Mice	110	22
NZW/N Mice	100	20
*P/JN Mice	100	10

* Mouse strains with characterized defects for tumor cytotoxicity

**Response of cells from C3H/HeN mice defined as 100%

inoculated intradermally with <u>L</u>. <u>tropica</u> amastigotes. Groups of
3-5 mice were inoculated in one hind footpad with 10^5 <u>L</u>. <u>tropica</u>
amastigotes. Depth (mm) of the developing lesion was measured
weekly for 12 weeks and compared to the uninoculated footpad
(Fig. 2).

Resistant strains developed lesions of less than 2 mm, and
lesions resolved by 12 weeks. In contrast, susceptible strains
developed lesions of greater than 3 mm that did not resolve.
BALB/c mice were most susceptible to <u>L</u>. <u>tropica</u> infection; the
infected foot became necrotic, metastatic lesions developed, and
visceralization of the infection to spleen and liver occurred.
Three other strains failed to resolve leishmanial lesions. The
mouse strains segregated into 2 groups as shown in Table 2.

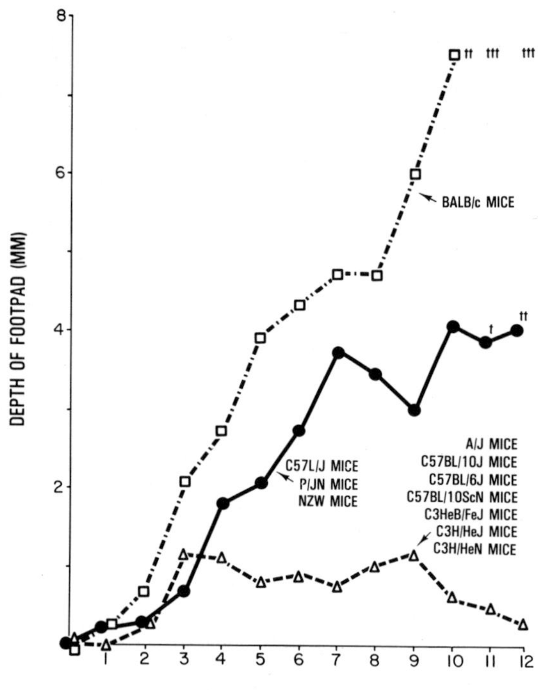

Figure 2.   L. tropica infection of inbred strains of mice. Mice
were inoculated with 10^5 amastigotes of L. tropica in
one hind footpad. Depth (mm) of inoculated and of the
control uninoculated footpads were measured with
calipers each week for 12 weeks. Results were expressed
as the mean difference in depth of inoculated and uninoculated footpads for groups of 3-7 mice of each
strain.

     All strains susceptible to in vivo challenge with L. tropica
were those whose macrophages did not respond to lymphokine signals
in vitro for intracellular killing of Leishmania. Resistance to
in vivo infection with L. tropica, then, may depend upon appropriate
macrophage response to lymphokine activation signals.

     To further analyze macrophage function in a susceptible mouse
strain, we treated macrophages from resistant C3HeB/FeJ mice and
susceptible P/N mice with different concentrations of lymphokines
after infection with L. tropica (Fig. 3). Even at high concentrations of lymphokines, P/MN macrophages were minimally responsive to
lymphokine signals that induce intracellular killing. Microbicidal
activity was only 20%. In contrast, 85-90% of C3HeB/FeJ macrophages were free of intracellular parasites by 72 hr.

Table 2. <u>L. tropica</u> Infection of Inbred Mouse Strains

Resistant	Susceptible
*A/J	BALB/cJ
AL/N	BALB/cN
BDP/N	BALB/cDub
B10D2	
CBA/N	C57L/J
C3H/HeN	
*C3H/HeJ	NZW/N
C3HeB/FeJ	
C57BL/6J	*P/J
C57BL/10J	*P/N
*C57BL/10ScN	

* Mouse strains with characterized macrophage defects

To determine which lymphokine(s) induce the limited intracellular killing observed in P/N macrophages, we fractionated lymphokine supernatants on Sephadex G-100 (Fig. 4). Three peaks of intracellular killing activity were observed for C3HeB/FeJ macrophages treated with lymphokine fractions whose apparent molecular weights were 130,000, 45,000, and ≤10,000 daltons. P/N macrophages responded to high molecular weight lymphokines for intracellular killing, but levels of this activity were 3-4 fold less than C3HeB/FeJ macrophages. P/N macrophages did not respond to lymphokine signals for intracellular killing present in either the 45,000 MW or the ≤10,000 MW lymphokine regions.

Susceptibility to <u>L. tropica</u> infection in inbred mouse strains could be correlated with the inability of macrophages from these strains to respond to lymphokine activation signals for intracellular killing of the parasite. Macrophages from one susceptible strain, P/N, responded to only one of three intracellular killing lymphokines, and the response of these macrophages was a fraction of that obtained with macrophages from a resistant mouse strain. Several possibilities exist: (1) macrophages from susceptible mouse strains are less responsive to lymphokine activation signals, or (2) macrophages are capable of responding to activation signals, but lack effector mechanisms induced by two of the three intracellular lymphokines which mediate killing. These possibilities

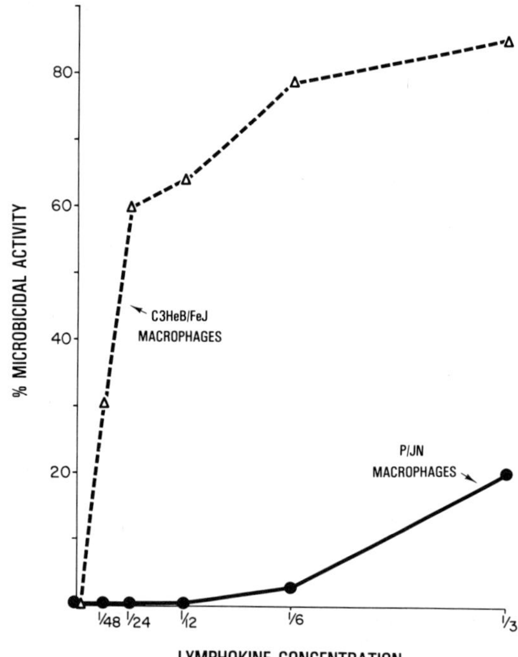

Figure 3.   Dose response of macrophages treated with lymphokines
            after infection. Macrophages were obtained from P/N
            and C3HeB/FeJ mice. Cells were infected with L. tropica
            (1 amastigote/macrophage) and were treated with various
            concentration of lymphokines for 72 hr after infection.
            Samples were observed microscopically for percent in-
            fected macrophages at 72 hr. Microbicidal activity was
            determined by the formula:

$$100 \times \frac{(\% \text{ infected control macrophages} - \% \text{ infected treated macrophages})}{\% \text{ infected control macrophages}}$$

are currently under analysis, both in P/N mice and in other strains
susceptible to L. tropica infection.

ACKNOWLEDGMENT

     In conducting research described in this report, the investi-
gators adhered to the Guide for Laboratory Animal Facilities and
Care as promulgated by the Committee of the Guide for Laboratory
Animals and Care of the Institute of Laboratory Animal Resources,
National Academy of Sciences, National Research Council.

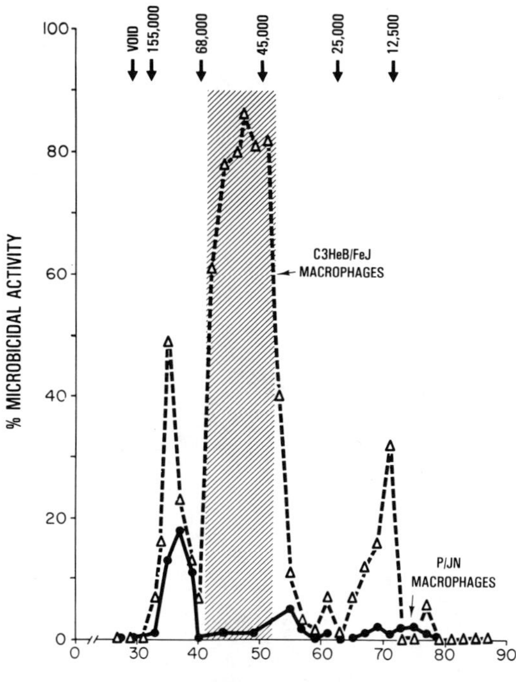

Figure 4. Fractionation of lymphokine supernatants. Lymphokines
were fractionated by Sephadex G-100 column chroma-
topography, and fractions were assayed at 1/12 dilution
for capacity to induce intracellular killing of L.
tropica in C3HeB/FeJ and P/N macrophages.

REFERENCES

1. Chang, K.P., and Dwyer, D.M. (1976). Sci. 193:678.
2. Chang, K.P., and Dwyer, D.M. (1976). Sci. 193:687.
3. Fowles, R.E., Fajardo, I.M., Leibavitch, J.L., and David, J.R.
        (1973). J. Exp. Med. 148:393.
4. Godal, T., Rees, R.J.W., and Lamvik, J.O. (1971). Clin. Exp.
        Immunol. 8:625.
5. Borges, J.S., and Johnson, W.D. (1975). J. Exp. Med. 141:483.
6. Nogiera, N., and Cohn, Z.A. (1978). J. Exp. Med. 148:288.
7. Nacy, C.A., and Meltzer, M.S. (1979). J. Immunol. 123:2544.
8. Nacy, C.A., Meltzer, M.S., Leonard, E.J., and Wyler, D.J.
        (submitted).
9. Boraschi, D., and Meltzer, M.S. (1979). Cell. Immunol. 45:188.

VIRUS-IMMUNE T CELLS AND MONOCLONAL ANTIBODIES IN THE MOUSE

INFLUENZA MODEL

Peter C. Doherty, Neil Greenspan, A. Dwight Lopes and
Walter Gerhard

The Wistar Institute, 36th and Spruce Streets
Philadelphia, PA   19104

INTRODUCTION

Evidence has been accumulating over the past 10 years that a
major factor in recovery from most acute virus infections is the
cellular immune response mediated by T cells which are, when tested
in vitro, cytotoxic for virus-infected target cells (reviewed in 1
and 2). The recent development of techniques for producing theoretically unlimited amounts of virus-specific monoclonal antibodies has
raised the possibility that such reagents may prove to be of great
therapeutic value (3,4). We wish to summarize here our current
understanding of the possible interactions between monoclonal
antibodies and virus-immune T cells in the influenza model. It
will be obvious to the reader that these experiments are still at
a very early stage, though some useful insights are emerging.

FOCUSING OF T CELLS ONTO CELL SURFACES

The characteristics of the virus immune T cell response have
recently been reviewed in greater (1) and lesser (2,5) detail.
The central point is that the T cells, which function together
with effectors from the monocyte-macrophate series to eliminate
virus infected cells from solid tissues and mucosal surfaces (6,7)
are focused onto the surface of the infected target by the need to
see both virus and H-2 determinants (5). This major histocompatibility complex (MHC) restriction phenomenon, seen operationally
as a need for H-2K or H-2D (or HLA-A or HLA-B) compatibility
between the T cell and the virus-infected target, serves to ensure
that the lymphocytes responsible for eliminating organ sites of
virus growth are not distracted by interacting with free virus in

plasma, or other extracellular fluid spaces. There is no convincing support for the idea that virus-immune cytotoxic T lymphocytes
(CTL) will bind free virus.

Even so, there is now good evidence that the cytotoxic T cell
must see virus on the cell surface. Target cells can be made _in
vitro_ by direct exposure to virus and H-2 antigens incorporated
into single liposomes (8,9). The need for recognition of, for
instance, the vesicular stomatitis virus (VSV) G protein has been
demonstrated using ts mutants (10). It is thus worth asking whether
antibody to virus can block the interaction of virus-immune T cells
with virally-modified cells. This is especially important if we
are considering introducing large amounts of monospecific antibody
into the _in vivo_ situation.

SPECIFICITY OF INFLUENZA-IMMUNE T CELLS

All of the experiments that we will be describing involve
mice infected with influenza A viruses, so it is first necessary
to briefly summarize the characteristics of the influenza A virus-
immune CTL response. The main point is that there is a major divergence in specificity patterns for CTL and antibody. Distinct subtypes of the influenza A viruses have been characterized on the
basis that antibodies (serum or monoclonal) to determinants
expressed on the virus hemagglutinin (H, HA) or neuraminidase (N)
molecules, which are present on the surface of the virion, dis-
criminate between them (reviewed in 11). For instance, there is
no cross-neutralization between the H1N1 and H3N2 subtypes of the
influenza A viruses. However, antibodies to the carbohydrate
moiety on the H glycoproteins show extensive cross-reactivity for
the complete range of influenza A and influenza B viruses. Also,
the internal virus proteins, matrix (M) and neucleoprotein (NP),
are very similar for all the influenza A viruses, though there is
no cross-reactivity with comparable components purified from the
influenza B viruses.

The situation for influenza A virus-immune CTL is that mice
primed with any influenza A virus develop potent effectors which
lyse targets infected with all influenza A, but not influenza B,
viruses. A subset of these CTL is specific (subtype-specific) for
the particular H antigen used for immunization (12-16), while the
majority of the T cell population is highly cross-reactive (type
specific). Some of these type-specific CTL may recognize shared
components on the viral HA molecule, as targets can be made by
incorporating purified HA glycoprotein into the cell membrane using
liposome technology (17). Analysis with monoclonal antibodies has
also shown that there are significant amounts of the internal NP
component on the cell-surface, which could also serve as a cross-
reactive viral determinant recognized by CTL (18).

However, the situation as currently understood, is that both
the majority of the antibody made in a mouse infected with an
influenza A virus, and the CTL populations generated, are recognizing the viral HA molecule. The antibodies distinguish readily
between subtypes, while most of the T cells will interact with H-2
compatible targets expressing any influenza A (but not influenza B)
HA glycoprotein. Similar patterns have also been found for man and
the rat (19-21).

THE PROTECTIVE EFFECTS OF MONOCLONAL ANTIBODIES

An intravenous dose of 50 µg of HA-specific monoclonal antibody given to BALB/c mice prior to aerosol challenge with 100 ID_{50}
of the PR8 (HON1) influenza A virus will prevent infection (22).
Antibodies of all classes (with the exception that IgE and IgD have
not yet been tested), binding to any of the antigenic determinants
on the viral HA molecule (23) have been used to prevent infection.
However, all mice became infected when pre-treated with monoclonal
antibodies reactive to other influenza proteins. Also, administration of as much as 300 µg HA-specific antibody after virus challenge
in most cases failed to stop the infection.

More success has been achieved for post-exposure treatment
using mice infected intracerebrally (i.c.) with 500 i.c. ID_{50} of the
A/WSN influenza virus. Such mice, which generally die about 5 to
7 days after i.c. challenge, can be completely protected by intravenous administration of 500 µg (but not 100 µg) of A/WSN-specific
monoclonal antibody given 2 to 3 days following initial exposure
(24). Preliminary experiments indicate that the protective
capacity of a particular monoclonal antibody is related to the
potency of that antibody to mediate hemagglutination-inhibition
(specific for viral HA) (W.G. & P.D., unpublished).

Direct inoculation of an influenza A virus into the brain
leads to generalized virus growth in cells of the meninges,
ependyma and choroid plexus (25). The barrier between blood and
cerebrospinal fluid (CSF) seems to break down (for immunoglobulin,
Ig) between 48 and 72 hours after injection of virus, apparently
as a direct result of virus-induced damage. The exogenously
administered monoclonal antibody then floods into the CSF and,
presumably, neutralizes the virus and prevents further spread of
the infectious process throughout the brain. The serum/CSF ratios
for an irrelevant Ig, which does not bind to virus, are generally
higher than those found for an antibody specific for A/WSN. This
no doubt reflects the fact that virus-infected cells in the brain
are acting as an immunoabsorbent, removing free antibody in the
CSF. It is also intriguing that post-exposure (2 days) inoculation
of complement-fixing (in vitro) antibody specific for A/WSN neither
promotes, nor prevents, the subsequent breakdown of the blood-CSF
barrier.

There are a number of possible reasons why exogenously-administered monoclonal antibody, when given post-exposure, might be more effective in eliminating virus from the brain than from the lung (considering that the blood-CSF barrier is breached). It may simply be easier to contain an infectious process occurring in a solid tissue, as compared with a mucosal surface. Also, lung cells may produce much more infectious virus than those in the brain. However, we should emphasize that we have not yet done controlled experiments which directly compare the two systems: the apparent divergence may simply reflect the use of different protocols.

BLOCKING OF SUBTYPE-SPECIFIC CTL WITH MONOCLONAL ANTIBODY

Some monoclonal antibodies which bind to the viral HA molecule will, when incorporated into the in vitro assay system, partially block T cell-mediated lysis of virus-infected target cells (26). The inhibition is seen preferentially for the subtype-specific CTL. A comparable situation (27) is found for T cell stimulation in vivo: mice given virus followed by monoclonal antibody (3 to 48 hours later) generate effector CTL populations which are skewed towards the cross-reactive, rather than the subtype specific, pattern of lysis. The two situations are probably analogous. The substantially subtype specific CTL, which recognize the portion of the viral HA characterized as a variable on the basis of antibody binding, are more likely to be blocked(at both the effector and stimulator level) by the presence of that antibody, than are the type-specific effectors, which show a much broader cross-reactivity pattern than antibody (12-16,26,27).

The comparability of the in vitro and in vivo systems also holds for the observation that one monoclonal antibody (H9D3) partially blocks influenza-immune CTL effector function mapping to $H-2D^d$, but not to $H-2K^d$ (27,28). Inoculation of H9D3 into BALB/c (K^dD^d) and B10.A(5R) (K^bD^d) mice given PR8 (H1N1) influenza virus 3 hours previously causes the resultant CTL response (measured 5 days later) to be much more cross-reactive in character. This is manifested as equivalent lysis of PR8 (H1N1) and NT60 (H3N2)-infected P815 (K^dD^d) cells. The same treatment protocol had absolutely no effect on the CTL response of C3H.OH (K^dD^k) mice measured using the P815 target, though another monoclonal antibody (H2-4B3) given in the same way reproduced the effect seen for H9D3 with the BALB/c and B10.A(5R) strains.

The divergence of the blocking effect at $H-2K^d$ and $H-2D^d$ with the H9D3 monoclonal antibody may be due to the fact that the subtype-specific CTL operating at these two loci are reactive to different regions of the viral HA molecule. Perhaps (if the basic requirement for T cell recognition is that the viral HA and

the H-2 glycoproteins are associated on cell membrane) H9D3 binds
to a determinant much more remote from the point of association
between HA and H-2K^d than that found for HA and H-2D^d. If one
argues this "interaction antigen" model, it seems unlikely that
H9D3 absolutely prevents the association of H-2D^d and the viral
HA; for if this were so, stimulation and effector function for
the type-specific CTL would also be inhibited, which is not the
case.

SUMMARY

     Very large doses of HA-specific monoclonal antibodies given to
mice infected with influenza A viruses can, at least in some
experimental systems, be of considerable therapeutic value (24).
Such treatment may cause a partial inhibition of T cell effector
function by blocking the stimulation in vivo of the more HA subtype-specific CTL (27). Comparable effects are seen for the lytic
interaction measured in vitro by the ^{51}Cr release assay (26,28).
However, development of the cross-reactive, type-specific CTL
effectors is not inhibited by such antibody treatments. It may,
in fact, be somewhat enhanced (27). The naturally (and experimentally) evolved virus-recombinant technology available for the
influenza A viruses (11) has allowed us to dissect this effect in
relation to antigen binding specificity. This same situation,
namely that many cells recognize determinants on virus-coded
molecules that are remote from the sites binding antibody, is
probably equally true for other viruses. Perhaps these cross-
reactive T cells are recognizing an 'altered-self' (29) antigen
that is partly virus and partly H-2. Alternatively, the T cell
(which has to deliver a lytic 'message' to the target) may focus
preferentially on more conserved viral determinants which are
close to the plasma membrane.

     Whatever the explanation for the existence of these cross-
reactive, type-specific CTL, it is highly satisfactory (from the
viewpoint of the Darwinian teleologist) that neither their generation, nor their effector function, is readily blocked by even very
large amounts of antibody. Some viruses spread readily from cell
to cell in the presence of specific Ig. Antibody blocking of T
cell-mediated lysis in such a situation could obviously have
disastrous consequences. The immunological surveillance function
of T cells is not likely to be completely inhibited by other
components of the immune response.

     Also, it seems that T cell responses have not constituted a
selective force driving influenza virus variability, even though
they play a central role in the termination of the infectious
process. The selective function must be ascribed to antibody (4).
In fact, an effective T cell response may be central to the

successful long-term, ecological relationship between virus and
host (30).

ACKNOWLEDGMENTS

     The experimental work described here was funded by grants
AI 14162 and NS 11036 from the Public Health Service, and by The
Medical Scientist Training Program (NIH 6m 07170).

REFERENCES

1.   Zinkernagel, R.M. and Doherty, P.C. (1977) Adv. Immunol. 27,
         51.
2.   Doherty, P.C. (1970) Prog. Immunol. 4, 563.
3.   Koprowski, H., Gerhard, W., and Croce, C.M. (1977) Proc.
         Natl. Acad. Sci. (USA) 74,2985.
4.   Gerhard, W.U., Yewdell, J., Frankel, M.E., Lopes, A.D. and
         Staudt, L. (1980) in "Monoclonal antibodies", R.H. Kennett,
         T.J. McKearn and K.B. Bechtol, eds., Plenum Press, New
         York, p. 317.
5.   Doherty, P.C., and Bennink, J.R. (1981) Fed. Proc. 40, 218.
6.   Blanden, R.V. (1974) Transplant. Rev. 19, 56.
7.   Yap, K.L., Ada, G.L. and McKenzie, I.F.C. (1978) Nature 273,
         238.
8.   Finberg, R., Mescher, M. and Burakoff, S.J. (1978) J. Exp.
         Med. 148, 1620.
9.   Hale, A.H. and Ruebush, M.H. (1980) J. Immunol. 125, 1569.
10. Zinkernagel, R.M. and Rosenthal, K.L. (1980) Immunol. Rev. 58,
         134.
11. Kilbourne, E.D. (1975) "The Influenza Viruses and Influenza",
         Academic Press, New York.
12. Effros, R.B., Doherty, P.C., Gerhard, W., and Bennink, J.R.
         (1977) J. Exp. Med. 145, 557.
13. Doherty, P.C., Effros, R.B. and Bennink, J.R. (1977) Proc.
         Nat. Acad. Sci. (USA) 74, 1209.
14. Zweerink, H.J., Courtneidge, S.A., Skehel, T.J., Crumpton, M.
         and Askonas, B.A. (1977) Nature 267, 354.
15. Lu, L.Y. and Askonas, B.A. (1980) Nature 288, 164.
16. Braciale, T.J., Andrew, M.E. and Braciale, V.L. (1981) J. Exp.
         Med. 153, 910.
17. Koszinowski, U.H., Allen, H., Gething, M.J., Waterfield, M.D.
         and Klenk, H.D. (1980) J. Exp. Med. 151, 945.
18. Yewdell, J., Frank, E. and Gerhard, W. (1981) J. Immunol. 126,
         1814
19. McMichael, A.J. and Askonas, B.A. (1978) Eur. J. Immunol.
         8, 705.
20. Biddison, W.E., Shaw, S. and Nelson, D.L. (1979) J. Immunol.
         122, 660.

21. Marshak, A., Doherty, P.C. and Wilson, D.B. (1977) J. Exp.
          Med. 146, 1773.
22. Lopes, A.D. (1981) Ph.D. Thesis, University of Pennsylvania.
23. Gerhard, W., Yewdell, J., and Frankel, M.E. (1981) Nature
          290, 713.
24. Doherty, P.C. and Gerhard, W. (1981) J. Neuroimmunol. 1, in
          press.
25. Mims, C.A. (1960) Brit. J. Exp. Path. 46, 586.
26. Effros, R.B., Frankel, M.E., Gerhard, W. and Doherty, P.C.
          (1979) J. Immunol. 123, 1343.
27. Greenspan, N. (1981) Ph.D. Thesis, University of Pennsylvania.
28. Frankel, W.E., Effros, R.B., Doherty, P.C. and Gerhard, W.U.
          (1979) J. Immunol. 123, 2438.
29. Zinkernagel, R.M. and Doherty, P.C. (1974) Nature 251, 547.
30. Doherty, P.C. (1980) in "Strategies of Immune Regulation",
          E. Sercarz and A. Cunningham, eds., Academic Press, New
          York, p. 103.

ESCAPE FROM IMMUNE SURVEILLANCE DURING PERSISTENT VIRUS INFECTION

Neal Nathanson and John R. Klein

Dpeartment of Microbiology, School of Medicine
University of Pennsylvania, Philadelphia, Pennsylvania
19104

INTRODUCTION

Persistent virus infections present a paradox; there exists
a potent array of antiviral immune defenses, and yet there are many
naturally occurring instances of persistent virus infection in
animals and in humans. In this short review we will examine the
diverse strategies whereby persistent viruses escape immune surveillance, and illustrate these mechanisms by reference to a few
selected examples. More detailed information is available in a
number of symposia and compendia (1-5).

ANTIVIRAL IMMUNE DEFENSES

To appreciate the paradox of virus persistence, a brief summary
of anti-viral immune defenses is necessary. The life cycle of
viruses is divided into two phases, the intracellular replicative
phase, and the extracellular vegetative phase where mature virions
pass from infected to uninfected cells. Different anti-viral
mechanisms operate against each phase.

<u>Virion</u>. Neutralizing antibody is the classical defense against
mature virions. Studies with monoclonal antibodies against influenza, rabies, bunya, and reoviruses indicate that a single protein
in the envelope or outer capsid is responsible for neutralization
(6-9). Neutralizing antibody coats the virion but does not destroy
it; rather it prevents attachment to cells, a process which mainly
involves one specific viral protein and a specific molecular
receptor on the surface of susceptible cells.

Virus coated with antibody may be further inactivated by several alternative routes. If the antibody has a complement-binding site and if the virion is enveloped, the virion may be irreversibly destroyed by virolysis (10-12). Alternatively, in the presence of macrophages bearing Fc receptors, the antibody coated virion may be phagocytosed. In vivo, virions in the circulation are removed much more rapidly in the presence of plasma antibody, a process classically called opsonization (13).

<u>Virus-infected cell</u>. Virus-infected cells play a key role in the dynamics of an ongoing infection because the lability of mature virions, even in the absence of antibody, requires a constant production of new virus particles to maintain or increase virus titer. If the decilife of mature virions is 2-4 hours, then there is essentially complete turnover of virus particles many times each day. Interruption of intracellular replication would lead very rapidly to virus clearance.

There are several immune mechanisms which can directly attack and destroy virus-infected cells. All of these depend on the fact that infected cells carry on their surface virus-coded neoantigens. Many enveloped RNA viruses mature by budding across membranes, and the insertion of viral glycoproteins into the plasma membrane is an integral part of virus assembly and maturation (14). Other viruses attach to plasma membranes as the first step in cellular infection, so that infected cells carry on their surfaces the external antigens of the virus.

There are three well-documented immune mechanisms whereby a cell bearing viral antigen can be destroyed. (i) Antibody and complement can lyse cells, by the classical complement pathway (15). (ii) Cytolytic T lymphocytes, in the absence of antibody or accessory cells, can lyse virus-infected targets (16-17). (iii) Null cells, binding antiviral antibody through Fc receptors, can acquire specificity and be armed to attack virus-infected targets, a process known as antibody-dependent cell-mediated cytolysis (18).

Two other immune mechanisms for interruption of viral replication also exist. (iv) Lymphocytes mediating delayed-type hypersensitivity, through secreted lymphokines, can attract and activate macrophages which will destroy virus-infected cells in the vicinity (19-20). (v) Immune interferon (INF-γ) is secreted by immune T lymphocytes activated by viral antigen, and can suppress viral synthesis by interrupting the intracellular replication process (21-22).

<u>Experimental evidence for immune clearance in vivo</u>. The cumulative impact of the specific mechanisms described above is dramatically illustrated in experimental models of acute virus infection. Clinical and experimental evidence suggests that

cellular mechanisms may be of particular importance for certain
classes of viruses such as herpes and pox viruses, while antibody
is particularly important for picornaviruses (23-24). However,
deletion of either T or B cells can markedly potentiate a number
of acute viral infections (25-27), and it seems likely that both
arms of the immune response play a synergistic role in coping with
a primary viral infection.

PERSISTENT VIRUS INFECTION

     Many naturally occurring examples of persistent virus infec-
tion in animals and humans have now been described. Some of these
examples have been studied in detail while for others information
is still sketchy. The cumulated data demonstrate that there are
a number of distinct strategies whereby a persistent virus can
escape immune elimination. Some of the best defined mechanisms
are set forth in Table I, although it must be recognized that
certain infections subsumed under a single category may differ
considerably. Each group will be discussed separately followed by
a brief description of one or two representative examples.

<u>Group I: Hyporesponse</u>

     Infections of this kind (Table 2) resemble "high dose"
immunological tolerance. That is, there are large amounts of

Table I. Escape from immune surveillance in persistent virus
          infection. Summary of mechanisms[*]

Group	Immune Response	Viral Genome Expression	Mechanism
I Hyporesponse	- or ±	+ or ++	"Tolerance" or Non-immunogenicity
II Nonexpressor	+	- or +	Latency-Activation
III Breakthrough	+ to ++	± to +	Infectious immune complexes or intercellular bridges

[*] Immune response refers to measurable levels of free antibody in
serum or other fluids. Viral genome expression refers to titer-
able levels of infectious virus in blood or tissue homogenates.
Mechanisms are described in the text.

Table 2. Escape from immune surveillance in persistant virus
          infection. Group I: Immune hyporesponse (selected
          examples)[*]

Immune Response	Agents and Immunogenicity	Specific Examples or Agents	Natural Host	References
−	Spongiform agents	Scrapie	Sheep	2
		Kuru	Human	2
	Non-immunogenic	Creutzfeldt-Jacob	Human	2
±	Conventional viruses	Hepatitis B	Human	14-1
		Rubella	Human	15-1
	Immunogenic	Lymphocytic choriomeningitis	Mouse	30
		Neurotropic murine leukemia	Mouse	31-33

* Conventional agents listed induce persistent infections only
  under limited conditions which often include exposure during
  the fetal or neonatal period. See text for a fuller discussion.

infectious virus which can be readily titrated over a long period
of time. This group of infections can be divided into two
categories.

     (a) Immunogenic conventional viruses which initiate a
persistent "tolerant" infection only if introduced into the ovum,
the fetus, or immediately after birth. Antibody may be made, but
is complexed by an excess of antigen, so that it can only be
detected by indirect methods. Likewise, cellular immune responses
are minimal or absent, where they have been explored.

     (b) Apparently nonimmunogenic, unconventional spongiform
agents, i.e. the causative agents of scrapie of sheep, transmiss-
ible mink encephalopathy (probably scrapie in mink), of kuru,
and of Creutzfeldt-Jacob disease of humans (2).

     <u>Neurotropic murine leukemia virus</u>. In a search for naturally
occurring murine leukemia viruses in wild mice, Gardner and his
associates (31) isolated a large series of lymphoma-producing
retroviruses. These agents were exogenous, i.e., infectious virus
was produced and was transmitted to uninfected mice. A subset of

these agents was found to induce a progressive paralytic disease
of mice in addition to lymphomas. The neurotropic agents were
ecotropic (replicated only in mouse cells) and N-tropic (replicated
preferentially in mice bearing the n allele at the Fv-1 genetic
locus). Since this allele is recessive, only homozygous mice
were fully susceptible. Paralytic disease occurred only when mice
were inoculated immediately after birth. Furthermore, inoculation
of large virus doses (over 10^4 xc/cell plaque forming units)
shortened incubation periods from 150 to 15 days (32).

After intracerebral virus injection of newborn mice, titers
rise to about 10^5 pfu per gram in many tissues and remain at this
level for the duration of infection (32). No neutralizing or other
antibody has been detected in the serum. Cellular immune responses
have not yet been studied. Histologically, an unusual lesion is
seen in the brain and spinal cord, consisting of multiple vacuoles,
destruction of neurons (particularly anterior horn motoneurons),
with no evidence of inflammation (33). Electron-microscopically,
vacuoles are seen in neurons, with intracytoplasmic budding and
accumulation of virus particles. Such infected neurons eventually
undergo necrosis.

It is not yet understood why this group of murine leukemia
viruses exhibit neuronotropism, nor is the mechanism of cell death
with this relatively noncytocidal agent explained. However, there
is a clear correlation between maximal virus replication and the
occurrence of paralytic lesions. These conditions are only met if
the mice inoculated as newborns fail to mount an effective immune
response, so that virus replication can proceed unchecked.

Group II: Nonexpressor

Certain viruses, particularly herpesviruses and retroviruses,
can be perpetuated in selected host cells which do not permit
expression of the viral genetic information. Nonexpressor cells
are not making viral proteins and therefore are not subject to
lysis (in the case of herpes, a lytic virus) and do not carry viral
antigens on the cell surface (in the case of retroviruses which
mature by budding from the plasma membrane). Such nonexpressor
cells can perpetuate the viral genome for the life of the cell,
which may survive as long as the individual (for example, nerve
cells). In many of the infections listed in Table 3, a few nonexpressor cells may from time to time become permissive, with
activation of the viral genome. Although the permissive cell may
shortly die, the release of infectious virions can disseminate
the infection.

Viral genetic information is usually perpetuated in the form
of double-stranded DNA. Herpesviruses, which can establish

Table 3. Escape from immune surveillance in persistant virus
          infection. Group II: Nonexpressor cells evade immune
          surveillance (selected examples)

Virus Class	Virus	Natural Host	Latency Site	Reference
Herpes	Herpes simplex	Human	Neuron	34-35
	Varicella zoster	Human	Neuron	36
	Epstein Barr	Human	B lymphocyte	37-38
Retro	Visna	Sheep	Macrophage or lymphocyte	39-40
	Murine leukemia	Mouse	All cells	4,45
	Equine infectious anemia	Horse	Macrophage or lymphocyte	46

latency in endstage non-replicating cells, may not integrate their
DNA into host DNA. Retroviruses, which can perpetuate in dividing
cells, integrate some copies of the provirus into host DNA. The
mechanisms of host cell restriction and of subsequent expression
are not well understood.

An active immune response to viral antigens is usually seen
in this kind of persistent infection. The immune response can
cope with any free virus and limits the spread of infection
following activation of the viral genome. However, nonexpressor
cells are always present and these remain exempt from immune
surveillance.

Visna. Visna (also called maedi or progressive pneumonia) is
a natural infection of sheep with a retrovirus (4,39,40). This
virus is exogenous (not carried as host genetic information) and
is transmitted as a respiratory infection between adult sheep or
from ewe to lamb. The virus is fully competent, does not carry
a transforming gene, is incapable of transforming cultured cells,
and does not cause tumors in vivo.

Infected Icelandic sheep have minimal titers of free infectious
virus and explanation and cocultivation with sheep fibroblasts is
often necessary for virus isolation. Studies by Haase and Brahic
(41), using in situ hybridization, have shown that proviral DNA
can be detected in many cells which do not make detectable amounts
of viral antigen. Restriction occurs at several points in the
replicative cycle, since there is a reduction in the number of

copies of viral DNA, RNA, and protein, by comparison with permissive cells in culture.

Sheep raise neutralizing antibody and antibody can be detected against both glycoprotein and the major internal p30 antigen (39). As measured by lymphoblast transformation, a cellular immune response also occurs. In the blood, virus can be isolated from explanted buffy coat cells, probably both lymphocytes and monocytes. However, free infectious virus has never been isolated from the plasma even in fetal or immunosuppressed sheep (39). The viral genome is perpetuated in nonexpressor cells which can escape an apparently potent immune response.

In visna, a second mechanism for escape from the immune response is also seen, namely the occurrence of antigenic variants (42). Variants are probably not essential for virus persistence but may play a role in local bursts of virus replication which have been postulated to precede the irregular occurrence of focal central nervous system (CNS) lesions.

The CNS lesions of visna are inflammatory infiltrates which may progress to destructive demyelinating foci resembling the plaques of multiple sclerosis (43). It seems likely that these changes are immunologically mediated (44), in view of the minimal levels of virus and the relatively noncytocidal nature of the agent. Immunosuppression eliminates the early lesions but definitive experiments to prevent late severe demyelination are not logistically feasible.

<u>Group III: Breakthrough</u>

There are a few persistent infections which appear to violate expectation since the infection continues to spread or infectious virus can be constantly isolated in the presence of an active antiviral immune response. As indicated in Table 4, at least two distinct mechanisms are in operation.

<u>Infectious immune complexes</u>. Infectious immune complexes can be demonstrated in certain examples, notably Aleutian disease of mink (48). Aleutian disease is caused by parvovirus, and infected mink are induced to synthesize extraordinary amounts of antiviral antibody, with a readily demonstrable oligo- or monoclonal globulin pattern in the serum. Although all free virus is complexed to this antibody, at least some of the complexes remain infectious, as demonstrated by transmission to susceptible Aleutian mink.

Theiler's virus (48,49) is another example where a highly lytic picornavirus persists in the spinal cord in the presence of high titers of neutralizing antibody in both serum and tissue.

Table 4.   Escape from immune surveillance in persistent virus
           infection. Group III: Breakthrough of infection in
           the face of overt immune response (selected examples)[*]

Mechanism	Antibody Titer	Virus & Class	Natural Host	Level Infectious Virus	Ref.
Infectious immune complexes	++	Aleutian disease (parvovirus)	Mink	++	47
	+	Theiler's (picornavirus)	Mouse	+	49
	+	Lactic dehydrogenase (flavivirus)	Mouse	++	51
Inter-cellular bridge	++	Subacute sclerosing panencephalitis: measles (paramyxovirus)	Human	±	54

* In these infections there is an overt immune response with antibody titers at normal or supranormal levels. Infectious virus
  can be isolated or infection continues to spread from cell to
  cell even though virus cannot be isolated (SSPE). See text for
  more detailed description.

Nevertheless, virus can be isolated from the spinal cord over
many months. In this instance, the titer is reduced by about
1000-fold below the level seen during the acute phase of infection,
and the residual infectivity (10-100 pfu per gm) may represent
the non-neutralizable fraction seen in kinetic neutralization
tests (53).

     Intercellular bridges. A quite distinct mechanism of per-
sistence is operative in subacute sclerosing panencephalitis
(SSPE), where a defective virus variant appears to spread by
intercellular bridges, created by virus-induced fusion of closely
approximated plasma membranes (54,55). SSPE is initiated by an
apparently uncomplicated measles infection and develops on the
average 7 years later as a rare complication following approximately one per 100,000 primary infections. After onset of symptoms, SSPE progresses to death over a period of 6-18 months, so
that the infectious agent is able to spread, but only slowly.
Although viral nucleocapsids and viral antigens can be readily

demonstrated in the brain, it is very difficult to isolate free
infectious virus, and in many instances brain explants continue
to exhibit the viral antigens but never yield a transmissable
agent. In some instances transmission can only be achieved by
the intracerebral inoculation of whole cells from the explant
culture.

SSPE patients produce very high titers of neutralizing and
other antibodies in both serum and spinal fluids, but the virus
can persist and spread in the presence of these supranormal
immune responses.

Recent studies (54,55) suggest that SSPE is caused by a group
of variant measles viruses which are defective in their synthesis
of the matrix protein which is essential for assembly of mature
infectious virus. Under these circumstances no competent infectious
virions bud into the extracellular phase and viral genomes can
probably be transmitted only by fusion of the plasma membrane of
an infected cell to an adjacent uninfected one, a process specifically mediated by the hemagglutinating glycoprotein of measles
virus. This putative mechanism would escape immune surveillance,
but permit a very slow spread of infection within the brain.

SUMMARY

In spite of the impressive array of immune defense mechanisms
which cooperate in virus clearance, there are a surprising number
of naturally occuring persisting infections in animals and humans.
Study of these examples demonstrates a considerable variety of
strategies which can be employed to escape immune elimination.
Many of these examples of persistence would not be predictable
from a study of anti-viral immunity. The study of virus persistence
has, in fact, enhanced our understanding of anti-viral immune
surveillance and of its limitations.

ACKNOWLEDGMENT

This work was supported in part by USPH grants NS 16010,
NS 01780, and IA 18085.

REFERENCES

1. Kimberlin, R.H., ed. (1976). Slow virus diseases of animals
         and man. American Elsevier, New York.

2. Prusiner, S.B. and Hadlow, W.J., eds. (1979). Slow transmiss-
    able diseases of the nervous system. Academic Press,
    New York.
3. Tyrrell, D.A.J., ed. (1979). Aspects of slow and persistent
    virus infections. M. Nijhoff, The Hague.
4. Stevens, J.G., Todaro, G.J. and Fox, C.F., eds. (1978).
    Persistent viruses. Academic Press, New York.
5. Youngner, J.S., ed. (1977). Persistent virus infections.
    Microbiology - 1977. Schlessinger, D., ed. American
    Society for Microbiology, Washington, p. 433.
6. Gerhard, W., Yewdell, J., Frankel, M.E. and Webster, R. (1981).
    Antigenic structure of influenza virus hemagglutinin
    defined by hybridoma antibodies. Nature (London) 290:
    713.
7. Flamand, A., Wiktor, T.J. and Koprowski, H. (1980). Use of
    hybridoma monoclonal antibodies in the detection of
    antigenic differences between rabies and rabies-related
    virus proteins: II. The glycoproteins, J. Gen. Virol.,
    48:105.
8. Gonzalez-Scarano, F., and Nathanson, N. (1981). Monoclonal
    antibodies which neutralize LaCrosse bunyavirus are
    directed against the GI glycoprotein. J. Virol.:
    submitted.
9. Weiner, H.L., Ault, K.A. and Fields, B.N. (1980). Inter-
    action of reovirus with cell surface receptors. I.
    Murine and human lymphocytes have a receptor for the
    hemagglutinin of reovirus type 3. J. Immunol. 124:2143.
10. Schleuderberg, A., Ajello, C. and Evans, B. (1976). Fate of
    Rubella genome ribonucleic acid after immune and non-
    immune virolysis in the presence of ribonuclease.
    Infect. Immun. 14:1097.
11. Welsh, R.M., Lampert, P.W., Burner, P.A. and Oldstone, M.B.A.
    (1976). Antibody-complement interactions with purified
    lymphocytic choriomeningitis virus. Virol. 73:59.
12. Oroszlan, S. and Nowinski, R.C. (1980). Lysis of retrovirus
    with monoclonal antibodies against viral envelope
    proteins. Virol. 101:296.
13. Silverstein, S. (1970). Macrophages and viral immunity.
    Semin. Hematol. 7:185.
14. Lodish, H.F., Zilberstein, A. and Porter, M. (1981).
    Synthesis and assembly of vesicular stomatitis virus
    and Sindla's virus glycoprotein. in: Perspectives in
    Virology, 11:31. Liss, Inc., N.Y., Editor: Morris
    Polland.
15. Witkor, T.J., Kuwert, E. and Koprowski, H. (1968). Immune
    lysis of rabies-infected cells. J. Immunol. 101:1271.
16. Doherty, P.C., Goetz, D., Trinchieri, G. and Zinkernagel,
    R.M. (1976). Models for recognition of virally modified
    cells by immune thymus derived lymphocytes. Immuno-
    genetics, 3:517.

17. Sissons, J.G.P. and Oldstone, M.B.A. (1980). Killing of virus
        infected cells by cytotoxic lymphocytes. J. Infect.
        Dis., 142:114.
18. Shore, S.L., Cromeans, T.L. and Norrild, B. (1979). Early
        damage of herpes-infected cells by antibody dependent
        cellular cytotoxicity: relative roles of virus-specific
        cell-surface antigens and input virus. J. Immunol. 123:
        2239.
19. Leung, K.N. and Ada, G.L. (1980). Production of DTH in the
        mouse to influenza virus: comparison of conditions for
        stimulation of cytotoxic T cells. Scand. J. Immunol.
        12:129.
20. Leung, K.N. and Ada, G.L. (1980). Two T-cell populations
        mediating delayed-type hypersensitivity to murine
        influenza infection. Scand. J. Immunol. 12:481.
21. DeMaeyer, E. and DeMaeyer-Guignard, J. (1979). Interferons,
        in: Comprehensive Virology, 15:205. Plenum Press, N.Y.
        Editors: Fraenkel-Conrat, H. and Wagner, R.R.
22. Johnson, H.M. (1981). Interferon and host defense systems.
        This symposium.
23. Fraenkel-Conrat, H. and Wagner, R.R. Editors (1979). Virus-
        Host interactions, Comprehensive Virology, 15.
24. Milla, A., Morse, W.W. III, Winkelstein, J. and Nathanson, N.
        (1978). Role of antibody in recovery from experimental
        rabies. I. Effect of deplection of B and T cells. J.
        Immunol. 120:321.
25. Nathanson, N., Johnson, E.D., Camenga, D.L., and Cole, G.A.
        (1974). Immunosuppression and experimental viral
        infection: the dual role of the immune response. E.
        Neter and F. Milgrom, eds. The immune system and infec-
        tious diseases. Karger, Basel. p. 76.
26. Nathanson, N. and Cole, G.A. (1970). Immunosuppression and
        experimental virus infections. Adv. Virus Res. Academic
        Press, New York. 16:397.
27. Nathanson, N. and Cole, G.A. (1971). Immunosuppression: a
        means for assessing the role of the immune response in
        acute viral infections. Fed. Proc. 31:1831.
28. Vyas, G.N., Cohen, S.N. and Schmid, R., eds. (1978). Viral
        hepatitus. Franklin Institute Press, Philadelphia.
29. Rawls, W.E. (1974). Viral persistence in congenital rubella.
        Prog. Medical Virology. 18:273.
30. Anonymous. International symposium on arenaviral infections
        of public health importance. Bull. W.H.O. 52:381.
31. Gardner, M.B. (1978). Type C viruses of wild mice: character-
        ization and natural history of amphotropic, ecotropic
        and zenotropic MuLV. Curr. Topics Microbiol. Immun.
        79:112.

32. Brooks, B.R., Swarz, J.R. and Johnson, R.T. (1980). Spongi-
     form polioencephalomyelopathy caused by a murine retro-
     virus. Laboratory Investigation 43:480.
33. Oldstone, M.B.A., Lampert, P.W., Lee, S. and Dixon, F.J. (1977).
     Pathogenesis of the slow disease of the central nervous
     system associated with WM 1504 E virus. Am. J. Pathology
     88:193.
34. Stevens, J.G. (1978). Latent herpetic infections in the central
     nervous system of experimental animals. in: Stevens, J.
     G., Todaro, G.J. and Fox, C.F., eds. Persistent Viruses,
     Academic Press, New York, p. 701.
35. Kucera, L.S. (1979). Herpes simplex virus-host interactions.
     CRC Critical Rev. Microbiol. 7:215.
36. Burnell, P.A. Varicella-zoster virus. Mandell, G.L., Douglas,
     R.G. and Bennett, J.A., eds. Principles and practice of
     infectious diseases. p. 1295.
37. Klein, G. (1978). EBV-persistence in human lymphoid and
     carcinoma cells. Stevens, J.G., Todaro, G.J. and Fox,
     C.F., eds. Persistent Viruses. Academic Press, New York,
     p. 27.
38. Snyder, B., Kintner, C.R. and Mark, W. (1979). The molecular
     biology of lymphotropic herpesvirus. Adv. Cancer Res.
     30:239.
39. Petursson, G., Martin, J., Georgsson, G., Nathanson, N. and
     Palsson, P.A. (1979). Visna. The biology of the agent
     and the disease. Tyrrell, D.A.J., ed. Aspects of slow
     and persistent infections. Martinus Nijhoff, The Hague,
     p. 165.
40. Petursson, G., Nathanson, N., Georgsson, G., Panitch, H. and
     Palsson, P.A. (1976). Pathogenesis of visna. I. Sequen-
     tial virological, serological and pathological studies.
     Lab. Investigation. 35:402.
41. Brahic, M., Stowring, L., Ventura, P. and Haase, A.H. (1981).
     Gene expression in visna virus infection. Nature, in
     press.
42. Narayan, O., Griffin, D.E. and Clements, J.E. (1978). Pro-
     gressive antigenic drift of visna virus in persistently
     infected sheep. in: Stevens, J.G., Todaro, G.J. and
     Fox, C.F., eds. Persistent viruses. Academic Press,
     New York, p. 663.
43. Georgsson, G., Nathanson, N., Palsson, P.A. and Petursson, G.
     (1976). The pathology of visna and maedi in sheep.
     in: Slow Virus Diseases of Animals and Man, Frontiers in
     Biology, R. Kimberlin, ed., North Holland Publishing Co.,
     Amsterdam, p. 61.
44. Nathanson, N.,Panitch, H., Palsson, P.A., Petursson, G. and
     Georgsson, G. (1976). Pathogenesis of visna. II. Effect
     of immunosuppression upon early central nervous system
     lesions. Lab. Investigation. 35:444.

45. Strayer, D.R. (1980). The nature and organization of retro-
       viral genes in animal cells. Virol. Mono. 17:1.
46. Crawford, T.B., Cheevers, W.P., Klevjer-Anderson, Pa. and
       McGuire, T.C. (1978). Equine infectious anemia: Virion
       characteristics, virus-cell interaction, and host
       responses. in: Stevens, J.G., Todaro, G.J. and Fox, C.
       F., eds. Persistent Viruses. Academic Press, New York,
       p. 727.
47. Porter, D.D. and Cho, H.J. (1980). Aleutian disease of mink:
       a model for persistent infection. in: H. Fraenkel-
       Conrat and R.R. Wagner, eds. Comprehensive Virology,
       16:233-256.
48. Lipton, H.L. (1978). The relationship of Theiler's mouse
       encephalomyelitis virus plaque size with persistent
       infection. in: Stevens, J.G., Todaro, G.J. and Fox, C.F.,
       eds. Persistent Viruses. Academic Press, New York,
       p. 679.
49. Lipton, H.L., Dal Canto, M.C. and Rabinowitz, S.G. (1977).
       Chronic Theiler's virus infection in mice. in: Micro-
       biology - 1977, D. Schlessinger, ed. American Society for
       Microbiology, Washington, 1977, p. 505.
50. Lipton, H.L. and Friedmann, A. (1980). Purification of
       Theiler's virus infection in mice. in: Microbiology -
       1977, D. Schlessinger, ed. American Society for Micro-
       biology, Washington, 1977, p. 505.
51. Riley, V. (1974). Persistence and other characteristics of
       the lactic dehydrogenase elevating virus (LDH-virus).
       Prog. Med. Virol. 18:198.
52. Brinton, M. (1981). Lactate dehydrogenase elevating virus.
       Foster, H.L., Small, J.D. and Fox, J.G., eds. The mouse
       in biomedical research. Academic Press, New York, in
       press.
53. Daniels, C.A. (1975). Mechanisms of virus neutralization.
       in: Notkins, A.L., ed. Viral Immunology and Immuno-
       pathology. Academic Press, New York, p. 79.
54. ter Meulen, V., Hall, W.W. and Kreth, H.W. (1978). Pathogenic
       aspects of subacute sclerosing panencephalitis. in:
       Persistent Viruses. Stevens, J.G., Todaro, G.J. and
       Fox, C.F., eds. Academic Press, New York, p. 615.
55. Hall, W.W. and Choppin, P.W. (1979). Evidence for lack of
       synthesis of the M polypeptide of measles virus in
       brain cells in subacute sclerosising panencephalitis.
       Virology. 99:443.

INFLUENCE OF VIRUSES ON CELLS OF THE IMMUNE RESPONSE SYSTEM

Herman Friedman, Steven Specter and Mauro Bendinelli

University of South Florida, College of Medicine
Tampa, Florida; and University of Pisa, Pisa, Italy

INTRODUCTION

Viruses are ubiquitous infectious agents involved in both overt clinical as well as subclinical or undetectable interactions with eukaryotic cells. It is now widely recognized that various viruses of man and experimental animals may affect the immune response mechanisms to a variable degree (1,6,10). Immunologic hyporeactivity associated with many viral diseases, however, is not entirely a consequence of physical distress. Immunosuppression may be the first sign of illness induced by many viruses. It often reaches levels unjustified by the severity of the disease and, in some instances, persists much longer than other pathologic manifestations. Depending upon the type of virus causing the disease, the reduced ability of an individual to mount a normal immune response may or may not be accompanied by a fall in the preexisting levels of immunity, may be limited to cell-mediated immune responses or extend to humoral ones as well, may be directed selectively to certain types of antigens or may bring about a deficit in selected parameters of the immune response or classes of immunoglobulins.

Evidence in favor of a specific role for viruses in influencing immune responses is emerging from many experimental models which permit some insight into the mechanisms involved. In most if not all instances immunologic impairment appears to be caused by functional alterations of immunocompetent cells caused by direct interactions of such cells with a virus or, possibly by substances produced by other cells as a consequence of viral infection (5,9). Indeed, available data accumulated from many laboratories has shown that many viruses have true immunodepressive properties. This

review will discuss some of the mechanisms involved and offer
illustrations of how some viruses interact with the immune defense
system.

CELLULAR BASIS FOR VIRUS INDUCED IMMUNOSUPPRESSION

The functional complexity of the immune response can be inter-
fered with at various levels. During virus infections, numerous
alterations of immunocompetent cell functions may occur. The known
effects of various viruses on distinct cell classes of the immune
system is shown in Tables 1 and 2. It is important to note that

Table 1. Differential effects of viruses on humoral vs cellular
          immune responses

| | Immune response system affected | |
Virus	Cell-mediated	Humoral
Adenovirus	+	+
Coxsackievirus	+	+
Cytomegalovirus	+	+
Dengue	+	+
Epstein-Barr	+	
Hepatitis A	+	
Hepatitis B	+	
Herpes simplex	+	+
Influenza	+	+
Lymphocytic choriomeningitis	+	+
Measles	+	+
Mumps	+	
Poliovirus	+	
Rabies	+	
Rubella	+	+
Smallpox	+	
Vaccinia	+	
Varicella-Zoster	+	
Yellow fever	+	

Table 2. Major effects of selected viruses on cells of the immune
system

| | Cell Type Affected | | |
| Virus Type | Macrophages | Lymphocytes | |
		T	B
Adenovirus		+	+
Cytomegalovirus	+		+
Dengue	+	+(±)	+
Echovirus	+	+(±)	
Epstein-Barr			+
Hepatitis B	+	+	
Herpes simplex	+	+(±)	+(±)
Influenza	+	+(±)	
LCM	+		
Measles	+	+	+
Mumps		+	+
Poliomyelitis	+	+(±)	
Retroviruses	+	+	+
Rubella	+	+	
Vaccinia	+	+(±)	
Varicella-Zoster	+		
Yellow fever	+	+	

functional alterations of these responses are not always accompanied
by extensive structural changes in lymphoid tissue. In fact,
although hyperplastic changes may occur as a rule, substantially
few if any histologic alterations develop in lymphoid tissue after
virus infections. Nevertheless, certain infections by viruses in
the lymphatic tissue involve necrotic lesions, usually limited in
their extent. These may resemble the cytopathogenic effects
characteristic of the infecting virus in vitro. For instance,
during measles infection typical polynucleated cells develop in
various lymphoid organs, beginning at the very early stage of
infection. Occasionally, variable degrees of cellular depletion
occur which appear to affect specific and distinct areas within the
lymphoid tissue, i.e., so-called thymus dependent areas.

The structural integrity of lymphoid tissue does not prevent immunocompetent cells from exhibiting profound manifestations of functional alterations when examined in vitro. For example, in vitro blastogenic responsiveness of lymphocytes to specific antigens, alloantigens and/or mitogens is often diminished in many viral diseases and may be readily reproduced in various experimental virus infections. Among the various modifications of immunocompetent cell behavior that have been described are those of lymphocyte migration activities. Modification of lymphocyte traffic has been documented only in certain specific experimental models. This has been tentatively attributed to cell surface changes with subsequent inability of the cells to interact with vessel endothelial cells. However, should altered cell traffic be shown not to be an experimental artifact, this might account for the marked fluctuations of peripheral lymphocyte counts which occur in many viral diseases which have so far remained largely unexplained. It is widely accepted that a lymphopenia occurs in certain virus infections, including those caused by measles, rubella, influenza, as well as by coxsackie and arboviruses. Although often transient, lymphopenia may be present from the early incubation stages of the disease and become quite pronounced, affecting a variety of cell types or, alternatively, selected cells such as T-lymphocytes. In evaluating the importance of these changes, it should be noted that continuous redistribution of immunocompetent cells among the lymphoid organs occurs through the lymphatic blood circulation, assuring homeostasis of the immune system, and permitting the interaction of immunocompetent cells among themselves and with antigen, resulting in normal development of the immune responses.

MECHANISMS OF VIRUS INDUCED IMMUNOSUPPRESSION

The immunocompetent cell alteration occurring in viral infections may be due to two different mechanisms: 1) a direct effect of the virus on the immune system or 2) intervention by soluble factors and/or lymphotoxins on the cells.

<u>Direct Virus Action</u> - During the acute phase of several virus diseases such as those caused by rubella, measles, dengue, etc., as well as following vaccination with many attenuated viruses, the specific virus may be isolated from peripheral lymphocytes. In the acute stage of some virus infections, such as those caused by cytomegalovirus, peripheral blood lymphocytes appear to be incapable of inducing complete infectious virus but do contain macromolecules of viral origin. The well-known persistance of Epstein-Barr virus within peripheral lymphocytes and the repeated isolation of viruses from lymph nodes several weeks after clinical recovery from measles indicates that association of virus with lymphoid cells may not be temporary and/or casual.

Viruses are capable of replicating within immunocompetent lymphoid cells. Studies of such viruses have permitted recognition of cellular types and sometimes specific subtypes that are susceptible to infection. In this regard, a wide variety of viruses have been studied in terms of cell types which they preferentially affect such as macrophages, either alone or in conjunction with uncharacterized lymphoid cells, which serve as a target of virus infection in the immune system. In other cases, lymphocytes alone or along with macrophages appear to be the preferential target for virus infection. Study of immunocompetent cell interaction with viruses has been aided by cell culture procedures in vitro and has been especially fruitful in permitting the observation that the outcome of an interaction between a virus and immunocompetent cell depends in a precise manner on the metabolic state of the target cells, possibly more so than on the virus type. Thus, macrophages become restrictive (incapable of replicating) for many viruses, once nonspecifically stimulated by intensive phagocytic stimuli, or become specifically activated through signals emitted by lymphocytes, i.e., lymphokines. In turn, the latter cells, if placed in contact with viruses while in the usual quiescent state, often result in an abortive infection, followed by incomplete synthesis of viral macromolecules, with little or no production of infectious progeny. In contrast, lymphocytes can be made to replicate large amounts of virus by blastogenic stimulation induced either by an appropriate antigen or mitogen.

Several classes of virus cause lymphocytes to undergo rapid clonal and/or blastogenic activation. This does not appear to be an experimental artifact since lymphocytes obtained from patients in the acute phase of measles or rubella and cultured in vitro show a similar high degree of "spontaneous" proliferation. Furthermore, patients with infectious mononucleosis show extremely large numbers of activated lymphocytes in their circulation. It is not yet known whether direct virus infection of lymphocytes is directly related to appearance of such functional alterations.

As a result of direct contact with the viruses in vitro, immunocompetent cells may develop defects similar to those observed following in vitro experimental or clinical infection.

Virus induced changes do not necessarily require that the virus replicate in an infected cell, since in some instances inactivated viruses or purified viral subunit fractions also result in similar alterations. Taken together, these observations suggest the possibility that varying degrees of immunosuppression observed in viral disease may be due to the ability of the virus to interact directly with cells of the immune system and to replicate selectively in those lymphoid cells which are more actively engaged in immune responses, i.e., stimulated to a higher degree of activity. Moreover, these observations emphasize a need for careful examination of the influences exerted by localized virus infections and the function of the lympho-reticular barriers adjacent to draining active sites of infection.

<u>Indirect Virus Effects</u> - Soluble substances toxic to lymphocytes have been repeatedly found in the serum of individuals during acute phases of various viral infections. Recently, it has been reported that serum factors which non-specifically inhibit cell-mediated immune responses may occur in patients with infections such as those caused by hepatitis or infectious mononucleosis virus. Although these factors do not appear to be viral components, their exact chemical nature and role have not been defined. Whether these substances are produced within the immune system itself is also not known. Tissues outside the immune system, under the influence of viruses, may release substances which affect immune responsiveness, as exemplified by fibroblast interferon which is endowed with distinct immunomodulatory properties. The significance of these interferon effects in virus induced immunosuppression is also not yet known.

<u>Immunosuppression by Leukemia Viruses</u> - Within the last decade or so, the most systematic study of immunosuppression by viruses has been performed using murine and other experimental animal leukemia viruses. For example, in the early and mid 1960's, it was noted that susceptible mice infected with oncornaviruses show a marked immunoderangement. Avian leukosis virus as well as feline leukemia viruses generally have similar immunoalterating properties. In the murine leukemia system, it was found that the leukemia virus complex as well as individual components, including the lymphatic leukemia virus, when injected individually or combined, into susceptible mice, alter immune function to antigens such as sheep erythrocytes, soluble serum protein antigens, etc. Also, allograft immune responses as well as a variety of in vitro cell mediated immune parameters, including MIF activity, alloantigen responsiveness, blastogenic responsiveness, and cellular reactivity of lymphocytes, are markedly altered. In experiments designed to determine mechanisms, it was observed that multifactoral systems are involved. For example, leukemia viruses appear to replicate preferentially in lymphoid cells, especially those which may have been sensitized previously with antigens or mitogens, either in vivo or in vitro. Interaction of virus with target lymphocytes alters their responsiveness to a wide variety of antigens, as well as their ability to interact with each other. For example, mice infected with Friend leukemia virus (FLV) show a rapid development of splenomegaly and hepatomegaly. However, before these changes become evident, immunocompetent cells in the spleen lose their ability to respond both in vivo and in vitro to test antigens such as sheep erythrocytes (Table 3). Marked immunodepression occurs prior to overt symptoms of disease and development of large amounts of viral material (2, 3).

Table 3. Effects of Friend leukemia virus on the antibody response
          of mouse spleen cells to sheep erythrocytes _in vivo_ and
          in vitro

Time in days after infection[a]	Spleen Weight (mg)	Antibody plaque response	
		In vivo[b]	In vitro[c]
Control (non-infected)	115 ± 16	1,260 ± 230	1,890 ± 230
+ 2	118 ± 28	630 ± 120	730 ± 310
+ 4	132 ± 42	280 ± 30	316 ± 60
+ 7	296 ± 150	160 ± 230	48 ± 12
+10	1,160 ± 230	98 ± 30	29 ± 10
+15	1,860 ± 530	110 ± 42	20 ± 8
+20	2,210 ± 450	48 ± 16	16 ± 5
+30	2,430 ± 310	20 ± 16	<10

[a] Group of BALB/c mice infected with 10^2 ID FLV on day indicated
before testing for antibody responsiveness.

[b] Mice immunized in vivo by i.p. injection with 10^8 SRBC 4 days
before testing for PFC.

[c] Spleen cells from infected mice cultured in vitro with 2×10^6
SRBC for 4 days before testing for PFC response.

     In vitro experiments have shown that purified FLV, when added
directly to cultures of normal murine spleen cells, depresses the
ability of the cells to respond to sheep erythrocytes (Table 4).
Although a "suppressor" cell activity appeared to be involved,
various studies showed that not only virus transformed cells, but
also cell-free virus preparations, when added to normal spleen
cells or co-cultured with such spleen cells in chambers containing
cell impermeable membranes, still resulted in marked suppression
(7). B-lymphocytes appeared to be usually affected, although
helper or regulatory T-cells were also affected, at least later
in the immune response and during the late phases of leukemia.

     Macrophages appeared to play an important role in the immune
derangement induced by FLV, since addition of normal peritoneal
exudate cells to virus-suppressed spleen cell cultures partially
restored immune responsiveness (8). Reagents which activate
macrophages, including endotoxins and non-specific immunomodulators,

Table 4. Effect of spleen cells, extracts or serum from FLV
          infected mice on antibody responsiveness of normal
          mouse spleen cell cultures immunized _in vitro_ with
          sheep erythrocytes

Preparation added to spleen cell cultures[a]	Antibody response of normal spleen cells[b]
None (Control)	1,680 ± 430
Normal spleen cells (10^2)	1,796 ± 382
FLV infected cells 10^{-4}	1,230 ± 415
10^{-5}	478 ± 150
10^{-6}	330 ± 97
FLV extract, 0.1 ml (unclarified)	1,730 ± 480
(purified)	460 ± 280
Normal serum	1,410 ± 310
FLV serum + 5 days	1,260 ± 240
+ 10 days	696 ± 150
+ 20 days	480 ± 79

[a] Indicated preparation from normal or FLV infected (100 ID_{50} 10 days earlier) mice added to 5 x 10^7 spleen cells from normal mice.

[b] Spleen cells tested for PFCs per 10^6 cells 4 days after in vitro immunization with 2 x 10^6 SRBC.

restored, either fully or partially, the immune responses of leukemia virus-suppressed splenocyte cultures. These and other experiments suggested that a leukemia virus may affect macrophages
directly, but also may have some effect on B-lymphocytes and
helper T-cells.

A portion of the immunomodulation occuring after FLV infection
appeared related to release of soluble immunoregulatory factors (4).
For example, spleen cells from mice infected with FLV, when separated
from target normal lymphoid cells by cell-impermeable membranes,
suppressed the immune response, not only because of release of
infectious virus, but also because of release of soluble substances.
This was shown by experiments in which cell-free preparations from
leukemia virus-infected spleen cells, when added to normal spleen
cells, suppressed both antibody responses and cytolytic reactivity.
The soluble supernatants obtained by ultracentrifugation and

treatment with ultraviolet light were as suppressive as virus
containing preparations. Soluble immunosuppressive factors were
obtained only late during the virus infection. Thus, it seemed
possible that both a direct effect by virus per se and an indirect
effect because of a soluble factor(s) from infected cells, mediated
immunoderangement. Antibody prepared to virus antigens neutralized
with the immunosuppressive properties of the virus, but not the
soluble factor(s) suggesting that a "chalone" or other mediators
from lymphocytes may be involved. The soluble factor(s) was
absorbable by bone-marrow cells but not thymocytes. Macrophages
also appeared to be involved in the immunosuppression induced by
the soluble factor(s), since normal peritoneal exudate cells, when
added to normal spleen cell cultures suppressed by the factor,
partially restored the immune response, similar to the restoration
observed when virus per se was used for immunosuppression. These
and similar experimental results with the FLV system showed that
the virus affected the immune responses of normal murine splenocytes by both direct and indirect mechanisms (2,3).

Selective replication of the leukemia virus in populations of
lymphocytes appeared mediated by specific receptors on the surface
of cells which reacted with the virus or virus antigen. Non-
infectious virus and viral derived antigens blocked the initial
interaction of infectious leukemia viruses and lymphoid cell
suspensions, inhibiting both replication of the virus and the
resulting immunosuppression. However, non-infectious virus
particles did not block the effect of the soluble immunomodulator
from other virus-infected cells. Thus, it appeared that the
initial interaction of a leukemia virus with lymphoid cells resulted
in penetration of the virus into the cells, followed by replication
of the virus and eventual release of soluble factors from the same
or other cells, which then further influenced the immune response
mechanism.

DISCUSSION

There appears to be little doubt that virus induced immunosuppression may be responsible for the decreased antimicrobial
defenses that accompany many viral infections, independently of
the presence of local tissue damage that might justify the decrease.
Such a condition reveals itself clinically as increased susceptibility to superinfection during many virus infections and reacti-
vation of latent infectious process. Such susceptibility often
disappears with recovery from virus infection, although occasionally the condition may last much longer, for example, more than
one year following measles or cytomegalovirus infection. Nevertheless, it is not yet known with certainty whether less specific
mechanisms may participate in alterations observed in other
clinical manifestations, i.e., parasitic and bacterial infections

and neoplasia per se also contribute to reduced host resistance.
Not clearly evaluated, for example, is a possible intervention of
hormonal disturbances or the influence of substances that may
leak through mucosal surfaces as a consequence of their invasion
by a virus.

A major unanswered problem concerning virus immunosuppression
is the role of such a suppression in the development of primary
infection by the virus and how this contributes to the pathological
consequences, including possibly tumor progression in experimental
animals. Possibly, because it had been taken for granted, and
certainly because the immunobiology of virus infection has been
exposed to detailed investigation only during relatively recent
years, this aspect still remains to be explored. It is clear,
however, that no true generalizations are possible at this time.
The contribution of immunopathologic mechanisms to virus induced
antigens is quite consistent. In some infections of lower animals,
both experimental and naturally acquired, pathologic lesions are
sustained solely because of the immune response against antigens
of the infecting virus, and there again at structures of the host
modified by the virus. Classic examples include the disorders
caused by lymphocytic chorimomeningitis virus infection of rodents.
Infection with this virus at an age when immunologic maturation is
complete or well underway results in rapid disease and death of
the animal unless drastic immunodepressive treatment is given,
whereas if the virus infection is acquired early in life, it remains
silent for many months until a feeble antibody response to viral
antigens causes deposition of immune complexes in the kidney.

In human disease similar clear-cut situations are not known,
except possibly in dengue hemorrhagic fever where pathologic disorders may be due entirely to a systemic immunocomplexemia
associated with virus infection. However, immunopathologic
phenomena are considered a consistent feature of Hepatitis B and
Yellow Fever virus infections, where they contribute to liver
damage. Similarly in measles and other viral exanthemas, immune
complexes are known to participate in cutaneous lesions. Furthermore, immune complexes are often found in the circulation during
many viral diseases and are believed to represent the pathogenic
basis for some complications, including polyarthritis nodosum
arthritis, urticaria and vasculitis, as well as disseminated
intravascular coagulation, a complication of Smallpox, Yellow
Fever, and other virus infections. In addition, immunopathologic
mechanisms triggered by virus infections have been postulated
in the genesis of certain chronic glomerulonephritis and other
autoimmune diseases.

It is possible that in some circumstances immunopathologic
impairment caused by a virus, even if it facilitates the spreading
of the virus and causes a delay in recovery, is nevertheless

beneficial to the patient by lowering the amount of immunologically
mediated damage caused by the virus. Thus, the depression of the
host's immune responsiveness may be of special importance to
continuation of a virus infection and, in many instances, may also
have some beneficial role to the individual in preventing certain
lethal consequences due to immune responses to virus altered cells.
In the case of leukemia and/or tumor-associated viruses, the
effective strategy of the host may not yet have evolved sufficiently
in such virus infections to diminish or abrogate the tumorigenic
reactions associated with infection. Only the immunosuppressive
responsiveness to the virus may be manifested and not the control
of the virus itself. Nevertheless, it seems reasonable to postu-
late that immunodepressive effects by viruses represent an important
element in the economy of viral infections and a possible prerequisite for other escape mechanisms to become effective for the
virus per se.

SUMMARY

     A wide variety of immune parameters may be influenced in vivo
and in vitro during virus infections. Some viruses appear to
specifically alter macrophage activity, while other viruses preferentially affect activity of lymphocytes, both B and T cells.
There are many mechanisms involved in immune derangement during
virus infection, including the possibility that viruses preferentially interact with selected classes of immunocompetent cells.
Alternately, some of the events associated with viral derangement
of the immune response may be associated with the release or
formation of factors by cells infected with a virus which affects
other cells of the immune system. Similarly, the development of
immune complexes resulting from the interaction of viruses or
their components with antibody directed to the virus may influence
the immune response, as well as induce other immunopathologic
manifestations. Many model systems have been studied, especially
those with viruses important to human infection. It is noteworthy
that leukemia viruses and, in some cases, other tumor associated
viruses may directly alter the immune response, usually before
other manifestations of the tumorigenic process is evident. Tumor
virus-induced immunomodulation is associated with direct effects
of a virus on immunocompetent cells per se as well as the formation
of soluble factors induced by the virus infection. These and other
considerations indicate that interactions of viruses with the host
immune defense system are an important consideration in understanding how virus infection is initiated and progresses, and how
such infection affects the host's physiologic activity, including
immunocompetence, per se.

REFERENCES

1. Bendinelli, M. (1981). Mechanisms and significance of immuno-
        depression in viral diseases. Clin. Immunol. Newsletter
        2:75:80.
2. Friedman, H. and Specter, S. (1981). Virus induced immuno-
        modulation. In: Advances in Immunopharmacology. J.
        Hadden, L. Chedid, P. Mullen and F. Spreafico (eds.)
        Pergammon Press, New York, p. 91-100.
3. Friedman, H., Specter, S., Farber, P. and Ceglowski, W. (1979).
        Interactions of viruses with lymphoid cells. In: Cell
        Biology and Immunology of Leukocyte Function. Academic
        Press, New York, p. 783-802.
4. Kateley, J.R., Kamo, I., Kaplan, G. and Friedman, H. (1974).
        Suppressive effect of leukemia virus-infected lymphoid
        cells on in vitro immunization of normal splenocytes.
        J. Natn. Cancer Inst. 53:1371-1378.
5. Proffitt, M.R. (1979). Virus-lymphocyte interactions:
        Implications for disease. Elsevier/North-Holland, New
        York.
6. Specter, S. and Friedman, H. (1978). Viruses and the immune
        response. Pharmac. Ther. A 2:595-622.
7. Specter, S., Patel, N. and Friedman, H. (1976). Immuno-
        suppression induced in vitro by cell-free extracts of
        Friend leukemia virus infected splenocytes. J. Natn.
        Cancer Inst. 56:143-147.
8. Specter, S., Patel, N. and Friedman, H. (1976). Peritoneal
        exudate cell induced restoration of antibody formation
        by leukemia virus suppressed spleen cell cultures.
        Proc. Soc. Exp. Biol. Med. 151:163-167.
9. Wainwright, W.H., Veltri, R.W. and Sprinkle, P.M. (1979).
        Abrogation of cell-mediated immunity by a serum blocking
        factor isolated from patients with infectious mono-
        nucleosis. J. Infect. Dis. 140:22-32.
10. Woodruff, J.F. and Woodruff, J.J. (1975). T lymphocyte
        interaction with virus and virus-infected tissues.
        Prog. Med. Virol. 19:120-160.

GENETICALLY CONTROLLED RESISTANCE TO VIRUSES

Margo A. Brinton

Wistar Institute
Philadelphia, PA   19104

INTRODUCTION

Historically, the first demonstrations of the existence of
host genes that could control resistance to disease induced by
an animal virus were reported independently by Lynch and Hughes
(1) and Webster and Clow (2) in 1936. Subsequently, a number of
other genes have been identified in various strains of inbred
mice, each of which confers resistance to a specific type of virus
infection (3,4,5). Since different families of viruses vary
greatly in their modes of replication, it would be expected a
priori that the various resistance gene products would also differ
in their mechanisms of action. Host resistance genes may act on
a particular virus infection at the level of receptors, intracellular replication, interferon inhibition, or the immune
response. A host resistance gene can therefore be an important
component in determining the outcome of a particular virus
infection. However, the severity of a virus-induced disease is
the result of the interaction between host resistance genes, viral
virulence genes, and the host defense system.

MURINE GENES INFLUENCING DISEASE INDUCED BY VIRUSES

The majority of studies of genetic resistance have been
carried out in various strains of inbred mice. Identification
of an increasing number of host genes controlling the specific
virus infections in inbred mice as well as in some wild populations (6) indicates the importance of such genetic effects. Genes
which modulate disease induced by nine different families of RNA
and DNA viruses in mice have thus far been identified. Among the

475

RNA viruses, genes which influence togavirus, orthomyxovirus,
paramyxovirus, coronavirus, and picorna virus infections have been
reported (Table 1). However, genes controlling RNA tumor virus
infections have to date been the most extensively studied (Table
2). Genes affecting disease induction by herpes virus, papova
virus, and pox virus infections have also been reported (Table 3).
Although in almost all cases the mechanism of action of these
resistance genes is not understood, some generalizations can be
made from the available data. In some instances, a single locus
is responsible for the difference between susceptibility or
resistance to virus-induced disease. However, in other cases,
two or more unlinked loci are involved. A dominant allele may
code for either the resistant or the susceptible phenotype. Each
resistance gene or gene group functions specifically by influencing
infection of only a particular group of viruses. The various
resistance genes appear to be unique. When tested, the resistance
genes have been found to segregate independently (4). The majority
of virus-resistance genes identified in inbred mice do not map
to the H-2 major histocompatibility locus.

GENETICALLY CONTROLLED RESISTANCE TO FLAVIVIRUSES

     Resistance to flavivirus - induced encephalitis is an example
of a case where a single allele controls the expression of resistance. Historically, it was the first mammalian virus-resistance
gene identified (1,2). Based on available data, it seems most
likely that the flavivirus resistance gene acts on an intracellular
flavivirus replication step. Such factors as the age of the host,
its immune status, the degree of virulence of the infecting virus,
and the route of infection have been found to influence the phenotypic expression of the flavivirus resistance gene at the whole
animal level. However, no evidence has yet been reported that
indicates that any of these factors are involved in the specific
mechanism of resistance mediated by the product of the resistance
gene. Mice that possess the resistance gene support the replication of flaviviruses, but virus yields in their brain tissues are
2 or more logs lower and the spread of the infection is slower
than in susceptible mice (6,29,30). The _in vivo_ expression of
genetic resistance to flaviviruses requires an intact lymphoreticular system (29,31,32). Immunosuppression of resistant mice
before infection with cyclophosphamide, X-irradiation, or thymus
cell depletion converts a normally asymptomatic flavivirus
infection into a lethal one. Nevertheless, under such conditions
the onset of disease in resistant mice is delayed several days as
compared to susceptible mice, and the virus titers in moribund
resistant mouse brains is 2 or more logs lower than in susceptible
brains.

Table 1. Murine genes affecting disease induced by RNA viruses

Virus	Disease	Susceptible Prototype Strain	Resistant Prototype Strain	Most Inbred Strains	Number of Genes or Designation	Dominant Trait	Maps to H-2	Reference
Toga								
Flavi	Encephalitis	C3H	PRI	S	1	R	No	7,8
LDV	Polioencephalitis	C58	C3H	R	>1	R	No	9
Orthomyxo								
Influenza A	Pneumonia	A/J	A2G	S	Mx	R	No	10
Paramyxo								
Measles	Encephalitis	C3H	S/JL	S	>1	R	No	11
Corona								
MHV2	Hepatitis	PRI	C3H	R	1	S	No	12
MHV3	Persistence	C3H/S	C3H	R	2	?	No	13,14
JHM	Demyelination	C3H	S/JL	S	Rhv-1	R	No	15
					Rhv-2	S	No	
Picorna								
EMC/M	Diabetes	SWR/J	C57BL/6	R	1	R	No	16

LDV, lactate dehydrogenase elevating virus; MHV, mouse hepatitis, JHM, variant of MHV; EMC/M, encephalomyocarditis/M variant; R, resistant; S, susceptible.

Table 2. Murine genes affecting disease induced by RNA tumor viruses

Leukemia Virus	Disease or Outcome	Susceptible Prototype Strain	Resistant Prototype Strain	Most Inbred Strains	Designa-tion of Gene	Dominant Trait	Maps to H-2	Chromo-Some	Refer-ence
Ecotropic	Replication	NIH	BALB/c	E	Fv-1	R	No	4	17
FLV	Splenic foci	Many	C57BL/6	S	Fv-2	S	No	9	18
Endogenous	Virus Expression	AKR	NIH	R	Akv-1*	S	No	7	19
Endogenous	Leukemia	AKR	NIH	R	Akv-2*	S	No	16	19
GLV	Leukemia	$H-2^k$	$H-2^b$	R	Rgv-1	R	Yes	17	20
GLV	Leukemia	$H-2^k$	$H-2^b$	R	Rgv-2	R	No	4?*	21
FLV	Recovery from splenomegaly	$H-2^k$	$H-2D^b$	E	Rfv-1	R/S	Yes	17**	22
FLV	Recovery from splenomegaly	$H-2^k$	$H-2K^b$	E	Rfv-2	R/S	Yes	17**	23
FLV	Recovery from splenomegaly	BALB/c	C57/BL6	S	Rfv-3	R	?	?	24

R, resistant; S, susceptible/ R/S, intermediate phenotype; E, both genotypes equally represented in inbred strains; H-2, located on mouse chromosome 17; FLV; Friend leukemia virus; GLV, Gross leukemia virus.

* Akv-1 and Akv-2 are transcripts of virus genetic information.

** Probably map outside I region.

Table 3. Murine genes affecting disease induced by DNA viruses

Virus	Disease	Susceptible Prototype Strain	Resistant Prototype Strain	Number of Genes Involved	Dominant Trait	Maps to H-2	Reference
Herpes							
HSV-1	Encephalitis	A/J	C57BL/6	2+	R	No	25
CMV	Encephalitis	BALB/c	C3H/He	1+	S	Yes	26
Papova							
Polyoma	Tumors	AKR	C57BL/6	1+	S	?	27
	Runting	AKR	C57BL/6	1+	R	No	27
Pox							
Ectro-melia	Mousepox	CBA	C57BL	1+	?	?	28

R, resistant; S, susceptible; HSV, herpes simplex virus; CMV, cytomegalovirus.

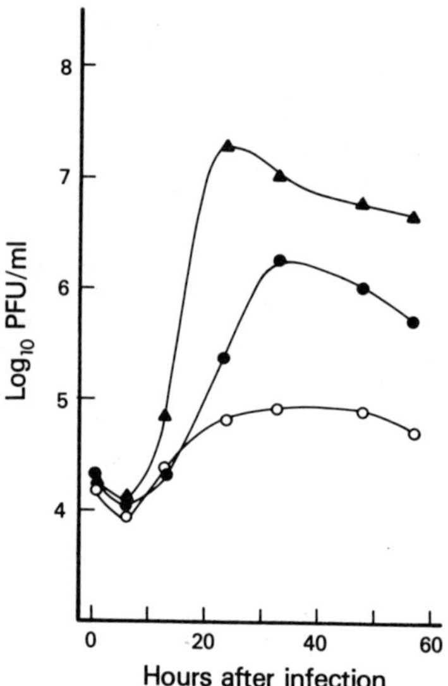

Figure 1.   Comparison of West Nile virus (WNV) replication in (O)
            resistant C3H/RV embryofibroblasts (●) susceptible
            C3H/HE embryofibroblasts, and (▲) BHK cells. Cultures
            were infected at a multiplicity of infection of 10.

        Cell cultures derived from various tissues obtained from
resistant mice produce lower yields of flaviviruses than do
comparable cultures of cells from susceptible animals (Fig. 1 and
Refs. 33,34). Unrelated viruses grow equally well in cultures
from the two types of mice. Virus adsorption and penetration
apparently occur equally well in resistant and susceptible cells,
since the same percentage of cells show perinuclear virus-positive
immunofluorescence in both types of culture by 6 to 8h after
infection. The development of the resistant C3H/RV mouse strain
which is congenic to C3H/HE has allowed comparative studies of
flavivirus resistance to be carried out against a low background
of unrelated variables. The 17D-vaccine strain of Yellow Fever
virus (YFV) and the E101 strain of West Nile virus (WNV) are the
flaviviruses used in our experiments.

        In the case of the gene in A2G mice which confers resistance
to influenza-induced disease, sensitivity to interferon inhibition
has been found to be specifically involved in resistance.
Injection of small amounts of antimouse interferon antibody
rendered the resistant A2G mice fully susceptible to influenza
virus-induced disease (35). Also, virus titers were increased
in the antibody-treated A2G mice, so that the amount of virus
produced was similar to that observed in susceptible A/J mice.
In contrast to what was observed with the A2G mice, injection of

Table 4. Effect of antibody to mouse interferon on the expression
of influenza and flavivirus resistance

Mouse Strain	Genotype	Virus[a]	Treatment[b]	Mortality	log 10 virus titer[c]
A/J	r/r	Influenza	NSG	8/8	6.0
			AIF	8/8	6.0
A2G	R/R	Influenza	NSG	0/8	3.7
			AIF	8/8	6.3
C3H/HE	r/r	Yellow Fever virus	NSG	4/4	5.7
			AIF	4/4	6.5
C3H/RV	R/R	Yellow Fever virus	NSG	0/4	2.4
			AIF	0/4	2.8

a. Mice were injected with 10^4 LD_{50} of strain A/Turkey/England
63 of influenza by the intraperitoneal route or $10^{3.7}$ PFU of
17D-Yellow Fever virus by the intracerebral route.

b. Sheep normal serum globulin (NSG) or sheep anti-mouse interferon globulin (AIF) were diluted 1 to 3 with PBS and 0.1 ml
was injected intravenously just before virus was injected.

c. Influenza titer expressed in EID_{50} per ml of blood; Yellow
Fever virus titer expressed as PFU per brain.

anti-interferon antibody had no effect on the flavivirus resistance
of C3H/RV mice (Table 4). Treatment with antibody to mouse interferon did allow an increase in virus production in both C3H/RV
and C3H/HE mice, but virus production in resistant mice remained
significantly lower than in susceptible mice (36). Results from
tissue culture experiments were similar. These data indicate
that interferon is not a necessary component in the expression
of flavivirus resistance.

Our recent studies indicate that the synthesis of flavivirus-
specific RNAs and proteins is less efficient in resistant cells
than in susceptible cells, indicating that the flavivirus resistance gene product acts intracellularly and affects an early step
in virus replication (37). Flaviviruses are small (40 to 60 nm
in diameter), round, enveloped RNA viruses, that contain a single-
stranded 40S genome of plus-strand polarity (38). This RNA is
the m-RNA for all virus-specific proteins. The replication of
these viruses takes place entirely in the cytoplasm.

The RNA contained in extracellular virions produced by flavi-
virus-infected resistant and susceptible cell cultures was compared.
RNA was extracted from pelleted virions that had been labeled with
[^{3}H] uridine from 4 to 29 h after infection and was sedimented
through 15 to 30% sucrose gradients. The infectivity of the resistant
cell virus sample was $10^{5.7}$ PFU and that of the susceptible virus
sample $10^{7.6}$ PFU. The radioactively labeled RNA contained in the
virus produced by the susceptible cell cultures was essentially
all normal size 40S RNA (Fig. 2). In contrast, the majority of
the labeled RNA from the virions grown in resistant cells was of
smaller size than 40S. The amount of labeled 40S RNA present in
the resistant cell virus was about 100-fold less than that in
susceptible cell virus. This correlated with the 100-fold lower
infectivity of the resistant cell virus sample.

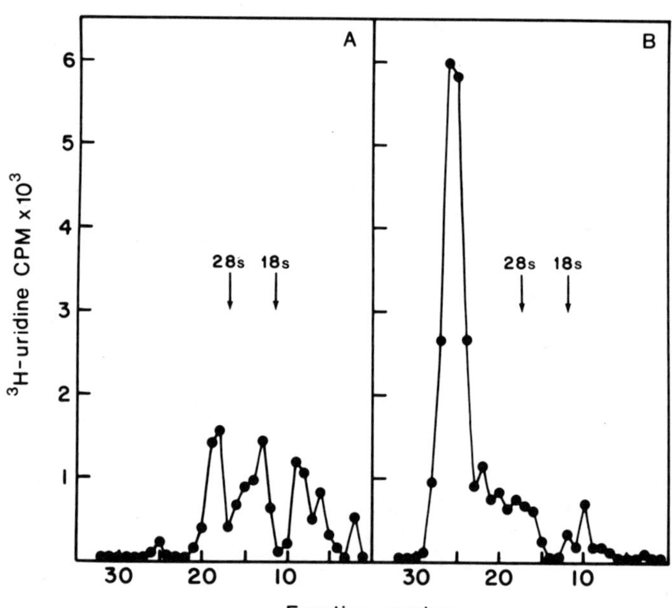

Figure 2.   Rate-zonal sedimentation patterns of WN virion-associated
            RNA in culture fluids from (A) resistant C3H/RV embryo-
            fibroblast cultures and (B) susceptible C3H/HE embryo-
            fibroblast cultures. Cultures were incubated with
            [^{3}H] uridine (20 μci/ml) from 4 to 29 h after infection.
            Virions in harvested culture fluids were pelleted, the
            RNA extracted and sedimented through 15 to 30% (wt/vol)
            sucrose gradients. Unlabeled BHK rRNAs were used as
            markers (37).

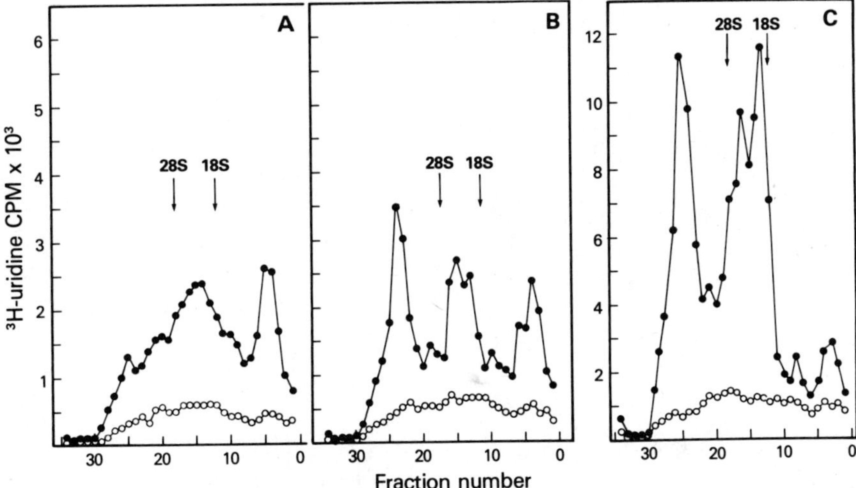

Figure 3.   Intracellular RNA synthesized between 24 and 25.5 h
             after infection with WNV in (A) resistant C3H/RV (B)
             susceptible C3H/HE and (C) BHK cultures. Actinomycin
             D (2 µg/ml) was added to culture fluids at 23 h, and
             [^{3}H] uridine (20 µci/ml) was added at 24 h after
             infection. Cell extracts were prepared in sodium
             dodecyl sulfate-containing buffer, and RNA extracted
             and sedimented in 15 to 30% (wt/vol) sucrose gradients.
             The position of the 18S and 28S rRNAs were determined
             from absorbance profiles. Symbols: (●), infected;
             (O), uninfected control (37).

     Analysis of intracellular actinomycin D-insensitive RNA
indicated that the synthesis of 40S RNA as measured by incorporation of [^{3}H] uridine was less efficient in resistant cells at all
times tested between 10 and 72h after infection (Fig. 3). This
difference was specific for flaviviruses, since RNA synthesis by
unrelated viruses such as sindbis, was found to be identical in
both susceptible and resistant cells. Since the 40S RNA is the
only viral mRNA present in flavivirus infected cells, a decrease
in the synthesis of this RNA would be expected to result in a
concomitant reduction in the level of viral protein synthesis.
The level of incorporation of [^{35}S] methionine into viral proteins
in resistant cells was found to be markedly less than that observed
in infected susceptible cells (37).

     Recently, we have isolated a mutant of WNV, designated WNV-RV,
that is able to replicate more efficiently in resistant cells than
standard WNV (Fig. 4). This mutant was isolated from the culture
fluid of a resistant C3H/RV cell culture persistently infected

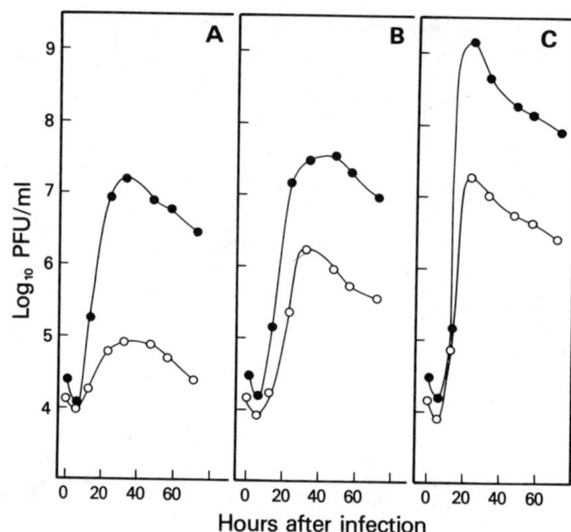

Figure 4. Comparison of replication of parental WNV (O) and mutant
WNV-RV (●) in cultures of (A) resistant C3H/RV embryofibroblasts, (B) susceptible C3H/HE embryofibroblasts,
and (C) BHK cells. Cultures were infected at a multi-
plicity of 10 (37).

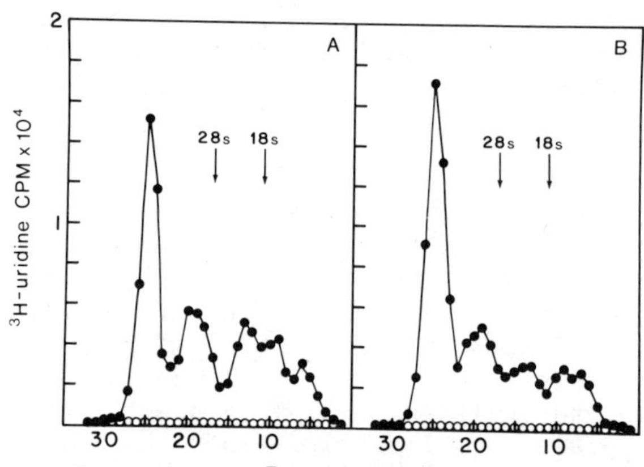

Figure 5. Rate-zonal sedimentation analysis of [³H] uridine-
labeled RNA associated with virions in culture fluids
from mutant WNV-RV infected (A) resistant C3H/RV embryofibroblasts and (B) susceptible C3H/HE embryofibroblasts.
Cultures were incubated with [³H] uridine (20 μci/ml)
from 4 to 29 h after infection. Virions in harvested
culture fluids were pelleted and the RNA was extracted
and sedimented through 15 to 30% (wt/vol) sucrose
gradients. Unlabeled BHK rRNAs were used as markers.
Symbols: (●), infected; (O), uninfected control (37).

with WNV for 25 weeks (37). Analysis of viral-specific RNA
synthesis after infection of cells with this mutant indicated that
the amount of [^{3}H] uridine incorporated into 40S virion RNA was
similar in WNV-RV produced by resistant and susceptible cells
(Fig. 5). However, peaks of smaller sized RNAs were observed in
mutant WN virions produced by both types of cells. Preliminary
information indicates that these smaller RNAs are WNV-specific.
It is not yet known whether the small RNAs observed in standard
WN virions produced by resistant cells (Fig. 2) and those in WNV-RV
virions are identical. If the small RNAs prove to be deleted forms
of virion RNA, then in infections with the WNV-RV mutant these
RNAs apparently have little autointerfering ability since high
titers of virus are produced.

Even though the WNV-RV mutant had acquired an enhanced ability
to replicate, it produced only a slight cytopathic effect in
resistant cell cultures and did not kill resistant animals. On
the other hand the WNV-RV mutant was found to be more cytopathic
for susceptible cells and more virulent for susceptible animals
than the parental WNV. The WNV-RV mutant is a unique tool for use
in further studies of the mode of action of the flavivirus resistance
gene. At present we can conclude that resistant cells alter the
amount and species of flavivirus RNA synthesized, which results
in the release of less infectious virus, and also that resistant
cells appear to be more resistant to cytopathic effects caused by
flavivirus replication.

ACKNOWLEDGMENTS

This work was supported by Public Health Service research
grant AI-14238 from the National Institute of Allergy and Infectious
Diseases and by contract DAMD-17-80-C-9108 from the U.S. Army
Medical Research and Development Command.

REFERENCES

1.   Lynch, C.J. and Hughes, T.P. (1936). Genetics 21:104.
2.   Webster, L.T. and Clow, A.D. (1936). J. Exp. Med. 63:827.
3.   Pincus, T. and Snyder, H.W. (1975). Viral Immunology and
        Immunopathology, ed. A. Notkins, Academic Press,
        New York, pp. 167.
4.   Bang, F.B. (1978). In Advances in Virus Research, 23:269.
5.   Brinton, M.A. and Nathanson, N. (1981). Epidemiologic Reviews,
        3, 115.
6.   Darnell, M.B., Koprowski, H. and Lagerspetz, M. (1974).
        J. Infect. Dis. 129:240.
7.   Webster, L.T. (1937). J. Exp. Med. 65:261.
8.   Sabin, A.B. (1952). Proc. Nat. Acad. Sci. USA,38:540.

9.   Martinez, D., Brinton, M.A., Tachovsky, T.G. and Phelps, A.H.
        (1980). Infect. Immun. 27:979.
10. Lindenmann, J. (1964). Proc. Soc. Exp. Biol. (N.Y.) 116:506.
11. Neighbour, P.A., Rager-Zisman, B., and Bloom, B.R. (1978).
        Infect. Immun. 21:764.
12. Bang, F.B., and Warwick, A. (1960). Proc. Nat. Acad. Sci.
        USA, 46:1065.
13. Virelizier, J.L., Dayan, A.D. and Allison, A.C. (1975).
        Infect. Immun. 12:1127.
14. Levy-LeBlond, E., Oth, D., and Dupuy, J.M. (1979). J. Immunol.
        122:1359.
15. Stohlman, S.A., and Frelinger, J.A. (1980). In Genetic Control
        of Natural Resistance to Infection and Malignancy,
        E. Shamene, P.A.L. Kougshaven and M. Landy, eds.,
        Academic Press, New York, pp. 247.
16. Onodera, T., Yoon, J.W., Brown, K.S. and Notkins, A.L. (1978).
        Nature, 274:693.
17. Pincus, T., Hartley, T.W. and Rowe, W.P. (1971). J. Exp.
        Med. 133:1234.
18. Lilly, F. (1970). J. Nat. Cancer Inst. 45:163.
19. Rowe, W.P. (1972). J. Exp. Med. 136:1272.
20. Lilly, F., Duran-Reynolds, M.L., and Rowe, W.P. (1975).
        J. Exp. Med. 141:882.
21. Lilly, F. (1966). J. Nat. Cancer Inst. 22:631.
22. Chesebro, B., Wehrly, K. and Stimpfling, J.H. (1974). J. Exp.
        Med. 140:1457.
23. Chesebro, B., and Wehrly, K. (1978). J. Immunol. 120:1081.
24. Doig, D., and Chesebro, B. (1979). J. Exp. Med. 150:10.
25. Lopez, C. (1975). Nature 258:152.
26. Chalmers, J. (1980). In Genetic Control of Natural Resistance
        to Infection and Malignancy, E. Shamene, P.A.L. Kougshaven,
        and M. Landy, eds., Academic Press, New York, pp. 283.
27. Chang, S.S. and Hildemann, W.H. (1964). J. Nat. Cancer Inst.
        33:303.
28. Kees, U. and Blanden, R.V. (1976). J. Exp. Med. 143:450.
29. Goodman, G.T. and Koprowski, H. (1962). J. Cell Comp.
        Physiol. 59:333.
30. Hanson, B. and Koprowski, H. (1969). Microbios. 1B:51.
31. Jacoby, R.O. and Bhatt, P.N. (1976). J. Infect. Dis. 134:158.
32. Jacoby, R.O., Bhatt, P.N., and Schwartz, A. (1980). J. Infect.
        Dis. 141:617.
33. Darnell, M.B. and Koprowski, H. (1974). J. Infect. Dis. 129:
        248.
34. Webster, L.T. and Johnson, M.S. (1941). J. Exp. Med. 74:489.
35. Haller, O., Arnheiter, H., Gresser, I. and Lindenmann, J.
        (1979). J. Exp. Med. 149:601.
36. Brinton, M.A. (1981). In Natural Resistance to Viruses and
        Tumors, O. Haller, Ed., Curr. Top. Microbiol. Immunol.
        92, pp. 1.

37. Brinton, MA. (1981). J. Virol. 39, 413.
38. Westaway, E.G. (1980). In The Togaviruses, Biology, Structure,
         and Replication. W. Schlesinger, ed., Academic Press,
         Inc., New York, pp. 531.

MACROPHAGE OXIDATIVE METABOLISM: A DEFENSE MECHANISM AGAINST

VIRUS INFECTION?

Bracha Rager-Zisman, Celia F. Brosnan, and
Barry R. Bloom

Departments of Microbiology, Immunology and Pathology
Albert Einstein College of Medicine, Bronx, New York
10461

There is ample evidence that macrophages (MØ) are intrinsic-
ally resistant to infection by a number of viruses in vitro and
are important in providing resistance to virus infection in vivo
(1). In neither case are the mechanisms of viral resistance
engendered by macrophages understood at the molecular level.

A number of potentially effective molecular mechanisms in
phagocytic cells capable of killing intracellular bacteria, fungi
and protozoa have been identified, including: i) low intracellular pH; ii) cationic proteins; iii) lysosomal enzymes, and
iv) lysozyme (2). Particular interest has focused on the role
of oxidative cytocidal mechanisms (3). One of the early events
associated with phagocytosis in mononuclear phagocytes is a
"respiratory burst", which is a rapid sequence of biochemical
events leading from increase in oxygen uptake to the formation of
highly reactive oxygen metabolites, such as the superoxide anion
(O_2^-), hydrogen peroxide (H_2O_2), hydroxyl radical (.OH) and singlet
oxygen ($^1O_2$) (3). Enhanced killing of parasites and fungi by
activated macrophages has been shown to be dependent on phagocytosis associated with this respiratory burst (4). In addition,
mouse primary macrophages activated by BCG or LPS, which are known
to produce elevated levels of O_2^- and H_2O_2, showed increased microbicidal capacity against trypanosomes, toxoplasma and candida (4).

Continuous cloned macrophage-like cell lines offer certain
advantages over primary cells for studying macrophage functions
because: 1) they are homogeneous cloned cells that are free from
other contaminating cell types, and 2) they grow rapidly in vitro
and can be cultured in sufficient quantities.

We have, over the past several years, used continuous macrophage-like cloned cell lines derived from a reticulum cell sarcoma, J774 (5). We have selected for a clone, clone 16, which under appropriate stimulation with phorbol myristate acetate (PMA) or aggregated IgG produces a respiratory burst which generates O_2^- and H_2O_2 in amounts of the same order of magnitudes as primary activated macrophages (6). In order to provide a model for the study of the role of oxidative cytocidal mechanisms, we have selected a variant, clone C3C, from the parental clone 16 cells, using nitroblue tetrazolium (NBT) as a selective agent, which under comparable conditions, lacks the ability to produce significant amounts of O_2^- and H_2O_2 (6).

Monolayers of the macrophage-like cell lines, clones 16 and C3C, were infected at different multiplicities with vesicular stomatitis virus (VSV). Both lines were found to be susceptible to VSV (Table 2). Within 24-28 hrs after infection, both cell lines were completely lysed by the virus.

In the experiments summarized in Fig. 1, clone 16 or variant C3C were stimulated for 10 minutes with PMA prior to virus infection, and then infected at several multiplicities. At low multiplicities there was a significant reduction (on the order of 2-3 logs) in the yield of VSV from the PMA-treated clone 16 cells. No reduction was found in yields of VSV in clone C3C similarly treated with PMA. Virus infection of clone 16 cells led to the formation of multinucleate giant cells, which was augmented with PMA. This cytopathic effect was abolished by the addition of anti-VSV serum.

The effect of PMA on VSV yields was expressed within a short critical time period. Stimulation with PMA was effective only

Table 1. Properties of Macrophage-Like Clones

	Parental Clone 16	Variant Clone C3C
Glucose-1-C oxidation	+	−
NBT reduction	+	−
O_2^- production	+	−
H_2O_2 production	+	−
Phagocytosis	+	+
Fc-receptor	+	+
Plasminogen activator	+	+

Table 2. Susceptibility of MØ Clones 16 and C3C to Infection
          with VSV

Clone	MOI	VSV yields pfu/5x10^{5}MØ[a]	%VSV Infectious Centers		Calculated VSV Yields pfu/single MØ
			+PMA	−PMA	
16	0.1	6 x 10^7	4	4	3 x 10^3
	0.01	3 x 10^7	1	1	6 x 10^3
C3C	0.1	5 x 10^7	1	1	1 x 10^4

[a]Three or four day cultures of clone 16 or C3C were harvested,
washed, and suspended in fresh medium at a cell density of
5 x 10^5 in 1 ml. One ml of this cell suspension was dispersed
into the wells of 16 mm cluster dishes, and was incubated at
37° in 5% CO_2. After monolayers were formed (2-3 hrs later)
the cells were infected with different amounts of VSV (Indiana,
HRC serotype).

when PMA was added either 10 minutes prior to, simultaneous with,
or 10 minutes following virus infection. Treatment at earlier
or later intervals had no effect on virus yields in clone 16.
The optimum dose of PMA that was effective and nontoxic appeared
to be 1 µg/ml.

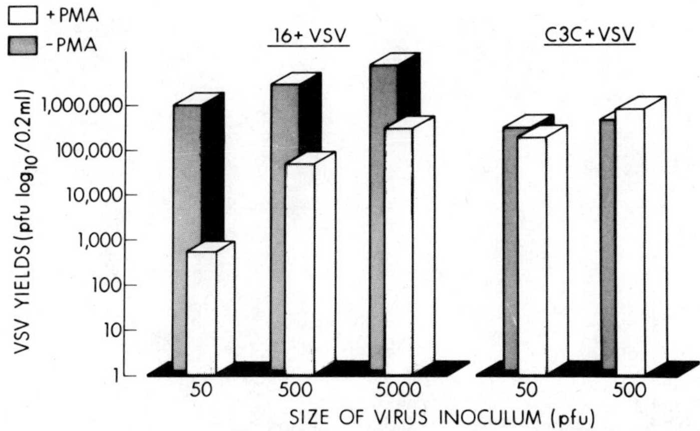

Figure 1.   Effect of PMA on VSV yields. Clone 16 or C3C were
            plated into wells of 16 mm cluster dishes. 2-3 hrs
            later the culture was stimulated with 1 µg/ml of PMA
            10 minutes prior to VSV infection. After 24 hrs, cell
            culture supernatants were collected and VSV yields
            were measured.

Since VSV infection in permissive cells has been shown to result in inhibition of host cell metabolism (7), it was of interest to ascertain whether the oxidative cytocidal mechanisms could be inactivated as a consequence of VSV infection. Accordingly, clone 16 cells, uninfected or infected by VSV, at a MOI of 3 or 1 for 3 hrs, was stimulated by PMA. Glucose-1-(^{14}C) oxidation via the hexose manophosphate (HMP) shunt (Fig. 2) and H_2O_2 production detected by cytochrome C peroxidase (ccp) assay (Fig. 2) were measured (6). Under these conditions virus infection appeared to have no effect on glucose oxidation or H_2O_2 production, either with suboptimal or optimal concentration of PMA. As expected, uninfected or infected virus clone C3C cells exhibited minimal levels of glucose-1-(^{14}C) oxidation or H_2O_2 production following stimulation with PMA.

Since it was found that clone 16 stimulated with PMA released approximately 0.13 nmol/min/10^6 of H_2O_2 (8), it was possible to test indirectly the role of H_2O_2 in reducing the yields of VSV by carrying out the experiment of PMA stimulation and virus infection in the presence of catalase which catalyzes the breakdown of H_2O_2 to O_2.

Direct evidence for the viricidal activity of H_2O_2 was obtained by exposing VSV to a H_2O_2 generating system consisting of glucose and glucose oxidase in Krebs Ringer phosphate buffer

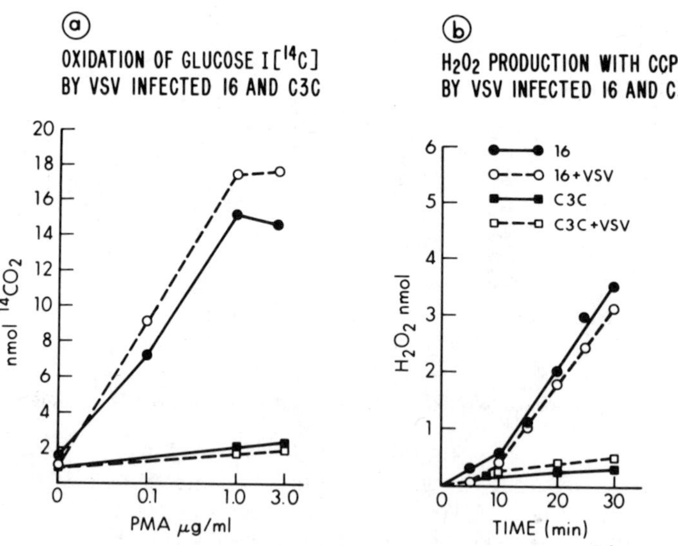

Figure 2.   Effects of VSV infection on glucose-1-(^{14}C) oxidation via HMP shunt and H_2O_2 production detected by ccp assay were measured.

Table 3. Catalase Inhibition of VSV Inactivation by PMA

| | VSV titers (pfu/0.2 ml) | |
Treatment	Clone 16	Clone C3C
None	4×10^5	2.5×10^5
1 µg PMA	4×10^1	7.5×10^5
1 µg PMA + 1 mg catalase[a]	1×10^3	7.5×10^5

[a] PMA stimulation and virus infection were carried out in the presence of 1 mg of catalase.

(KRPB). As shown in Fig. 3, reduction in viral infectivity was detected only with low initial numbers of infectious virus. With higher amounts of virus little reduction in VSV titers could be detected.

In this assay the LD_{50} (5–10×10^3 pfu) of VSV killing by H_2O_2 was at the level of 18.2 nmol/ml/min. Assuming that 5×10^5 clone 16 cells could produce 0.065 nmol/min/ml (8) and using our standard inactivation curve it can be estimated that approximately 35–50 pfu could be inactivated by the amount of H_2O_2 produced by clone 16 after stimulation with PMA.

It therefore appears that H_2O_2 has the ability to kill only a limited number of pfu. The simplest assumption is that since

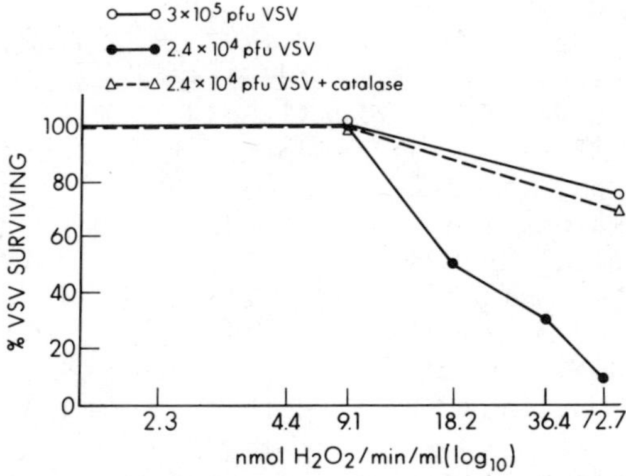

Figure 3. VSV was incubated for 1 hr at 37°C in 1 ml of KRPB containing 4.5×10^4 M glucose and various amounts of glucose oxidase. The rate of H_2O_2 production in this system was independently assayed with ccp.

reduction in yields is seen only at low multiplicities of infection, H_2O_2 produced by macrophages probably acts on extracellular input virus, rather than on the intracellular viral progeny.

This hypothesis was tested using the infectious centers assays. If viruses were killed intracellularly following PMA stimulation, one would expect to find a diminution in infectious centers in clone 16 but not in C3C cells treated with PMA. On the other hand, if only input virus were killed extracellularly, one might expect a diminution in yields in PMA-stimulated clone 16 cells at low MOI, but one would not expect a diminution in infectious centers. The results summarized in Table 2 show that only 1-4% of clone 16 cells produced infectious centers. Addition of PMA to VSV infected cells had no effect on the number of infectious centers. These results suggest first that in cells capable of producing H_2O_2 PMA does not impart resistance to intra-cellular VSV, and secondly that there must exist other mechanisms than the oxidative one which protect macrophages from infection by virus.

ACKNOWLEDGMENT

Supported by USPHS grant AI07118 and grant 1003 from the MS society.

REFERENCES

1. Bloom, B.R. and Rager-Zisman, B. (1975). In: Viral Immunity
         and Immunopathology. Ed. by , A. L. Notkins, Academic
         Press, New York, pp. 113.
2. Klebanoff, S.J. and Hamon, C.B. (1975). In: Mononuclear
         Phagocytes in Immunity, Infection and Pathology. Ed.
         by R. Van Furth, Blackwell Scientific Pub. Ltd., pp. 507.
3. Babior, B.M. (1978). Oxygen-dependent microbial killing by
         phagocytes. N. Engl. J. Med. 298-659.
4. Sasada Masataka and Johnson, R.B. Jr. (1980). Macrophage
         microbicidal activity correlation between phagocytosis-
         associated oxidative metabolism and the killing of
         candida by macrophages. J. Exp. Med. 152:85.
5. Ralph, P. (1980). In: Mononuclear Phagocytes. Ed. by, R. van
         Furth. Martinus Nijhoff, The Hague, The Netherlands.
6. Damiani, G., Kiyotaki, C., Soeller, D., Sasada, M., Peisach,
         J. and Bloom, B.R. (1980). Macrophage variants in
         oxygen metabolism. J. Exp. Med. 152:808.
7. Fenner, F., McAuslan, B.R., Mims, C.A., Sambrook, J. and
         White, D.O. (1974). In: The Biology of Animal Viruses.
         Academic Press, pp. 261.

8. Tanaka, Y., Kiyotaki, C., Tanowitz, H. and Bloom, B.R. (1981).
    Microbicidal mechanisms of macrophages II. (Submitted
    for publication).

INTERFERON-INDUCED AUGMENTATION OF NATURAL KILLER CELL ACTIVITY

BY SPLENOCYTES FROM LEUKEMIA VIRUS IMMUNOSUPPRESSED MICE

Steven Specter, Mauro Bendinelli, William I. Cox and
Herman Friedman

Department of Medical Microbiology and Immunology
University of South Florida, College of Medicine
Tampa, Florida 33612

INTRODUCTION

Interactions of Friend leukemia virus (FLV) and immune
responsiveness in mice has been studied extensively over the past
decade (4). In most circumstances FLV has been shown to depress
cellular and humoral immunity. However, the role of these
depressed responses in the pathogenesis of this virus infection
is not known. In this regard, our laboratory recently began
investigation of natural killer (NK) cell activity in FLV infected
BALB/c mice.

Initial studies indicated that neither FLV infected spleen
cells, nor a cell line induced by Friend virus, GM 979, are
susceptible to NK cell cytolysis (unpublished observations).
Subsequently the ability of spleen cells from FLV infected mice
to lyse the NK sensitive target cell, YAC-1, was studied. Cytotoxicity assays using release of ^{51}Cr were performed after either
4 or 18 hours incubation at 37°C as previously described (1,3).

RESULTS

BALB/c mice infected intraperitoneally with approximately 10^3
ID_{50} FLV were tested at varying times post-infection ranging from
4 to 40 days. Effector to target ratios of 12, 25, 50 and 100:1
were tested. NK activity was reduced in the spleen cells of
infected mice as early as 4 days after infection, and the reduction
increased with progression of infection (Table 1). Although
splenomegaly is very notable beyond 10 days post-infection, the
cause of decreased NK activity was not believed to be dilution

Table 1. Decreased natural killer cell activity in FLV infected
          spleen cells

| Cell Donors[a] | Percent Specific Cytotoxicity[c] | | | |
| | E:T Ratio[b] | | | |
	25	50	100	200
Control mice	19	28	40	42
FLV infected				
- 4 days	14	22	28	34
- 6 days	9	15	19	25
- 8 days	7	15	19	25
-11 days	11	19	25	20
-18 days	4	3	3	2
-25 days	3	3	2	3

[a] Two or three spleens from mice infected with 10^3 ID_{50} FLV were
pooled at each time indicated, and spleen cells were used in an
18 hour cytotoxicity assay. Experiments were repeated at least
3 times and the standard deviations were consistently under
5 percent.

[b] 10^4 YAC-1 target cells were incubated with the appropriate
number of spleen cells

[c] % Specific cytotoxicity =

$$\frac{\text{cmp experimental release} - \text{cpm spontaneous release}}{\text{cpm maximum release} - \text{cpm spontaneous release}} \times 100$$

of NK cells by non-reactive tumor cells in the FLV-infected spleens.
This interpretation is based on the observation that reduced NK
responsiveness is also seen at 4 days post-infection, when spleens
are virtually normal in size. Interferon was capable of enhancing
NK activity in FLV infected mice, however, restoration of NK
activity to control levels was not observed (Table 2). The
suppressive nature of the FLV infected spleen cells was investigated in co-culture studies. The data in Table 3 show that the
NK activity of normal BALB/c spleen cells could be diminished by
FLV infected spleen cells from 11 to 25 days post-infection. This
result also supports the hypothesis that active suppression, rather
than a dilution of effector cells, is responsible for the decreased
NK activity during FLV infection. Reactivity in the suppressed
co-cultures was readily enhanced by interferon, although cultures
containing FLV-infected cells were not as reactive as control
cultures (Table 4).

Table 2. Effect of interferon on the natural killer cell activity
          of spleen cells from FLV-infected mice for YAC-1 targets

	Cell Donor	Percent Specific Cytotoxicity[b]				
		Interferon Concentration (U/ml)[a]				
		0	10^2	3×10^2	10^3	3×10^3
Exp. 1	Control mice	21				42
	FLV -11 days	6				11
	-16 days	4				5
Exp. 2	Control mice	14	13	18	20	26
	FLV -17 days	-1	1	3	6	8

[a] Mouse fibroblast interferon was purchased from Calbiochem as a
stock of 10^5 U/ml.

[b] % Cytotoxicity had a standard deviation of $\leq 5\%$ in all cases.
Effector:target ratio = 50:1.

Table 3. Suppression of natural killer cell activity by FLV
          infected spleen cells

Cells Added[a]	Percent of Control Response[b]			
	E:T Ratio			
	0.5	1	2	4
Controls (cultured)[c]	91	90	90	90
FLV infected				
- 11 days	85	70	66	64
- 18 days	55	44	35	27
- 18 days (cultured)[c]	70	53	40	32
- 25 days	60	51	36	23

[a] 10^4 YAC-1 cells were incubated for 18 hours with 5×10^5 normal
BALB/c spleen cells and the appropriate number of suppressor
cells from FLV infected mice.

[b] Control cultures with no suppressor cells added had cytotoxicity
responses of 21 ± 3%.

[c] Cultured cells were placed in petri dishes and pre-cultured for
24 hours at 37°C. Uninfected spleen cells incubated in this
manner are non-reactive and are referred to as "filler cells".

Table 4.   Enhancement of natural killer cell activity of normal
           spleen cells by interferon in the presence of suppressive
           FLV-infected spleen cells

| | | Percent Specific Cytotoxicity | | |
| | | Suppressor:Effector Ratio | | |
Cells Added[a]	Interferon[b]	0	0.5	1
Controls (cultured)	−	25 ± 4	30 ± 3	28 ± 4
FLV minus 14 days	−	N.D.	18 ± 7	14 ± 4
Controls (cultured)	+	15 ± 4	52 ± 3	53 ± 2
FLV minus 14 days	+	N.D.	38 ± 4	37 ± 1

[a] 10^4 YAC-1 cells were incubated for 18 hours with 10^6 normal BALB/c
spleen cells and the appropriate number of suppressor cells from
FLV infected mice.

[b] Cultures were incubated in the presence or absence of 10^3 U/ml
fibroblast interferon for the duration of the experiment.

FLV infection was noted to severely depress splenic NK cell
activity. This phenomenon has also been reported for Moloney
sarcoma virus infected mice (2). This is in contrast to many
acute virus infections which have been reported to enhance NK
activity, probably via induction of interferon (5). The FLV-
induced suppression appears to be an active phenomenon, since
transfer of infected cells to cultures of normal spleen cells
results in depressed NK activity. Enhancement of NK activity in
control spleen cell cultures, FLV-infected spleen cell cultures,
and control-suppressor FLV-infected spleen cell co-cultures was
achieved using mouse fibroblast interferon. The increase in
activity due to interferon was proportional to the baseline NK
activity in unstimulated cultures. These data suggest that the
FLV-induced suppression is not due to depletion of interferon
induceable NK precursor cells. The nature of suppression of NK
activity by FLV is not yet delineated, but is currently under
investigation. It has been hypothesized, however, that NK cells
have an important role in immunologic surveillance. Thus, the
ability of FLV to suppress this activity may be important in the
pathogenesis of this virus and its resultant neoplasm.

REFERENCES

1. Brunner, K.T., Mavel, J., Cerotinni, J.C. and Chapius, B. (1968).
       Quantitative assay of the lytic action of immune lymphoid
       cells on ^{51}Cr - labelled allogeneic target cells in vitro;
       inhibition by isoantibody and by drugs. Immunol. 14:181-
       195.
2. Herberman, R.B., Holden, H.T., Varesio, L., Taniyama, T.,
       Puccetti, P., Kirchner, H., Gerson, J., White, S. and
       Keisari, Y. (1980). Immunologic reactivity of lymphoid
       cells in tumors. Contemp. Topics Immunobiol. 10:61-78.
3. Kiessling, R., Klein, E., Pross, H. and Wigzell, H. (1975).
       Natural killer cells in the mouse. II. Cytotoxic cells
       with specificity for mouse Moloney leukemia cells.
       Characteristics of the killer cell. Eur. J. Immunol. 5:
       117-121.
4. Specter, S. and Friedman, H. (1978). Viruses and the immune
       response. Pharmac. Ther. A 2:595-622.
5. Welsh, R.M., Jr. and Doe, W.F. (1980). Cytotoxic cells induced
       during lymphocytic chorromeningitis virus infection of
       mice; Natural killer cell activity in cultured spleen
       leukocytes concomitant with T-cell dependent immune
       interferon production. Infect. Immunity 30:473-483.

IMMUNOLOGICAL COMPARISON OF OCULAR DISEASE INDUCED BY TWO STRAINS

OF HERPES VIRUS OF DIFFERENT VIRULENCE

R. M. Nagy, R. McFall and P. Dixon

Wills Eye Hospital Research Division
Philadelphia, Pennsylvania   19107

INTRODUCTION

In previous reports (4,5) we discussed the immunopotentiating
or immunosuppressive role of adherent cells from regional draining
lymph nodes and spleen during the course of experimental Herpes
simplex virus (HSV) keratitis induced by the highly virulent H-4
strain of HSV. Further, we reported that carrageenan sensitive
cells (macrophages) induced differential effects in lymphocyte
responsiveness to the mitogens phytohaemagglutinin (PHA), concanavalin A (Con A) and pokeweed (PWM), as well as to specific HSV
antigens. In a variety of disease states, modulation of immune
function by macrophages has been attributed to prostaglandins.
The evidence implicating prostaglandins in immune regulation is
based on the ability of a prostaglandin synthetase inhibitor,
Indomethacin, to enhance the mitogenic response of lectin
stimulated lymphocytes.

The purpose of the present study was to compare certain
immunological mechanisms during the course of experimental HSV
keratitis induced by two widely divergent strains of virus.
Modulation of helper cell function (PHA responsive lymphocytes),
or suppressor/cytotoxic cell function (Con A responsive lymphocytes), as well as the response of lymphocytes to PWM and specific
virus antigens was monitored. Additionally, macrophage and
prostaglandin modulation of the immune response during the
course of ocular disease was evaluated.

METHODS

Both corneas of 66 New Zealand White rabbits were anesthetized, abraded, and infected with 5000 $TCID_{50}$ of either the highly virulent H-4 strain, or the less virulent HF strain of HSV. In addition, both corneas of 10 rabbits were anesthetized, abraded and control cytosol was topically applied. On days 3, 7, 10, 14, 17 and 21 post inoculation, rabbits were sacrificed and the pre-auricular and cervical lymph nodes (RDLN) were removed, and single cell suspensions were prepared for testing via the lymphoblast transformation assay. Two-fold dilutions of three different T cell mitogens, as well as specific HSV antigens, were tested in triplicate, to obtain maximal levels of lymphoproliferative responses.

For each rabbit, 5×10^5 cells were tested in each assay as: 1) whole suspension; 2) whole + 100 µgm/ml of carrageenan; or 3) whole + 10 µgm/ml of Indomethacin. PHA was tested at final concentrations of 200-25 µgm/ml; PWM at 100-25 µgm/ml, and Con A at 10-1.5 µgm/ml. HSV and control antigens were tested at 120-30 µgm/ml.

For HF infected rabbits for each day tested, N = 6,6,6,5,4 and 4 respectively. For H-4 infected rabbits for each day tested N = 6,6,8,5,6 and 4 respectively. For control rabbits N = 10.

The data obtained from lymphocyte transformation assays was pooled for each day, for each mitogen or antigen tested. Each experimental point represents the ΔCPM ^{3}H-thymidine incorporation determined as follows: mean ^{3}H-thymidine incorporation of triplicate test (antigen or mitogen stimulated) cultures minus the mean ^{3}H-thymidine incorporation of triplicate control antigen stimulated cultures.

A modification of the method of Williams for measuring tritiated thymidine incorporation by mitogen and antigen stimulated lymphocytes was used in this study (9).

RESULTS

As shown in Figs. 1 through 4, cyclic immune reactivity by RDLN lymphocytes is apparent during the course of HSV ocular disease induced by the less virulent HF strain and the highly virulent H-4 strain of HSV. Decreased levels of helper cell activity (PHA responsive lymphocytes) are apparent during the course of ocular infection with the highly virulent strain (H-4) of HSV (Fig. 1). The decreased levels of helper activity are shown both in whole cell populations, as well as those populations treated with carrageenan which is toxic to macrophages. Additionally, decreased levels of suppressor/cytoxic cell activity (Con A

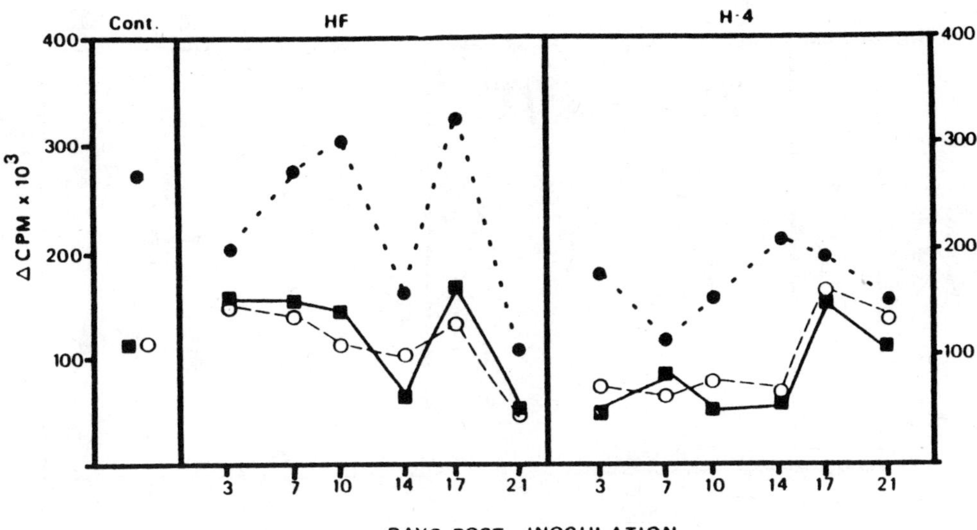

Figure 1.   PHA induced proliferative responses of Whole (■);
Carrageenan (●) and Indomethacin (O) treated RDLN
lymphocytes.

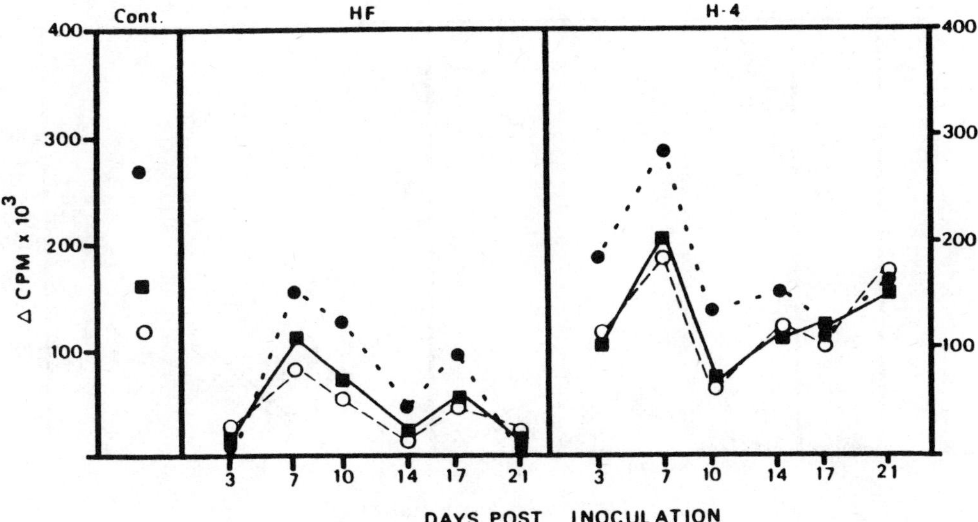

Figure 2.   PWM induced proliferative responses of Whole (■);
Carrageenan (●) and Indomethacin (O) treated RDLN
lymphocytes.

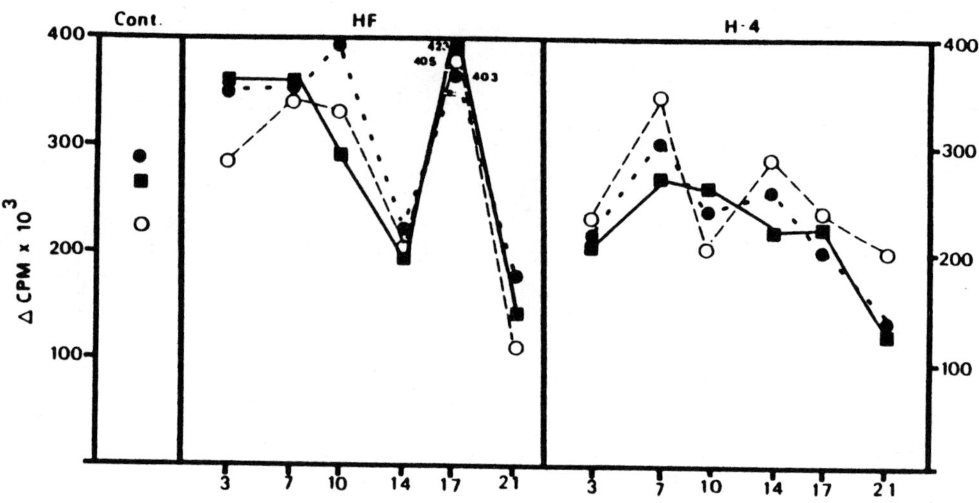

Figure 3. Con A induced proliferative responses of Whole (■); Carrageenan (●), and Indomethacin (○) treated RDLN lymphocytes.

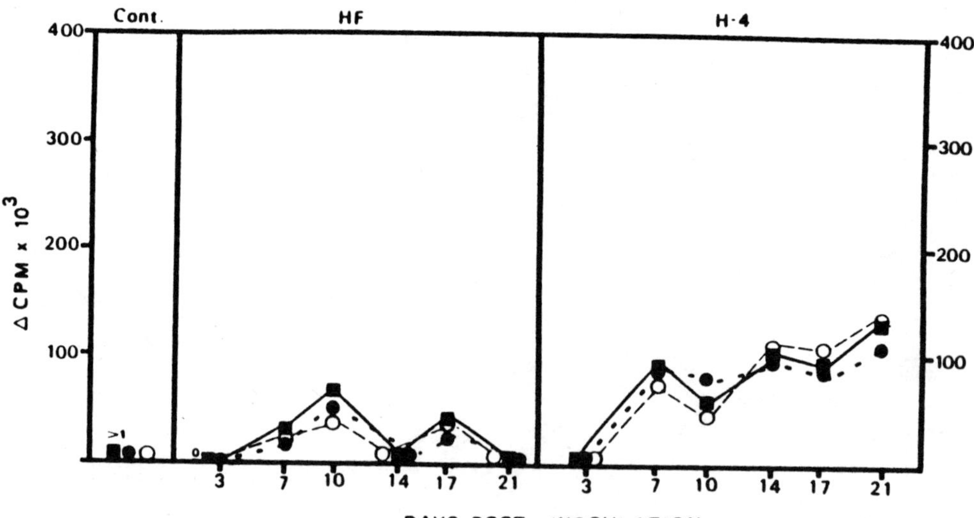

Figure 4. HSV antigen induced proliferative responses of Whole (■); Carrageenan (●), and Indomethacin (○) treated RDLN lymphocytes.

responsive lymphocytes) are evident during the course of ocular
infection with the highly virulent strain (H-4) of HSV (Fig. 3).
Interestingly, significant modulation of Con A responsive lymphocytes by macrophages is not demonstrated. PWM responsive lymphocyte levels are generally greater throughout the course of ocular
infection with the highly virulent strain of HSV (Fig. 2).
Immunopotentiation of PWM lymphocyte responsiveness after
carrageenan treatment of macrophages is evident during the course
of ocular disease induced by either strain of HSV. In contrast,
prostaglandin associated immunosuppression was minimal after
ocular infection with either virus strain (with the exception of
Con A responsive lymphocytes during H-4 induced HSV keratitis)..

Specific HSV responsive lymphocytes attained higher levels
of reactivity after ocular infection with the highly virulent H-4
strain of HSV (Fig. 4). Additionally, peak reactivities of HSV
specific responsive lymphocytes during the cycles of the immune
response differ during the course of ocular infection with the
two strains of HSV.

Splenic lymphocyte responsiveness to mitogens and antigens
was greatly reduced when compared to RDLN lymphocyte responsiveness (data not shown). Immunopotentiation of Con A responsive
lymphocyte populations was demonstrated after addition of carrageenan or Indomethacin to cultured splenic lymphocytes on days 3,
7, 14 and 17 post infection with either strain of HSV. Immunopotentiation of PHA responsive lymphocytes was demonstrated
after addition of carageenan to cultured spleen cells, but not
after the addition of Indomethacin. Lymphocyte responses to
HSV were near control levels on all days tested after ocular
infection with the H-F strain. Specific HSV responsive lymphocytes were, however, more variable on all days tested post H-4
induced ocular infection.

DISCUSSION

Most studies of HSV ocular infections focus attention on a
variety of immune defects, as well as, genetic susceptibility of
the host. Bloomfield and Lopez (1) have suggested the possibility
of at least three defects in homeostatic mechanisms during the course
of HSV ocular infections. More recently Wander, Centifano and
Kaufman (8) suggested that although host factors are important,
patterns of herpetic ocular disease may be attributed at least
partially, to the differing biological behavior and virulence of
certain strains of HSV.

In the present investigation, we compared certain immunological
mechanisms during ocular infection with two widely divergent strains
of HSV. The HF strain of HSV presents with a mild, self limiting

epithelial keratitis, while the H-4 strain presents with a mild
epithelial keratitis which progresses to stromal edema, necrotizing
stromal keratitis and uveitis. The data shown (Figs. 1 through 4)
suggests that certain immunological disturbances occur during HSV
ocular disease induced with the virulent H-4 strain of HSV. These
disturbances may include too few helper T cells, too few suppressor
T lymphocytes, as well as possible defects in macrophage regulatory
function. Interestingly, although ocular infection with H-4 HSV
is somewhat immunosuppressive, immunogenic properties of this
virus remain intact.

Other studies have indicated the existence of various "check
and balance" mechanisms that may be essential for insuring a proper
immune response (2,6). These mechanisms require the collaboration
of various cellular and humoral compartments of the immune system.
It may be that virulent HSV strains, as well as certain tumors
and parasites (3,7) have the capacity to subvert the immune system
and thereby offset these delicate check and balance mechanisms.
Evasion of the host immune surveillance could insure their
survival.

REFERENCES

1. Bloomfield, S.E. and Lopez, C. (1979). Wellcome Trends in
        Ophthalmology 1:4-5.
2. Cantor, H., Hugenberger, J., McVoy-Bondrean, L., Eardley, D.,
        Kemp, J., Shen, F.W. and Gershon, R.K., (1978).
        J. Exp. Med. 148:871-877.
3. Jaywardena, A.N. and Waksman, B.H. (1977). Nature (London)
        265:539-540.
4. Nagv, R.M., Dixon, P. and McFall, R.C. (1979). Assoc. Res.
        Vis. Opthalmol. Spring Suppl. p. 57.
5. Nagy, R.M., Dixon, P. and McFall, R.C. (1980). Associ. Res.
        Vis. Opthalmol. Spring Suppl. p. 31.
6. Smith, R.T. and Landy, M. (1970). Eds. Immune Surveillance,
        Academic Press, N.Y. 1-85.
7. Stutman, O. (1975). Adv. Cancer Res. 22:261-422.
8. Wander, A.H., Centifano, Y.M. and Kaufman, H.E. (1980).
        Arch. Ophthalmol. 98:1458-1461.
9. Williams, R.M. (1973). Cell. Immunol. 9:435-444.

CELLULAR PROCESSING OF THE LARGE GLYCOPROTEIN OF LACROSSE VIRUS

(FAMILY BUNYAVIRIDAE); IMPLICATIONS FOR VIRION ASSEMBLY AND HOST

DEFENSE

David H. Madoff and John Lenard

UMDNJ-Rutgers Medical School, Dept. of Physiology and
Biophysics, Piscataway, NJ 08854

INTRODUCTION

LaCrosse virus (LAC) is a major causative agent of mosquito-
borne human encephalitis in the USA (1). LAC is an enveloped RNA
virus containing two glycoproteins, G1 and G2 (2). Based on
electron micrographs of LAC-infected cells, it is thought that
LAC derives its lipid envelope by budding from host smooth membranes rather than from the plasma membrane (3). We have demonstrated that the LAC G1 protein accumulates in a juxtanuclear region
(probably Golgi) and not at the plasma membrane of the infected
cell. Further evidence suggests that G1 is inefficiently
modified in Golgi (4).

RESULTS AND DISCUSSION

LAC- and VSV-infected baby hamster kidney (BHK) cells that
were labelled with ^{35}S-methionine between 4 and 6 h post-infection
(pi) were treated with either chymotrypsin (2 mg/ml in PBS-def
containing 2 mM EDTA; Sigma Type II) or control buffer and analyzed
by SDS-PAGE. The conditions for proteolysis were sufficient to
completely degrade accessible G1 and nucleocapsid (N) protein
(data not shown). The relative intensities of glycoprotein and
nucleocapsid bands were determined by densitometry of an auto-
radiogram of the dried gel, and the % protease resistance of the
labelled glycoprotein population was calculated as in Table 1.
The results demonstrate that <20% of LAC G1 and >50% of VSV G
protein are sensitive to external protease. Also, in LAC-infected
cells that were enzymatically radioiodinated, LAC G1 was not
measurably labelled (Figure 1C) even though G1 in isolated virions

Table 1. Proteolysis of Viral Glycoproteins in Infected Cells
          Following Steady-State Labelling

Virus	% Protease Resistance[1]
LAC	82.8 ± 8.1
VSV	45.6 ± 13.9

[1] Calculated from densitometry of GI (G) and N bands in SDS gels
of chymotrypsin treated or control cells (see text) as follows:
$[(G1/N)_{Treated}/(G1/N)_{Control}] \times 100\%$.

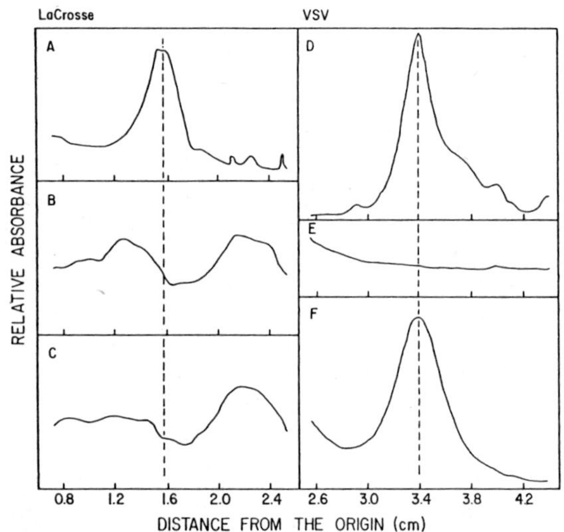

Figure 1. Surface iodination of infected BHK cells. LAC-(C),
           VSV-(F), and mock-infected cells (B, E) were iodinated,
           and analyzed by SDS-PAGE. Individual lanes of the auto-
           diogram of the dried gel were scanned on a Joyce Loebl
           microdensitometer. Only the region co-migrating with
           authentic LAC G1 or VSV G protein is presented. (A) [35]S-
           labelled LAC-infected cell; (B) Iodinated mock-infected
           cell; (C) Iodinated LAC-infected cell; (D) ^{35}S-labelled
           VSV-infected cell; (E) Iodinated mock-infected cell;
           (F) Iodinated VSV-infected cell.

was readily labelled (data not shown). Under these conditions, the
VSV G protein in VSV-infected cells was heavily labelled (Fig. 1F).

LAC-infected cells were doubly stained with anti-LAC G1 antibody plus RITC-goat anti-rabbit IgG, and either FITC-Ricin or
FITC-Con A as described previously (5,6). The LAC glycoprotein
was localized to a juxtanuclear region of the cell that coincides
with FITC-Ricin but not FITC-Con A staining (Data not shown).
Since Ricin stains Golgi and Con A stains rER (6), our data suggests
that LAC G1 accumulates in Golgi within the infected cell. This is
consistent with the proteolysis and iodination data (Table 1 and
Fig. 1) which indicate that only a small percentage of the cell-
associated G1 is expressed on the cell surface at any time.

Golgi-mediated oligosaccharide modification of the LAC G1
protein was studied using the enzyme Endo β-N-acetylglucosamini-
dase H (Endo H). Endo H removes mannose-rich but not complex
oligosaccharides from glycoproteins (7). Endo H resistance is
acquired only after passage through and modification within the
Golgi apparatus (8). The results in Fig. 2 show that the virion-

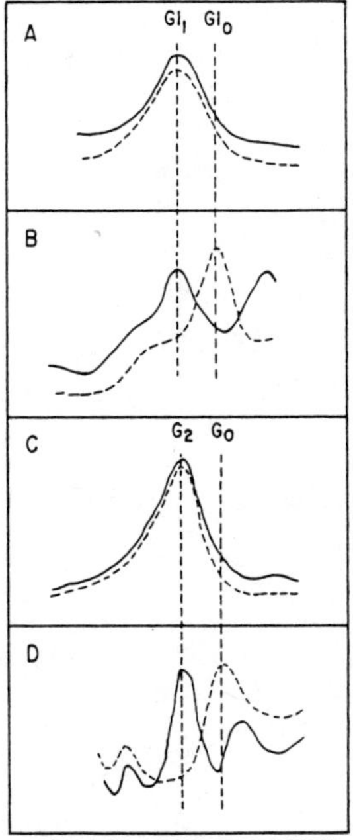

Figure 2. Endo H Resistance of the Virion-Associated Glycoproteins
           of LAC and VSV. Purified virus (A,C) and cells pulse-
           labelled for 5 min (B,D) were treated with Endo H (---)
           or control buffer (——). The SDS-PAGE profile of the
           region co-migrating with authentic virus glycoproteins
           is presented. (A) LAC virus; (B) Pulse-labelled LAC-
           infected cells; (C) VSV; (D) Pulse-labelled VSV-infected
           cells.

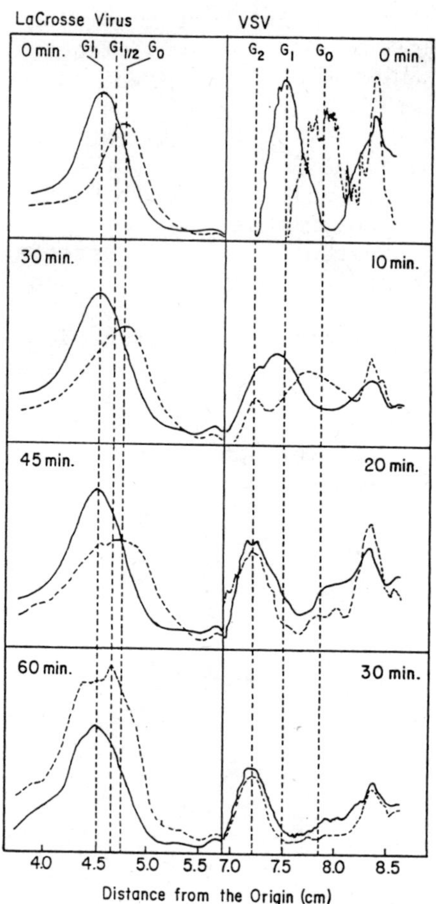

Figure 3. Endo H Sensitivity of LAC and VSV Glycoproteins Following
           Pulse. Infected cells were pulse-labelled for 5 min
           at 5 h pi and chased as in Figure 2. At each time
           point the cells were treated with Endo H (---) or
           control buffer (———) and analyzed as in Figure 2.

associated LAC G1 and VSV G proteins are resistant to Endo H, while
the neoly synthesized proteins are both Endo H sensitive.
Therefore, LAC G1 must pass through the Golgi prior to appearance in
extracellular virus. Pulse-labelled infected cells were treated
with Endo H at different times during the chase period and analyzed
by SDS-PAGE for Endo H resistant (Gl_1 for LAC; G_2 for VSV) and
Endo H sensitive (Gl_0 for LAC; G_0 for VSV) glycoprotein species
(Fig. 3). The VSV G protein was completely converted from an
Endo H sensitive (G_0) to an Endo H resistant (G_2) form within 20
min of synthesis. In contrast, only 50% of the LAC G1 pool was converted to an Endo H resistant form (Gl_1) as late as 45-60 min after

synthesis, and furthermore, a glycoprotein of intermediate sensitivity ($G1_{1/2}$) was observed. Some of the oligosaccharides of $G1_{1/2}$ must be of the complex type and the others must be mannose-rich. The existence of this intermediate form is consistent with the slow modification and accumulation of LAC G1 in Golgi.

The covalent addition of fatty acid to the LAC G1 protein was studied by incubating LAC- and VSV-infected cells with ^{3}H-palmitate and ^{35}S-methionine. Cells and extracellular virus were isolated and their proteins analyzed by SDS-PAGE (Data not shown). The extent of fatty acid modification of cell- and virion-associated glycoproteins was determined by calculating the ^{3}H/^{35}S ratio in LAC G1 and G2 and VSV G proteins (Table 2). The relative extent of fatty acid attachment to the VSV G protein is virtually identical for virion- and cell-associated species, whereas the degree of fatty acid attachment to cell-associated LAC G1 protein is lower than the corresponding virion-associated glycoprotein is lower than the corresponding virion-associated glycoprotein (Table 2). Therefore, under steady state conditions, the cell-associated LAC G1 protein is relatively deficient in fatty acid, while the cell-associated VSV G protein is identical to the virion-associated glycoprotein in this respect. Since the intracellular site of fatty acid attachment is in the Golgi (9), the data imply that the predominant cell-associated LAC G1 population is blocked in its ability to become fatty acylated, while the VSV G protein is readily processed in the Golgi to its fatty acylated form. It has been suggested that the fatty acid modification is required before a glycoprotein is transported from Golgi to the plasma membrane (10). It is therefore possible that LAC G1 protein is arrested in the Golgi due to its lack of fatty acid.

Table 2. Fatty Acid Attachment to the Glycoproteins of LAC Virus and VSV

Virus	Protein	^{3}H-Palmitate/^{35}S-Methionine Ratio[1]	
		Cell-Associated	Virion Associated
LAC	G1	0.7	2.7
	G2	*	2.1
VSV	G	3.1	3.3

* Cannot be determined.

[1] Calculated from SDS-PAGE profile.

It is interesting that extracellular virions contain a highly fatty
acylated Gl which may either be derived from a subset of the Golgi
Gl pool or from the small proportion of Gl that is processed to
the cell surface. (Note: None of the evidence to date excludes
the possibility that LAC buds at the plasma membrane. The electron
micrographic data is inconclusive since the actual budding event
has not been visualized.)

In summary, we have demonstrated that the viral glycoproteins
in LAC-infected cells accumulate in Golgi and not at the plasma
membrane. The mechanism of this localization may involve inefficient
addition of fatty acid, thus preventing the transport of the glycoproteins to the plasma membrane. Slow processing of Gl in Golgi
is reflected in its delayed acquisition of Endo H resistance. The
assembly of LAC virions must occur in Golgi or at the plasma membrane in such a way that fatty acylated, Endo H resistant Gl is
incorporated. If budding occurs at the plasma membrane, then fatty
acylation and transport of the glycoproteins to the plasma membrane
represents a rate-limiting step. Therefore, the cell surface is
minimally altered by viral glycoproteins during the infectious
process and LAC-infected cells would be expected to escape detec-
tion by the host immune system.

REFERENCES

1.   Bishop, D.H.L. and Shope, R.E. (1979). Bunyaviridae. In:
          Comprehensive Virology, Vol. 14. H. Fraenkel-Conrat
          and R.R. Wagner, Eds. Plenum Press, New York, Chapter
          1, pp. 1-156.
2.   Obijeski, J.F., Bishop, D.H.L., Murphy, F.A. and Palmer, E.L.
          (1976). Structural proteins of LaCrosse virus. J.
          Virol. 19(3):985-997.
3.   Murphy, F.A., Whitfield, S.G., Coleman, P.H., Calisher, C.H.,
          Rabin, E.R., Jenson, A.B., Melnick, J.L., Edwards, M.R.
          and Whitney, E. (1968). California group arboviruses:
          Electron microscope studies, Exp. Mol. Path. 9:44-56.
4.   Madoff, D.H. and Lenard, J. (1982). A membrane glycoprotein that
          that accumulates intracellularly: Cellular processing
          of the large glycoprotein of La Crosse virus. Cell. 28,
          821-829.
5.   Saraste, J., von Bonsdorff, C.-H., Hashimoto, K., Kääriäinen, L.
          and Keränen, S. (1980). Semliki forest virus mutants
          with temperature-sensitive transport defect of envelope
          proteins. Virology. 100:229-245.
6.   Vitranen, I., Ekblom, P. and Laurila, P. (1980). Subcellular
          compartmentalization of saccharide moieties in cultured
          normal and malignant cells. J. Cell Biol. 85:249-434.

7. Robbins, P.W., Hubbard, S.C., Turco, S.J. and Wirth, D.F.
          (1977). Proposal for a common oligosaccharide intermed-
          iate in the synthesis of membrane glycoproteins. Cell
          12:893-900.
8. Rothman, J.E. and Fries, E. (1981). Transport of newly
          synthesized vesicular stomatitis viral glycoprotein
          to purified Golgi membranes. J. Cell Biol. 89:162-168.
9. Schmidt, M.F.G. and Schlessinger, M.J. (1980). Relation of
          fatty acid attachment to the translation and matura-
          tion of vesicular stomatitis and Sindbis virus membrane
          glycoproteins. J. Biol. Chem. 255(8):3334-3339.
10. Zilberstein, A., Snider, M.D., Porter, M. and Lodish, H.F.
          (1980). Mutants of vesicular stomatitis virus blocked
          at different stages in maturation of the viral glco-
          protein. Cell 21:417-427.

CMV AND RENAL ALLOGRAFT SURVIVAL

R. H. Kerman, R. Conklin, D. Cahall, C. T. Van Buren
and B. D. Kahan

Departments of Surgery and Laboratory Medicine
University of Texas Medical School, Houston, Texas

INTRODUCTION

A majority of renal transplant recipients become actively infected with cytomegalovirus. (CMV) (1). Most patients displaying
CMV antibody preoperatively experience infections due to reactiva-
tion of latent virus. Primary CMV infections occur in preoperatively seronegative recipients due to organs from seropositive
donors or to blood transfusions (2). CMV infection has serious
consequences, including acute allograft rejection and patient
death (1). Although CMV infection in renal allograft recipients
has been thought to be associated with rejection and graft loss,
there is little information concerning the relationship between
tissue typing for HLA A, B and DR antigens, CMV infection, and
renal allograft survival (3). In the present study the effect of
CMV infection on the success of renal transplantation was assessed
by serologic analysis of recipients before and following transplantation. Recipients were grouped based on the degree of incom-
patibility for HLA A, B and DR antigens with their donors.

MATERIALS AND METHODS

Eighty-three renal transplant recipients, including 58 cadaveric (CAD) and 25 living related donor (LRD) grafts, were studied
for the association between CMV infection and graft survival. All
donors and recipients were ABO compatible. HLA A, B and DR tissue
typing was performed using the microlymphocytotoxicity technique
with NIH and Terasaki tissue typing trays (4,5). All transplants
were performed following a negative comprehensive immunological
cross match (6). The CMV status was determined by immunofluorescent

517

reactivity of patient sera against infected human diploid fibroblasts or by growth of CMV directly from peripheral blood leukocytes, tissue samples or urine (7). A patient was considered
seropositive when the serum titer was greater than 1:16. All
patients were followed for at least one year post-transplantation.
The Student "t" test was employed to assess the statistical
significance of differences between groups.

RESULTS AND DISCUSSION

      Sixty-three patients (44 CAD and 19 LRD) were seropositive
(CMV titer > 1:16) and 20 (14 CAD and 6 LRD) were seronegative prior
to transplantation. Thirteen of the 20 seronegative patients
became seropositive within the first year postoperatively. The
overall cadaveric graft survival of 50% was the same, whether
patients were seropositive or negative preoperatively. Recipients
with fewer than 2 mismatched HLA A and B antigens had a 65% one
year graft survival compared to 51% for recipients with 2 or more
HLA A and B mismatched antigens (p < 0.05). Patients engrafted
with a 0-1 DR antigen mismatched kidney had 68% graft survival
compared to 46% for one year graft survival for recipients of a
2 DR mismatch (p < 0.05). Moreover, patients who were seropositive
pretransplantation, and received well matched grafts (less than 2
HLA A and B mismatches), demonstrated a significantly better one
year graft survival (p < 0.05) than pretransplant seronegative
recipients (71% vs. 50%, respectively).

      A four-fold increase in CMV antibody titer postoperatively,
a finding which was deemed consistent with CMV infection, occurred
in 21 of 76 postoperatively seropositive patients (8 primary and
13 recurrent infections). There were three grafts lost among 10
of 21 patients displaying seroaugmented CMV infections and not
treated for rejection, whereas 4 grafts were lost in 11 of 21
patients who were diagnosed and treated for rejection episodes
concomitant with CMV infections. Since 14 of 21 grafts (67%)
were intact and functioning following CMV infection, we concluded
that rejection accompanied by graft loss does not necessarily
follow CMV infection. Indeed, the data suggest that patients
with CMV infections may be less prone to rejection episodes and
graft loss since patients who did not experience CMV infections
during this same time period (30/62, 48%) bore a lesser fraction
of functional grafts (p < 0.05).

REFERENCES

1. Simmons, R.L., Matas, A.J., Rattazzi, L.C., Balfour, Jr., H.H.,
        Howard, R:J. and Najarian, J.S. (1977). Clinical
        characteristics of the lethal cytomegalovirus infection
        following renal transplantation. Surgery, 82:537-546.

2. Andrus, C.H., Betts, R.F., May, A.G. and Freeman, R.B. (1976).
    Cytomegalovirus infection blocks the beneficial effect
    of pretransplant blood transfusions on renal allograft
    survival. Transplantation 28:451-456.

3. May, A.G., Betts, R.F., Freeman, R.B. and Andrus, C.H. (1978).
    An analysis of cytomegalovirus infection and HLA antigen
    matching on the outcome of renal transplantation. Ann.
    Surg. 187:110-117.

4. Staff, Transplantation and Immunology Branch (1979). NIH
    microlymphocytotoxicity technique, p. 39, In: NIAID
    Manual of tissue typing techniques, Ray, J.G. (ed.)
    Government Printing Office, Washington, D.C. (NIH
    publication No. 80-545).

5. Ayoub, G., Park, M.S., Terasaki, P.I. and Iwaki, Y. (1980).
    B. cell antibodies and cross-matching. Transplantation
    29:227-229.

6. Kerman, R.H., Kahan, B.D. (in press). Immunological evaluation
    of transplant rejection: Pre-and postoperative indices
    detecting immune responsiveness, In: Hayry, P. (ed.):
    Present and Future Trends in Clinical Transplantation,
    Annals of Clinical Research.

7. Reynolds, D.W., Stagno, S. and Alford, C.A. (1980). Laboratory
    diagnosis of cytomegalovirus infection, p. 425, In:
    Diagnostic Procedures for Viral, Rickettsial and
    Chlamydial Infections, Lennette, E. and Schmidt, N.V.
    (eds.) Amer. Pub. Hlth. Assoc., Washington, D.C.

TO MARTYRS

The Bard of Berryville*

Their eyes are pink, their fur is white

When treated well they rarely bite

Their tails provide a place to seize

To pick them up or give i.v.s

They're small in size, live ten per cage

That keeps one rat of early age

And should we need to raise our own

They do know how if left alone

They take three weeks to incubate

And then give birth to more than eight

The pups grow up in thirty days

And can be used in many ways

They give their lives to help us learn

The fact of life for which we yearn

All tests are done with groups of size

So we may seek a Nobel prize

So let us toast throughout the house

That tool of science, the martyred mouse

*AKA L. Joe Berry, Department of Microbiology, University of
Texas, Austin, TX   78712

This poem was part of Dr. Berry's address at the banquet held
with the conference.

CONTRIBUTORS

Aber, V.R., MRC Unit for Laboratory Studies of Tuberculosis,
    Royal Postgraduate Medical School, London, W12 OHS,
    UNITED KINGDOM.

Actor, Paul, Smith, Kline & French Laboratories, 1500 Spring
    Garden Street, Philadelphia, Pennsylvania 19101, U.S.A.

Andrew, P.W., MRC Unit for Laboratory Studies of Tuberculosis,
    Royal Postgraduate Medical School, London W12 OHS,
    UNITED KINGDOM.

Baba, Tohru, Department of Pathology, University of Connecticut
    Health Center, Farmington, Connecticut 06032, U.S.A.

Barile, Michael F., Division of Bacterial Products, Bureau of
    Biologics, Food and Drug Administration, Bethesda,
    Maryland 20205, U.S.A.

Bendinelli, Mauro, University of South Florida, College of
    Medicine, 12901 N. 30th Street, Tampa, Florida 33612, U.S.A.

Berry, L. Joe, Department of Microbiology, University of Texas
    at Austin, Austin, Texas 78712, U.S.A.

Biddison, William E., Immunology Branch, National Cancer
    Institute, National Institutes of Health, Bethesda,
    Maryland 20205, U.S.A.

Bloom, Barry R., Departments of Microbiology, Immunology and
    Pathology, Albert Einstein College of Medicine, Bronx,
    New York 10461, U.S.A.

Brinton, Margo, Wistar Institute, 36th and Spruce Streets,
    Philadelphia, Pennsylvania 19104, U.S.A.

Brosnan, Celia F., Departments of Microbiology, Immunology and
    Pathology, Albert Einstein College of Medicine, Bronx,
    New York 10461, U.S.A.

Bullock, Ward E., University of Cincinnati Medical Center,
     231 Bethesda Avenue, Cincinnati, Ohio 45267, U.S.A.

Byrne, Gerald I., Cornell University Medical College, 525 East
     68th Street, New York, New York 10021, U.S.A.

Cahall, D., University of Texas Medical School, Houston, Texas
     77030, U.S.A.

Carrow, Emily, Department of Microbiology and Immunology, Tulane
     University School of Medicine, 1430 Tulane Avenue, New Orleans,
     Louisiana 70112, U.S.A.

Cebra, John J., Department of Biology, University of Pennsylvania,
     Leidy Laboratory G6, Philadelphia, Pennsylvania 19004, U.S.A.

Cohen, Stanley, Department of Pathology, University of Connecticut
     Health Center, Farmington, Connecticut 06032, U.S.A.

Collins, Frank M., Trudeau Institute, P.O. Box 59, Saranac Lake,
     New York 12983, U.S.A.

Conklin, R., University of Texas Medical School, Houston, Texas
     77030, U.S.A.

Cox, William I., University of South Florida, College of Medicine,
     12901 N. 30th Street, Tampa, Florida 33612, U.S.A.

DeChatelet, Lawrence R., Bowman Gray School of Medicine, Winston-
     Salem, North Carolina 27103, U.S.A. (Deceased)

Dixon, P., Wills Eye Hospital Research Division, Philadelphia,
     Pennsylvania 19107, U.S.A.

D'Silva, Helen, University of Connecticut Health Center,
     Farmington, Connecticut 06032, U.S.A.

Dodge, G.R., Wyeth Laboratories, Inc., P.O. Box 8299, Philadelphia,
     Pennsylvania 19101, U.S.A.

Doherty, Peter C., Wistar Institute, 36th and Spruce Streets,
     Philadelphia, Pennsylvania 19104, U.S.A.

Domer, Judith, Department of Microbiology, Tulane University
     School of Medicine, 1430 Tulane Avenue, New Orleans,
     Louisiana 70112, U.S.A.

Eisenstein, Toby K., Temple University Medical School, Department
     of Microbiology and Immunology, 3400 N. Broad Street,
     Philadelphia, Pennsylvania 19140, U.S.A.

CONTRIBUTORS

Farrell, Jay P., Department of Pathobiology, University of
Pennsylvania,3800 Spruce Street, Philadelphia, Pennsylvania
19104, U.S.A.

Forget, Adrien, University of Montreal, CP 6125-A, Montreal,
Quebec, CANADA

Friedman, Herman, Chairman, Department of Medical Microbiology,
University of South Florida, College of Medicine, 12901 N.
30th Street, Tampa, Florida 33612, U.S.A.

Goidl, Edmond A., Department of Medicine, Cornell University
Medical College, 1300 York Avenue, New York, New York 10021,
U.S.A.

Goren, Mayer B., National Jewish Hospital, 3800 E. Colfax Avenue,
Denver, Colorado 80206, U.S.A.

Grappel, Sarah F., Smith, Kline & French Laboratories, 1500 Spring
Garden Street, Philadelphia, Pennsylvania 19101, U.S.A.

Greenberg, L.E., National Jewish Hospital and Research Center,
Denver, Colorado 80206, U.S.A.

Greenspan, Neil, The Wistar Institute, 36th and Spruce Streets,
Philadelphia, Pennsylvania 19104, U.S.A.

Gros, Philippe, Montreal General Hospital Research Inst., Montreal,
Quebec, CANADA.

Grun, James L., Department of Microbiology, Hahnemann Medical
College, 230 N. Broad Street, Philadelphia, Pennsylvania
19102, U.S.A.

Hank, J.A., Departments of Human Oncology and Pediatrics, University
of Wisconsin-Madison, K4-423 CSC, 600 Highland Avenue, Madison,
Wisconsin 53792, U.S.A.

Hashemi, Shahab, Hahnemann Medical College, Department of Pathology,
230 N. Broad Street, Philadelphia, Pennsylvania 19102, U.S.A.

Haverly, Anne L., Department of Immunology, Walter Reed Army
Institute of Research, Washington, D.C. 20012, U.S.A.

Henry, Robin R., Department of Immunology, Walter Reed Army
Institute of Research, Washington, D.C. 20012, U.S.A.

Jackett, P.S., MRC Unit for Laboratory Studies of Tuberculosis,
Royal Postgraduate Medical School, London, W12 OHS, UNITED
KINGDOM.

Jerrells, Thomas R., Walter Reed Army Institute of Research,
    Bldg. 40, Room B091, Washington, D.C. 20012, U.S.A.

Johnson, Howard, Department of Microbiology, University of Texas
    Medical Branch, Galveston, Texas 77550, U.S.A.

Jones, Thomas C., Department Medicine, Cornell University Medical
    College, 1300 York Avenue, New York, New York 10021, U.S.A.

Kahan, B., University of Texas Medical School, Houston, Texas
    77030, U.S.A.

Katz, Sheila M., Hahnemann Medical College, Department of Pathology,
    230 N. Broad Street, Philadelphia, Pennsylvania 19102, U.S.A.

Kerman, Ronald H., University of Texas Medical School, Department
    of Surgery, 6431 Fannin, Houston, Texas 77030, U.S.A.

Killar, Loran, Temple University Medical School, Department of
    Microbiology, 3400 N. Broad Street, Philadelphia, Pennsylvania
    19140, U.S.A.

Kirkpatrick, C.H., National Jewish Hospital and Research Center,
    Denver, Colorado 80206, U.S.A.

Kirsh, Richard, Department of Tumor Biology, Smith, Kline Corpora-
    tion, 1500 Spring Garden Street, Philadelphia, Pennsylvania
    19101, U.S.A.

Klein, John R., Department of Medical Microbiology, University of
    Pennsylvania, Philadelphia, Pennsylvania 19104, U.S.A.

Komisar, Jack L., Department of Biology, University of Pennsylvania,
    Leidy Laboratory G6, Philadelphia, Pennsylvania 19104, U.S.A.

Kongshavn, Patricia A.L., Department of Physiology, McGill
    University, Montreal, Quebec, CANADA

Laveck, Moira A., Department of Microbiology, Uniformed Services
    University, School of Medicine, Bethesda, Maryland 20014, U.S.A.

Leake, Eva S., Department of Microbiology and Immunology, The
    Bowman Gray School of Medicine of Wake Forest University,
    Winston-Salem, North Carolina 27103, U.S.A.

Lenard, John, UMDNJ-Rutgers Medical School, Department of
    Physiology and Biophysics, Piscataway, New Jersey 08854, U.S.A.

Leonard, Edward J., National Cancer Institute, National Institutes
    of Health, Bethesda, Maryland 20205, U.S.A.

Leunk, Robert D., Michigan State University, East Lansing,
    Michigan 48824, U.S.A.

Lopes, A. Dwight, The Wistar Institute, 36th and Spruce Streets,
    Philadelphia, Pennsylvania 19104, U.S.A.

Lowrie, D.B., MRC Unit for Laboratory Studies of Tuberculosis,
    Royal Postgraduate Medical School, London W12 OHS, UNITED
    KINGDOM.

McFall, R.C., Wills Eye Hospital Research Division, Philadelphia,
    Pennsylvania 19107, U.S.A.

McPhail, Linda C., The Bowman Gray School of Medicine, Winston-
    Salem, North Carolina 27103, U.S.A.

Madoff, David H., UMDNJ-Rutgers Medical School, Department of
    Physiology and Biophysics, Piscataway, New Jersey 08854, U.S.A.

Mason, U.G., III, National Jewish Hospital and Research Center,
    Denver, Colorado 80206, U.S.A.

Meltzer, Monte S., National Cancer Institute, National Institutes
    of Health, Bethesda, Maryland 20205, U.S.A.

Mergenhagen, Stephen E., National Institute of Dental Research,
    9000 Rockville Pike, Bethesda, Maryland 20205, U.S.A.

Metcalf, Eleanor S., Uniformed Services University School of
    Medicine, 4301 Jones Bridge Road, Bethesda, Maryland 20014,
    U.S.A.

Moon, Robert J., Department of Microbiology and Public Health,
    Michigan State University, East Lansing, Michigan 48824, U.S.A.

Moore, Robert N., Department of Microbiology, University of
    Tennessee, Knoxville, Tennessee 37916, U.S.A.

Murray, Henry W., Cornell Univeristy Medical College, 525 East
    68th Street, New York, New York 10021, U.S.A.

Myrvik, Quentin N., Department of Microbiology and Immunology,
    Bowman Gray School of Medicine, Winston-Salem, North Carolina
    27103, U.S.A.

Nacy, Carol A., Walter Reed Army Institute of Research, Department
    of Immunology, Washington, D.C. 20012, U.S.A.

Nagy, R.M., Wills Eye Hospital Research Division, Philadelphia,
    Pennsylvania 19107, U.S.A.

Nathanson, Neal, Department of Medical Microbiology, University of
     Pennsylvania, Philadelphia, Pennsylvania 19104, U.S.A.

O'Brien, Allison D., Uniformed Services University School of
     Medicine, 4301 Jones Bridge Road, Bethesda, Maryland 20014,
     U.S.A.

Okamura, Noburu, Tokyo Medical and Dental University, School of
     Medicine, Tokyo, JAPAN

Oster, Charles N., Walter Reed Army Institute of Rsearch, Depart-
     ment of Immunology, Washington, D.C. 20012, U.S.A.

Osterman, J.V., Walter Reed Army Institute of Research, Bldg. 40,
     Room B091, Washington, D.C. 20012, U.S.A.

Pagano, Joseph, President, Eastern Pennsylvania Branch, American
     Society for Microbiology, Smith, Klein & French Laboratory,
     1500 Spring Garden Street, Philadelphia, Pennsylvania 19101,
     U.S.A.

Pappas, Michael G., Walter Reed Army Institute of Research, Depart-
     ment of Immunology, Washington, D.C. 20012, U.S.A.

Polansky, Michael J., Smith, Kline Corporation, 1500 Spring Garden
     Street, Philadelphia, Pennsylvania 19101, U.S.A.

Poste, George, Director of Research, Smith, Kline & French
     Laboratories, 1500 Spring Garden Street, Philadelphia,
     Pennsylvania 19101, U.S.A.

Rager-Zisman, Bracha, Departments of Microbiology, Immunology and
     Pathology, Albert Einstein College of Medicine, Bronx,
     New York 10461, U.S.A.

Ramstedt, Urban, Department of Immunology, Biomedicum Centrum,
     University of Uppsala, Uppsala, SWEDEN.

Rappaport, R.S., Wyeth Laboratories, Inc., P.O.Box 8299,
     Philadelphia, Pennsylvania 19101, U.S.A.

Root, Richard K., Department of Internal Medicine, Yale University
     School of Medicine, 333 Cedar Street, New Haven, Connecticut
     06510, U.S.A.

Rose, Faina V., Department of Biology, University of Pennsylvania,
     Leidy Laboratory G6, Philadelphia, Pennsylvania 19104, U.S.A.

Scott, Phillip A., Department of Pathobiology, University of
     Pennsylvania, 3800 Spruce Street, Philadelphia, Pennsylvania
     19104, U.S.A.

Shirley, Pamela S., The Bowman Gray School of Medicine, Winston-
    Salem, North Carolina 27103, U.S.A.

Skamene, Emil, Montreal General Hospital Research Institute, 1650
    Cedar Avenue, Montreal, Quebec H3G 2A4, CANADA

Sondel, P.M., Departments of Human Oncology and Pediatrics, Univer-
    sity of Wisconsin-Madison, K4-423 CSC, 600 Highland Avenue,
    Madison, Wisconsin 53792, U.S.A.

Specter, Steven, University of South Florida, College of Medicine,
    12901 N. 30th Street, Tampa, Florida 33612, U.S.A.

Spitznagel, John K., Chairman, Department of Microbiology, 501
    Woodruff Memorial Building, Emory University, Atlanta,
    Georgia 30322, U.S.A.

Steeg, Patricia S., National Institute of Dental Research, 9000
    Rockville Pike, Bldg. 30, Room 327, NIH, Bethesda, Maryland
    20205, U.S.A.

Stevenson, Mary M., Montreal General Hospital Research Institute,
    Montreal, Quebec H3G 1A4,CANADA.

Suko, Matsunobu, Department of Pathology, University of Connecticut
    Health Center, Farmington, Connecticut 06032, U.S.A.

Sultzer, Barnet M., Department of Microbiology and Immunology,
    Downstate Medical Center, Brooklyn, New York, U.S.A.

Taylor, Ben A., Jackson Laboratory, Bar Harbor, Maine, U.S.A.

Tenner-Racz, Klara, Department of Microbiology and Immunology, The
    Bowman Gray School of Medicine of Wake Forest University,
    Winston-Salem, North Carolina 27103, U.S.A.

Trischmann, Thomas M., Tropical Medicine Center, John Hopkins
    University, 615 N. Wolfe Street, Baltimore, Maryland 21205,
    U.S.A.

Van Buren, C., University of Texas Medical School, Houston, Texas
    77030, U.S.A.

Watson, Susan R., University of Cincinnati Medical Center, 231
    Bethesda Avenue (Rm. 7168), Cincinnati, Ohio 45267, U.S.A.

Weidanz, William P., Department of Microbiology, Hahnemann Medical
    College, 230 N. Broad Street, Philadelphia, Pennsylvania 19102,
    U.S.A.

Wing, Edward J., Montefiore Hospital, 3459 Fifth Avenue, Pittsburgh,
    Pennsylvania 15213, U.S.A.

Yen, S.S., National Jewish Hospital and Research Center, Denver,
    Colorado 80206, U.S.A.

Yoshida, Takeshi, Department of Pathology, University of Connecti-
    cut Health Center, Farmington, Connecticut 06032, U.S.A.

SUBJECT INDEX